STUDENT STUDY GUIDE

MARK J. COMELLA

JOHN D. CUTNELL
KENNETH W. JOHNSON
Southern Illinois University at Carbondale

with the assistance of
ANDREW ELBY
University of California-Berkeley

to accompany

PHYSICS

FOURTH EDITION

JOHN D. CUTNELL
KENNETH W. JOHNSON
Southern Illinois University at Carbondale

JOHN WILEY & SONS, INC.
NEW YORK • CHICHESTER • WEINHEIM • BRISBANE • SINGAPORE • TORONTO

ISBN 0-471-16411-9

Printed in the United States of America

10 9 8 7 6 5 4 3 2 1

Printed and bound by Courier Kendallville, Inc.

TO THE STUDENT

This study guide has been written to accompany *Physics*, fourth edition, by John D. Cutnell and Kenneth W. Johnson. It is designed to assist you in understanding the concepts and in developing the problem-solving skills necessary to master a first course in college physics. It is not intended that the Study Guide be a replacement for the text. It should be viewed as an additional resource to be used *after reading* the text. Each chapter of this study guide is organized as follows:

- **Preview** The *Preview* provides a brief synopsis of the chapter contents.
- **Quick Reference** The *Quick Reference* contains a list of *Important Terms*, a list of *Equations* and a summary of the main concepts and principles encountered in the chapter.
- **Discussion of Selected Sections** This section provides discussions of most of the sections found in the text with the order of topics following that of the text. The topics discussed are usually the main topics in the chapter, as well as topics that students generally find most difficult. These discussions include a large number of quantitative examples to supplement those found in your text.
- **Practice Problems** The *Practice Problems* are provided to give you additional practice at solving "single concept" problems.
- **Helpful Suggestions** The *Helpful Suggestions* provide useful tips that will aid you both in problem solving and in your conceptual understanding of physics.
- **Everyday Physics** *Everyday Physics* applies the concepts in the chapter to practical situations. Often, this section describes simple experiments and demonstrations that you can perform and explain using basic physics.
- **Chapter Quiz** The *Chapter Quizzes* are included to help you diagnose any weak areas in the concepts of the chapter.
- **Solutions and Answers** This section contains detailed solutions to the *Practice Problems* and answers to the *Chapter Quizzes.*
- **MCAT Review Problems** This section contains problems designed to help students review for the MCAT exam that is required as part of the admission process to medical school. The physical sciences portion of the exam differs radically from many other science tests. Most of the questions are associated with passages. A typical passage describes an experiment, makes an argument, or presents some information, often covering material you have never seen. The questions test whether you can use your physics knowledge to interpret the experiment, settle the argument, or understand the new material. Knowing lots of formulas will *not* help. Instead, you need to understand the *concepts* . Many passages deal with multiple concepts simultaneously. Since these review problems do the same thing, they might also help you prepare for *regular* physics exams, not just the MCAT. In order to test your ability to handle real-life journal articles and medical textbooks, the MCAT contains a few poorly written passages. To prepare you for this ordeal, some of these review problems

intentionally present information unclearly or too quickly. Sometimes, the vocabulary and symbols on the exam (and in these review problems) differ slightly from those in your text. Fortunately, by avoiding panic, you can sort through the confusion and answer the questions, most of which do not require you to understand all the "dirty details" of the passage. Since you may not bring calculators, graphing paper, or other aides to the MCAT, the exam questions do not require difficult calculations. Often, approximations can get you the answer quickly.

In addition, this study guide includes four *Sample Examinations* with answers, and advice on *How to Take a Physics Course*.

HOW TO TAKE A PHYSICS COURSE

Often students entering their first course in physics anticipate having a "hard time". Much of what they have heard from friends and relatives about physics courses leads them to believe that physics can not be learned well by the average person. This is simply not true. The secret to success in a physics course is to learn the material day by day. Trying to "cram" physics the night before an exam is what leads to the "hard time" in the course. Physics is a subject which requires time for the concepts to be absorbed and understood.

Physics is NOT learned by memorizing equations and then trying to find the right numbers to plug into them. Physics is learned by using it day after day to solve problems and by thinking about the concepts and relating them to everyday experience.

How to Learn Physics

Learning physics is a building process much like building a house. You must first build a good foundation and learn to use the proper tools. The foundation in physics consists of a set of laws and principles. The tools are mathematics and problem solving skills. If you build a good foundation, the rest of the structure will be straight and sound. If the foundation is weak, the whole structure may collapse.

After the foundation is laid, everything else rests on it. The same is true of the foundation that you will build in physics. If a principle or concept is not well understood, it will come back to plague you later. Learn everything well as you encounter it. The following will aid you in your effort.

1. **Attend lecture faithfully**. Physics professors are prone to include material which is either not in the text or modified to some extent.

2. **Take GOOD notes**. Good note taking skills are learned by practice in the classroom. Good note taking does not mean writing down everything that is put on the board or appears on the overhead transparencies. Good note taking requires that you are already familiar with the topic by reading the text in advance of the lecture. The material presented in the lecture can then be digested properly and your notes will consist of the digested material in the form of outlines of the major concepts and problem solving methods. Do not write down all of the little facts that you already know unless they add something to your understanding.

3. **Use your notes**. A good way to study is to compile the day's notes and the textbook material into a written discussion of the concepts and problem solving methods. Pretend that you are explaining the material to someone as you write. Also, try out your understanding by teaching the material to a sympathetic friend or fellow physics student. This "teaching" is a very effective way to learn. Indeed, physics professors know physics so well because they teach.

4. **Learn the laws, principles, and terminology of physics**. Avoid memorizing verbatim from the book. Simply knowing the words is useless, you must gain an understanding of the meaning. Restating the laws, principles, and definitions in **your own words** aids in the understanding of their meaning. Memorization is not learning, however, learning sometimes involves memorization. Frequently used equations and constants should be memorized. Actually, most of the equations which you will need, you will automatically remember after doing a lot of homework problems.

5. **Solve problems**. Solving lots of problems from the text and the study guide helps you to develop a deeper understanding of the concepts. In fact, it is often argued, that if you can't solve physics problems, then you don't know physics. After solving several problems of a particular type you will develop "physical intuition". This means that you can analyze a problem and predict the results before you actually solve it.

Problem solving skills are developed by practice and often vary from student to student. There is no cookbook approach to problem solving. However, some general guidelines to problem solving can be written down. Use the following guidelines to help you develop your problem solving technique.

a. Understand the problem. This requires that you carefully read the problem and determine exactly what is asked for in the problem. An invaluable aid to understanding a problem is a labeled diagram of the situation presented in the problem. This can be as simple as a quick sketch of the objects involved and their relationships to one another. Include all of the information given in the problem on your sketch.

b. Determine the physical principles involved in the problem. What kind of a problem is it? Mechanics, electricity, etc. or a combination? What physical principles apply in this problem? Newton's laws, conservation of energy, conservation of momentum, etc.?

c. Try to gain an general understanding of what the results should be like before you do calculations. Does the ball go left or right? Up or down? Does the car speed up or slow down?

d. Write the mathematical expressions of the relevant laws and principles in the notation of your particular problem. Take advantage of any zero quantities to simplify the expressions. DO NOT plug in any other numbers.

e. After you have the mathematical expressions ask yourself, "how can I manipulate these mathematical equations to find what I am asked to find? Students often become stuck at this point because they look at the equations and realize that they don't know a quantity which appears to be needed, say, the mass of an object. DON'T worry. Solve the equations and obtain an algebraic form for the solution. Often the unknown quantity, such as the mass, will cancel out. If it doesn't, THEN worry about how to find it. In any case, you have a general solution to the problem. All you need now are some specific numbers to "plug" in.

f. Plug in your numbers to get a specific answer for your problem. Since you already have the general solution, you may plug in different numbers to get the specific solution to a new problem without having to do all of the work again.

g. Interpret your solution. Do the units work out right? Does the answer make sense? A simple example of an answer which does not make sense is when your mathematics tells you that a ball rises when dropped and you know the object will fall.

How to Study for Exams

1. If you have **studied** your physics **each day** and **solved** a good number of **problems**, you should need little additional preparation for exams. The night before an exam, you might review the important concepts, refresh your memory on some problem solving techniques, go to bed early and get a good night's sleep. If you don't know the material, then cramming will be of little help. Cramming means trying to memorize far more material than possible in the short time available with very little understanding of any of it. This kind of studying may help a little on questions involving pure recall, but none on questions involving reasoning and deep thought. DON'T CRAM!

2. More preparation is required for the final exam even if you have studied day by day, particularly, if the final is comprehensive. This is because so much time has elapsed since you have studied some of the material that it may have become hazy in your mind. Plan your study to include a review spread out over several days. If you learned the material well in the first place, it will quickly come back to you. Cramming for a final is even less effective than it is for other exams. Again, DON'T do it.

How to Use the Text and the Study Guide

Aside from your instructor, the textbook is the most important source of information available to you. Read the text daily, paying particular attention to the examples worked out in the text. Then try some of the chapter problems on your own. If you need additional help, consult this study guide. It has several worked out example problems and practice problems with the solutions given. It is a good idea to try to solve the practice problems on your own before looking at the solutions. Use the solutions only to check to see if you have done the problem right or to get you started again if you become stuck on a problem.

The text has a large amount of information in the appendices which can be of great help if you know what is there. Spend some time browsing through the appendices. Don't try to memorize all that you find there, but keep in mind what is available. Later, when you need a constant or conversion, etc. you will recall that these are in the appendices and you will not have spend a great deal of time trying to find them. The Study Guide does not have the information contained in the appendices of the text. The authors of the Study Guide are expecting that you will use both books together.

The Study Guide has chapter quizzes to help you diagnose any weak areas in your understanding of the concepts of the chapter. These quizzes are not meant to be representative of the tests which will actually be given in your particular course. They are diagnostic only. The answers to the quizzes are given at the end of the chapter. Take these quizzes and check your progress.

CONTENTS

Chapter

CHAPTER 1
Introduction and Mathematical Concepts

PREVIEW

In this chapter you will be introduced to the physical units most frequently encountered in physics. After completion of the chapter you will be able to convert between these systems of units and use them as an aid in problem solving.

Also, after a review of trigonometry, you will become acquainted with vectors and the methods of vector addition. Upon completion of this material you should be able to add and subtract vectors graphically, decompose vectors into their components and use the components to reconstruct the vectors.

QUICK REFERENCE

Important terms

Scalar quantity
A quantity which can be described by a single number.

Vector quantity
A quantity which can be adequately described by a number (magnitude) and a direction.

Vector components
Two perpendicular vectors which added together produce the original vector.

Systems of Units

System	Length	Mass	Time
SI	meter (m)	kilogram (kg)	seconds (s)
BE	foot (ft)	slug (sl)	seconds (s)
CGS	centimeters (cm)	gram (g)	seconds (s)

Trigonometry

$$\sin\theta = \frac{h_o}{h} \quad (1.1) \qquad \theta = \sin^{-1}\left(\frac{h_o}{h}\right) \quad (1.4)$$

$$\cos\theta = \frac{h_a}{h} \quad (1.2) \qquad \theta = \cos^{-1}\left(\frac{h_a}{h}\right) \quad (1.5)$$

$$\tan\theta = \frac{h_o}{h_a} \quad (1.3) \qquad \theta = \tan^{-1}\left(\frac{h_o}{h_a}\right) \quad (1.6)$$

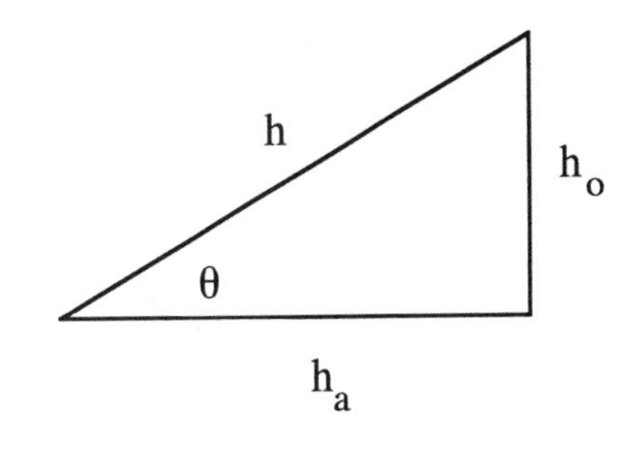

$$h^2 = h_o^2 + h_a^2 \quad (1.7)$$

Graphical Addition of Vectors

Vectors may be added (or subtracted) by placing them head to tail.

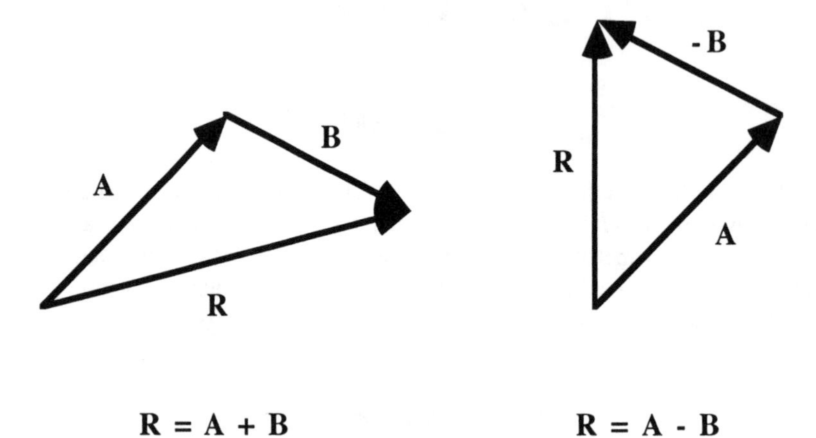

Components of a Vector

The following expressions refer to the vector diagram shown below.

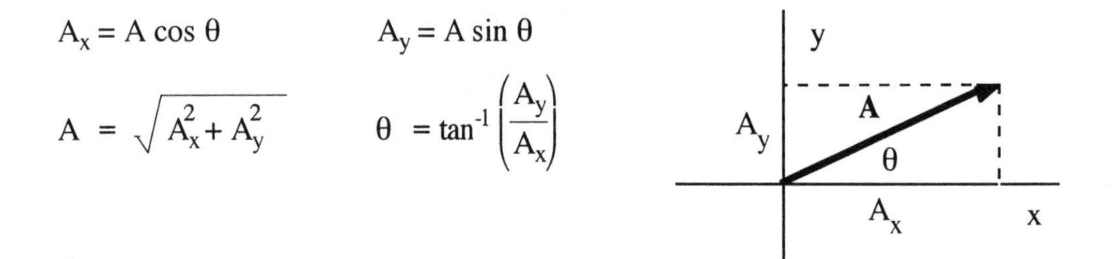

$$A_x = A\cos\theta \qquad A_y = A\sin\theta$$

$$A = \sqrt{A_x^2 + A_y^2} \qquad \theta = \tan^{-1}\left(\frac{A_y}{A_x}\right)$$

Addition of Vectors Using Components

If **C** = **A** + **B**, then the components of **C** are:

$$C_x = A_x + B_x \qquad C_y = A_y + B_y$$

and the magnitude and direction of **C** are:

$$C = \sqrt{C_x^2 + C_y^2} \qquad \theta = \tan^{-1}\left(\frac{C_y}{C_x}\right)$$

where θ is measured counterclockwise from the + x axis.

DISCUSSION OF SELECTED SECTIONS

1.3 The Role of Units in Problem Solving

Units are very important in the study of physics in that all physical quantities have units. These may be either the *base units* of length (m), mass (kg), and time (s), or *derived units* such as the joule (kg m^2/s^2).

When used in algebraic expressions, the units which accompany the numbers can be used to check not only the accuracy of the calculation, but also the validity of the equation. For this reason, the units will always be displayed along with the numbers in this study guide. You are encouraged to do the same in your solutions to problems. Remember, if the units do not work out, your solution is not right either.

Example 1 *Manipulating Units and Converting Between Systems*
Convert 5.00 mi/h to m/s.

Conversion factors:

$$1 \text{ mi} = 5280 \text{ ft} \qquad 1 \text{ ft} = 0.305 \text{ m}$$
$$1 \text{ km} = 1000 \text{ m} \qquad 1 \text{ h} = 3600 \text{ s}$$

Each of the equalities above can be used to form a fraction or *conversion factor* that is equal to unity (i.e., equal to 1). In multiplying by unity we do not change the value of the physical quantity; we are merely expressing the same quantity in a different set of units. One side of the equality will appear in the numerator and the other side will appear in the denominator of the fraction. The specific choice will be made so that the unwanted units cancel and the desired units appear in the final answer.

$$5.00 \text{ mi/h} = 5.00 \frac{\text{mi}}{\text{h}}\left(\frac{5280 \text{ ft}}{1 \text{ mi}}\right)\left(\frac{0.305 \text{ m}}{1 \text{ ft}}\right)\left(\frac{1 \text{ h}}{3600 \text{ s}}\right) = 2.24\frac{\text{m}}{\text{s}}.$$

Notice that the first conversion factor changes miles to feet, the second changes feet to meters, and the third conversion factor changes hours to seconds. Again notice that in each conversion factor, the numerator is equal to the denominator and, therefore, the conversion factor equal to unity.

Example 2 *Another Look at Converting Units*
Convert 2.00 mi^2 to ft^2.

We will start with the fact that [illegible] 5280 ft. In setting up the conversion factor as a fraction equal to unity, however, we must square the fracti[illegible] that it converts square miles (mi^2) to square feet (ft^2).

$$2.00 \text{ mi}^2 = 2.00 \text{ mi}^2\left(\frac{5280 \text{ ft}}{1 \text{ mi}}\right)^2 = 5.58 \times 10^7 \text{ ft}^2$$

Notice that the conversion factor is still equal to unity ($1^2 = 1$). Also notice that the final answer is expressed in scientific notation to preserve the correct number of significant figures; three in this case.

Example 3 *Using Units to Check Equations*
An equation which may result from the application of the conservation of energy principle is

$$\frac{1}{2}\,mv^2 = mgh$$

where m is a mass with the units of kg, v is a velocity with the units of m/s, g is an acceleration with the units of m/s^2 and h is a height with the units of m.

This equation may be checked for validity by simply substituting the units into it and manipulating them as if they were numbers. If the units on the right and left side of the equation do not match, the equation is definitely NOT VALID.

$$(kg)\left(\frac{m}{s}\right)^2 = (kg)\left(\frac{m}{s^2}\right)(m)$$

$$(kg)\left(\frac{m^2}{s^2}\right) = (kg)\left(\frac{m^2}{s^2}\right)$$

$$\frac{kg\ m^2}{s^2} = \frac{kg\ m^2}{s^2}$$

In this case, the units on each side of the equation do match and the equation may be a valid one. This procedure does NOT guarantee that the equation has any "true" meaning, however.

By the way, the above combination of units appear so often in physics that they are given the special name of joules (J). This is the unit of ENERGY.

1.4 Trigonometry

Most likely, you have already become acquainted with the basic trigonometric functions in a high school or college course. These can be found in the Quick Reference section of this book.

Many times the use of the "trig" functions in physics involve finding the length of one side of a right triangle when you know one other side and one of the acute angles. This is particularly important in finding the components of a vector which will be studied in the next section.

Other variations are, of course, possible and useful. For instance, if two sides of a right triangle are known, then the trigonometric functions can be used to find the angle.

Example 4 *Finding one side of a right triangle if one other side and an angle are known*
An observer, whose eyes are 6.0 ft above the ground, is standing 105 ft away from a tree. The ground is level, and the tree is growing perpendicular to it. The observer's line of sight with the tree top makes an angle of 20.0° above the horizontal. How tall is the tree?

The observer's eye, the tree top, and a point on the trunk are the vertices of a right triangle. The height of the tree is

$$H = 6.0\text{ ft} + h_o$$

Now h_o can be found from the right triangle by using equation (1.3)

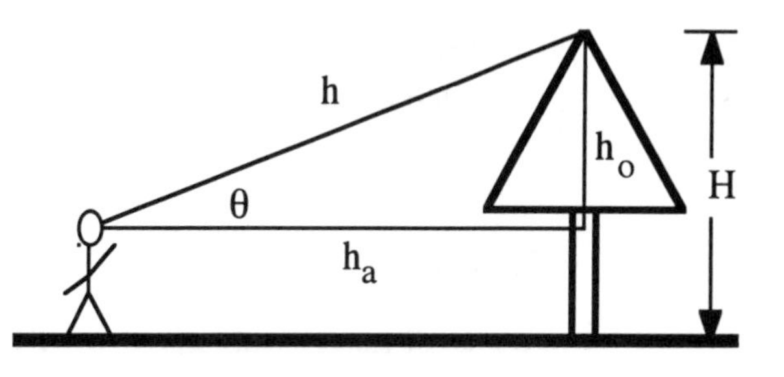

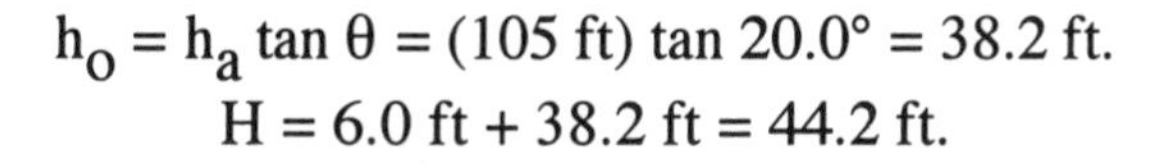

$$h_o = h_a \tan\theta = (105\text{ ft})\tan 20.0° = 38.2\text{ ft}.$$
$$H = 6.0\text{ ft} + 38.2\text{ ft} = 44.2\text{ ft}.$$

Another common need in physics is to find one side of a triangle when the other two sides are known but no acute angle is given. Then you should use the Pythagorean theorem (1.7) or one of its variants.

Example 5 *Using the Pythagorean theorem*
In the previous example, it is desired to know the straight line distance from the person's eye to the top of the tree. The Pythagorean theorem gives

$$h = \sqrt{h_o^2 + h_a^2} = \sqrt{(105\text{ ft})^2 + (38.2\text{ ft})^2} = 112\text{ ft}.$$

1.6 Vector Addition and Subtraction

If vectors are colinear, then they may be added or subtracted by simply adding or subtracting their magnitudes. The directions of the vectors are usually specified by calling a vector pointing to the right (or up) positive and a vector pointing to the left (or down) negative. Then the sign of the resultant vector tells which way it points.

Example 6 *Adding and subtracting colinear vectors*
Find the resultant, **C**, of the following vectors (u is an arbitrary unit).

In both cases the vectors are colinear. One could imagine placing the vectors **A** and **B** in a "tail-to-head" fashion. In case 1, both **A** and **B** point in the same direction. The resultant vector will have a magnitude equal to the length of **A** plus the length of **B**. Since both vectors are directed to the right, the resultant will also point to the right. In case 2, the vectors **A** and **B** point in opposite directions. If they are placed in a tail-to-head fashion it will be clear that part of **A** is canceled by **B**. The resultant will have a magnitude that is equal to the magnitude of **A** minus the magnitude of **B**. Since **A** has the greater magnitude, the resultant will point in the direction of **A**. The resultant **C** is shown below for both cases.

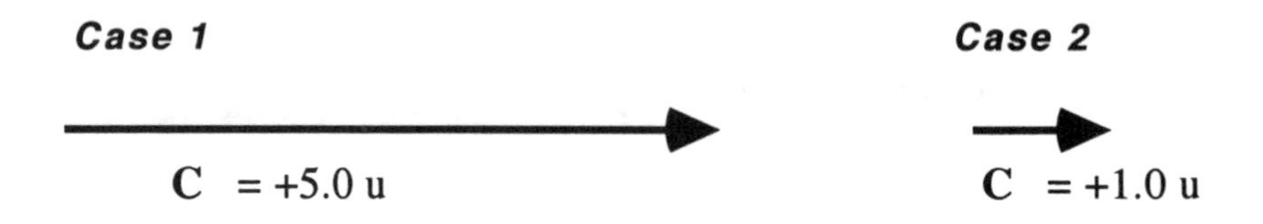

When vectors are not colinear but perpendicular, they may be added by using the Pythagorean theorem and the tangent function. This method yields the magnitude and direction of the resultant vector as a number and an angle.

Example 7 *Adding perpendicular vectors*
Find the resultant, **C**, of the vectors shown where **A** = + 2.0 u and **B** = + 5.0 u.

The resultant is **C** = **A** + **B** as shown in the diagram. Notice that it was constructed by placing the vectors **A** and **B** in a "tail-to-head" fashion and then connecting the "head" of **A** to the "tail" of **B**.

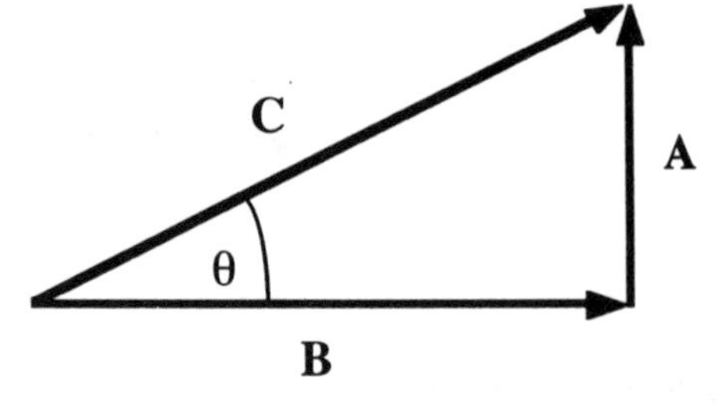

The length of the resultant vector, C, is given by the Pythagorean theorem to be

$$C = \sqrt{A^2 + B^2}$$

$$C = \sqrt{(2.0\ \text{u})^2 + (5.0\ \text{u})^2}$$

$$C = 5.4\ \text{u}.$$

The direction of **C**, as specified by the angle, θ, is given by (1.6) to be

$$\theta = \tan^{-1}\left(\frac{A}{B}\right) = \tan^{-1}\left(\frac{2.0\ \text{u}}{5.0\ \text{u}}\right) = 22^{\circ}.$$

Now **C** = 5.4 u, 22° counterclockwise from **B**.

1.7 The Components of a Vector

As you might infer from the preceding example, any vector, **C**, may be expressed as the sum of two perpendicular vectors. When these perpendicular vectors are placed along the x and y axes of a Cartesian coordinate system, they are referred to as the x and y components (C_x and C_y) of the vector.

The components of a vector may be found by using a vector diagram similar to the above and equations (1.1) and (1.2). Please keep in mind that if equations (1.1) and (1.2) are used to find the components of a vector, the angle, θ, is defined as being measured from the x axis.

Example 8 *Finding the components of a vector*
Find the components of the vector **A** = 5.0 u, θ = 30 °.

An application of (1.2) to the vector shown in the figure gives

$$A_x = A\cos\theta = (5.0\text{ u})\cos 30°$$
$$A_x = 4.3\text{ u}.$$

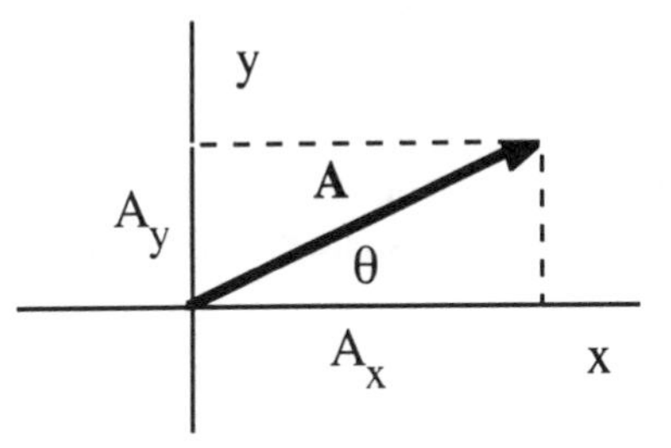

A similar application of (1.1) gives

$$A_y = A\sin\theta = (5.0\text{ u})\sin 30°$$
$$A_y = 2.5\text{ u}.$$

Sometimes you will encounter situations where you will need to find the x and y components of a vector, but the angle given, ϕ, is NOT measured from the x axis. In this case, the components may NOT be given by

$$A_x = A\cos\phi \quad \text{and} \quad A_y = A\sin\phi.$$

You now have two choices. You may either find the angle, ϕ, in terms of the angle, θ, or you may try to find a different right triangle in which to apply equations (1.1) and (1.2).

Example 9 *Finding the components of a vector when the angle is not measured from the +x axis*
Find the x and y components of the vector, **A**, whose magnitude is 10.0 u and which makes an angle of 45° with the +y axis. Refer to the figure shown.

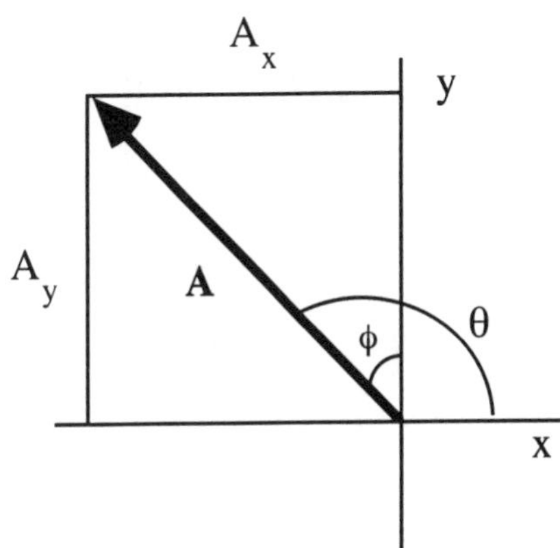

We may use
$A_x = A\cos\theta$ and $A_y = A\sin\theta$ if we can find the angle, θ.

From the figure it is seen that

$$\theta = 90° + \phi \text{ so that}$$
$$\theta = 90° + 45° = 135°.$$

Now

$$A_x = (10.0\text{ u})\cos 135°$$
$$A_x = -7.07\text{ u}$$
$$A_y = (10.0\text{ u})\sin 135°$$
$$A_y = +7.07\text{ u}.$$

We may also apply equations (1.1) and (1.2) to the right triangle shown in the figure to obtain

$$A_x = - A \sin \phi = - (10.0 \text{ u}) \sin 45° = - 7.07 \text{ u}$$

and

$$A_y = + A \cos \phi = + (10.0 \text{ u}) \cos 45° = + 7.07 \text{ u}.$$

1.8 Addition of Vectors by Means of Components

Since like components of two or more vectors are colinear, they may simply be added, as in example 6, to give the corresponding component of the resultant vector. In this way, all of the components of the resultant vector may be found. The magnitude and direction of the resultant vector may then be determined by equations (1.3) and (1.7).

Example 10 *Adding vectors by the component method*
Add the vectors **A**, **B**, and **C** shown in the figure using the component method. A = 5.0 m, B = 7.0 m and C = 4.0 m.

An application of equations (1.1) and (1.2) to each of the triangles shown in the figure gives

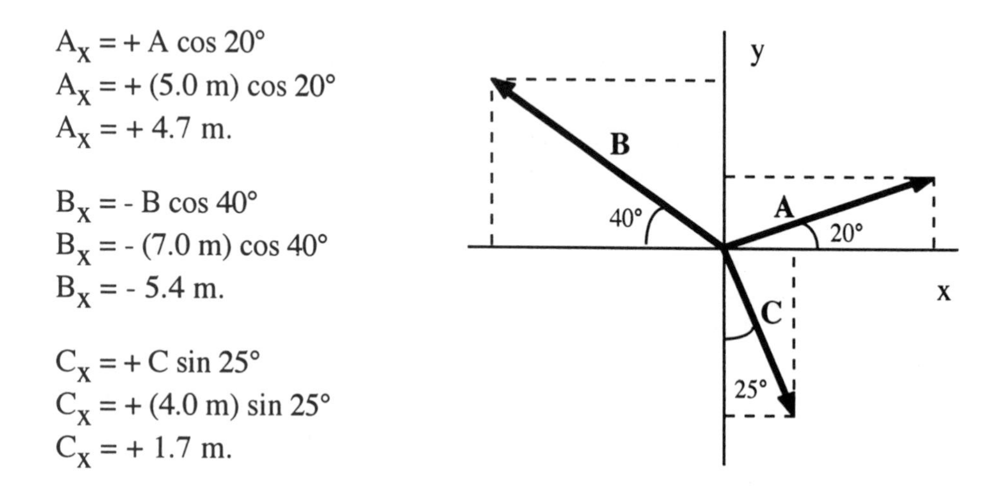

$$A_x = + A \cos 20°$$
$$A_x = + (5.0 \text{ m}) \cos 20°$$
$$A_x = + 4.7 \text{ m}.$$

$$B_x = - B \cos 40°$$
$$B_x = - (7.0 \text{ m}) \cos 40°$$
$$B_x = - 5.4 \text{ m}.$$

$$C_x = + C \sin 25°$$
$$C_x = + (4.0 \text{ m}) \sin 25°$$
$$C_x = + 1.7 \text{ m}.$$

The x component of the resultant vector is

$$R_x = A_x + B_x + C_x$$

$$R_x = 4.7 \text{ m} - 5.4 \text{ m} + 1.7 \text{ m}$$

$$R_x = 1.0 \text{ m}.$$

Repeating the above for the y component gives

$$A_y = + A \sin 20° = + (5.0 \text{ m}) \sin 20° = + 1.7 \text{ m}.$$

$$B_y = + B \sin 40° = + (7.0 \text{ m}) \sin 40° = + 4.5 \text{ m}.$$

$$C_y = - C \cos 25° = - (4.0 \text{ m}) \cos 25° = - 3.6 \text{ m}.$$

The y component of the resultant vector is

$$R_y = A_y + B_y + C_y = 1.7\text{ m} + 4.5\text{ m} - 3.6\text{ m}$$

$$R_y = 2.6\text{ m}.$$

Now the magnitude of the resultant can be found from the Pythagorean theorem

$$R = \sqrt{R_x^2 + R_y^2}$$

$$R = \sqrt{(1.0\text{ m})^2 + (2.6\text{ m})^2} = 2.8\text{ m}$$

The angle that R makes with the +x axis is

$$\theta = \tan^{-1}(R_y/R_x) = \tan^{-1}(2.6\text{ m}/1.0\text{ m}) = 69°.$$

PRACTICE PROBLEMS

The following problems are provided to give you additional practice solving single concept problems. Some work space has been left for you after or to the right of the problem. The solutions to these problems will be found at the end of this chapter.

1. How many significant figures are in the numbers

    a. 2.03

    b. 0.0054

    c. 2.8×10^5

    d. 1500

    e. 0.23×10^{-4}

2. Convert the following into the indicated units.

    a. 5.00 m to ft

    b. 1.2 ft/s to m/s

    c. 60.0 mi/h to km/h

    d. 9.80 m/s^2 to ft/s^2

    e. 535 kg to slugs

3. A basketball coach insists that his players be at least 180.0 cm tall. Would a player of height 5 ft 11.5 in tall qualify for the team?

4. A football field is 100.0 yards long. Express this distance in millimeters.

5. Check the following equations for possible validity.
   a. $x^2 = 1/2\ gt$ where x is in m, g is in m/s^2, t is in s.

   b. $v^2 = 2\ ax$ where v is in m/s, a is in m/s^2 and x is in ft.

6. A right triangle has a side of length 3.5 m. The angle opposite the 3.5 m side is 25°. Find the length of the other sides.

7. A 500.0 m tall building casts a shadow 800.0 m long over level ground. What is the sun's elevation angle above the horizon?

8. A right triangle has two sides of length 25 ft and 15 ft. Find the length of the hypotenuse and all angles.

9. A bridge 50.0 m long crosses a chasm. If the bridge is inclined at an angle of 20.0° to the horizontal, what is the difference in height between the two ends?

10. A displacement vector of magnitude 5.0 m points in an easterly direction. A second displacement vector points north and has a magnitude of 9.7 m. Find the magnitude and direction of the vector sum.

11. An electric field vector, **E**, has a magnitude of 1.0 newtons per coul (N/C) and makes an angle of 33° CCW from the +x axis of a Cartesian coordinate system. Find the components of **E**.

12. A magnetic field vector, **B**, is oriented 65° clockwise from the -y axis. It has a magnitude of 0.010 tesla (T). Find the x and y components of **B**.

13. A vector, **A**, has a magnitude of 10.0 u and points 21.0° north of west. A second vector, **B**, has a magnitude of 5.2 u and points 47.0° east of south. Find the magnitude and direction of the sum by the component method.

14. A car drives 2 km west, then 8 km south, and then 10.0 km at an angle 53° north of east. Find the car's final displacement. (magnitude and direction)

HELPFUL SUGGESTIONS

1. When expressing a vector quantity for an answer, be sure to specify both its magnitude and direction.

2. Vectors may be moved from place to place, providing that you maintain the same length and direction.

3. Calculating the components of a vector as $A_x = A \cos \theta$ and $A_y = A \sin \theta$ is correct **only if** θ is an angle measured with respect to the x axis.

4. When giving the answer to a problem, ALWAYS include the units. The units can be used as a tool for checking whether or not the answer is "dimensionally" correct.

5. Try to avoid just "plugging in" numbers into equations. Try to understand the ideas and concepts rather than just memorizing equations.

6. When you obtain your solution, always ask yourself, "is the solution reasonable?", "does it make sense?", and "are the units consistent?"

EVERYDAY PHYSICS

1. If you want to lay out a garden, build a foundation or anything which should have 90 degree corners, use a 3-4-5 right triangle. Use a string to roughly define two adjacent sides of your garden, measure 3 ft along one side and 4 ft along the other side and mark the location of each point. Measure from mark to mark and adjust the angle between the strings until the measurement yields exactly 5 ft. The strings will then be 90 degrees from each other. For more accuracy you may want to use a 6-8-10 right triangle or even a 9-12-15 right triangle.

2. Most of us are familiar with the SI units of meter and kilogram, but the British units of foot and pound are deeply ingrained since they are the ones we use in everyday life. Become more familiar with the SI units by lifting 1 kg of a common substance like sugar. Also, measure several common objects like your car in both meters and feet.

3. Determine the height of a tall object by measuring its shadow and estimating the angle of elevation of the sun. If it is possible, find the actual height of the object and compare with your results. Can you think of ways to determine the angle more accurately?

CHAPTER QUIZ

This quiz has been provided to help you diagnose possible weak areas in the understanding of the key chapter concepts. The answers can be found at the end of the chapter in this guide.

1. How many significant figures are in the answer of 1.20 - 1.01?
   a. one b. two c. three d. four

2. How many significant figures are in the result of 1600/2.80?
   a. one b. two c. three d. four

3. The derived unit $(kg^2\ m^2)/(kg\ m\ s^2)$ is equivalent to
   a. $(kg^2)/s^2$ b. $(kg\ m/s^2)$ c. $(kg\ m)/s$ d. $(kg\ m^2)/m$

4. How many feet are in 25 m?
   a. 82 b. 0.12 c. 0.31 d. 3.28

5. If you are given one angle and the length of the side opposite in a right triangle, which trig function would allow you, in a single step, to find the length of the hypotenuse?
   a. sine b. cosine c. tangent d. any

6. If you are given one angle and the length of the hypotenuse of a right triangle, which trig function would you use to find the side adjacent to the angle?
   a. sine b. cosine c. tangent d. any

7. Vectors may be added graphically by placing them
   a. tail to tail. b. parallel. c. tail to head. d. head to head.

8. Which of the following is an example of a scalar?
   a. force b. volume c. displacement d. velocity

9. The parts of a vector which lie in perpendicular directions are called
   a. magnitudes b. resultants c. components d. scalars

10. If the angle, less than 90°, specifying the direction of a vector is given as measured from the y axis, then the x component will ALWAYS involve what trig function?
    a. sine b. cosine c. tangent d. any

11. Which of the following is NOT a vector quantity?
    a. time b. velocity c. force d. displacement

12. Two displacement vectors have magnitudes of 8 m and 12 m , respectively. When these two vectors are added the magnitude of the sum
    a. is 20 m
    b. is 4 m.
    c. is larger than 20 m.
    d. could be as large as 20 m or as small as 4 m.

13. A physical quantity is calculated from the formula $K = 3\pi\ c\ (a^2 + b^2)$ where a, b, and c are all lengths. What is the dimension of K?
    a. $[L]$
    b. $[L]^2$
    c. $3\pi\ [L]^2$
    d. $[L]^3$

SOLUTIONS AND ANSWERS

Practice Problems

---

1. a. 2.03 has **THREE** significant figure.

   b. 0.0054 has **TWO** significant figures. Note that it can be written as $5.4 \text{ X} 10^{-3}$.

   c. $2.8 \text{ X } 10^5$ has **TWO** significant figures.

   d. 1500 has **TWO** significant figures since both zeros are in doubt.

   e. $0.23 \text{ X } 10^{-4}$ has **TWO** significant figures.

---

2. a. We have

$$5.00 \text{ m} = (5.00 \text{ m})(3.28 \text{ ft}/1 \text{ m}) = \mathbf{16.4 \text{ ft}}.$$

   b. Also

$$1.2 \text{ ft/s} = (1.2 \text{ ft/s})(0.305 \text{ m}/1 \text{ ft}) = \mathbf{0.37 \text{ m/s}}.$$

   c. We know

$$60.0 \text{ mi/h} = (60.0 \text{ mi}/1 \text{ h})(5280 \text{ ft}/1 \text{ mi})(0.305 \text{ m}/1 \text{ ft})(1/1000 \text{ km}/1 \text{ m}) = \mathbf{96.6 \text{ km/h}}.$$

   d. In this case

$$9.80 \text{ m/s}^2 = (9.80 \text{ m/s}^2)(3.28 \text{ ft}/1 \text{ m}) = \mathbf{32.1 \text{ ft/s}^2}.$$

   e. Finally

$$535 \text{ kg} = (535 \text{ kg})(6.85 \text{ X } 10^{-2} \text{ sl}/1 \text{ kg}) = \mathbf{36.6 \text{ sl}}.$$

---

3. The conversion looks like

$$180.0 \text{ cm} = (180.0 \text{ cm})/(1 \text{ in}/2.54 \text{ cm}) = 70.9 \text{ in} = 5 \text{ ft } 10.9 \text{ in. } \mathbf{YES}$$

---

4. In this case

$$100.0 \text{ yds} = (100.0 \text{ yds})(3 \text{ ft/yd})(12 \text{ in/ft})(2.54 \text{ cm/in})(10 \text{ mm/cm}) = \mathbf{91\ 440 \text{ mm}}.$$

---

5. a. Substitute the units into the equation to see if both sides match.

$$\text{m}^2 = (\text{m/s}^2)(\text{s}) = \text{m/s}$$

   The units on each side do NOT match so **the equation is not valid**.

   b. Similarly,

$$\text{m}^2/\text{s}^2 = (\text{m/s}^2)(\text{m}) = \text{m}^2/\text{s}^2.$$

   The units DO match so the **equation MAY be valid**.

---

6. The hypotenuse is given by (1.1)

$$h = (3.5 \text{ m})/\sin 25° = 8.3 \text{ m}.$$

   The side adjacent is given by (1.2),

$$h_a = (8.3 \text{ m})\cos 25° = \mathbf{7.5 \text{ m}}.$$

---

7. The shadow, the building, and a line drawn from the top of the building to the corresponding point on the end of the shadow form a right triangle. The angle between this latter line and the horizontal is the elevation angle of the sun and is

$$\theta = \tan^{-1} (500.0 \text{ m}/800.0 \text{ m}) = \mathbf{32.01°}.$$

---

8. The length of the hypotenuse is given by (1.7),

$$h^2 = (25 \text{ ft})^2 + (15 \text{ ft})^2 ,$$

$$h = \mathbf{29\ ft}.$$

The angle opposite the 25 ft side is given by (1.6),

$$\theta = \tan^{-1} (25/15). \quad \theta = \mathbf{59°}.$$

The angle opposite the 15 ft side is

$$\theta = \tan^{-1} (15/25) = \mathbf{31°}.$$

Note that the angles add to 90° as they should.

---

9. We are given the hypotenuse and one angle in a right triangle. (1.1) gives the side opposite the angle to be

$$h_o = (50.0 \text{ m}) \sin 20.0° = \mathbf{17.1\ m}.$$

---

10. The vectors are perpendicular; hence the magnitude of the vector sum can be found from the Pythagorean theorem.

$$\text{magnitude} = \mathbf{11\ m.}$$

The angle the vector makes with the east-west line (+x) is

$$\theta = \tan^{-1} (9.7/5.0) = \mathbf{63°\ N\ of\ E}.$$

---

11. The x component is

$$E_x = (1.0 \text{ N/C}) \cos 33° = \mathbf{0.84\ N/C}.$$

The y component is

$$E_y = (1.0 \text{ N/C}) \sin 33° = \mathbf{0.54\ N/C}.$$

---

12. The angle as measured from the +x axis is $\theta = 270° - 65° = 205°$. Then

$$B_x = (0.010 \text{ T}) \cos 205° = \mathbf{-\ 0.0091\ T}$$

and

$$B_y = (0.010 \text{ T}) \sin 205° = \mathbf{-\ 0.0042\ T}.$$

---

13. The x components are:

$$A_x = -(10.0 \text{ u}) \cos 21.0° = -9.34 \text{ u}, \quad B_x = (5.2 \text{ u}) \sin 47.0° = 3.80 \text{ u}, \quad R_x = -\ 5.54 \text{ u}.$$

The y components are:

$$A_y = (10.0\text{ u}) \sin 21.0° = 3.58\text{ u}, \quad B_y = -(5.2\text{ u}) \cos 47.0° = -3.55\text{ u}, R_y = 0.03\text{ u}.$$

The magnitude of the resultant vector is found from the Pythagorean theorem to be

$$R = \sqrt{(-5.54\text{ u})^2 + (0.03\text{ u})^2} = \mathbf{5.54\text{ u}.}$$

The angle is

$$\theta = \tan^{-1}(0.03/5.54) = \mathbf{0.310°\ N\ of\ W.}$$

---

14. Let **A** be the westerly displacements, **B** the southerly displacement, and **C** be the remaining displacement. Taking east and north to be positive, the east-west (x) components are:

$A_x = -2$ km, $B_x = 0$ km, $\quad C_x = +(10.0\text{ km}) \cos 53° = 6$ km. $\quad R_x = +4$ km.

The north-south (y) components are:

$A_y = 0$ km, $\quad B_y = -8$ km, $\quad C_y = +(10.0\text{ km}) \sin 53° = +8$ km. $\quad R_y = 0$ km.

The magnitude of the resultant vector is given by the Pythagorean theorem to be

$$R = \sqrt{(4\text{ km})^2 + (0\text{ km})^2} = \mathbf{4\text{ km}}$$

and the direction is

$$\theta = \tan^{-1}(0/4) = 0°.\ \text{That is, } \mathbf{EAST}.$$

Quiz answers

1. b
2. b
3. b
4. a
5. a
6. b
7. c
8. b
9. c
10. a
11. a
12. d
13. d

MCAT REVIEW PROBLEMS

When an airplane flies, its total velocity with respect to the ground is

$$\mathbf{v}_{total} = \mathbf{v}_{plane} + \mathbf{v}_{wind},$$

where $\mathbf{v}_{plane}$ denotes the plane's velocity through motionless air, and $\mathbf{v}_{wind}$ denotes the wind's velocity. Crucially, all the quantities in this equation are vectors. The magnitude of a velocity vector is often called the "speed."

Consider an airplane whose speed through motionless air is 100 meters per second (m/s). To reach its destination, the plane must fly east.

The "heading" of a plane is the direction in which the nose of the plane points. So, it is the direction in which the engines propel the plane.

---

1. If the plane has an eastward heading, and a 20 m/s wind blows towards the southwest, then the plane's speed is:

   A. 80 m/s **B.** more than 80 m/s but less than 100 m/s

   C. 100 m/s **D.** more than 100 m/s

---

2. The pilot maintains an eastward heading while a 20 m/s wind blows northward. The plane's velocity is deflected
from due east by what angle?

   A. $\sin^{-1} \frac{20}{100}$ **B.** $\cos^{-1} \frac{20}{100}$ **C.** $\tan^{-1} \frac{20}{100}$ **D.** none of the above

---

3. Let ϕ denote the answer to question 2. The plane in question 2 has what speed with respect to the ground?

   A. $(100 \text{ m/s})\sin\phi$ **B.** $(100 \text{ m/s})\cos\phi$

   C. $\frac{100 \text{ m/s}}{\sin\phi}$ **D.** $\frac{100 \text{ m/s}}{\cos\phi}$

---

4. Because the 20 m/s northward wind persists, the pilot adjusts the heading so that the plane's total velocity is eastward. By what angle does the new heading differ from due east?

   A. $\sin^{-1} \frac{20}{100}$ **B.** $\cos^{-1} \frac{20}{100}$ **C.** $\tan^{-1} \frac{20}{100}$ **D.** none of the above

---

5. Let θ denote the answer to question 4. What is the total speed, with respect to the ground, of the plane in question 4?

   A. $(100 \text{ m/s})\sin\theta$ **B.** $(100 \text{ m/s})\cos\theta$ **C.** $\frac{100 \text{ m/s}}{\sin\theta}$ **D.** $\frac{100 \text{ m/s}}{\cos\theta}$

ANSWERS TO MCAT REVIEW PROBLEMS

---

1. **B.** In all of these problems, a good diagram gets you half way to the answer. For this plane, $\mathbf{v}_{\text{plane}} = 100$ m/s eastward, and $\mathbf{v}_{\text{wind}} = 20$ m/s towards the southwest. To represent the total velocity, add these two vectors:

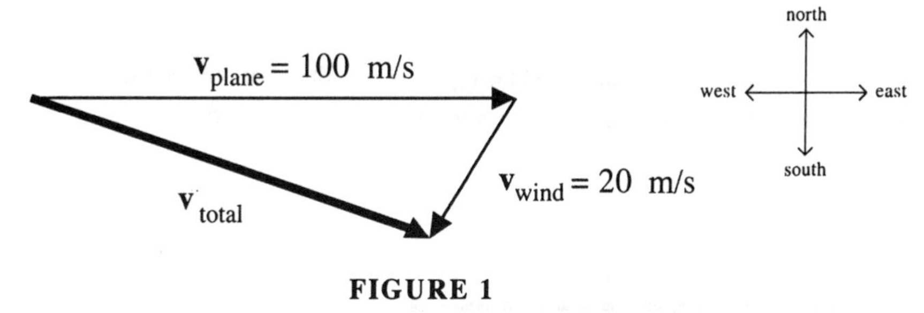

FIGURE 1

Just by looking at this rough diagram, you can see that the wind slows down the plane. So, the answer must be A or B. Since the wind does not blow westward, it does not directly oppose the plane's motion. Therefore, the wind cannot slow the plane down by a full 20 m/s from the plane's "default" velocity of 100 m/s.

---

2. **C.** Because the pilot maintains an eastward heading, the engines still push the plane due east. So, $\mathbf{v}_{\text{plane}} =$ 100 m/s eastward. But the wind blows at $\mathbf{v}_{\text{wind}} = 20$ m/s northward, thereby knocking the plane off course by angle ϕ.

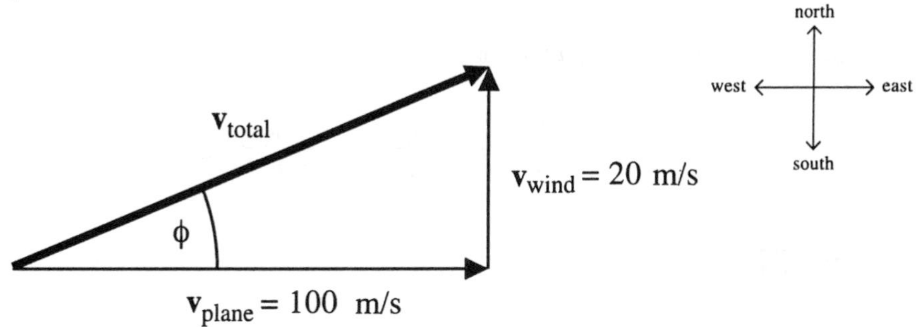

FIGURE 2

In this triangle, we know the two legs but not the hypotenuse. Specifically, in units of m/s, the "opposite" leg has length $h_o = 20$, while the "adjacent" leg has length $h_a = 100$. Therefore, by the definition of tangent,

$$\tan\phi = \frac{h_o}{h_a} = \frac{20}{100}.$$

Take the inverse tangent of both sides to get $\phi = \tan^{-1}\frac{20}{100}$.

---

3. **D.** By looking at figure 2, and remembering the definition of cosine, we get

$$\cos\phi = \frac{h_a}{h} = \frac{v_{\text{plane}}}{v_{\text{total}}} = \frac{100 \text{ m/s}}{v_{\text{total}}}.$$

Multiply through by v_{total}, and then divide through by cos ϕ, to get

$$v_{total} = \frac{100\text{ m/s}}{\cos\ \phi}$$

---

4. **A.** The pilot now adjusts the heading so that $\mathbf{v}_{total}$ points due east. Since the wind knocks the plane northward, the plane must head slightly southward, to "offset" the wind:

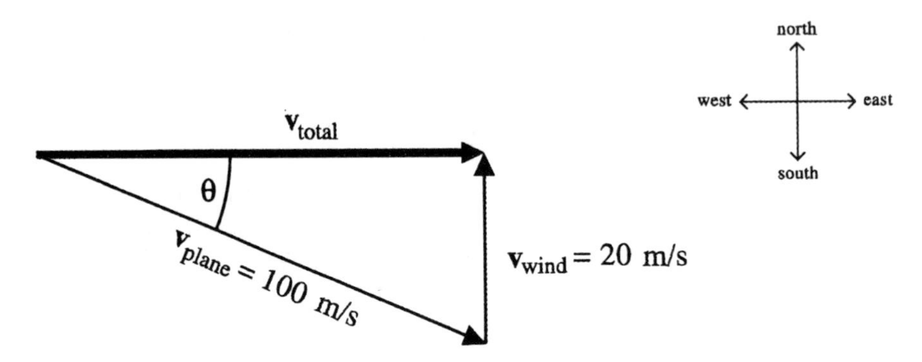

FIGURE 3

Here, $\mathbf{v}_{plane}$ is the hypotenuse. From figure 3, we get

$$\sin\theta = \frac{h_o}{h} = \frac{v_{wind}}{v_{plane}} = \frac{20\text{ m / s}}{100\text{ m / s}},$$

and hence, $\theta = \sin^{-1}\frac{20}{100}$.

---

5. **B.** Again from figure 3, the hypotenuse has length $h = v_{plane}$ = 100 m/s, while the adjacent leg has length $h_a = v_{total}$. So,

$$\cos\theta = \frac{h_a}{h} = \frac{v_{total}}{v_{plane}} = \frac{v_{total}}{100\text{ m / s}}$$

Multiply through by 100 m/s to get

$$(100\text{ m/s})\cos\theta = v_{total}.$$

CHAPTER 2
Kinematics in One Dimension

PREVIEW

In this chapter you will learn how to describe the motion of an object in one dimension, that is, straight line motion. You will be introduced to the concepts of displacement, speed, velocity, and acceleration. Equations for straight line motion with constant acceleration are developed, with applications including freely falling bodies.

QUICK REFERENCE

Important Terms

Mechanics
The branch of physics that deals with the motion of objects and the forces that change it.

Kinematics
A branch of mechanics which describes the motion of an object without explicit reference to the forces that act on the object.

Displacement
A vector representing the change in the position of an object, drawn from the initial to the final position. The most common units for displacement are meters (m), centimeters (cm), or feet (ft).

Average Speed
The distance (a scalar quantity) an object moves in a given amount of time divided by the time. Units are (m/s), (cm/s), or (ft/s).

Average Velocity
The displacement (a vector quantity) of an object divided by the elapsed time. Units are the same as average speed.

Average Acceleration
The change in the velocity (a vector quantity) of an object divided by the elapsed time. Units are (m/s^2), (cm/s^2), or (ft/s^2).

Instantaneous Velocity
The rate of change of displacement. Units are (m/s), (cm/s), or (ft/s).

Instantaneous Speed
The magnitude of the instantaneous velocity vector. Units are the same as instantaneous speed.

Instantaneous Acceleration
The rate of change of velocity. Units are (m/s^2), (cm/s^2), or (ft/s^2).

Free Fall
An idealized motion, in which air resistance is neglected and the acceleration is nearly constant. This acceleration is due to gravity and taken to be 9.80 m/s^2 or 32.2 ft/s^2 at the surface of the earth.

Kinematic Equations for One Dimensional Motion (Constant Acceleration)

$$v = v_0 + a t \qquad (2.4) \qquad\qquad x = \frac{1}{2}(v_0 + v)\, t \qquad (2.7)$$

$$x = v_0 t + \frac{1}{2} a t^2 \qquad (2.8) \qquad\qquad v^2 = v_0^2 + 2 a x \qquad (2.9)$$

DISCUSSION OF SELECTED SECTIONS

2.1 Displacement

The displacement is a **vector quantity** whose magnitude represents the shortest distance between the initial and final points, and whose direction points from the initial to the final position.

Example 1
A car travels 3.00 km due west, then 4.00 km due south, as shown in the figure. What is the car's displacement? The displacement vector, **s**, is the vector sum of vectors **A** and **B**, that is $\mathbf{s} = \mathbf{A} + \mathbf{B}$.

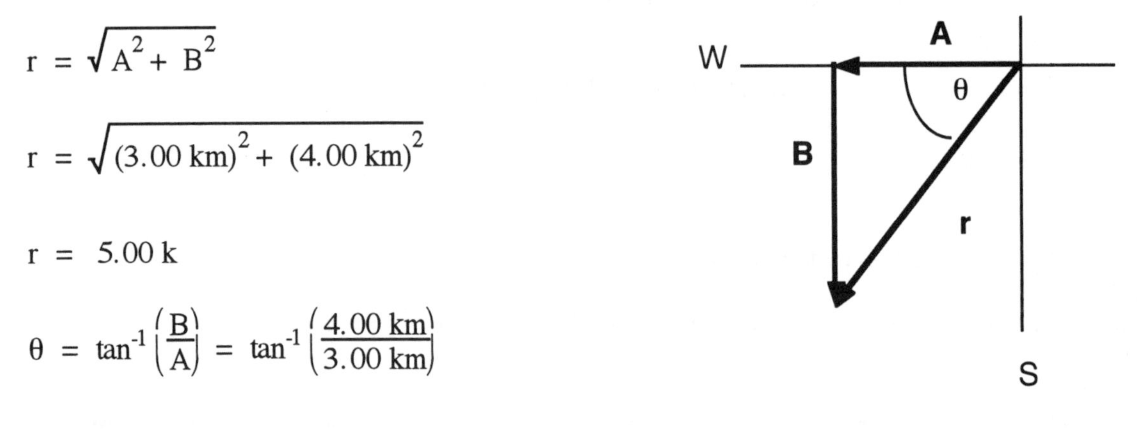

$$r = \sqrt{A^2 + B^2}$$

$$r = \sqrt{(3.00\text{ km})^2 + (4.00\text{ km})^2}$$

$$r = 5.00\text{ k}$$

$$\theta = \tan^{-1}\left(\frac{B}{A}\right) = \tan^{-1}\left(\frac{4.00\text{ km}}{3.00\text{ km}}\right)$$

$$\theta = 53.1°\text{ South of West.}$$

2.2 Speed and Velocity

The **average speed** is a scalar quantity defined as the distance traveled divided by the amount of time taken to travel this distance.

Example 2
If, in Example 1, the car takes 10.0 minutes to travel from its initial to its final position, what is its average speed?

$$\text{Average Speed} = \frac{\text{Distance}}{\text{Elapsed Time}}$$

$$= \left(\frac{7.00\text{ km}}{10.0\text{ min}}\right)\left(\frac{1000\text{ m}}{1\text{ km}}\right)\left(\frac{1\text{ min}}{60\text{ s}}\right) = 11.7\text{ m/s.}$$

The **average velocity** is a vector quantity defined as the displacement divided by the elapsed time. In general, the average speed is not the same as the magnitude of the average velocity. There is one special case where these two quantities will be equal. When an object travels in a straight line and does *not* change the direction of motion, the average speed of the object over any time interval will be equal to the magnitude of the average velocity over the same time interval.

Example 3
Find the average velocity of the car in Examples 1 and 2.

$$\bar{\mathbf{v}} = \frac{\text{Displacement}}{\text{Elapsed Time}} = \left(\frac{5.00\text{ km}}{10.0\text{ min}}\right)\left(\frac{1000\text{ m}}{1\text{ km}}\right)\left(\frac{1\text{ min}}{60\text{ s}}\right) = 8.33\text{ m/s at }53.1^{\circ}\text{ S of W.}$$

Notice that the average speed and the magnitude of the average velocity are **not** the same.

The **instantaneous velocity** indicates both how fast and in what direction an object is moving at a particular instant of time. To determine the instantaneous velocity we measure the time Δt it takes the object to travel a very small displacement $\Delta \mathbf{r}$. The instantaneous velocity is then given by

$$\mathbf{v} = \lim_{\Delta t \to 0} \frac{\Delta \mathbf{x}}{\Delta t} \qquad (2.3)$$

The notation $\Delta t \to 0$ essentially means that we want to isolate an instant of time and find the velocity during that instant. The **instantaneous speed** is simply the magnitude of the instantaneous velocity.

2.3 Acceleration

Acceleration is any change in velocity for an object divided by the time interval over which the change occurs. This means a change in MAGNITUDE and/or DIRECTION. We define the **average acceleration** of an object as

$$\bar{\mathbf{a}} = \frac{\mathbf{v} - \mathbf{v}_0}{t - t_0} = \frac{\Delta \mathbf{v}}{\Delta t} \qquad (2.4)$$

Example 4
A car traveling at 10 m/s in a straight line speeds up to 30 m/s in 5.0 s. What is the average acceleration?

$$\bar{\mathbf{a}} = \frac{\mathbf{v} - \mathbf{v}_0}{\Delta \mathbf{t}} = \frac{30\text{ m/s} - 10\text{ m/s}}{5.0\text{ s}} = \frac{+20\text{ m/s}}{5.0\text{ s}} = +4\frac{\text{m/s}}{\text{s}} = +4\text{ m/s}^2$$

We have assumed that the direction in which the car moves is the + direction, and the + sign in the above result indicates the acceleration is in this same direction.

The **instantaneous acceleration** can be defined by analogy with the instantaneous velocity:

$$\mathbf{a} = \lim_{\Delta t \to 0} \frac{\Delta \mathbf{v}}{\Delta t} \qquad (2.5)$$

2.4 Equations of Kinematics for Constant Acceleration

The motion of an object traveling with constant acceleration along a straight line can be described by equations (2.4), (2.7), (2.8), and (2.9), as listed in the **Quick Reference** section. Note that these equations are valid only for uniform or constant acceleration. If the acceleration is changing, these equations cannot be used. One exception is if the motion can be broken up into segments in which the acceleration is constant for each segment. The kinematic equations can then be applied to each segment separately.

Example 5
Starting from rest, a car accelerates for 8.0 s to a final speed of 16 m/s. Find the acceleration of the car and the distance traveled during this time interval.

Using equation (2.4) with the initial velocity equal to zero we obtain

$$a = (v - v_0)/t = (16\ \text{m/s} - 0\ \text{m/s})/(8.0\ \text{s}) = +2.0\ \text{m/s}^2.$$

To find the distance traveled use equation (2.8)

$$x = v_0 t + (1/2)\,at^2 = 0 + (1/2)\,(2.0\ \text{m/s}^2)\,(8.0\ \text{s})^2 = 64\ \text{m}.$$

Note that this last result can also be obtained using equation (2.7)

$$x = (1/2)\,(v_0 + v)t = (1/2)\,(0\ \text{m/s} + 16\ \text{m/s})\,(8.0\ \text{s}) = 64\ \text{m}.$$

2.6 Freely Falling Bodies

In the absence of air resistance, it is found that all bodies near the surface of the earth fall vertically with the same acceleration. The acceleration of a freely falling body is called the **acceleration due to gravity**, and its magnitude is given by the symbol 'g'. This acceleration is always directed downward, toward the center of the earth. Near the surface of the earth g is approximately constant and has a value of

$$g = 9.80\ \text{m/s}^2 \quad \text{or} \quad 32.2\ \text{ft/s}^2.$$

Example 6
A ball is dropped from the top of a 123 m high cliff. Ignoring air resistance find the final velocity of the ball, and the amount of time it takes for it to hit the ground below.

The final velocity can be obtained using equation (2.9) and taking the direction of motion as the y direction.

$$v = \sqrt{v_0^2 + 2\,a\,y} = \sqrt{(0\ \text{m/s})^2 + 2\,(+9.80\ \text{m/s}^2)\,(123\ \text{m})}$$

$v = \mathbf{49.1\ m/s}.$

The time of flight can be obtained using equation (2.4)

$$v = v_0 + a\,t \quad \text{or}$$

$$t = \frac{v - v_0}{a} = \frac{49.1\ \text{m/s} - 0\ \text{m/s}}{+9.80\ \text{m/s}^2} = \mathbf{5.0\ s}.$$

2.7 Graphical Analysis of Velocity and Acceleration

Consider an object moving in a straight line at a constant speed. The position, x, is plotted as a function of time, t, in the following graph.

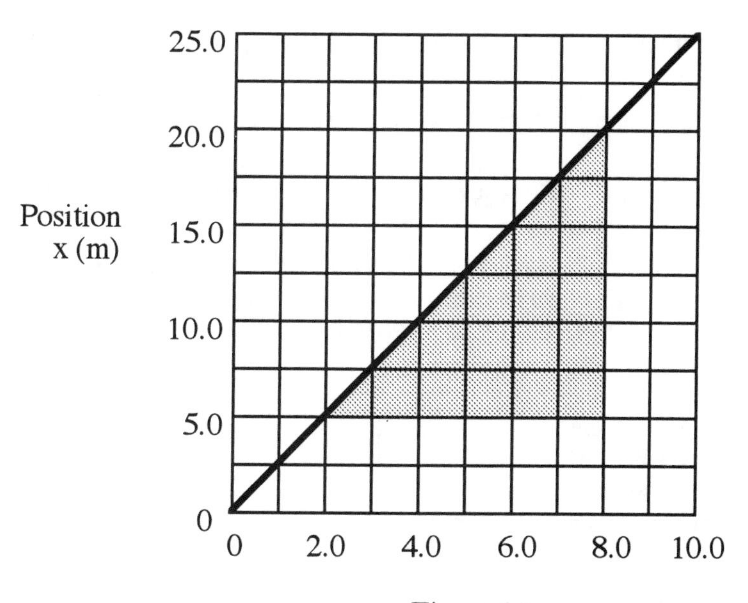

We can see from the graph that the position is increasing linearly with time, i.e., the graph is a straight line. Between the times 2.0 and 8.0 s note that the object travels from 5.0 to 20.0 m. The change in time is Δt = 6.0 s, while the change in position is Δx = +15.0 m. The ratio of $\Delta x/\Delta t$ is called the **slope** of the straight line. That is

$$\text{Slope} = \frac{\Delta x}{\Delta t} = \frac{+15.0 \text{ m}}{6.0 \text{ s}} = +2.5 \text{ m/s}$$

Note that the slope has units of velocity (m/s). In fact, the slope of the position vs. time graph does represent the average velocity of the moving object, as evidenced by the fact that the above equation is identical to equation (2.2). In the real world, however, the situation may not be quite as simple. Consider the following example.

Example 7 *Determining the slope of a position vs. time graph*
Consider the following position vs. time graph and determine the velocity over each segment.

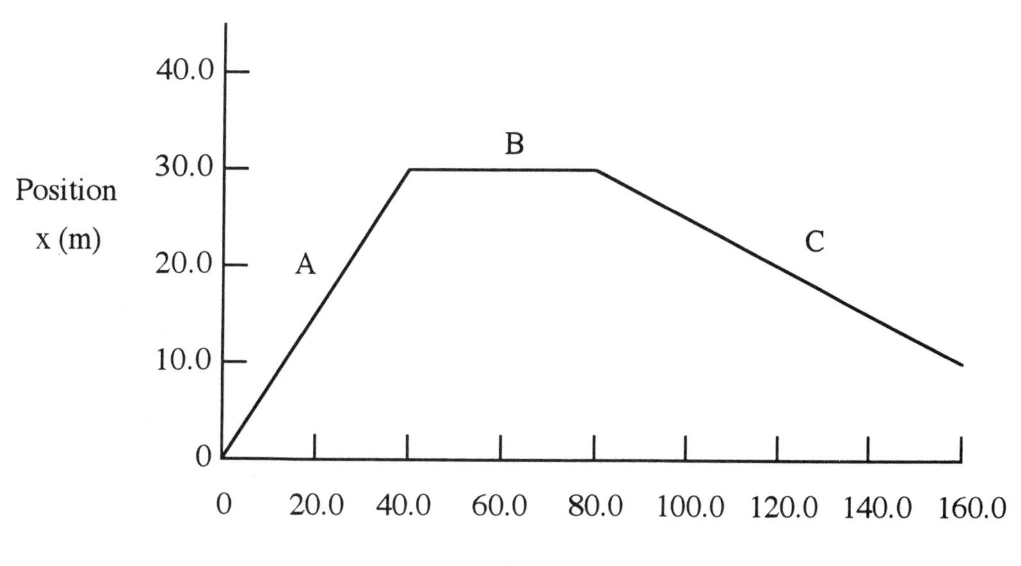

Segment A: $v = \Delta x/\Delta t = (+30.0\ m)/(40.0\ s) = +0.750\ m/s.$

Segment B: $v = \Delta x/\Delta t = (0\ m)/(20.0\ s) = 0\ m/s.$

Segment C: $v = \Delta x/\Delta t = (-20.0\ m)/(80.0\ s) = -0.250\ m/s.$

In segment A the object is traveling with a positive velocity of 0.75 m/s. Over segment B the object does not change position, so its velocity is zero. Over segment C the slope of the graph is negative and hence the velocity is negative, i.e., the object is now moving in the negative direction.

Example 8 *Accelerated motion, slope of a tangent line*
Using the x vs. t graph shown at the right, we can determine the velocity of the object at say, t = 40 s. We see that the velocity is changing with time. The velocity at any instant of time can be determined by measuring the slope of the tangent to the curve at that point.

This will then provide us with a value of the **instantaneous velocity** at the point in question.

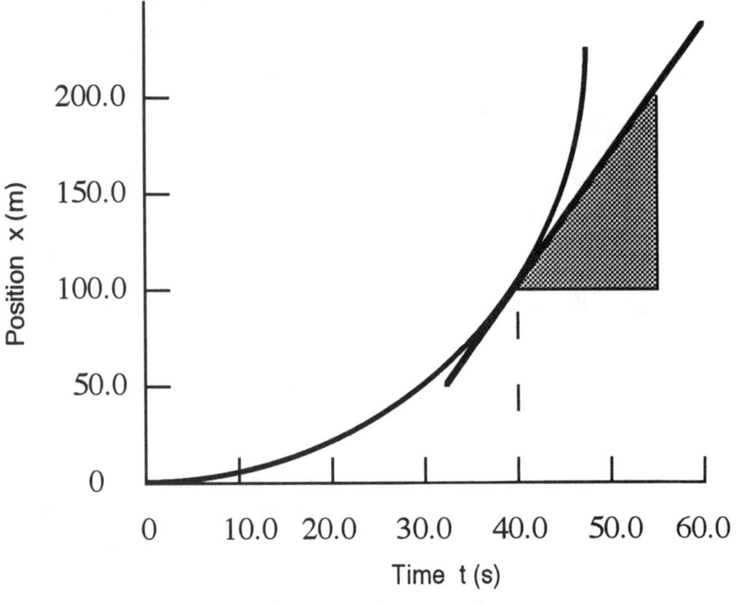

The slope at t = 40.0 s can be determined by referring to the shaded triangle. The slope is

$$v = \text{slope} = \Delta x/\Delta t = (200.0\ m - 100.0\ m)/(55.0\ s - 40.0\ s) = (+100.0\ m)/(15.0\ s) = +6.67\ m/s.$$

If we were to plot a graph of velocity vs. time for the data in the drawing for Example 8 we would obtain the graph shown at the right.

The slope of the above graph yields

$$\begin{aligned}\text{Slope} &= \Delta v/\Delta t \\ &= (6.7\ m/s - 2.0\ m/s)/(40.0\ s - 12.0\ s) \\ &= (+4.7\ m/s)/(28.0\ s) = +0.17\ m/s^2.\end{aligned}$$

Notice that the slope of the velocity vs. time graph represents the acceleration of the object over this time interval.

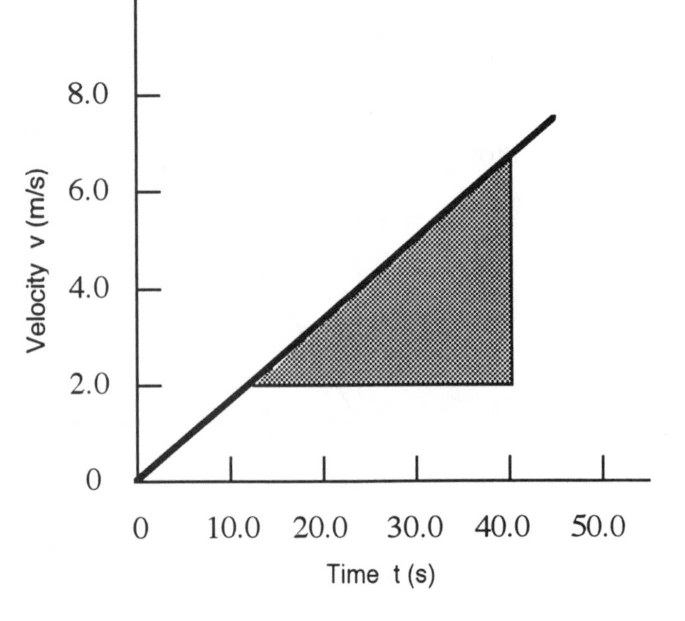

PRACTICE PROBLEMS

1. A car travels 10.0 km due north, then 8.00 km due east, the total trip taking 30.0 minutes. Find (a) the displacement of the car, (b) the average speed (in m/s), and (c) the average velocity (in m/s).

2. A bicycle accelerates to a velocity of 25.0 mi/h in 5.00 s. The average acceleration over this time interval is 4.00 ft/s^2. What was the initial velocity of the bike?

3. A train accelerates from rest to a speed of 25.0 m/s in 10.0 s. Assuming the acceleration to be constant over this interval find (a) the magnitude of the acceleration, and (b) the distance traveled during this interval.

4. A drag racer, starting from rest, travels a quarter mile in 6.0 s. What is the racer's speed (in mi/h) at the end of the race?

5. Suppose a car traveling at 12.0 m/s sees a traffic light turn red. After 0.510 s have elapsed, the driver applies the brakes, and the car decelerates at 6.20 m/s^2. What is the stopping distance of the car, as measured from the point where the driver first notices the red light?

6. A car starts from rest heading due east. It first accelerates at 3.0 m/s^2 for 5.0 s and then continues without further acceleration for 20.0 s. It then brakes for 8.0 s in coming to rest. (a) What is the car's velocity after the first 5.0 seconds? (b) What is the car's acceleration over the last 8.0 s interval? (c) What is the total displacement?

7. A ball is thrown upward with an initial velocity of 20.0 m/s. What maximum height does the ball reach?

8. A stone is thrown vertically downward from a 2.00×10^2 m high cliff at an initial velocity of 5.00 m/s. (a) How long does it take for the stone to reach the base of the cliff? (b) What is the stone's final velocity?

9. A sports car, picking up speed, passes between two markers in a time of 4.1 s. The markers are separated by 120 m. All the while, the car accelerates at 1.8 m/s^2. What is its speed at the second marker?

10. A heavy ball is dropped into a lake from a height of 30.0 m above the water. It hits the water with a certain velocity and continues to sink to the bottom of the lake at this same constant velocity. It reaches the bottom of the lake 10.0 s after it was dropped. How deep is the lake?

11. Using the position-time graph shown below, draw the corresponding velocity-time graph.

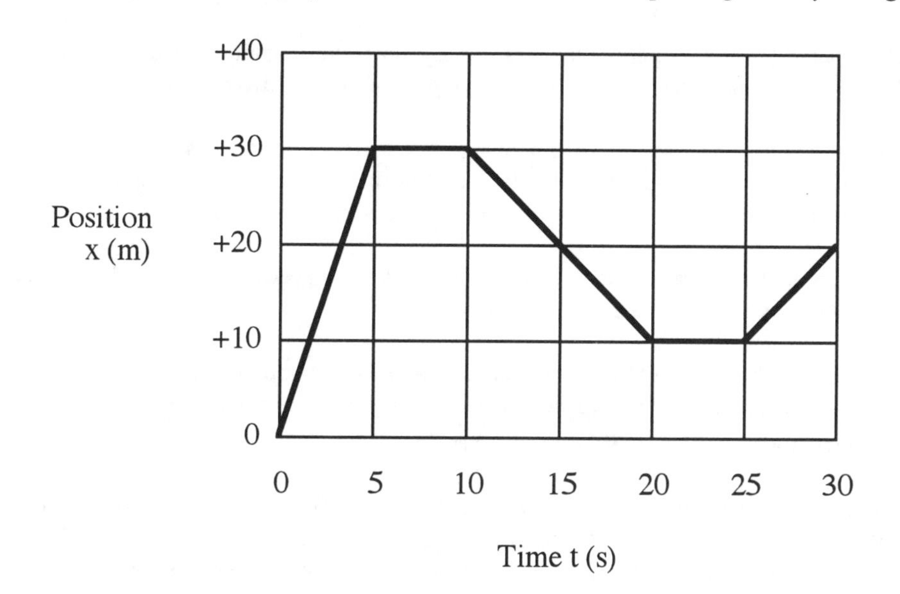

12. A snowmobile moves according to the velocity-time graph shown below. What is the snowmobile's average acceleration during each of the segments labeled A, B, and C?

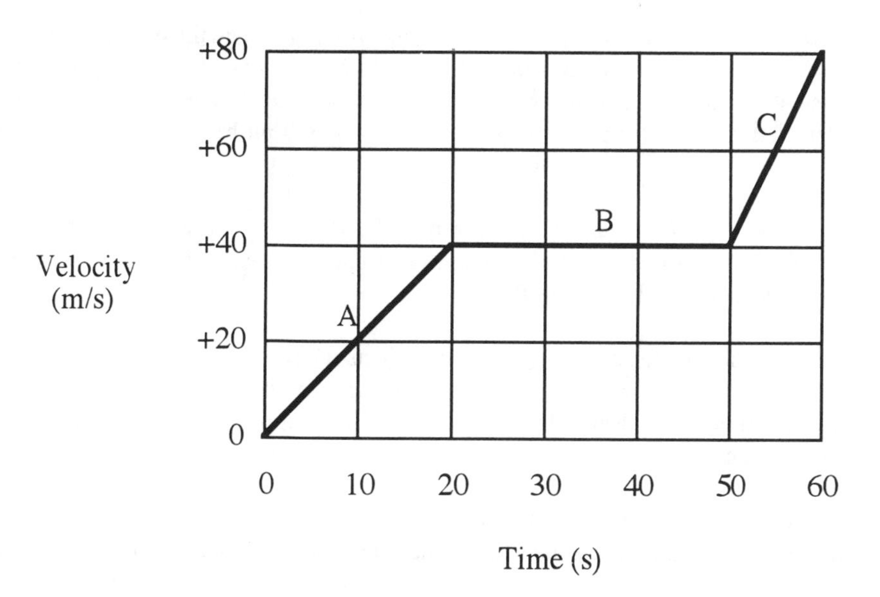

HELPFUL SUGGESTIONS

1. The **velocity** is a vector quantity which requires that you specify both the magnitude and direction. The **instantaneous speed** is a scalar quantity and represents the magnitude of the **instantaneous** velocity.

2. An object may be traveling at a constant speed but with a changing velocity. Consider a satellite in a circular orbit around the earth. It travels with a steady speed of about 18 000 mi/h but is constantly changing direction since it travels in a circle. It is therefore changing velocity because it is changing direction.

3. In one dimension, the direction of the displacement, velocity, and acceleration vectors is denoted by + and - signs. It is therefore important to pay attention to the sign of these vectors.

4. It is possible for an object to have zero velocity but non-zero acceleration. For example, an object thrown upward has zero velocity when it reaches its maximum vertical height. However, its acceleration is non-zero since it is constantly subject to the acceleration of gravity, 9.80 m/s^2, directed downward.

5. The kinematic equations of motion, equations (2.4), (2.7), (2.8), and (2.9), can **only** be used if the acceleration is in fact constant or uniform. If the acceleration is continually changing, they cannot be used **unless** the motion of the object can be broken up into segments, the acceleration being constant over each segment.

6. For freely falling objects near the earth's surface, the acceleration of gravity always has a magnitude of 9.80 m/s^2 or 32.2 ft/s^2 and is directed toward the center of the earth. Air resistance is usually neglected although this would not be the case in real life.

EVERYDAY PHYSICS

1. An example of the difference between speed and velocity is the following: you jog around the block, a distance of a half mile, in 3 minutes. You've traveled 0.5 mi in 3 min (1/20 of an hour) so your average speed was 10 mi/h. However, since you finish at the same point that you started, your displacement is zero. Since velocity is displacement divided by the elapsed time, the average velocity for your jog was 0 mi/h! You've certainly gone no place fast (or slow).

2. Suppose you wanted to find out how deep a well was. Drop a rock into the well and time how long it takes for you to hear the sound. Since you know the speed of sound (343 m/s) and the kinematic equations of motion for constant acceleration, you can calculate the depth of the well.

3. Try dropping objects of different weight from the same height (avoid objects which would be strongly affected by air resistance, such as a sheet of paper or a feather). You will notice that all the objects dropped from the same height reach the ground at the same time. This demonstrates that the acceleration of gravity is indeed the same for all objects, regardless of their weight.

4. You can determine how far away lightning is by counting the number of seconds between when you see the lightning flash and when you hear the thunder. Since sound travels at about 343 m/s (about 1100 ft/s), it takes sound about 5 seconds to travel a mile. Since the flash of light is seen almost instantly (the speed of light is 300 000 000 meters per second), every five second delay between the flash and the sound represents a distance of one mile.

5. Think of some examples of acceleration in your everyday life, not only such obvious things as speeding up your car using the accelerator or slowing down using the brakes (i.e., decelerating) but also changing direction (like going around a turn).

CHAPTER QUIZ

1. A car travels for 30.0 min in a straight line for a distance of 50.0 miles. What is its average speed?
   a. 1.67 mi/hr b. 25.0 mi/h c. 100 mi/h d. 150 mi/h

2. A plane flies 300 km due north, then 400 km due east. What is the magnitude of its displacement?
   a. 100 km b. 500 km c. 700 km d. 1000 km

3. A box slides across a horizontal floor in a straight line. It starts with a speed of 5 m/s and slows down to 2 m/s after 3 seconds. What is the average acceleration over this interval?
   a. -1 m/s^2 b. +1 m/s^2 d. -3 m/s^2 d. -9 m/s^2

4. A ball is thrown upward. Neglecting air resistance, what is the instantaneous acceleration at the top of its path?
   a. It is zero.
   b. It is 9.80 m/s^2 downward.
   c. It is 9.80 m/s^2 upward.
   d. It is less than 9.80 m/s^2 downward.

5. An object starts from rest and undergoes constant acceleration. In the first second it travels 5.00 meters. How far does it travel in the **next** second?
   a. 5.00 m b. 10.0 m c. 15.0 m d. 20.0 m

6. Ball A is dropped from the top of a tall building. Ball B is thrown downward from the same building. Both balls start at the same time. When they reach the bottom of the building which statement is correct?
   a. B has a greater velocity and acceleration than A.
   b. B has a greater velocity but A has a greater acceleration.
   c. B has a greater velocity but its acceleration is the same as A's.
   d. Not much can be said unless we know the height of the building.

7. The slope of a position vs. time graph gives
   a. average acceleration.
   b. average velocity.
   c. instantaneous acceleration.
   d. instantaneous velocity.

8. The slope of a velocity vs. time graph gives
   a. average acceleration.
   b. average velocity.
   c. instantaneous acceleration.
   d. instantaneous velocity.

9. Two cars are accelerating on the road. It is observed that the distance between the cars is increasing. What does this imply about their accelerations?
   a. It implies that the leading car has the greater acceleration.
   b. It implies that the trailing car has the smaller acceleration.
   c. It implies that both cars are accelerating at the same rate.
   d. It implies nothing about their accelerations.

10. An arrow is launched vertically upward at a speed of 50 m/s. What is the arrow's speed when it returns to the ground? Ignore air resistance.
    a. greater than 50 m/s.
    b. less than 50 m/s.
    c. 50 m/s.
    d. Impossible to say without knowing the maximum height reached by the arrow.

11. An object is thrown upward from the top of a 10.0 m high building at a speed of 10.0 m/s. How fast is the object moving when it reaches the bottom of the building?
    a. 10.0 m/s b. 14.0 m/s c. 17.2 m/s d. 24.0 m/s

SOLUTIONS AND ANSWERS

Practice Problems

---

1. a. The total displacement is the vector sum of the individual displacements. Since these displacements are perpendicular, the vector sum is given by the Pythagorean theorem.

$$r^2 = (10.0 \text{ km})^2 + (8.00 \text{ km})^2$$
$$r = 12.8 \text{ km}.$$

The angle is found using

$$\tan \theta = (10.0 \text{ km}/8.00 \text{ km})$$
$$\theta = \mathbf{51.3° \text{ N of E.}}$$

b. Avg. Speed = Distance/Time = (18.0 km/30.0 min)(1000 m/1 km)(1 min/60 s) = **10.0 m/s**.
c. Avg. Velocity = Displacement/Time = (12.8 km/30.0 min)(1000 m/1 km)(1 min/60 s) = **7.11 m/s.**

---

2. The average acceleration is given by equation (2.4),

$$a = \Delta v/\Delta t = (v - v_0)/\Delta t$$

or

$$v_0 = v - at$$

where

$$v = (25.0 \text{ mi/h})(5280 \text{ ft}/1 \text{ mi})(1 \text{ h}/3600 \text{ s}) = 36.7 \text{ ft/s}.$$

Then

$$v_0 = 36.7 \text{ ft/s} - (4.00 \text{ ft/s}^2)(5.00 \text{ s}) = \mathbf{16.7 \text{ ft/s.}}$$

---

3. a. Using equation (2.4) we have

$$a = (v - v_0)/\Delta t = (+25.0 \text{ m/s})/(10.0 \text{ s}) = \mathbf{+2.50 \text{ m/s}^2.}$$

b. Using equation (2.7) we have

$$x = (1/2)(v + v_0)t = (1/2)(25.0 \text{ m/s} + 0)(10.0 \text{ s}) = \mathbf{125 \text{ m.}}$$

---

4. Using $v_0 = 0$, $x = (0.250 \text{ mi})(5280 \text{ ft/mi}) = 1320 \text{ ft}$, $t = 6.0$ s and equation (2.7) we solve for v to get

$$v = 2x/t = 2(1320 \text{ ft})/(6.0 \text{ s}) = (440 \text{ ft/s})(1 \text{ mi}/5280 \text{ ft})(3600 \text{ s}/1 \text{ h}) = \mathbf{3.0 \times 10^2 \text{ mi/h.}}$$

---

5. The motion should be divided into two segments, segment 1 is constant velocity and segment 2 is decelerating.

$$x_1 = v_0 t = (12.0 \text{ m/s})(0.510 \text{ s}) = 6.12 \text{ m};$$
$$x_2 = (v^2 - v_0^2)/2a = -(12.0 \text{ m/s})^2/[2(-6.20 \text{ m/s}^2)] = 11.6 \text{ m}.$$

The stopping distance is therefore

$$x = x_1 + x_2 = 6.12 \text{ m} + 11.6 \text{ m} = \mathbf{17.7 \text{ m.}}$$

---

6. a. $a = 3.0\ m/s^2$, $t = 5.0\ s$, $v_0 = 0$. Equation (2.4) gives

$$v = v_0 + at = (3.0\ m/s^2)(5.0\ s) = \mathbf{15\ m/s\ East.}$$

b. $v_0 = 15\ m/s$, $v = 0$, $t = 8.0\ s$. Equation (2.4),

$$a = (v - v_0)/t = (-15\ m/s)/(8.0\ s) = \mathbf{1.9\ m/s^2\ West.}$$

c. The distance is then

$$x = x_1 + x_2 + x_3 = (1/2)a_1t_1^2 + v_2t_2 + + v_2t_3 + (1/2)a_3t_3^2 = \mathbf{4.0 \times 10^2\ m.}$$

---

7. $v_0 = 20.0\ m/s$, $v = 0$, $a = -9.80\ m/s^2$. Equation (2.9) is

$$v^2 = v_0^2 + 2ax$$

Solving for x we obtain

$$x = -v_0^2/2a = -(20.0\ m/s)^2/[2(-9.80\ m/s^2)] = \mathbf{20.4\ m.}$$

---

8. a. $v_0 = 5.00\ m/s$, $a = 9.80\ m/s^2$, $x = 2.00 \times 10^2\ m$. Equation (2.8): $x = v_0t + (1/2)at^2$.
Substituting yields the quadratic equation

$$4.90\ t^2 + 5.00\ t - 2.00 \times 10^2 = 0$$

which has a solution

$$t = \mathbf{5.90\ s.}$$

b. Since $v^2 = v_0^2 + 2ax = (5.00\ m/s)^2 + 2(9.80\ m/s^2)(2.00 \times 10^2\ m)$

then

$$v = \mathbf{62.8\ m/s.}$$

---

9. $a = 1.8\ m/s^2$, $x = 120\ m$, $t = 4.1\ s$. Equation (2.8),

$$x = v_0t + (1/2)at^2$$

yields

$$v_0 = [x - (1/2)at^2]/t = 25.6\ m/s.$$

Equation (2.4), gives

$$v = v_0 + at = 25.6\ m/s + (1.8\ m/s^2)(4.1\ s) = \mathbf{33\ m/s.}$$

---

10. Find the velocity with which the ball hits the water and the time it takes to do so. Since $v_0 = 0$ we have

$$x = (1/2)at^2$$

which gives

$$t = \sqrt{(2x/a)} = \sqrt{2(30.0m)/(9.80m/s^2)} = 2.47\ s.$$

Now

$$v = at = (9.80\ m/s^2)(2.47\ s) = 24.2\ m/s.$$

The time it takes to sink is

$$t' = 10.0\ s - 2.47\ s = 7.5\ s.$$

The depth of the lake is therefore found from

$$\text{Depth} = vt' = (24.2\ m/s)(7.5\ s) = \mathbf{180\ m.}$$

11.

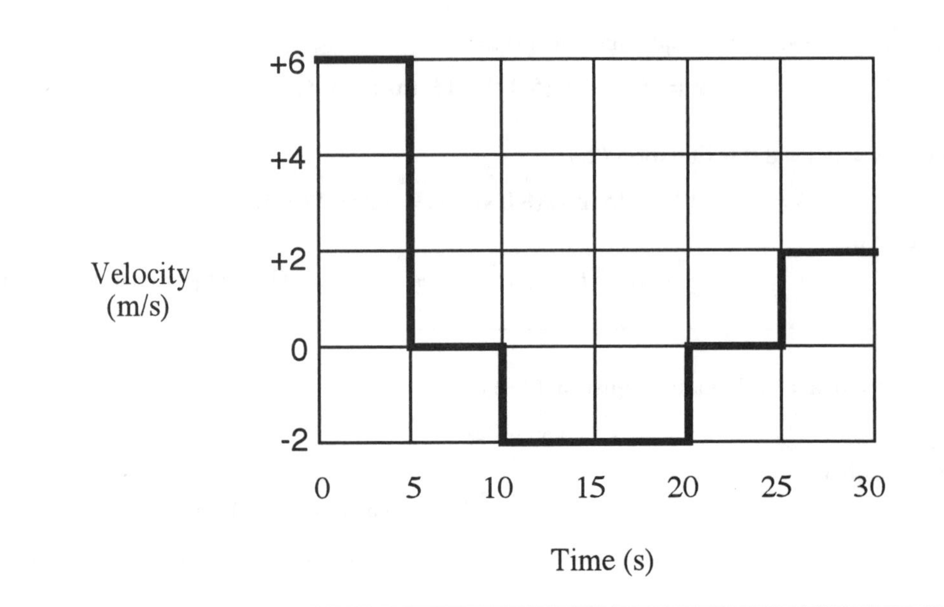

---

12. During segment:

A; $a = (40 \text{ m/s})/(20 \text{ s}) = \mathbf{2\ m/s^2}$.

B; $a = (0 \text{ m/s})/(30 \text{ s}) = \mathbf{0\ m/s^2}$.

C; $a = (40 \text{ m/s})/(10 \text{ s}) = \mathbf{4\ m/s^2}$.

Quiz answers

1. c
2. b
3. a
4. b
5. c
6. c
7. d
8. c
9. d
10. c
11. c

MCAT REVIEW PROBLEMS

During a car crash, the more rapidly a person decelerates, the more likely she is to be injured. A large deceleration is dangerous, even if it lasts for a short time.

Airbags are designed to decrease the magnitude of the deceleration. Before the airbag inflates, the driver continues forward at constant speed. But once the airbag inflates, the driver decelerates gradually, instead of getting thrown into the windshield or steering wheel.

Three different models of airbags were tested using identical crash test dummies. Sensors measured the velocity of the crash test dummy as a function of time, when the car crashed at 20 m/s into a brick wall. The following velocity vs. time graphs resulted. Time $t = 0$ is the moment the car crashes.

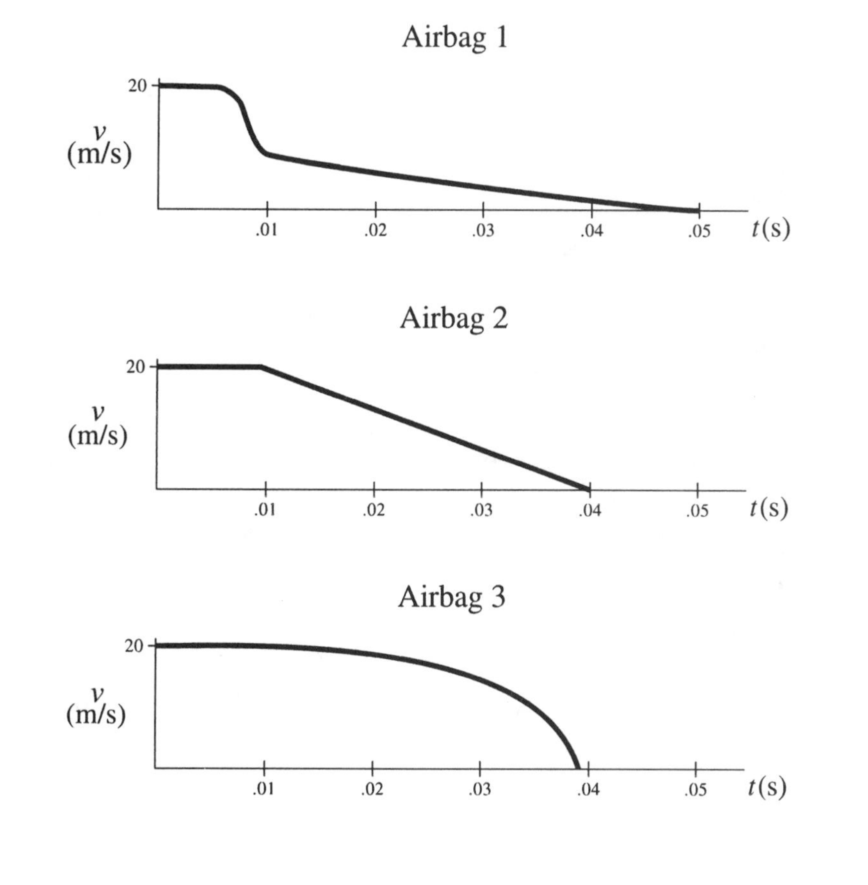

1. All three airbags are the same size and shape. Which one inflates most quickly?

A. Airbag 1 **B.** Airbag 2

C. Airbag 3 **D.** We cannot determine the answer from the given information.

2. Let a_{max} denote the largest instantaneous acceleration that the crash test dummy experiences during the crash. The best airbag is the one for which $|a_{max}|$ is as small as possible. Which airbag is best?

   A. Airbag 1 **B.** Airbag 2

   C. Airbag 3 **D.** We cannot determine the answer from the given information.

---

3. For airbag 2, which of the following graphs best represents the *position* of the crash test dummy as a function of time? Let $x = 0$ be the dummy's position at time $t = 0$.

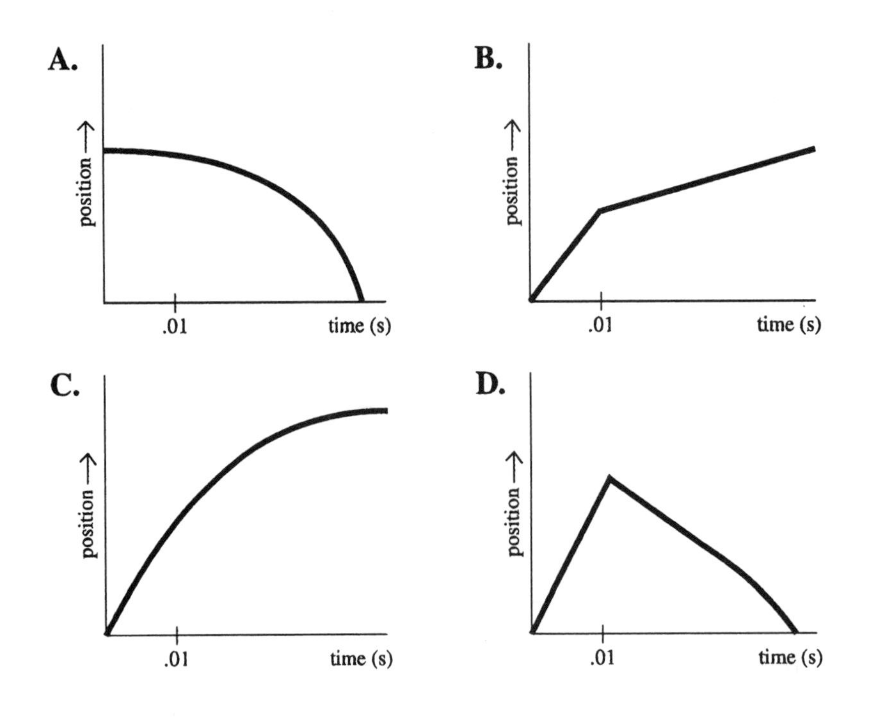

---

4. For airbag 2, approximately how much distance does the dummy cover between the moment the car crashes and the moment the dummy first makes contact with the airbag?

   A. 0.2 m **B.** 0.4 m **C.** 0.6 m **D.** 1.0 m

ANSWERS TO MCAT REVIEW PROBLEMS

---

1. **A.** According to the problem, the crash test dummy doesn't start decelerating (slowing down) until the airbag inflates. So, we know that the airbag has finished inflating when the crash test dummy's velocity first dips below 20 m/s. For airbag 1, this happens before t = 0.01 s. By contrast, the other airbags aren't inflated until t = 0.01 seconds (for airbag 2) or later (for airbag 3).

---

2. **B.** Since acceleration is the rate of change of velocity with time, it corresponds to the slope of the v vs. t graph. Here, the bigger the (negative) slope, the bigger the deceleration. So, for airbag 1, the dummy experiences a huge deceleration just before time t = 0.01 s. For airbag 3, the dummy feels a huge deceleration right before t = 0.04 s. By contrast, the dummy using airbag 2 never feels such a dramatic deceleration; that graph never gets as steep.

   Some students choose A, because airbag 1 causes the smallest *average* deceleration. But even though this deceleration is the smallest on average, it still gets huge just before t = 0.01 s. A human driver might suffer severe injury at that time.

---

3. **C.** Between t = 0 and t = 0.01 s, the dummy "coasts" forward at constant velocity. So, its position steadily increases, corresponding to an upward-sloped straight line on the position vs. time graph. (A flat line would correspond to no motion, i.e., zero velocity.)

   Between t = 0.01 s and t = 0.04 s, the dummy does *not* move backwards. It keeps moving forward, but at a slower and slower rate. Therefore, its position graph does not "come down." Instead, the graph continues to "go up," but at a decreasing rate—i.e., decreasing slope.

---

4. **A.** The dummy first makes contact with the airbag at t = 0.01 s, the moment the dummy starts to slow down. So, this problem wants the distance covered by the dummy between t = 0 and t = 0.01 s. During that interval, the dummy travels at constant speed, v = 20 m/s. There's no acceleration. So, we need not use a fancy kinematic formula. We can just use distance = rate × time:

$$\Delta x = v\Delta t = (20 \text{ m/s})(0.01 \text{ s}) = 0.2 \text{ m}.$$

CHAPTER 3
Kinematics in Two Dimensions

PREVIEW

In this chapter the concepts of displacement, velocity, and acceleration are applied to motion in two dimensions. You will learn to describe motion along a curved line, such as a car moving on a race track, a ball flying through the air, or a spacecraft in deep space. The kinematic equations for displacement, velocity, and acceleration developed in Chapter 2 will be applied here to describe projectile motion and relative velocity.

QUICK REFERENCE

Important Terms

Projectile Motion
An object moving in two dimensions which experiences only the acceleration due to gravity. Air resistance is assumed to be negligible and there is no acceleration in the x direction, i.e., $a_x = 0$. In the y direction on earth we have $a_y = -9.80\ m/s^2$. A baseball or football thrown from one person to another is a good example.

Relative Velocity
The velocity of a moving object can take on different values depending on the point of view of the observer. That is, the value of the velocity must be measured 'relative' to a particular observer.

Kinematic Equations for Two Dimensional Motion (Constant Acceleration)

$$v_x = v_{0x} + a_x t \quad (3.3a) \qquad v_y = v_{0y} + a_y t \quad (3.3b)$$

$$x = \frac{1}{2}(v_{0x} + v_x)\, t \quad (3.4a) \qquad y = \frac{1}{2}(v_{0y} + v_y)\, t \quad (3.4b)$$

$$x = v_{0x} t + \frac{1}{2} a_x t^2 \quad (3.5a) \qquad y = v_{0y} t + \frac{1}{2} a_y t^2 \quad (3.5b)$$

$$v_x^2 = v_{0x}^2 + 2\, a_x x \quad (3.6a) \qquad v_y^2 = v_{0y}^2 + 2\, a_y y \quad (3.6b)$$

DISCUSSION OF SELECTED SECTIONS

3.2 Equations of Kinematics in Two Dimensions

In Chapter 2 we emphasized that displacement, velocity, and acceleration were vector quantities. In one-dimensional motion the direction of these vectors was simply denoted by a positive or negative sign. In the more complex realm of two dimensions we must be careful to treat both the x and y motion independently, with the understanding that they will add together as vectors.

In treating two-dimensional motion we must be careful to view the x and y components of the motion as separate but related quantities. We will see that the x part of the motion occurs exactly as it would if the y part did not occur at all. Similarly, the y part of the motion occurs exactly as it would if the x part of the motion did not exist. Therefore, two sets of kinematic equations are needed to describe the full two dimensional motion.

Example 1 *Components of the velocity vector*
A ball is traveling at a constant velocity of 20.0 m/s at an angle of 30.0 degrees with respect to the x axis. How far does the ball travel in the x and y directions in 1 minute? See the diagram shown below.

First find the x and y components of the velocity.

$$v_x = v \cos\theta = (20.0 \text{ m/s})\cos 30.0^\circ = 17.3 \text{ m/s}.$$

$$v_y = v \sin\theta = (20.0 \text{ m/s})\sin 30.0^\circ = 10.0 \text{ m/s}.$$

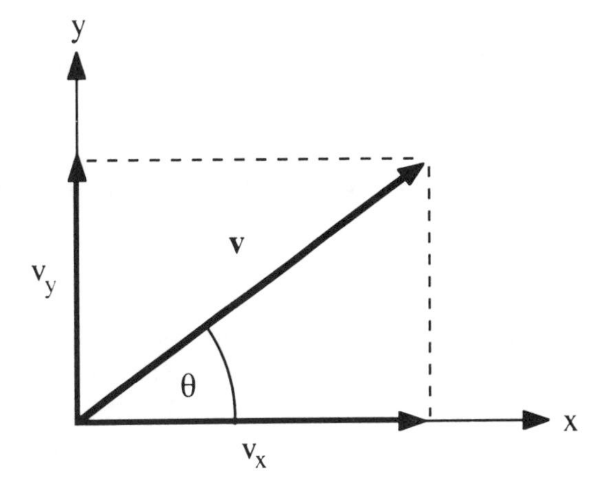

Since there is no acceleration, the x and y displacements are given by equations (3.5a) and (3.5b) with $v_{0x} = v_x$ and $v_{0y} = v_y$. We therefore have,

$$x = v_x\, t = (17.3 \text{ m/s})\,(60 \text{ s}) = 1040 \text{ m}.$$
$$y = v_y\, t = (10.0 \text{ m/s})\,(60 \text{ s}) = 600 \text{ m}.$$

3.3 Projectile Motion

When we throw a ball, or fire a bullet from a gun, the motion is one in which only the vertical component of the velocity changes with time. The horizontal velocity component remains constant (assuming that air resistance is negligible). We refer to this type of motion as **projectile motion**. The equations describing the trajectory of such a projectile are equations (3.3) through (3.6). Consider the following example.

Example 2
A football is kicked with an initial velocity of 82 ft/s at an angle of 53° with respect to the horizontal. How high does the football go and how far down the field does it land?

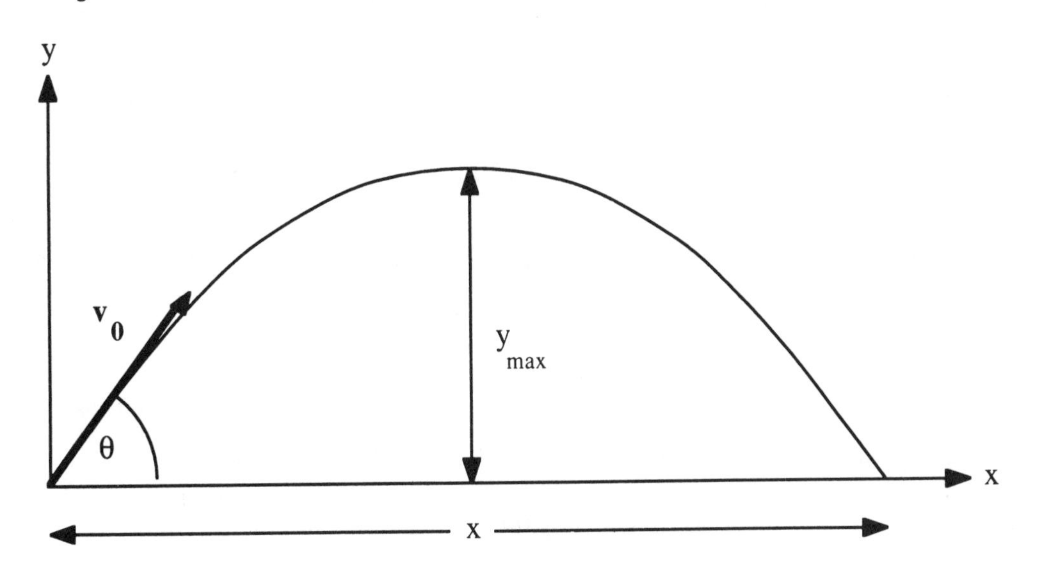

The x and y motions can be treated separately. Begin by looking at the motion in the y direction. We have

$$v_{0y} = v_0 \sin\theta = (82 \text{ ft/s}) \sin 53^\circ = 65 \text{ ft/s}; \quad a_y = -32.2 \text{ ft/s}^2; \quad v_y = 0.$$

Notice we have chosen the final velocity to be zero. In effect, we are looking at the first half of the motion in the y direction, i.e., the portion of the motion in which the ball is rising. When the ball reaches its maximum height, $v_y = 0$ and $y = y_{max}$. Using equation (3.6b) we can solve for this maximum height y,

$$y_{max} = y = \frac{-v_{0y}^2}{2a_y} = \frac{-(65 \text{ ft/s})^2}{2(-32.2 \text{ ft/s}^2)} = 66 \text{ ft}$$

To find the distance the football travels in the x direction, we need to know the time of flight for the ball. We can find the time it takes for the ball to rise using equation (3.3b), i.e.,

$$t = -v_{0y}/a = -(65 \text{ ft/s})/(-32.2 \text{ ft/s}^2) = 2.0 \text{ s.}$$

The total time of flight is twice this value, or 4.0 seconds. Now we can find the x distance using equation (3.5a) and noting that $a_x = 0$ and $v_{0x} = v_0 \cos\theta = (82 \text{ ft/s}) \cos 53° = 49 \text{ ft/s}$. We have

$$x = v_{0x}t = (49 \text{ ft/s})(4.0 \text{ s}) = 2.0 \text{ X } 10^2 \text{ ft.}$$

So the ball rose 66 ft in the air, stayed up 4.0 s (hang time), and traveled 2.0 X 10^2 ft.

Example 3

A jet fighter is traveling horizontally with a speed of 111 m/s at an altitude of 3.00 x 10^2 m, when the pilot accidentally releases an outboard fuel tank. (a) How much time elapses before the tank hits the ground? (b) What is the speed of the tank just before it hits the ground? (c) What is the horizontal distance traveled by the tank?

(a) The y-direction data is; $y = -3.00 \text{ x } 10^2$ m, $v_{0y} = 0$, $a_y = -9.80 \text{ m/s}^2$. Using equation (3.5b) and solving for t,

$$t = \sqrt{\frac{2y}{a_y}} = \sqrt{\frac{2(-3.00 \text{ x } 10^2 \text{ m})}{(-9.80 \text{ m/s}^2)}} = 7.82 \text{ s.}$$

(b) The y component of the velocity as it strikes the ground can be found using equation (3.3b),

$$v_y = v_{0y} + a\,t = 0 + (-9.80 \text{ m/s}^2)(7.82 \text{ s}) = -76.6 \text{ m/s.}$$

The speed as it hits the ground can be obtained by calculating the vector sum of the x and y components. The x component of the velocity is constant at 111 m/s. Therefore,

$$v_{tot} = \sqrt{v_x^2 + v_y^2} = \sqrt{(111 \text{ m/s})^2 + (-76.6 \text{ m/s})^2} = 135 \text{ m/s.}$$

(c) The horizontal distance can be obtained using equation (3.5a) with $a_x = 0$ and $v_{0x} = 111$ m/s,

$$x = v_{0x}\, t = (111 \text{ m/s})(7.82 \text{ s}) = 868 \text{ m.}$$

Note that the initial velocity, v_{0x}, plays no role in determining the time required for the fuel tank to hit the ground. Even if the jet were moving faster, the tank would still take 7.82 s to reach the ground. However, the speed on impact and the horizontal distance would be affected by v_{0x}.

3.4 Relative Velocity

The velocity measured for a particular object depends on the observer who is making the measurement. For example, suppose a ground based observer is measuring the velocity of two cars, each moving in the same direction. Car A is moving at 45 km/h relative to the ground, while car B is moving at 55 km/h relative to the ground. The observer in car A, however, measures the velocity of car B as 10 km/h relative to himself. Car B measures A's velocity as -10 km/h in his reference frame. So each observer obtains different values for the velocities because each is making measurements from different points of reference.

Example 4 *Relative velocity in one dimension*
A passenger in a moving train is walking towards the dining car in the front of the train. The train is moving at a speed of 21 ft/s and the passenger is walking at 4 ft/s relative to the train. How far will the passenger move in 10.0 seconds relative to (a) the train, and (b) a ground based observer.

Use the following symbols to represent the different relative velocities;

v_{TG} = velocity of the train relative to the ground.
v_{PT} = velocity of the passenger relative to the train.
v_{PG} = velocity of the passenger relative to the ground.

We can see that the three velocities are related in the following manner;

$$v_{PG} = v_{PT} + v_{TG} = 4 \text{ ft/s} + 21 \text{ ft/s} = 25 \text{ ft/s}.$$

The distances measured in each case are therefore,

(a) $s_{PT} = v_{PT}t = (4 \text{ ft/s})\ (10.0 \text{ s}) = 40 \text{ ft}.$
(b) $s_{PG} = v_{PG}t = (25 \text{ ft/s})\ (10.0 \text{ s}) = 250 \text{ ft}.$

When dealing with relative velocities in two dimensions it is important to emphasize the vector nature of the velocities. For example, if a boat is sailing across a stream, and the stream is flowing, both the boat's and the stream's velocity must be considered. Similarly, the velocity of an airplane in flight is effected by the velocity of the wind. The following example illustrates this effect.

Example 5 *Relative velocity in two dimensions*
A boat whose speed is 10.0 mi/h in still water travels across a river that is 2.00 miles wide. The river current is 5.00 mi/h directed parallel to the river bank. How long does it take for the boat to cross the river and how far downstream will the boat arrive?

First find the velocity of the boat relative to the shore. From the diagram we can see that

$$\mathbf{v}_{BS} = \mathbf{v}_{BW} + \mathbf{v}_{WS}$$

Since the vectors $\mathbf{v}_{BW}$ and $\mathbf{v}_{WS}$ are perpendicular to each other, the magnitude and direction of v_{BS} can be found from the Pythagorean theorem.

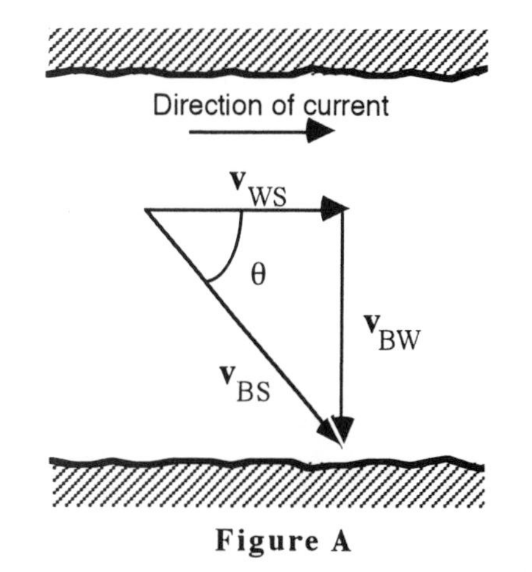

Figure A

$$v_{BS} = \sqrt{v_{BW}^2 + v_{WS}^2}$$

$$= \sqrt{(10.0 \text{ mi/h})^2 + (5.00 \text{ mi/h})^2}$$

$$= 11.2 \text{ mi/h}.$$

To find the amount of time and downstream distance we need to know the angle θ shown in Figure A. We have

$$\theta = \tan^{-1}\left(\frac{v_{BW}}{v_{WS}}\right) = \tan^{-1}\left(\frac{10.0 \text{ mi/h}}{5.00 \text{ mi/h}}\right) = 63.4^{o}.$$

In order to find the time taken for the trip we need the total distance traveled. We see that this distance is

$$r = \frac{2.00 \text{ mi}}{\sin\theta} = \frac{2.00 \text{ mi}}{\sin 63.4^{o}} = 2.24 \text{ mi}.$$

Note that the angle θ that $\mathbf{v}_{BS}$ makes with $\mathbf{v}_{WS}$ (in Figure A) is the same angle θ that **r** makes with **x** (in Figure B).

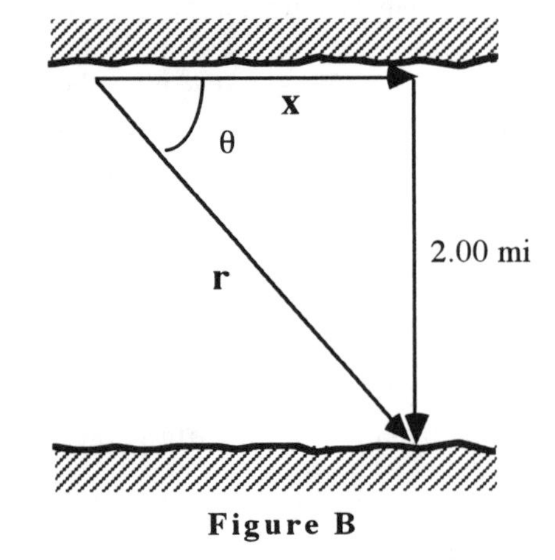

Figure B

Therefore, the time taken can be obtained using

$$t = \frac{r}{v_{BS}} = \frac{2.24 \text{ mi}}{11.2 \text{ mi/h}} = 0.20 \text{ h}.$$

The distance the boat lands downstream can be found from

$$x = \frac{2.00 \text{ mi}}{\tan\theta} = \frac{2.00 \text{ mi}}{\tan 63.4^{o}} = 1.00 \text{ mi}.$$

PRACTICE PROBLEMS

1. A pitcher throws a baseball at a speed of 90.0 mi/h towards home plate, which is located 60.0 ft away. How far has the ball dropped when it reaches the plate? Assume the ball leaves the pitcher's hand traveling horizontally.

2. An arrow is shot at an initial velocity of 25.0 m/s at an angle of 30.0 degrees with respect to the horizontal. (a) What is the maximum height reached by the arrow? (b) How far does the arrow travel in the horizontal direction in returning to the same level from which it was shot?

3. A plane is traveling at a velocity of 2.0×10^2 m/s at an angle of 45 degrees below the horizontal. It releases a bomb at an altitude of 1.0×10^3 m, which strikes a target on the ground. (a) How far was the plane from the target when the bomb was released? (b) How long did it take the bomb to strike the target after being released? (c) What was the bomb's speed at the moment of impact?

4. A diver springs upward from a three-meter board. At the instant she contacts the water her speed is 8.90 m/s and her body makes an angle of 75.0° with respect to the surface of the water. (a) Determine her initial velocity, both magnitude and direction. (b) How much time does she spend in the air?

5. A ball rolls off the top of a stairway with a horizontal velocity of 2.0 m/s. The steps are each 20.0 cm high and 20.0 cm wide. Which step will the ball hit first?

6. When a cannon is aimed at an angle of 45° above the horizontal, a cannon ball lands 1.0×10^2 m down range. What was the muzzle velocity of the cannon ball?

7. The compass of an aircraft indicates it is headed due north, and its airspeed indicator shows that it is moving through the air at 120 mi/h. If a wind of 50.0 mi/h blowing from west to east suddenly arises, what is the velocity of the aircraft relative to the earth?

8. An airplane pilot wishes to fly due south. A wind of 25 km/h is blowing toward the west. If the flying speed of the plane is 300.0 km/h (its speed in still air), in what direction should the pilot head?

9. Two trains approach a railroad station, one from the north at 30.0 m/s, the other from the east at 40.0 m/s. Find the velocities of (a) the railroad station and (b) the south-bound train; relative to the west-bound train.

10. A boat can travel at a speed of 20.0 km/h relative to the water. The boat sails across a river (perpendicular to the river bank) and reaches a point 5.00 km directly across from its launch point. The water is flowing at a speed of 8.00 km/h parallel to the river bank. (a) At what angle must the boat steer to reach its destination? (b) How much time is required for the boat to make the trip?

HELPFUL SUGGESTIONS

1. For a projectile, in the absence of air resistance, the velocity in the x direction always continues unchanged. In the y direction near the earth the acceleration is always **g**, directed downward.

2. For a projectile at the top of its path, $v_y = 0$, but v_x = constant.

3. The x and y motions for a projectile must be treated separately, since they are independent motions. However, the time of flight, t, will be common to both sets of equations.

4. When a velocity is given it should always be resolved into its components, which can then be used independently in the kinematic equations. If the solution requires you to state a final velocity, the velocity vector can be reconstructed from the x and y components.

5. The time that a projectile spends in the air is governed by the y variable; not the x variable. Once the time of flight has been determined, however, it can then be used in the x equations.

6. In many relative velocity problems the velocities form right triangles. However, if this is not the case, the law of sines or cosines must be applied.

EVERYDAY PHYSICS

1. Projectile motion can be observed in many everyday situations. Throwing a baseball or football are obvious examples. Shooting a rifle or an arrow are also examples of projectile motions, although these are harder to observe because of the high initial velocities involved. However, the effects of gravity must be accounted for when you aim at a target, i.e., you need to aim a little above the target to compensate for this effect.

2. You can observe projectile motion for the simple case of a ball rolling off a table. Notice that the ball never lands right at the base of the table, but always some distance away. Try giving the ball different initial velocities and see how the landing distance varies. How can you determine the initial velocity of the ball by measuring the height of the table and the distance the ball lands from the base of the table?

3. A good way to observe projectile motion is with a garden hose. Hold the hose at different angles and watch the resulting flow of water. Which angle causes the water to reach the greatest horizontal distance? What measurements do you need to make in order to calculate the nozzle velocity of the water?

4. In real life, we really can't ignore the effects of air resistance and friction for moving objects. How would air resistance affect our treatment of projectile motion? What assumptions do we need to modify regarding the x and y motions? How would the kinematic equations of motion be affected?

CHAPTER QUIZ

1. Rock A is dropped from rest from the top of a building. Rock B is thrown horizontally from the top of the same building at the same time. In the absence of air resistance compare the time it takes for each rock to hit the ground below.
   a. Both objects take the same time to reach the ground.
   b. Rock A strikes the ground first.
   c. Rock B strikes the ground first.
   d. It is impossible to say which reaches first without knowing the building's height.

2. A tennis ball is hit into the air and moves along an arc. Neglecting air resistance, where along the arc is the speed of the ball a minimum?
   a. At its launch point. b. At the top of its arc. c. At the landing point.

3. A ball moving at 3.0 m/s rolls off a table and strikes the floor 0.30 s later. What is its speed on impact?
   a. 3.0 m/s b. 4.2 m/s c. 5.9 m/s d. 9.0 m/s

4. An arrow is fired with an initial velocity of 20.0 m/s at an angle of 37.0° with the horizontal. How long does the arrow stay in the air?
   a. 1.22 s b. 2.46 s c. 3.26 s d. 4.08 s

5. Two cars travel towards each other on the same road, one at 45 mi/h, the other at 55 mi/h. They are initially separated by 150 miles. How long will it take for the cars to meet?
   a. 1.5 h b. 6.0 h c. 15 h d. 0.67 h

6. A boat is traveling upstream at 10.0 mi/h. The current moves at 3.0 mi/h. How fast is the boat moving relative to the shore?
   a. 13 mi/h b. 10 mi/h c. 7 mi/h d. 3.3 mi/h

7. A boat which can travel at 6 km/h wants to go directly across a stream whose current is 3 km/h. At what angle must the boat steer (relative to the shore) to accomplish this?
   a. 30° b. 40° c. 60° d. 90°

8. A plane can fly at a speed of 150 m/s in still air. When a cross wind (a wind directed perpendicular to the plane) blows at 80.0 m/s, what is the plane's speed relative to the ground?
   a. 70 m/s b. 130 m/s c. 170 m/s d. 230 m/s

9. A railroad flatcar is equipped with a missile launcher that points straight up. If the flatcar travels at 10 m/s and a missile is launched at an initial speed of 100 m/s, at what angle (relative to the vertical) is the missile launched as seen by a stationary ground observer?
   a. 6° b. 10° c. 30° d. 90°

10. A golf ball is hit and makes an angle of 45° with the horizontal as it leaves the tee. The ball lands 200.0 yards away and was in the air for 10 s. What was the ball's initial speed?
    a. 28 ft/s b. 60 ft/s c. 85 ft/s d. 200 ft/s

11. A plane traveling at 300 m/s (horizontally) drops a bomb which hits its target 10 s later. How high was the plane flying?
    a. 30 m b. 300 m c. 500 m d. 3000 m

SOLUTIONS AND ANSWERS

Practice Problems

---

1. The horizontal speed is

$$v_x = (90.0 \text{ mi/h})(1\text{h}/3600 \text{ s})(5280 \text{ ft}/1 \text{ mi}) = 132 \text{ ft/s}.$$

The time to travel 60.0 ft is

$$t = x/v_x = (60.0 \text{ ft})/(132 \text{ ft/s}) = 0.455 \text{ s}.$$

$$y = v_{0y}t + (1/2)a_yt^2 = 0 + (1/2)(32.2 \text{ ft/s}^2)(0.455 \text{ s})^2 = \mathbf{3.33 \text{ ft.}}$$

---

2. a. The initial velocity is

$$v_{0y} = v_0 \sin 30.0° = (25.0 \text{ m/s})(0.500) = 12.5 \text{ m/s};$$

Use equation (3.6b) and solve for y

$$y = -v_0^2/2a_y = -(12.5 \text{ m/s})^2/[2(-9.80 \text{ m/s}^2)] = \mathbf{7.97 \text{ m.}}$$

b. Find the time from $v_y = v_{0y} + at$

$$t = -v_{0y}/a_y = -(12.5 \text{ m/s})/(-9.80 \text{ m/s}^2) = 1.28 \text{ s}.$$

Finally, we have

$$x = v_{0x}t = v_0 \cos 30.0° \, t = (25.0 \text{ m/s}) \cos 30.0° \, (2.56 \text{ s}) = \mathbf{55.4 \text{ m.}}$$

---

3. Using equation (3.6b) we have

$$v_y^2 = v_{0y}^2 + 2a_yy = [(2.0 \times 10^2 \text{ m/s}) \sin 45°]^2 + 2(-9.80 \text{ m/s}^2)(-1.0 \times 10^3\text{m}),$$

Which yields $v_y = 2.0 \times 10^2$ m/s. Now, $v_x = v_0 \cos 45° = 141$ m/s. The time can be found from equation (3.3b),

$$t = (v_y - v_{0y})/a = (2.0 \times 10^2 \text{ m/s} - 1.4 \times 10^2 \text{ m/s})/(9.80 \text{ m/s}^2) = 6.1 \text{ s}.$$

a. $x = v_{0x}t = (141 \text{ m/s})(6.1 \text{ s}) = 860 \text{ m}$, $y = 1.0 \times 10^3$ m,

$$s^2 = x^2 + y^2 = (860 \text{ m})^2 + (1.0 \text{ X } 10^3 \text{ m})^2, \text{ so that } s = \mathbf{1300 \text{ m.}}$$

b. We have already seen that $t = \mathbf{6.1 \text{ s.}}$

c. Finally,

$$v^2 = v_x^2 + v_y^2 = (141 \text{ m/s})^2 + (2.0 \times 10^2 \text{ m/s})^2 \text{ so } v = \mathbf{240 \text{ m/s.}}$$

---

4. a. At a height of 3.00 m above the water find v_o. We know

$$v_y = v \sin 75.0° = (8.90 \text{ m/s})\sin 75.0° = 8.60 \text{ m/s}.$$

From equation (3.6b) we have

$$v_{0y}^2 = v_y^2 - 2a_y y = (8.60 \text{ m/s})^2 - 2(9.80 \text{ m/s}^2)(3.00 \text{ m}) \rightarrow v_{0y} = 3.89 \text{ m/s}.$$

Also,

$$v_{0x} = v_x = v \cos 75.0° = (8.90 \text{ m/s}) \cos 75.0° = 2.30 \text{ m/s} \rightarrow v_o^2 = v_{ox}^2 + v_{0y}^2 \text{ so that}$$

$$v_0 = \mathbf{4.52 \text{ m/s}}.$$

The direction is obtained from

$$\theta = \tan^{-1}(v_{0y}/v_{0x}) = \tan^{-1}[(3.89 \text{ m/s})/(2.30 \text{ m/s})] = \mathbf{59.4°}.$$

b. The time to go up is $t_{up} = -v_{0y}/a = 0.397$ s; the time down is $t_{down} = v_y/a = 0.878$ s. So the total time is

$$t = t_{up} + t_{down} = 0.40 \text{ s} + 0.88 \text{ s} = \mathbf{1.28 \text{ s}}.$$

---

5. In order to determine which step the ball lands on, the ball must fall at least a distance equal to its horizontal displacement. For example, at $x = 0.2$ m,

$$t = x/v_x = (0.2 \text{ m})/(2.0 \text{ m/s}) = 0.1 \text{ s},$$
$$y = (1/2)a_y t^2 = 0.05 \text{ m}.$$

So the ball goes past the first step. At x' = 0.8 m (4th step), t' = 0.4 s and y' = 0.78 m, so it goes past the 4th step. However, at x" = 1.0 m (5th step), t" = 0.5 s, y" = 1.23 m. So the ball lands on the **fifth step.**

---

6. The time taken to travel half the distance in the x direction, using equation (3.5a), is

$$t = x/(v_0 \cos 45°) = (50.0 \text{ m})/(v_0 \cos 45°).$$

The time it takes for the ball to reach its maximum height is the also this value of t, and using equation (3.3b) with $v_y = 0$,

$$t = v_{0y}/a = v_0 \sin 45°/(9.8 \text{ m/s}^2).$$

Equating the two expressions for the time, and noting that cos 45° = sin 45° yields,

$$(v_0 \cos 45°)^2 = (50.0 \text{ m})(9.8 \text{ m/s}^2).$$

Thus,

$$v_0^2 = (50.0 \text{ m})(9.8 \text{ m/s}^2)/(\cos^2 45°)$$
$$v_0 = \mathbf{31 \text{ m/s}}.$$

---

7. A vector diagram shows the

$$v^2 = v_P^2 + v_W^2 = (120 \text{ mi/h})^2 + (50.0 \text{ mi/h})^2$$
$$v = \mathbf{130 \text{ mi/h}}.$$

The direction is

$$\theta = \tan^{-1}(v_P/v_W) = \tan^{-1}[(120 \text{ mi/h})/(50.0 \text{ mi/h})] = \mathbf{67° \text{ North of East}}.$$

---

8. Since the final direction of motion and the direction of the wind are perpendicular, we can find the angle between the plane and wind, i.e.,

$$\theta = \cos^{-1}(v_W/v_P) = \cos^{-1}[(25 \text{ km/h})/(300.0 \text{ km/h})] = \mathbf{85° \text{ S of E.}}$$

---

9. a. We can write $\mathbf{v}_{RW} = \mathbf{v}_{RG} + \mathbf{v}_{GW}$ where R → railroad station, W → west bound train, G → ground. So

$$v_{RW} = 0 + 40 \text{ m/s} = \mathbf{40 \text{ m/s East.}}$$

b. We have $\mathbf{v}_{SW} = \mathbf{v}_{SG} + \mathbf{v}_{GW}$. Since $\mathbf{v}_{SG}$ and $\mathbf{v}_{GW}$ are perpendicular, we can write $v_{SW}^2 = v_{SG}^2 + v_{GW}^2$. So,

$$v_{SW}^2 = (30.0 \text{ m/s})^2 + (40.0 \text{ m/s})^2$$
$$v_{SW} = \mathbf{50.0 \text{ m/s.}}$$

Also,

$$\theta = \tan^{-1}(v_{SG}/v_{GW}) = \mathbf{36.9° \text{ S of E.}}$$

---

10. a. A vector diagram shows that the angle at which the boat must steer is $\theta = \cos^{-1}(v_S/v_B)$, where the velocity v_S is that of the stream and v_B is that of the boat. So,

$$\theta = \cos^{-1}[(8.00 \text{ km/h})/(20.0 \text{ km/h})] = \mathbf{66.4°.}$$

b. The velocity across the river is

$$v^2 = v_B^2 - v_S^2 = (20.0 \text{ km/h})^2 - (8.00 \text{ km/h})^2$$
$$v = 18.3 \text{ km/h.}$$

Since it's 5.00 km across the river, the time to get across is

$$t = s/v = (5.00 \text{ km})/(18.3 \text{ km/h}) = \mathbf{0.273 \text{ h.}}$$

Quiz answers

1. a	5. a	9. a
2. b	6. c	10. c
3. b	7. c	11. c
4. b	8. c	

MCAT REVIEW PROBLEMS

A student performs an experiment to determine how the range of a ball depends on the velocity with which it is released. The "range" is the distance between where the ball lands and where it was released, assuming it lands at the same height from which it was released.

In each trial, the student uses the same baseball, and launches it at the same angle. Table 1 shows the experimental results.

TABLE 1

TRIAL	launch speed (m/s)	range (m)
1	10	8.0
2	20	31.8
3	30	70.7
4	40	122.5

Based on this data, the student then hypothesizes that the range, R, depends on the initial speed, v_0, according to the following equation:

$$R = Cv_0^n,$$

where C is a constant, and n is another constant.

---

1. Based on this data, the best guess for the value of n is

   A. 1/2 **B.** 1 **C.** 2 **D.** 3

---

2. The student speculates that the constant C depends on:

   I. The angle at which the ball was launched.
   II. The ball's mass.
   III. The ball's diameter.

   If we neglect air resistance, then C actually depends on

   A. I only **B.** I and II **C.** I and III **D.** I, II, and III

---

3. The student performs another trial in which the ball is launched at speed 5.0 m/s. Its range is approximately:

   A. 1.0 meters **B.** 2.0 meters **C.** 3.0 meters **D.** 4.0 meters

4. Let θ denote the angle of the ball's initial velocity, as measured from the horizontal. Neglect air resistance. At the peak (highest point) of its trajectory, the ball's speed is:

A. 0 **B.** $v_0 \sin\theta$ **C.** $v_0 \cos\theta$ **D.** v_0

---

5. For trial 2, which of the following graphs best represents the vertical component of the ball's velocity as a function of time, assuming upward is positive?

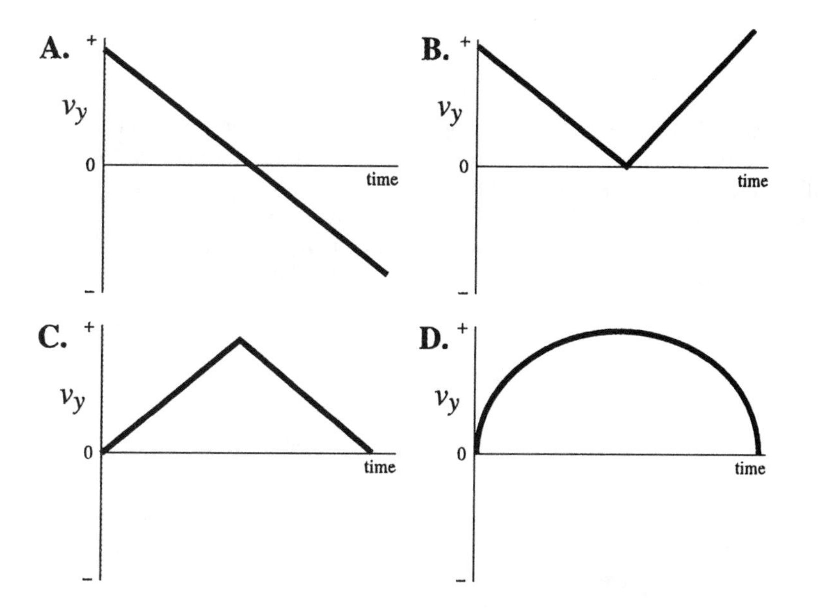

ANSWERS TO MCAT REVIEW PROBLEMS

---

1. **C.** If n equals 1/2, then we'd need to quadruple v_0 in order to double R. But here, quadrupling v_0 makes R increase much more dramatically. Similarly, if n equals 1, then doubling v_0 would double R. Again, that's not even close. But $n = 2$ fits the data well. If n equals 2, then doubling v_0 should quadruple R. So, between trials 1 and 2, R should jump from 8.0 m to 32 m. To good approximation, that's what happens. Similarly, when we triple v_0 (between trials 1 and 3), the range should increase by a factor of 9, from 8.0 m to 72 m. Again, the data closely matches this prediction. The other data points "work," too.

Notice that you can answer this question without remembering the relevant textbook formula. Table 1 supplies all the necessary information.

---

2. **A.** Neglecting air resistance, a projectile's motion depends *only* on the launch speed, launch angle, and gravity. If air resistance were present, then a bigger ball would travel less far, other things being equal. But we're neglecting air resistance. The mass also makes no difference in projectile motion. This is demonstrated by the fact that a tennis ball and a bowling ball, when dropped simultaneously from the same height, hit the ground at the same time.

---

3. **B.** As found in question 1, R depends on v_0^2. Using this insight, we can compare this new trial to trial 1. The initial velocity, v_0, is reduced by a factor of 2. Therefore, R goes down by a factor of $2^2 = 4$. When v_0 gets cut in half, R gets cut "in fourth."

---

4. **C.** If thrown straight up, the ball would stop at the peak, and A would be correct. But in general, only the *vertical* component of a projectile's motion vanishes at the peak. The horizontal motion keeps going. Intuitively, at the peak, the ball momentarily stops rising or falling, but keeps moving horizontally. Therefore, the total speed at the peak is the magnitude of the ball's horizontal velocity at the peak.

As emphasized in your textbook, the horizontal component of a projectile's velocity stays constant. That's because gravity affects only the vertical part of the motion. So, the horizontal velocity at the peak equals the ball's initial horizontal velocity. And the initial horizontal velocity is $v_0 \cos\theta$, as this diagram shows.

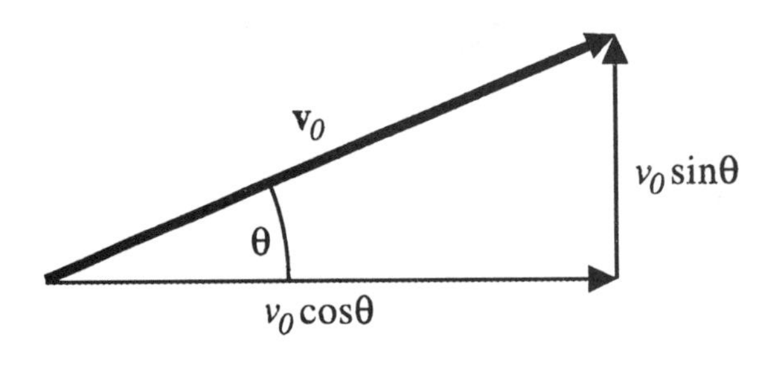

---

5. **A.** As the ball rises, its vertical velocity decreases until reaching zero at the peak. Intuitively, after getting launched, the ball rises quickly at first, but then rises more and more slowly until reaching the peak. So, in the first segment of the graph, the vertical velocity gradually decreases until hitting zero. And since gravity causes a constant acceleration, the velocity decreases at a steady rate, corresponding to a straight line on the velocity vs. time graph. By contrast, graph D would represent the ball's *height* vs. time, rather than its vertical velocity vs. time. But here, we want *vertical velocity* vs. time; and as the ball gets higher, it gets slower. This is true for all trials, not just trial 2.

At the peak, the ball "turns around" and starts falling, slowly at first, but then faster and faster. In other words, the velocity now increases. For this reason, many students choose graph B. But when the ball falls, its velocity is *negative* (assuming an upward velocity counts as positive). So, after the ball reaches its peak, the vertical velocity increases in the negative direction.

CHAPTER 4
Forces and Newton's Laws of Motion

PREVIEW

In this chapter you will begin the study of **dynamics**, that branch of physics which explains why objects accelerate. You will be introduced to the concept of **force**, and study Newton's laws of motion, which apply to all forces that occur in nature. You will learn how to construct free-body diagrams, and use them to analyze systems subject to such forces as gravity and friction. You will also apply Newton's laws of motion to solve a number of different problems. The applications will include both equilibrium and non-equilibrium problems.

QUICK REFERENCE

Important Terms

Force
The push or pull required to change the state of motion of an object, as defined by Newton's second law. It is a vector quantity with units of newtons (N), dynes (dyn), or pounds (lb).

Inertia
The natural tendency of an object to remain at rest or in uniform motion at a constant speed in a straight line.

Mass
A quantitative measure of inertia. Units are kilograms (kg), grams (g), or slugs (sl).

Inertial Reference Frame
A reference frame in which Newton's law of inertia is valid.

Free-Body Diagram
A vector diagram that represents all of the forces acting on an object.

Gravitational Force
The force of attraction that every particle of mass in the universe exerts on every other particle.

Weight
The gravitational force exerted by the earth (or some other large astronomical body) on an object.

Normal Force
One component of the force that a surface exerts on an object with which it is in contact. This component is directed normal, or perpendicular, to the surface.

Friction
The force that an object encounters when it moves or attempts to move along a surface. It is always directed parallel to the surface in question.

Tension
The tendency of a rope (or similar object) to be pulled apart due to the forces that are applied at either end.

Equilibrium
The state an object is in if it has zero acceleration. Mathematically, equilibrium means $\Sigma \mathbf{F} = 0$.

Apparent Weight
The force that an object exerts on the platform of a scale. It may be larger or smaller than the true weight, depending on the acceleration of the object and the scale.

Newton's Laws of Motion

First Law

An object continues in a state of rest or in a state of motion at a constant speed along a straight line, unless compelled to change that state by a net force. By "net" force we mean the **vector sum** of all of the forces acting on an object.

For example, consider a spaceship in deep space, isolated from any other object or force. If the ship is stationary, it will remain so. But if the ship is moving (its rocket engines are shut down) it will continue to move in a straight line with a constant speed. So if the ship were traveling into deep space at say, 100 000 mi/h, it would continue to move at this speed in a straight line, even without the rocket engines firing, until an outside force acted to stop or change its motion.

Second Law

When a net force $\Sigma\mathbf{F}$ acts on an object of mass m, the acceleration **a** that results is directly proportional to the net force and has a magnitude that is inversely proportional to the mass. The direction of the acceleration is the same as the direction of the net force. This statement is usually written as

$$\Sigma\mathbf{F} = m\mathbf{a} \qquad \text{or} \qquad \mathbf{a} = \Sigma\mathbf{F}/m \tag{4.1}$$

The symbol $\Sigma\mathbf{F}$ represents the net force, that is, the **vector sum** of all the forces acting on an object. This means that the components of the forces must be examined. For example, in two dimensions, equation (4.1) becomes

$$\Sigma F_x = ma_x \tag{4.2a}$$
$$\Sigma F_y = ma_y \tag{4.2b}$$

Equations (4.1) and (4.2) can be used to determine the **units** of force. They are

SI	kg m/s^2 = newton (N)
CGS	g cm/s^2 = dyne (dyn)
BE	slug ft/s^2 = pound (lb)

Third Law

Whenever one body exerts a force on a second body, the second body exerts an oppositely directed force of equal magnitude on the first body.

This is sometimes referred to as the "action-reaction" law, i.e., "for every action there is an equal, but opposite, reaction". An example is a spaceship firing its rocket engines. The engines eject hot gases out the rear of the rocket at very high velocity, this is the "action". The "reaction" is that the rocket accelerates in the forward direction.

Universal Gravitation

Every particle in the universe exerts an attractive force on every other particle in the universe. The force acting between two particles of masses m_1 and m_2 separated by a distance r has a magnitude given by

$$F = G\frac{m_1 m_2}{r^2} \tag{4.3}$$

where G is the universal gravitational constant, whose value (obtained experimentally) is given by

$$G = 6.673 \times 10^{-11}\ \text{N m}^2/\text{kg}^2.$$

The **weight** of an object on the earth is the gravitational force that the earth exerts on an object. This force is always directed downward, toward the center of the earth. For the earth (mass = M_E, radius = R_E) we can write

$$W = G\frac{M_E m}{R_E^2}$$

where the gravitational acceleration is $g = GM_E/R_E^2 = 9.80\ m/s^2$. The weight of an object can therefore be written

$$W = m g \qquad (4.5)$$

DISCUSSION OF SELECTED SECTIONS

4.3/4.4 Newton's Second Law of Motion and the Vector Nature of Newton's Second Law of Motion

Newton's second law, as expressed in equations (4.1) and (4.2), contains vectors. The equation $\Sigma\mathbf{F} = m\mathbf{a}$ means that we need to determine the net force acting on an object. The terminology, $\Sigma\mathbf{F}$, means that you must determine the vector sum (Σ = summation sign) of **all** the forces acting on an object. You should begin by drawing a diagram that represents the object and all the forces acting on it, a so-called **free-body diagram**.

Example 1 *A Free-Body Diagram*
A 2.0 kg object is subjected to a 6.0 N force acting in the x direction, and an 8.0 N force acting in the y direction. What is the acceleration (magnitude and direction) of this object?

The first diagram shows the mass m and forces acting on it. The first diagram is the free-body diagram where the mass is placed at the origin of the x-y coordinate system and the forces are shown acting from this point.

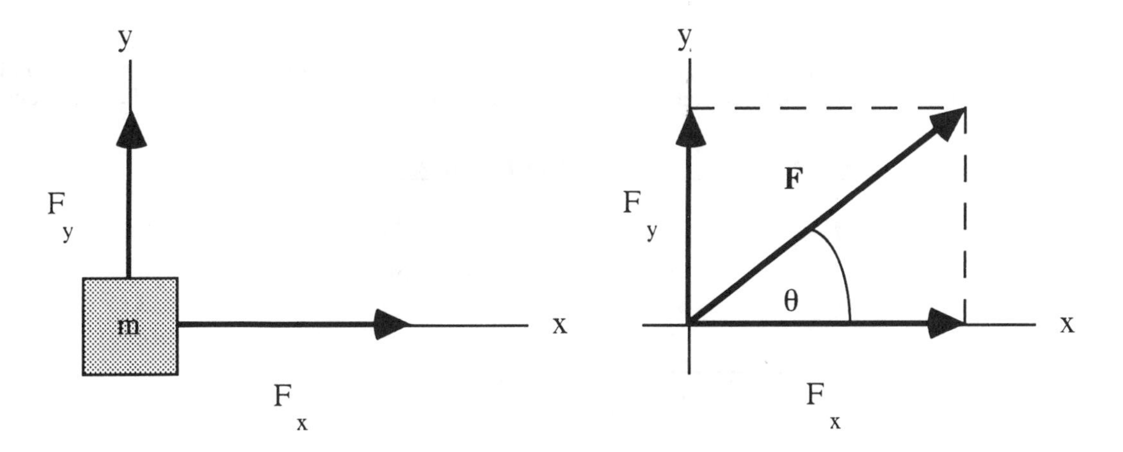

In order to find the acceleration of the object, **a**, we need to find the net force, **F**, acting on the object. Since the forces acting on m are perpendicular, we can find the net force using

$$F = \sqrt{F_x^2 + F_y^2} = \sqrt{(6.0\ N)^2 + (8.0\ N)^2} = 10\ N.$$

$$\theta = \tan^{-1}\left(\frac{F_y}{F_x}\right) = \tan^{-1}\left(\frac{8.0\ N}{6.0\ N}\right) = 53^\circ.$$

Since the acceleration is in the same direction as the net force, **a** is directed 53° up from the +x axis. Its magnitude is

$$a = F/m = (10\text{ N})/(2.0\text{ kg}) = 5\text{ m/s}^2.$$

In most cases, the forces acting on an object can have any orientation in space. The individual forces must be resolved into x and y components before the net force, and hence the acceleration, can be found. The next example illustrates this.

Example 2 *Finding the net force and acceleration*
Suppose a 1.8 kg object is subjected to forces as illustrated in the following diagram. The magnitudes of the forces are F_1 = 20.0 N, F_2 = 14.0 N, F_3 = 15.0 N. What is the net force and acceleration acting on the object?

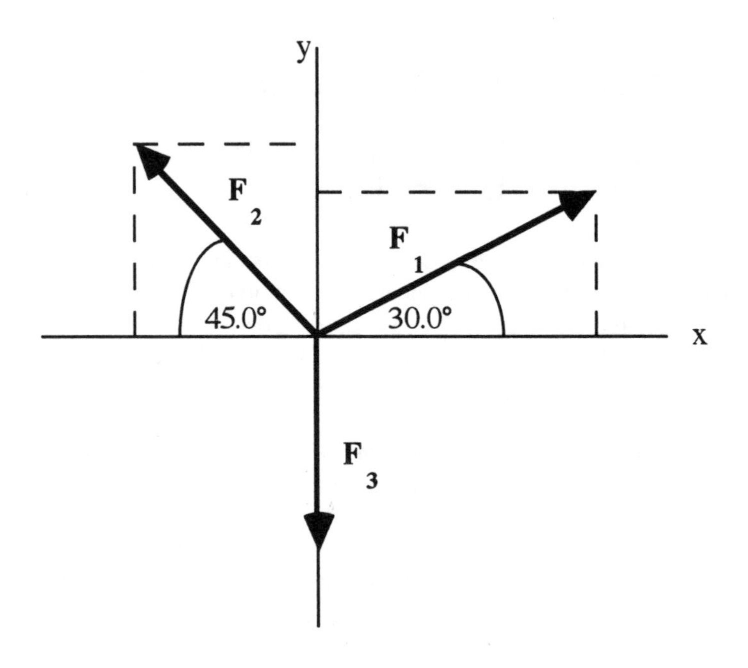

To find $\Sigma\mathbf{F}$ and the accelerations, we need to look at the components of the forces and use equations (4.2a) and (4.2b). That is, we need to sum the force components in the x direction (ΣF_x) and y direction (ΣF_y). We have

Force	x component
$\mathbf{F_1}$	$+(20.0\text{ N})\cos 30.0° = +17.3\text{ N}$
$\mathbf{F_2}$	$-(14.0\text{ N})\cos 45.0° = -9.9\text{ N}$
$\mathbf{F_3}$	0

$$\Sigma F_x = +17.3\text{ N} - 9.9\text{ N} = +7.4\text{ N}$$

Force	y component
$\mathbf{F_1}$	$+(20.0\text{ N})\sin 30.0° = +10.0\text{ N}$
$\mathbf{F_2}$	$+(14.0\text{ N})\sin 45.0° = +9.9\text{ N}$
$\mathbf{F_3}$	-15.0 N

$$\Sigma F_y = +10.0\text{ N} + 9.9\text{ N} - 15.0\text{ N} = +4.9\text{ N}$$

The accelerations in the x and y directions can now be obtained using ΣF_x and ΣF_y found above:

$$a_x = \frac{\Sigma F_x}{m} = \frac{+7.4\ N}{1.8\ kg} = 4.1\ m/s^2; \quad a_y = \frac{\Sigma F_y}{m} = \frac{+4.9\ N}{1.8\ kg} = 2.7\ m/s^2.$$

4.7 The Gravitational Force

Consider two particles of masses m_1 and m_2 separated by a distance r. Each mass exerts a force of attraction on the other mass. The force is directed along the line joining the particles and the force has a magnitude given by

$$F = G\frac{m_1 m_2}{r^2} \tag{4.3}$$

Example 3
Find the force between objects of mass $m_1 = 1.00 \times 10^3$ kg, $m_2 = 2.00 \times 10^3$ kg, separated by r = 3.00 m.

$$F = G\frac{m_1 m_2}{r^2} = (6.67 \times 10^{-11}\ N\ m^2/kg^2)\frac{(1.00 \times 10^3\ kg)(2.00 \times 10^3\ kg)}{(3.00\ m)^2} = 1.48 \times 10^{-5}\ N$$

Example 4 *The earth's gravity and weight*
The weight of an object on earth is defined as the gravitational force exerted by the earth on the object. Find the weight on earth of a 50.0 kg mass. Use $M_E = 5.98 \times 10^{24}$ kg, $R_E = 6.38 \times 10^6$ m for the earth's mass and radius.

$$W = G\frac{M_E m}{R_E^2} = (6.67 \times 10^{-11}\ N\ m^2/kg^2)\frac{(5.98 \times 10^{24}\ kg)(50.0\ kg)}{(6.38 \times 10^6\ m)^2} = 4.90 \times 10^2\ N$$

Note that according to Newton's second law, F = ma = mg, since g is the acceleration of gravity. Therefore,

$$g = \frac{GM_E}{R_E^2} = \frac{(6.67 \times 10^{-11}\ N\ m^2/kg^2)(5.98 \times 10^{24}\ kg)}{(6.38 \times 10^6\ m)^2} = 9.80\ m/s^2.$$

So the weight of an object on earth can be expressed as W = mg. This expression will prove to be quite useful.

Weight

The weight of an object is not the same as its mass. The mass of an object is the same regardless of where the object is located in the universe. The weight, being simply the force that gravity exerts on an object, depends on how close the object is to another massive object such as the earth. Close to the surface of the earth, an object of mass, m, has a weight of mg, where "g" is the acceleration due to gravity. The acceleration, g, close to the surface of the earth is

$$9.80\ m/s^2 \text{ or } 32.2\ ft/s^2.$$

4.8/4.9 The Normal Force and Static and Kinetic Frictional Forces

The normal force, $\mathbf{F_N}$, is the perpendicular force that a surface exerts on an object with which it is in contact. For an object at equilibrium on a flat surface (see figure below) we can see that $\mathbf{F_N} = \mathbf{W}$. The normal force is simply the surface's reaction to the forces pushing the object against the surface.

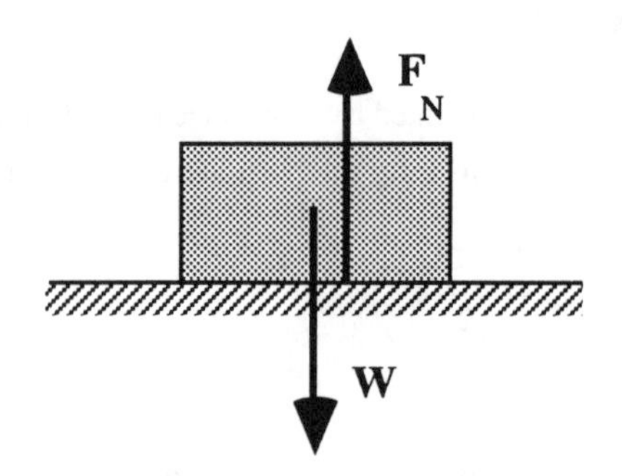

If an object is at equilibrium on an inclined plane, the magnitude of the normal force will be equal to the magnitude of a component of the object's weight, as is shown in the figure.

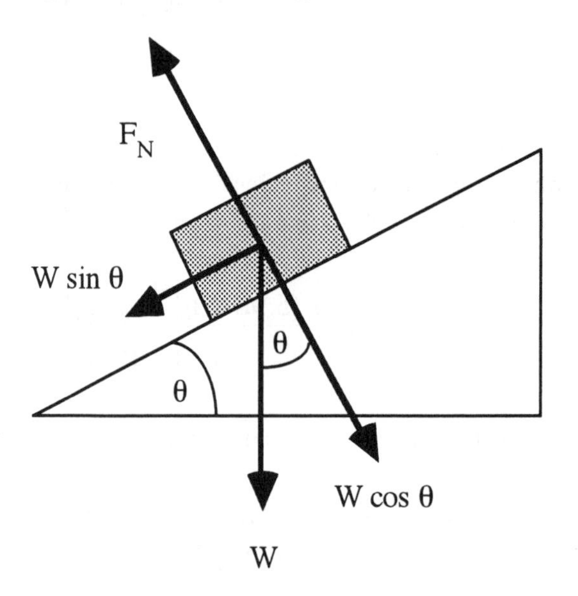

The magnitude of the normal force is $F_N = W \cos \theta$, which is the case even for a flat surface, i.e., for $\theta = 0$ since $\cos \theta = 1$, therefore $F_N = W$.

When an object is in contact with a surface, there is another force in addition to the normal force and gravity acting on it. This additional force is **friction**. When the object slides over the surface, the surface exerts a frictional force on the object. This is known as **kinetic friction**. However, a frictional force exists even if two objects are not in motion. This force is **static friction**.

An important characteristic of the frictional force is that its magnitude is proportional to the magnitude of the normal force. For example, an object sliding on a rough surface is subject to the force of kinetic friction, f_k, according to

$$f_k = \mu_k F_N \tag{4.8}$$

where μ_k is the **coefficient of kinetic friction**. The coefficient of friction can have values which are usually $0 \le \mu_k \le 1$.

Example 5
A 5.0 kg object is pulled over a rough horizontal surface by a 100.0 N force directed at an angle of 14.0° with respect to the surface (see figure). If the coefficient of kinetic friction is 0.40, find the acceleration of the object.

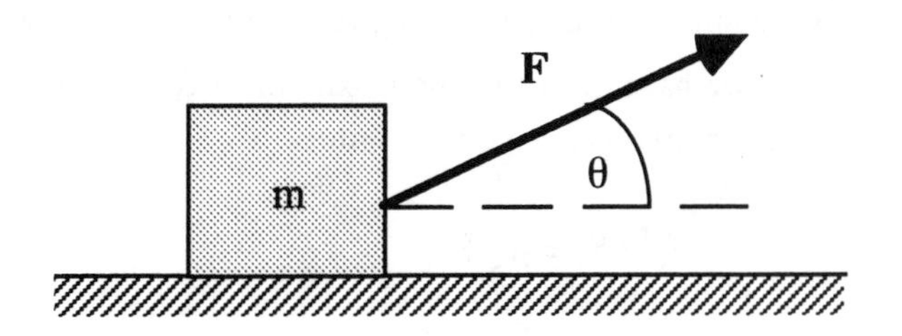

We begin by drawing a free-body diagram for this example. The forces which must be included are the applied force, **F**, the frictional force, $\mathbf{f_k}$, the normal force, $\mathbf{F_N}$, and the weight, $\mathbf{W} = m\mathbf{g}$. If the object is moving in the +x direction, we have

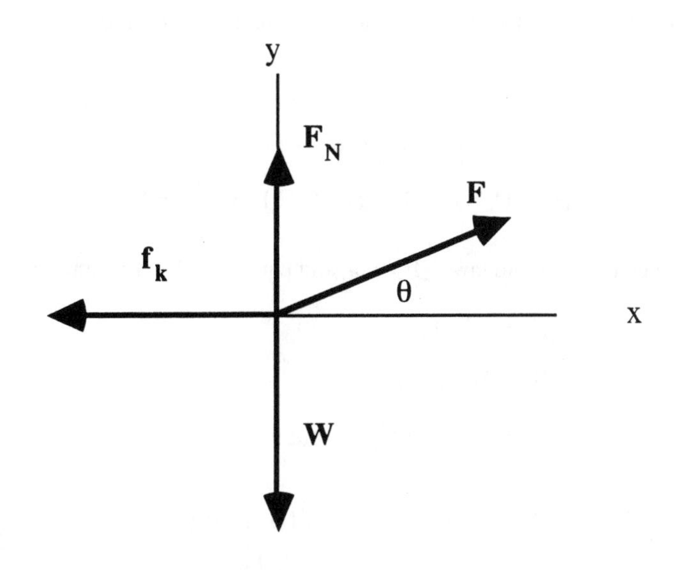

Look at the forces in the x and y directions:

y direction: $$F_N + F_y - W = 0$$

since there is no acceleration in the y direction, so

$$F_N = W - F_y = mg - F \sin \theta$$

$$F_N = (5.0 \text{ kg})(9.80 \text{ m/s}^2) - (100.0 \text{ N}) \sin 14.0° = 25 \text{ N}.$$

x direction: $$F_x - f_k = ma_x$$

since the object is accelerating in the x direction, so

$$F \cos \theta - \mu_k F_N = ma_x$$

$$(100.0 \text{ N}) \cos 14.0° - (0.40)(25 \text{ N}) = 87 \text{ N} = ma_x$$

so

$$a = (87\ \text{N})/m = (87\ \text{N})/(5.0\ \text{kg}) = 17\ \text{m/s}^2.$$

Suppose an object is resting on a flat surface. If we apply a small horizontal force, friction resists this force and the object remains at rest. This is the force of static friction, f_s. As we gradually increase the applied force, the object will eventually begin to move once we have overcome the maximum frictional force, f_s^{max}. If the coefficient of static friction is μ_s, the maximum static friction force can be written

$$f_s^{max} = \mu_s F_N \tag{4.7}$$

Example 6 *Static Friction*
A 10.0 kg block sits on a flat surface whose coefficient of static friction is $\mu_s = 0.60$ and whose coefficient of kinetic friction is $\mu_k = 0.40$. (a) What horizontal force is required to get the block to move? (b) If we continue to apply the same force as in part (a), what will the block's acceleration be?

(a) We can use equation (4.7) to find the force F necessary to overcome friction, that is,

$$F = f_s^{max}$$
$$F = \mu_s F_N = \mu_s mg$$
$$F = (0.60)(10.0\ \text{kg})(9.80\ \text{m/s}^2) = 59\ \text{N}.$$

(b) Using the above force, Newton's second law, $\Sigma F = ma$, and equation (4.8) for kinetic friction, we have

$$\Sigma F = F - f_k$$
$$\Sigma F = F - \mu_k F_N$$
$$\Sigma F = F - \mu_k mg = ma$$

Now

$$a = \frac{\Sigma F}{m} = \frac{F - \mu_k mg}{m} = \frac{59\ \text{N} - (0.40)(10.0\ \text{kg})(9.80\ \text{m/s}^2)}{10.0\ \text{kg}} = 2.0\ \text{m/s}^2$$

Example 7
A box weighing 147 N sits on a horizontal surface. The coefficient of static friction between the box and the surface is 0.70. If a force **P** is exerted on the box at an angle directed 37° below the horizontal, what must the magnitude of **P** be to get the box moving? The box and the associated free body diagram are shown below.

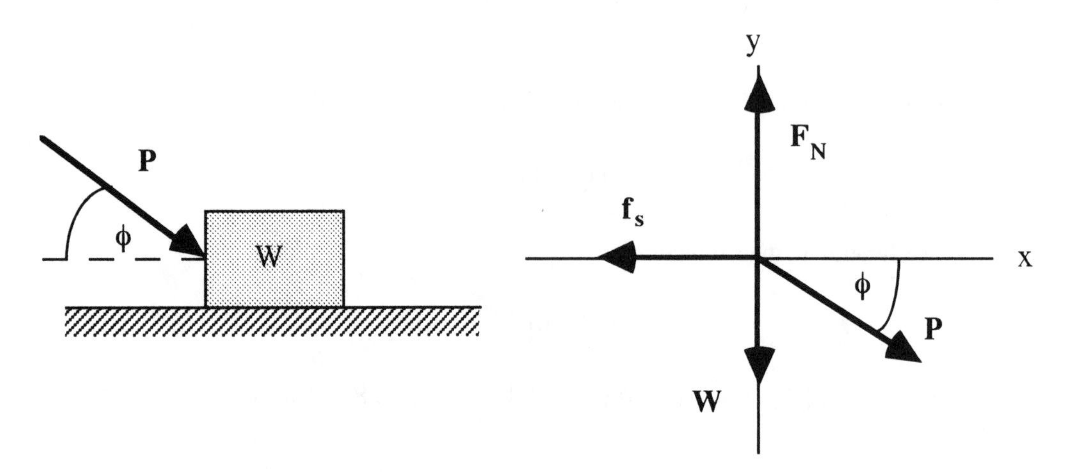

Use the component equations of Newton's second law, equations (4.2b) and (4.2a).

In the y direction

$$\Sigma F_y = ma_y$$

The box has no acceleration in the vertical direction so

$$F_N - P_y - W = 0$$

which gives

$$F_N = P \sin \phi + W \qquad (1)$$

In the x direction

$$\Sigma F_x = ma_x$$

The box has no acceleration in the horizontal direction so

$$P_x - f_s = 0$$

or

$$P \cos \phi - \mu_s F_N = 0$$

or

$$P \cos \phi - \mu_s(P \sin \phi + W) = 0 \qquad (2)$$

Solving (2) for P yields;

$$P = \frac{\mu_s W}{\cos \phi - \mu_s \sin \phi} = \frac{(0.70)\,(147\text{ N})}{\cos 37° - (0.70) \sin 37°} = 270\text{ N}.$$

4.10 The Tension Force

Forces are sometimes applied by ropes, cables, or wires that pull on an object. We usually assume that the rope is massless and that the magnitude of the tension, **T**, is the same throughout the rope.

Example 8 *The tension force*

A 12.0 kg object is pulled upward by a massless rope with an acceleration of 3.00 m/s^2. What is the tension in the rope?

The tension force is directed upward (+) while the weight of the object (mg) is directed downward (-). Since the object is being accelerated upward (+) we can write $\Sigma \mathbf{F} = m\mathbf{a}$ as

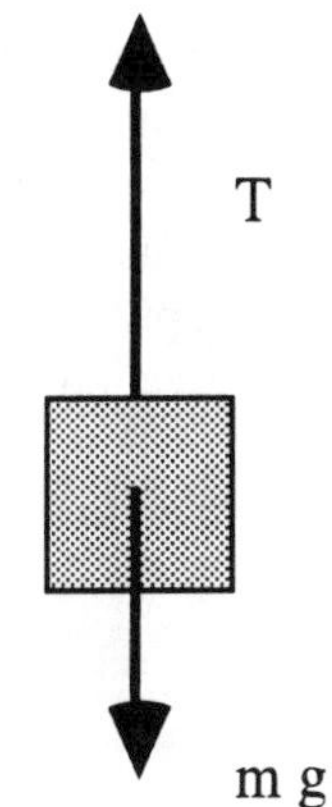

$$T - mg = ma$$

so

$$T = m\,(g + a)$$
$$T = (12.0\text{ kg})(9.80\text{ m/s}^2 + 3.00\text{ m/s}^2)$$
$$T = 154\text{ N}.$$

Example 9

Suppose the object in the previous example were accelerating downward at 3.00 m/s^2 instead of upward. What would the tension in the rope be in this case?

The only difference in this case would be that the acceleration is downward, so take $a = -3.00\ m/s^2$. We have,

$$T - mg = ma,$$

so that

$$T = m\,(g + a)$$
$$T = (12.0\ kg)\,(9.80\ m/s^2 - 3.00\ m/s^2)$$
$$T = 82\ N.$$

The situations depicted in examples 8 and 9 can be observed when you're in an elevator. Suppose you're on the bottom floor and the elevator starts accelerating upward. You feel as if you are being pushed downward, and the tension in the elevator cable increases. When you're on the top floor and the elevator begins accelerating downward, you feel as if you are being lifted off the floor and there is a corresponding decrease in the elevator cable tension. When the elevator is at rest, or moving with a steady speed, you feel "normal" and the cable tension would just be equal to the weight of you and the elevator.

4.11 Equilibrium Applications of Newton's Laws of Motion

An object that is at rest or traveling with a constant velocity has zero acceleration and is said to be in "equilibrium". This implies that the net force acting on the object is zero. In two dimensions we can write

$$\Sigma F_x = 0 \qquad (4.9a)$$

$$\Sigma F_y = 0 \qquad (4.9b)$$

There are five steps that can be followed in solving equilibrium problems. They are:

Step 1. Select an object, the "system", about which the most information is known. If two or more objects are connected, it may be necessary to treat each individually.

Step 2. Draw a "free-body" diagram for the system. As discussed previously, this is a drawing that represents the object and ALL the forces acting on it.

Step 3. Choose a convenient set of x, y axes for the object and resolve all the forces into components that point along these axes. By "convenient" we mean that you should choose your axes so that as many forces as possible point directly along the x or y axis. This will minimize the number of calculations needed.

Step 4. Apply equations (4.9a) and (4.9b) by setting the sum of the components of the forces in the x direction equal to zero and the sum of components in the y direction equal to zero.

Step 5. Solve the two equations obtained in step 4 for the desired unknown quantities. In many cases the equations must be solved simultaneously for the two unknowns.

The following are a number of examples employing the method just outlined.

Example 10 *A weight suspended by cables*
A 280 lb block is suspended by two cables, as shown below. Find the tension in each cable.

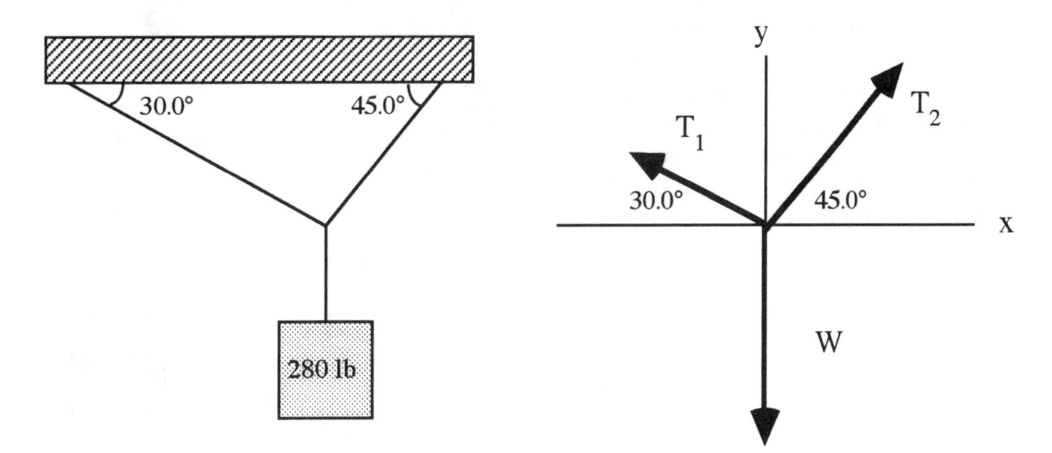

Choosing the box as our object, we have drawn the free-body diagram and chosen x and y axes, as shown. We now apply equations (4.9a) and (4.9b) using the following table

Force	x component	y component
$\mathbf{T_1}$	$-T_1 \cos 30.0$	$+T_1 \sin 30.0°$
$\mathbf{T_2}$	$+T_2 \cos 45.0°$	$+T_2 \sin 45.0°$
W	0	- 280 lb

Use Newton's second law

$$\Sigma F_x = -T_1 \cos 30.0° + T_2 \cos 45.0° = 0 \tag{1}$$

$$\Sigma F_y = +T_1 \sin 30.0° + T_2 \sin 45.0° - 280 \text{ lb} = 0 \tag{2}$$

Solve for T_1 in equation (1),

$$T_1 = T_2 \left(\frac{\cos 45.0°}{\cos 30.0°}\right) = 0.816\, T_2 \tag{3}$$

Substitute the result into equation (2),

$$(0.816)\, T_2 \sin 30.0° + T_2 \sin 45.0° = 280 \text{ lb}$$

Solving for T_2 then yields,

$$T_2 = \frac{280 \text{ lb}}{(0.816 \sin 30.0° + \sin 45.0°)} = 250 \text{ lb}$$

And finally, using this in equation (3) yields

$$T_1 = (0.816)\, T_2 = 200 \text{ lb.}$$

Example 11

Consider the following system in equilibrium. Determine the value of the weight, W, and the tension in the left wire.

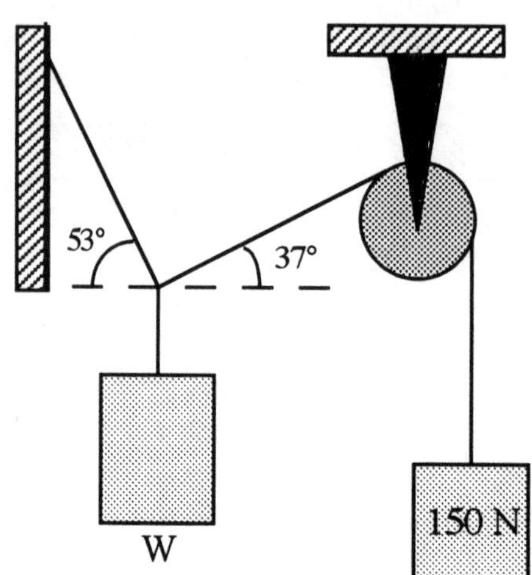

The free body diagram can be drawn in two different ways, as shown below. In (a), we can make the y axis vertical and draw in the forces. Perhaps a better choice would be to align the coordinate axes as shown in diagram (b). This is more convenient because two of the three forces now lie along the axes, and are thus resolved into x and y components automatically.

The equilibrium equations, (4.9a) and (4.9b) can now be written as

$$\Sigma F_x = 150\text{ N} - W \sin 37° = 0$$

which gives

$$W = (150\text{ N})/(\sin 37°) = 250\text{ N}.$$

and

$$\Sigma F_y = T - W \cos 37° = 0$$

which gives

$$T = W \cos 37° = 200\text{ N}.$$

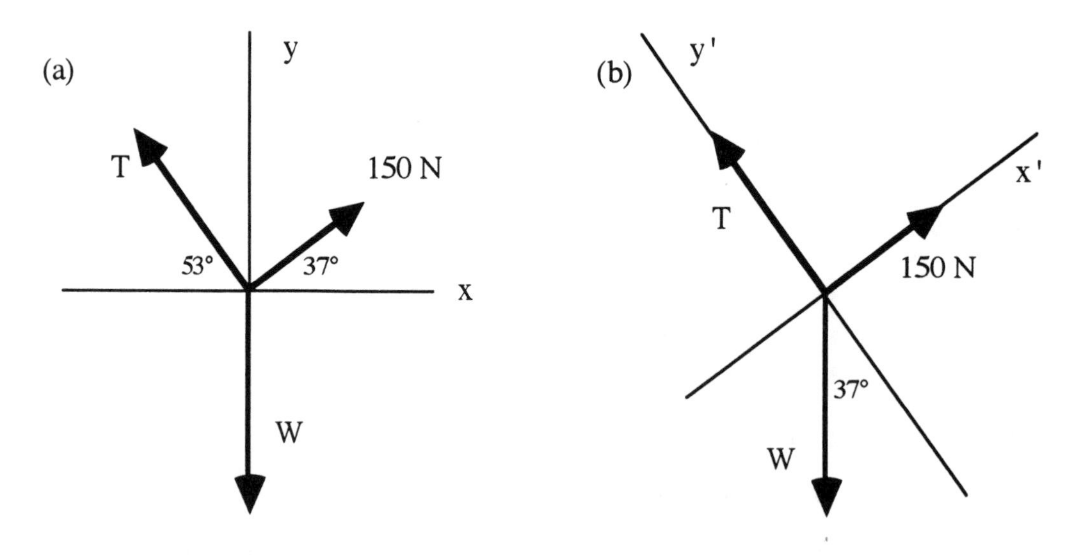

Notice that the choice of x and y axes in (b) made the equilibrium equations rather easy to solve for W and T. If we had chosen the axes as in drawing (a), the equations would have been a bit more difficult to solve, i.e.,

$$\Sigma F_x = (150\text{ N}) \cos 37° - T \cos 53° = 0$$

$$\Sigma F_y = (150\text{ N}) \sin 37° + T \sin 53° - W = 0.$$

Thus, the appropriate choice of coordinate axes can facilitate the solution quite a bit in most cases.

Example 12 *Object on an inclined plane*
Consider a mass M = 25.0 kg sitting on an inclined plane whose angle of inclination is θ = 30.0°. A massless cord connected to M passes over a frictionless pulley and is attached to a mass m = 15.0 kg. The mass M slides up the plane at a constant speed. Find the coefficient of kinetic friction between the mass M and the plane.

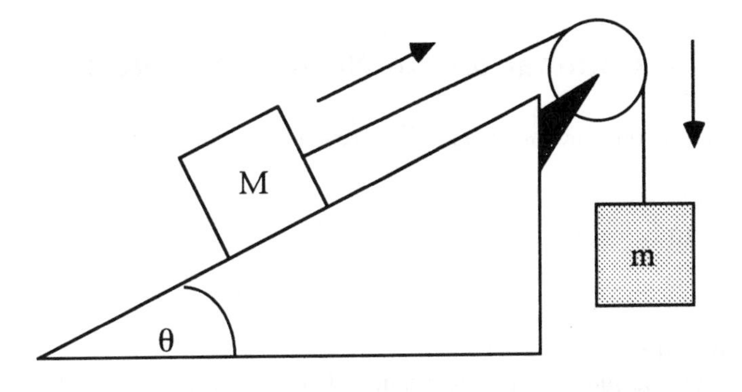

Since the masses move with constant velocity (a = 0), the system is in equilibrium. We need to look at each mass individually in order to determine the necessary quantities. The free-body diagrams for M and m look like

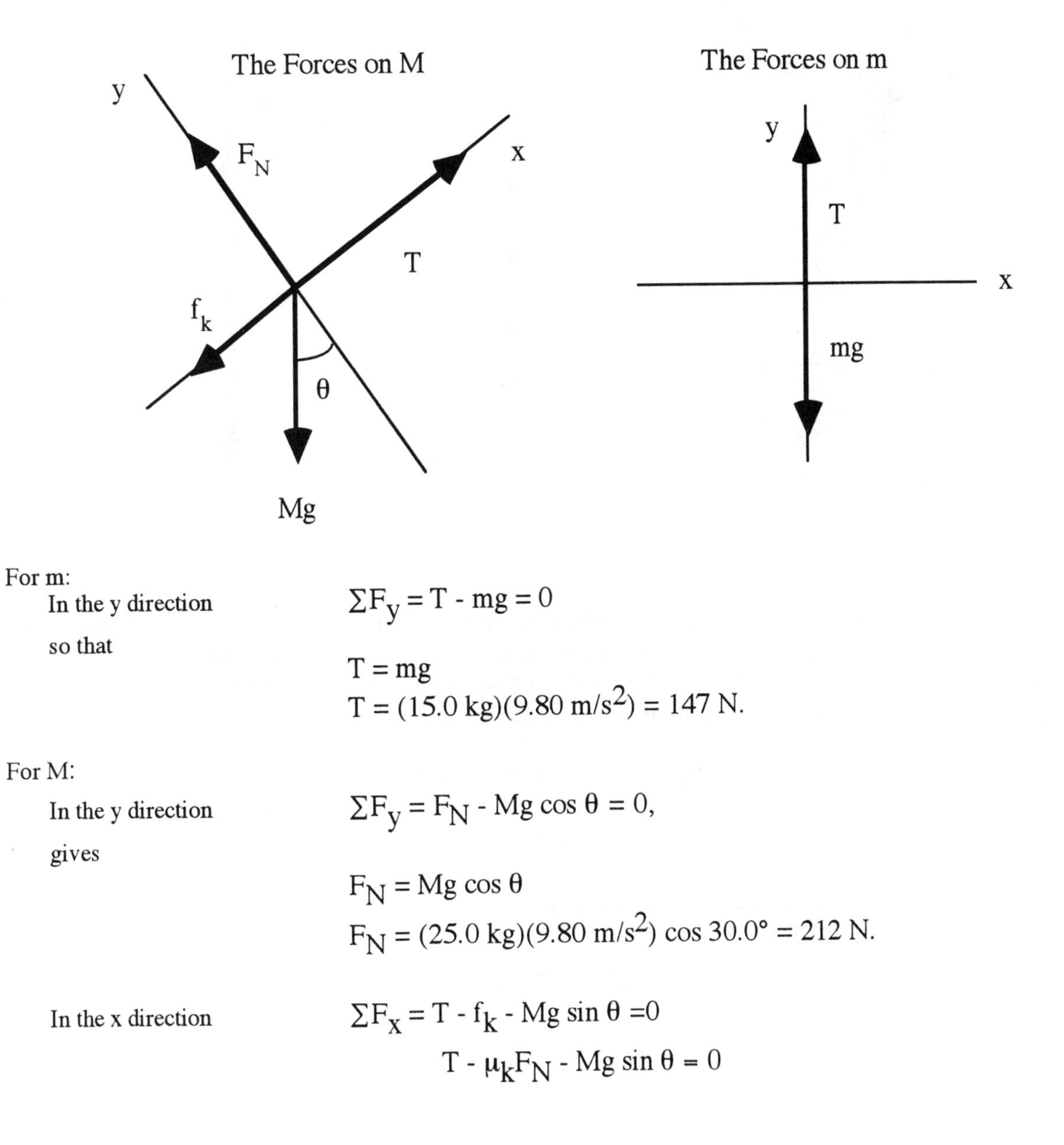

For m:

In the y direction $\quad \Sigma F_y = T - mg = 0$

so that

$$T = mg$$
$$T = (15.0\text{ kg})(9.80\text{ m/s}^2) = 147\text{ N}.$$

For M:

In the y direction $\quad \Sigma F_y = F_N - Mg\cos\theta = 0,$

gives

$$F_N = Mg\cos\theta$$
$$F_N = (25.0\text{ kg})(9.80\text{ m/s}^2)\cos 30.0° = 212\text{ N}.$$

In the x direction $\quad \Sigma F_x = T - f_k - Mg\sin\theta = 0$

$$T - \mu_k F_N - Mg\sin\theta = 0$$

Solving for μ_k yields

$$\mu_k = \frac{T - Mg \sin \theta}{F_N} = \frac{147 \text{ N} - (25.0 \text{ kg})(9.80 \text{ m/s}^2) \sin 30.0°}{212 \text{ N}} = 0.12.$$

4.12 Nonequilibrium Applications of Newton's Laws of Motion

Nonequilibrium occurs when the forces acting on an object are not balanced. In this case the object accelerates and Newton's second law is

$$\Sigma F_x = ma_x \quad (4.2a) \qquad \Sigma F_y = ma_y \quad (4.2b)$$

Example 13 *Atwood's Machine*
Two masses, m_1 and m_2 (assume $m_2 > m_1$), are connected by a massless string that passes over a massless, frictionfree pulley, as shown in the figure. Find the acceleration of the masses and the tension in the string.

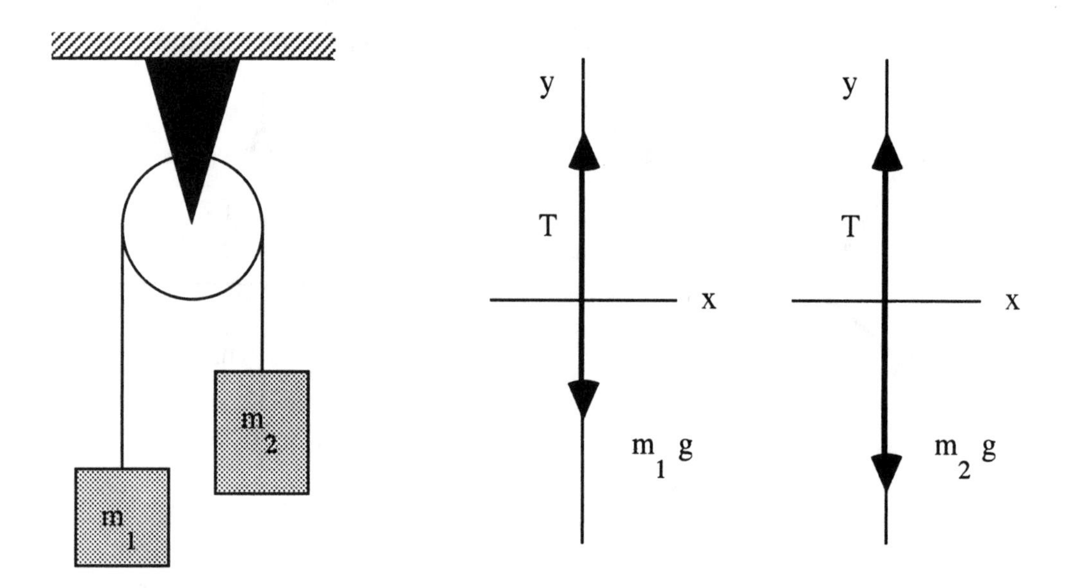

We need to look at each mass individually in order to obtain the two equations necessary to solve for the unknown quantities **a** and **T**. The free-body diagram for each mass is shown above. Since both masses are connected by the string, the magnitudes of their accelerations are the same. Since $m_2 > m_1$, m_1 has an upward acceleration and m_2 has a downward acceleration. We also assume that the tension in the string on each side of the pulley is the same.

For m_1 the acceleration is upward, so $T > m_1 g$ and we can therefore write,

$$T - m_1 g = m_1 a.$$

For m_2 the acceleration is down, so $m_2 g > T$. Therefore, we can write,

$$m_2 g - T = m_2 a.$$

Solving for T in the first equation and substituting into the second gives,

$$m_2g - (m_1g + m_1a) = m_2a.$$

Finally, solving for the acceleration yields,

$$a = \frac{(m_2 - m_1)}{(m_1 + m_2)} g$$

Substituting this expression for the acceleration into either of the mass equations, we can solve for the tension.

$$T = \frac{2m_1m_2}{(m_1 + m_2)} g$$

Example 14
Consider the following arrangement with M_1 = 15.0 kg, M_2 = 10.0 kg, and μ_k = 0.30. Find the acceleration of the system and the tension in the connecting cord. Free-body diagrams accompany the drawing.

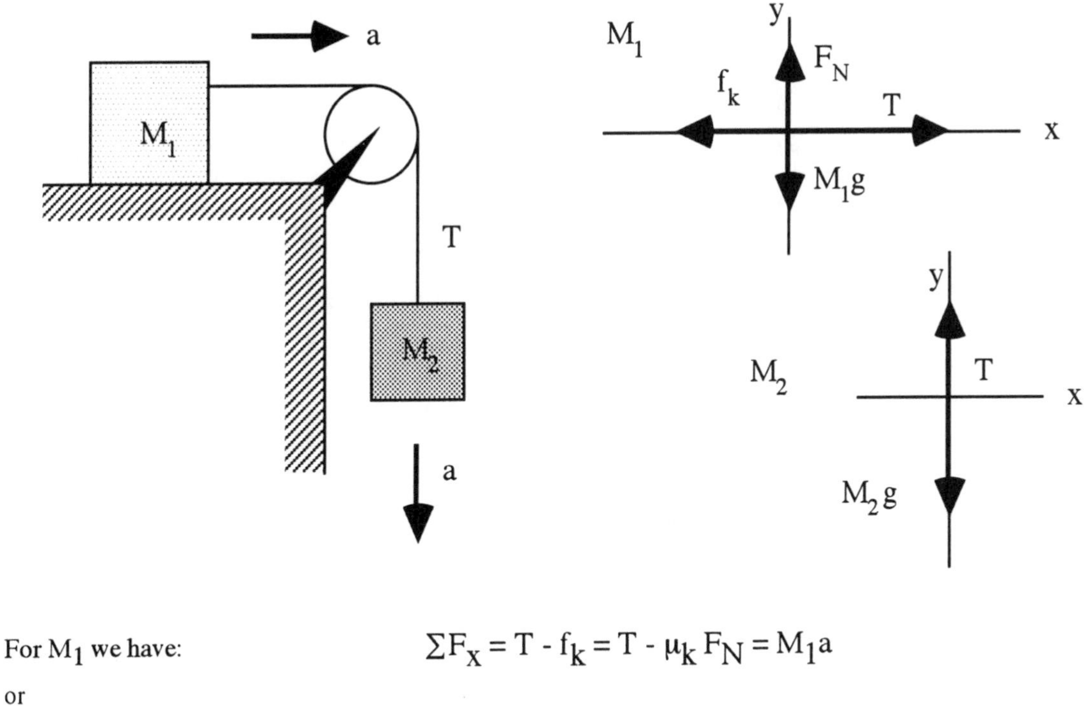

For M_1 we have: $$\Sigma F_x = T - f_k = T - \mu_k F_N = M_1a$$

or

$$T - \mu_k M_1 g = M_1a.$$

For M_2 we have: $$\Sigma F_y = M_2 g - T = M_2 a.$$

Solving for T in the second equation and substituting into the first yields:

$$M_2 g - M_2 a - {}_k M_1 g = M_1 a$$

or

$$(M_1 + M_2)\, a = (M_2 - {}_k M_1)\, g$$

The acceleration is therefore,

$$a = \frac{(M_2 - \mu_k M_1)g}{(M_1 + M_2)} = \frac{[10.0 \text{ kg} - (0.30)(15.0 \text{ kg})](9.80 \text{ m/s}^2)}{(15.0 \text{ kg} + 10.0 \text{ kg})} = 2.2 \text{ m/s}^2.$$

The tension can now be found by substituting this value of **a** into the equation for M_2.

$$T = M_2\,(g - a)$$

$$T = (10.0 \text{ kg})\,(9.80 \text{ m/s}^2 - 2.2 \text{ m/s}^2) = 76 \text{ N}.$$

PRACTICE PROBLEMS

1. A block of mass 4.0 kg rests on a horizontal surface. What horizontal force is required to accelerate the block at 5.0 m/s^2 if (a) there is no friction, and (b) if the coefficient of kinetic friction is 0.25?

2. An object of mass 10.0 kg is subjected to the following forces: 25.0 N due east, 50.0 N at an angle of 53.0° north of east, and 40.0 N due west. What is the acceleration of the object?

3. At what altitude above the earth's surface would the acceleration of gravity be 7.00 m/s^2? Use the values of the earth's mass and radius given in example 4.

4. Two 75 kg masses are separated by a distance of 1.0 m. A 5.0 kg object is placed 0.25 m from one of the masses (0.75 m from the other) along the line joining the masses. What is the gravitational force on the object?

5. Two blocks are connected by a cord on a horizontal surface. A force F pulls on the blocks, as shown below. Find the acceleration of the blocks and the tension in the connecting cord if (a) the surface is frictionless, and (b) the coefficient of kinetic friction is 0.20.

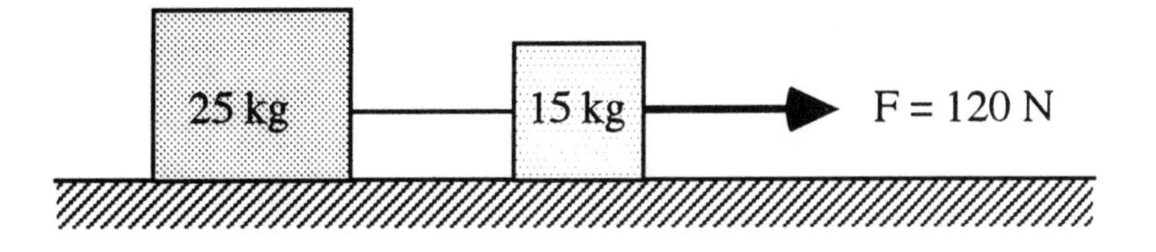

6. Traveling at a speed of 58.0 km/h, the driver of an automobile suddenly locks the wheels by slamming on the brakes. The coefficient of kinetic friction between the tires and the road is 0.720. How far does the car skid before coming to a halt? Ignore effects of air resistance.

7. You are standing on a scale in an elevator, the scale reads 176 lb when the elevator is at rest. The elevator begins to accelerate upward at 12.0 ft/s^2. (a) What is the reading on the scale? (b) Suppose the elevator were accelerating downward at 12.0 ft/s^2. What does the scale read now?

8. A 250 lb block rests on a horizontal table. The coefficient of kinetic friction between the block and table is 0.30. What force is required to pull the block at constant speed if (a) a horizontal rope is attached, or if (b) the rope makes an angle of 25° above the horizontal?

9. A space traveler weighs 130 lb on Earth. What will the traveler weigh on another planet whose radius is three times that of the earth and whose mass is twice that of the earth?

10. An object rests on an inclined plane. One end of the plane is lifted until the object begins to slide. If the angle at which the object begins to slide is 35°, what is the coefficient of static friction?

11. Find the tensions T_1 and T_2 in the following diagram.

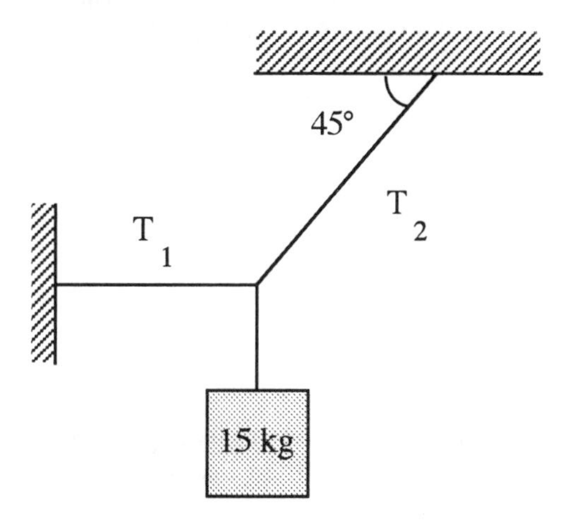

12. A fisherman is fishing from a bridge and using "10-lb test line". In other words, the line will sustain a maximum force of 10.0 lb without breaking. (a) What is the heaviest fish that can be pulled up vertically through the air when the line is reeled in at a constant speed? (b) Repeat part (a), assuming that the line is given an upward acceleration of 6.50 ft/s^2.

13. Find the acceleration of the system and the tension in the cord. Take m = 2.0 slugs and M = 3.5 slugs.

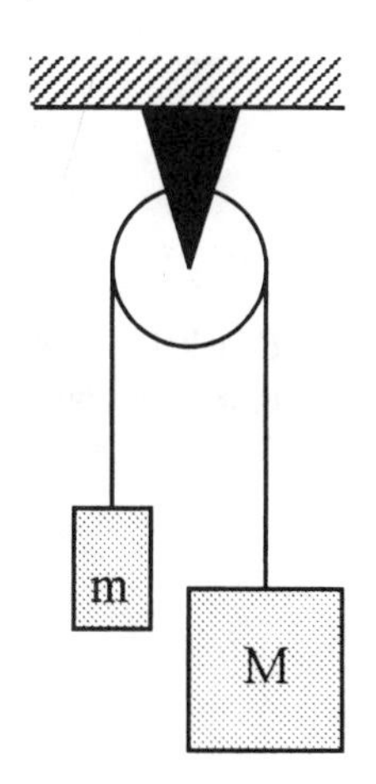

14. A 55.0 kg person is riding in a hot air balloon, and a scale shows this person's weight to be 549 N. The acceleration of gravity at the location of the balloon is known to be 9.79 m/s^2. Determine the magnitude and direction of the vertical component of the balloon's acceleration.

15. Find the acceleration of the system if (a) there is no friction, and (b) if the coefficient of kinetic friction is 0.33. Take M = 7.5 kg, and m = 2.5 kg.

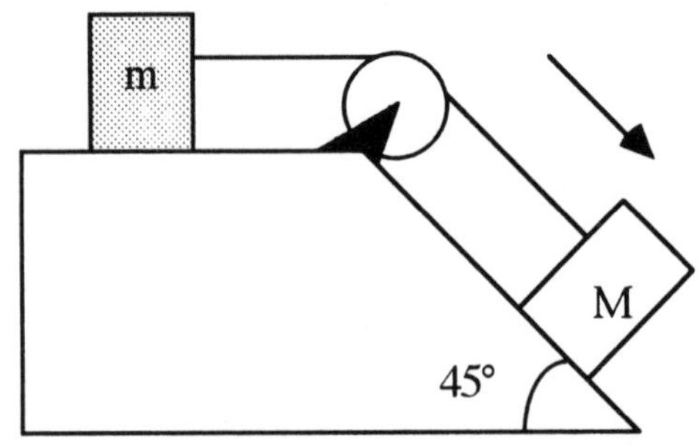

16. Starting from rest, a piece of ice slides down a 45.0° incline in twice the time it takes to slide down a frictionless 45.0° incline. What is the coefficient of kinetic friction between the ice and the incline?

17. What is the magnitude of the force **F** (assumed horizontal) required to push the block up the plane at a constant speed? Take the coefficient of kinetic friction to be 0.25.

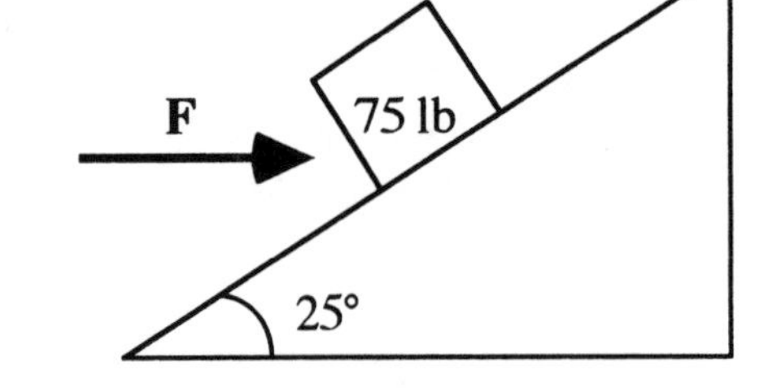

18. Block 1 weighs 96 lb. The coefficient of static friction between block 1 and the table is 0.50. Find the maximum weight of block 2 for which the system will remain in equilibrium.

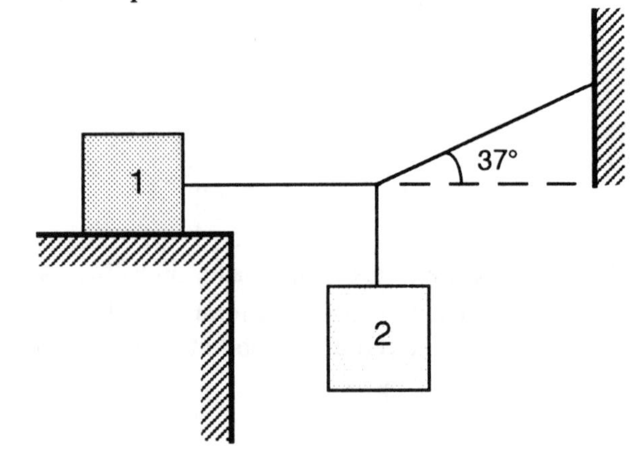

HELPFUL SUGGESTIONS

1. To remember the units of force (or any other quantity for that matter) use the defining equation, i.e., using $F = ma = (kg)(m/s^2) = kg\ m/s^2$ = newton.

2. Before drawing a free-body diagram, be sure to identify ALL the physical forces acting on an object.

3. Keep in mind that even though an object may be moving, there doesn't necessarily have to be a force acting directly in the direction of motion. (A component is sufficient.)

4. The gravitational force is ALWAYS a force of attraction. The gravitational force between two objects is such that the force acting on one object is equal in magnitude but opposite in direction to the force acting on the other object.

5. The direction of a frictional force is parallel to the contact surface and always acts in a direction so as to oppose the relative motion between the object and the surface.

6. The normal force is NOT always equal to an object's weight. If the object is on an inclined plane, or if additional forces have components that act perpendicular to a surface, F_N may be greater than or less than the weight W.

7. Tension forces are "pulling forces" and are therefore always directed away from the object on which they are pulling.

8. For a massless rope or string, the tension is the same everywhere in the rope, even if it is wrapped around a pulley (this assumes a frictionless and massless pulley). Thus, the rope exerts the same force on each side of the pulley.

9. Take some time to choose a "convenient" coordinate system for each individual problem. The best coordinate systems are those which reduce the number of forces for which components must be derived. Try to set up your axes so that as many forces act along the chosen axes as possible.

10. Remember that systems in "equilibrium" are not necessarily systems at rest. Any system moving with a constant velocity (**a** = 0) is also considered to be in equilibrium.

11. Follow the five steps outlined in the text that facilitate the application of Newton's laws. These are:

    Step 1: Choose an object or objects.
    Step 2: Draw the free-body diagram for the system.
    Step 3: Choose a convenient set of axes.
    Step 4: Apply Newton's second law ($\Sigma\mathbf{F} = m\mathbf{a}$).
    Step 5: Solve the equations obtained in step 4.

EVERYDAY PHYSICS

1. Consider Captain Kirk sitting on the bridge of the USS Enterprise. He tells Mr. Sulu to go to warp 5 and the ship instantly accelerates to a very high speed. If Newton's second law is still valid in the 23rd century, what would happen to Captain Kirk?

2. Many examples of action-reaction can be seen in everyday life. Blow up a balloon and let it go, shoot a gun and feel the recoil, step on the accelerator of your car and feel yourself pressed by your seat. These are just a few of many examples. Can you think of a few more?

3. You are riding upward on an elevator when it suddenly comes to a stop. Explain the "sinking" feeling you feel in your stomach.

4. One way that astronaut trainees experience weightlessness is in an airplane. The plane is flying level and suddenly goes into a steep dive and accelerates downward. For the brief period during the acceleration the astronauts are "weightless" and float about the airplane cabin. How can this be explained in terms of Newton's laws? What would expect the acceleration of the plane to be to make this situation occur?

5. You can tighten the head of a hammer by slamming the butt end of the handle down on a hard surface. Explain.

6. Imagine what would happen if the bullets in a rifle had a mass equal to the rifle itself. What would happen when you fired the rifle?

7. Supposed you moved from New York (at sea level) to Denver (one mile elevation). Would you gain or lose weight? How much weight would you gain or lose? (the earth's radius is about 4000 miles).

8. Think of some everyday examples of systems in equilibrium. For example, imagine the complicated structure of a long suspension bridge, or how a crane supported by cables and pulleys can lift a heavy object. Think about the muscles in your body and how they can control the function of bones and tissue.

9. Speculate on what the world would be like without friction. Consider simple things like walking, driving a car, riding a bicycle, etc. The compelling conclusion you'll reach is that nothing much would get done.

10. An example of "apparent weight" is when an astronaut takes off in a rocket. As the rocket begins lifting off the ground we sometimes say he is experiencing so many "g s", where g is the acceleration of gravity. If an astronaut weighs 150 lb and experiences 8 "g s" on liftoff, what is his or her apparent weight? Have you ever experienced a similar effect when you push down the accelerator in your car?

11. Which of the following are examples of systems in equilibrium and which are not?

    - A parachutist approaching the ground at constant velocity.
    - Fighting with a fish on the other end of a fishing line.
    - Driving around a curve in the car at constant speed.
    - Swinging on a swing.
    - A planet orbiting around the sun.
    - Walking up or down the stairs.
    - A ball at the exact top of its trajectory.

CHAPTER QUIZ

1. A net force **F** acts on a mass m and produces an acceleration **a**. What acceleration results if a force, 8**F** acts on a mass 4m?
   a. 32**a** b. 8**a** c. 4**a** d. 2**a**

2. Suppose an astronaut is taking a space walk to fix his spacecraft with a hammer. His life-line breaks and the jet packs on his backpack are out of fuel. How can he return safely to his spaceship (without someone else's help)?
   a. He can't return safely.
   b. He can throw his hammer at the spaceship in disgust.
   c. He can throw the hammer away from the spaceship.
   d. He can fling his arms around in circles.

3. A horizontal force of 10.0 N accelerates an 8 kg mass from rest at a rate of 1.0 m/s^2 in the positive x direction. What frictional force acts on the 8 kg mass?
   a. +2 N b. -2 N c. +8 N d. +80 N

4. A box is sliding along a horizontal surface with a constant acceleration of -3.3 m/s^2 as it comes to rest. Approximately what is the coefficient of kinetic friction between the box and surface?
   a. 0.20 b. 0.34 c. 0.66 d. 1.00

5. If the distance between the earth and moon were doubled, what would happen to the force of gravity between the earth and moon?
   a. It would double.
   b. It would be one-half as great.
   c. It would be four times as great.
   d. It would be one-fourth as great.

6. A force of 24 N acts on a 4 kg mass, originally at rest. In how many seconds will the body acquire a velocity of 54 m/s?
   a. 3 s b. 6 s c. 9 s d. 10 s

7. Assume that you are driving down a straight road at a steady speed. A small ball tied on the end of a string is hanging from the rear view mirror. Which way will the ball swing when you suddenly apply the brakes?
   a. forward b. backward c. to the left d. It wouldn't swing

8. If the mass and weight of an astronaut are measured on earth and on the moon, we will find that the masses are __________ and the weights are ____________.
   a. the same ... the same
   b. the same ... different
   c. different ... the same
   d. different ... different.

9. A string accelerates a 5.0 kg object upward at 4.0 m/s^2. What is the tension in the string?
   a. 29 N b. 49 N c. 69 N d. 676 N

10. A string is attached a 5.0-kg object which accelerates downward at 4.0 m/s^2. What is the tension in the string?
    a. 29 N b. 49 N c. 69 N d. 680 N

11. An object is sitting on a plane which can be inclined with respect to the horizontal. What happens to the normal force of the plane on the object as the angle of inclination increases?
    a. It increases b. It decreases c. It remains the same

12. A rock is suspended from a string and accelerates downward. It follows that the tension in the string is
    a. equal to the weight of the rock.
    b. less than the weight of the rock.
    c. greater than the weight of the rock.
    d. either greater or less than the weight, depending on the magnitude of the acceleration.

13. A mass m slides down a friction-free inclined plane of inclination angle θ. The force pulling the body down is
    a. mg b. mg sin θ c. mg cos θ d. zero

14. An object of mass M is pulled by a horizontal force along a surface at constant speed. If the coefficient of kinetic friction between the block and surface is μ, what is the tension in the rope pulling the object?
    a. $T = \mu g$ b. $T = Mg/\mu$ c. $T = Mg$ d. $T = \mu Mg$

15. What force is needed to lift a 64 lb object at an acceleration of 16 ft/s^2?
    a. 32 lb b. 48 lb c. 64 lb d. 96 lb

16. What force is required to lower a 64 lb object at an acceleration of 16 ft/s^2?
    a. 32 lb b. 48 lb c. 64 lb d. 96 lb

17. An object slides down an inclined plane (inclination angle $\theta = 37°$) at constant velocity. What is the coefficient of kinetic friction between the object and the plane?
    a. 0.40 b. 0.60 c. 0.75 d. 0.90

18. Which of the following statements about an object in equilibrium is false?
    a. The net force acting on the object MUST be zero.
    b. The body may be moving with constant velocity.
    c. The body MUST be at rest.
    d. The body can be moving at a steady speed.

19. A horizontal wire is attached between two supports. If an object of weight W is hung from the center of a wire, which "sags" a very small amount. Which statement best describes the tension in the wire?
    a. It increases by an amount W.
    b. It increases by an amount greater than W.
    c. It increases by an amount less than W.
    d. It can increase or decrease depending on the actual value of W.

20. The driver of a car moving at 49 m/s slams on the brakes and the car comes to a halt. If the coefficient of friction between the car and the road is 0.50, how long does it take for the car to come to rest?
    a. 5 s b. 10 s c. 25 s d. 98 s

21. For the situation described in question 20, how far does the car travel before coming to rest?
    a. 250 m b. 98 m c. 490 m d. 49 m

22. A 10.0 g bullet traveling 155 m/s strikes a wooden block and comes to rest 0.100 m inside the block. What was force exerted on the bullet by the block?
    a. 1.55 N b. 155 N c. 1200 N d. 1550 N

SOLUTIONS AND ANSWERS

Practice Problems

---

1. (a) Using equation (4.1),

$$F = ma = (4.0\ \text{kg})(5.0\ \text{m/s}^2) = \mathbf{2.0 \times 10^1\ N}.$$

(b) We now have two forces,

$$F - f_k = ma,$$

where

$$f_k = \mu_k F_N = \mu_k mg.$$

Therefore,

$$F = \mu_k mg + ma,$$

or

$$F = m(\mu_k g + a) = (4.0\ \text{kg})[(0.25)(9.80\ \text{m/s}^2) + 5.0\ \text{m/s}^2] = \mathbf{3.0 \times 10^1\ N}.$$

---

2. Use equations (4.2a) and (4.2b) to find the accelerations in the x and y directions. We have,

$$\Sigma F_x = ma_x$$

or

$$25.0\ \text{N} + (50.0\ \text{N})\cos 53.0° - 40.0\ \text{N} = (10.0\ \text{kg})\ a_x$$

so

$$a_x = (15.1\ \text{N})/(10.0\ \text{kg}) = \mathbf{1.51\ m/s^2}.$$

Also

$$\Sigma F_y = ma_y$$

or

$$(50.0\ \text{N})\sin 53.0° = 39.9\ \text{N} = (10.0\ \text{kg})\ a_y$$

then

$$a_y = (39.9\ \text{N})/(10.0\ \text{kg}) = \mathbf{3.99\ m/s^2}.$$

The acceleration **a** has magnitude a and direction θ given by

$$a^2 = a_x^2 + a_y^2$$

or a = **4.27 m/s²**. Also,

$$\theta = \tan^{-1}(a_y/a_x) = \mathbf{69.3°\ N\ of\ E}.$$

---

3. The acceleration is

$$a = GM_E/r^2,$$

so

$$r^2 = GM_E/a = (6.67 \times 10^{-11}\ \text{N m}^2/\text{kg}^2)(5.98 \times 10^{24}\ \text{kg})/(7.00\ \text{m/s}^2).$$

Solving gives

$$r = 7.55 \times 10^6\ \text{m}.$$

Now

$$r = R_E + h$$

Hence,

$$h = r - R_E = 7.55 \times 10^6\ \text{m} - 6.38 \times 10^6\ \text{m} = \mathbf{1.17 \times 10^6\ m}.$$

---

4. We can use equation (4.3),

$$F_1 = Gm_1M/r_1^2 = G(75 \text{ kg})(5.0 \text{ kg})/(0.25 \text{ m})^2 = 4.0 \times 10^{-7} \text{ N}.$$

The second mass gives

$$F_2 = G(75m \text{ kg})(5.0 \text{ kg})/(0.75 \text{ m})^2 = 4.4 \times 10^{-8} \text{ N}.$$

The resultant is

$$F = F_1 - F_2 = \mathbf{3.6 \times 10^{-7} \text{ N}}.$$

---

5. (a)

$$F = ma \text{ (where } m = m_1 + m_2 = 40 \text{ kg)}$$

so

$$a = F/m = (120 \text{ N})/(40 \text{ kg}) = \mathbf{3.0 \text{ m/s}^2}.$$

The tension is

$$T = m_1a = (25 \text{ kg})(3.0 \text{ m/s}^2) = \mathbf{75 \text{ N}}.$$

(b) Newton's second law requires

$$F - f_k = ma$$

or

$$F - \mu_k mg = ma$$

Then

$$a = (F - \mu_k mg)/m = \mathbf{1.0 \text{ m/s}^2}.$$

The tension is found from

$$T - f_{1k} = m_1a$$

or

$$T = m_1 (a + \mu_k g) = \mathbf{74 \text{ N}}.$$

---

6. The initial velocity is

$$v_0 = (58 \text{ km/h})(1 \text{ h}/3600 \text{ s})(1000\text{m}/1 \text{ km}) = 16.1 \text{ m/s}.$$

Only the friction force acts, so Newton's second law

$$f_k = ma$$

gives

$$a = - f_k/m = - \mu_k mg/m = - \mu_k g = - (0.720)(9.80 \text{ m/s}^2) = -7.06 \text{ m/s}^2.$$

Using equation (2.9) we have

$$v^2 = v_0^2 + 2as$$

or

$$s = -v_0^2/2a = -(16.1 \text{ m/s}^2)/[2(-7.06 \text{ m/s}^2)] = \mathbf{18.4 \text{ m}}.$$

---

7. (a) The reading on the scale will be the same as the normal force exerted by the platform of the scale. We have

$$F_N - W = ma,$$

so that

$$F_N = W + ma = 176 \text{ lb} + [(176 \text{ lb})/(32.2 \text{ ft/s}^2)](12.0 \text{ ft/s}^2) = \mathbf{242 \text{ lb}}.$$

(b)
$$W - F_N = ma$$
gives
$$F_N = W - ma = 176 \text{ lb} - (5.5 \text{ sl})(12.0 \text{ ft/s}^2) = \mathbf{110 \text{ lb}}.$$

---

8. (a) Since the acceleration is zero we have
$$F - f_k = 0$$
or
$$F = \mu_k F_N = (0.30)(250 \text{ lb}) = \mathbf{75 \text{ lb}}.$$

(b) The normal force is not equal to the weight in this case, instead
$$F_N + F_y = W$$
so that
$$F_N = W - F_y = W - F \sin 25°.$$
Therefore,
$$F \cos 25° = \mu_k (W - F \sin 25°),$$
solving for F we obtain
$$F = \mu_k W/(\cos 25° + \mu_k \sin 25°)$$
$$F = [(0.30)(250 \text{ lb})]/[\cos 25° + (0.30) \sin 25°)] = \mathbf{73 \text{ lb}}.$$

---

9. Using equation (4.4) we can find the ratio of the weights on the planet (P) and earth (E).
$$\frac{W_P}{W_E} = \frac{GM_Pm/R_P^2}{GM_Em/R_E^2} = \frac{M_PR_E^2}{M_ER_P^2} = \frac{2M_ER_E^2}{M_E(3R_E)^2} = \frac{2}{9}$$

The weight of the traveler on the planet is therefore,
$$W_P = (2/9)\, W_E = (2/9)\,(130 \text{ lb}) = \mathbf{29 \text{ lb}}.$$

---

10. Refer to the following figure

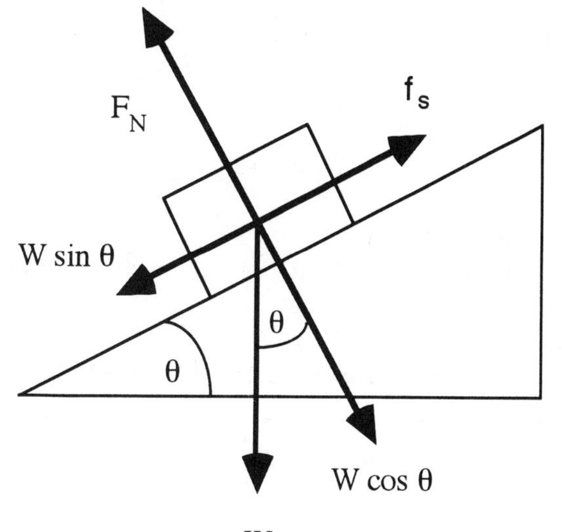

The force pulling the object down the plane is

$$F = W \sin\theta.$$

The frictional force is

$$f_s = \mu_s F_N = \mu_s (W \cos\theta).$$

Just as the block begins to slide we see that

$$F = f_s$$

so that

$$W \sin\theta = \mu_s W \cos\theta$$

Then

$$\mu_s = \sin\theta/\cos\theta = \tan\theta = \tan(35°) = \mathbf{0.70}.$$

---

11. Refer to the following free-body diagram.
In the y direction:

$$\Sigma F_y = T_{2y} - mg = 0$$

which gives

$$T_2 \sin 45° = mg$$

or

$$T_2 = mg/\sin 45° = (15\ \text{kg})(9.8\ \text{m/s}^2)/(\sin 45°) = \mathbf{210\ N}.$$

In the x direction:

$$\Sigma F_x = T_{2x} - T_1 = 0$$

which gives

$$T_1 = T_2 \cos 45°$$

$$T_1 = (210\ \text{N}) \cos 45° = \mathbf{150\ N}.$$

---

12. (a) Since there is no acceleration, Newton's second law requires that

$$T - W = 0$$

or

$$W = T = \mathbf{10.0\ lb}.$$

(b) Newton's second law with a non-zero acceleration, is

$$T - mg = ma$$

or

$$T = m(g+a),$$

solving for m yields

$$m = T/(g + a)$$
$$m = (10.0\ \text{lb})/(32.2\ \text{ft/s}^2 + 6.50\ \text{ft/s}^2) = 0.258\ \text{sl}.$$

Therefore, the weight of the fish is

$$W = mg = (0.258\ \text{sl})(32.2\ \text{ft/s}^2) = \mathbf{8.32\ lb}.$$

---

13. Using the results from example 13, we have

$$a = (M - m)g/(M + m)$$
$$a = (1.5 \text{ sl})(32 \text{ ft/s}^2)/(5.5 \text{ sl}) = \mathbf{8.7 \text{ ft/s}^2}.$$

The tension is then

$$T = 2Mmg/(M + m)$$
$$T = 2(3.5 \text{ sl})(2.0 \text{ sl})(32 \text{ ft/s}^2)/(5.5 \text{ sl}) = \mathbf{81 \text{ N}}.$$

---

14. Note that $F_N = 549$ N > mg = (55.0 kg)(9.79 m/s^2) = 539 N, so the acceleration is **UPWARD**. Using equation (4.6),

$$F_N = mg + ma$$

we have

$$a = (F_N - mg)/m = (549 \text{ N} - 539 \text{ N})/(55.0 \text{ kg}) = \mathbf{0.19 \text{ m/s}^2}.$$

---

15. (a) Newton's second law gives

$$\Sigma F = (M+m)a$$

where the weight is the only force that is pulling M down the plane. Therefore,

$$\Sigma F = Mg \sin 45° = (M+m)a$$

or

$$a = Mg \sin 45°/(M+m)$$
$$a = (7.5 \text{ kg})(9.8 \text{ m/s}^2) \sin 45°/(10.0 \text{ kg}) = \mathbf{5.2 \text{ m/s}^2}.$$

(b) With friction (on both surfaces),

$$\Sigma F = Mg \sin 45° - \mu_k(Mg \cos 45° + mg) = (M+m)a$$

which gives

$$a = \mathbf{2.7 \text{ m/s}^2}.$$

---

16. Frictionless plane:

$$a = g \sin 45.0°$$
$$a = (9.80 \text{ m/s}^2) \sin 45.0° = 6.93 \text{ m/s}^2;$$

and

$$s = (1/2)at^2.$$

With friction:

$$a' = g \sin 45.0° - \mu_k g \cos 45.0°$$

which gives

$$s = (1/2)a't'^2 = (1/2)a'(2t)^2$$
$$s = 2a't^2$$

or

$$a' = (1/4)a = 1.73 \text{ m/s}^2.$$

Now

$$\mu_k = (g \sin 45.0° - a')/(g \cos 45.0°)$$
$$\mu_k = [(9.80 \text{ m/s}^2) \sin 45.0° - 1.73 \text{ m/s}^2]/[(9.80 \text{ m/s}^2)(\cos 45.0°)] = \mathbf{0.75}.$$

---

17. The free-body diagram is shown below. We have:

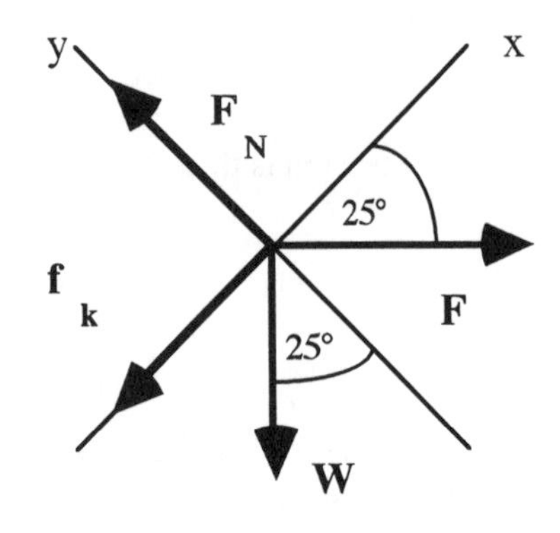

$\Sigma F_y = F_N - W \cos 25° - F \sin 25° = 0.$
$\Sigma F_x = F \cos 25° - f_k - W \sin 25° = 0.$

Since
$$f_k = \mu_k F_N = \mu_k(W \cos 25° + F \sin 25°)$$
we have:
$$F \cos 25° - \mu_k(W \cos 25° + F \sin 25°) - W \sin 25° = 0.$$

Solving for F we obtain:

$F = W(\sin 25° + \mu_k \cos 25°)/(\cos 25° - \mu_k \sin 25°) =$ **61 lb**.

---

18. The free-body diagram for block 2 is shown below.

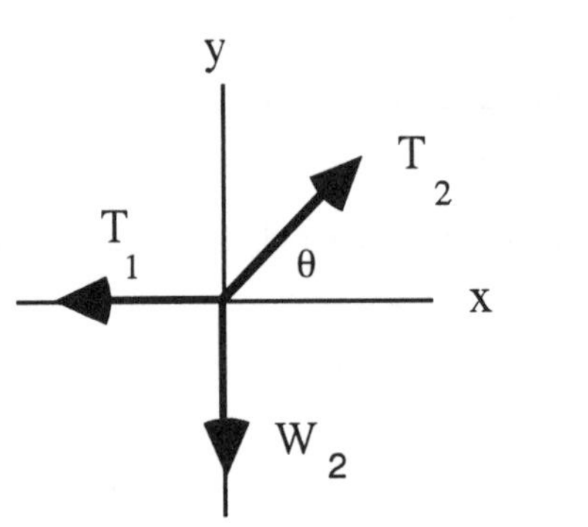

Block 2: $\Sigma F_x = T_{2x} - T_1 = T_2 \cos 37° - T_1 = 0$
so that,
$$T_2 = T_1/\cos 37°.$$

$$\Sigma F_y = T_{2y} - W_2 = T_2 \sin 37° - W_2 = 0$$
so that,
$W_2 = T_2 \sin 37° = T_1 \sin 37°/\cos 37°.$

Similarly for block 1: $T_1 - f_s = 0$
which gives
$$T_1 = \mu_s F_N$$
$$T_1 = \mu_s W_1 = (0.50)(96 \text{ lb}) = 48 \text{ lb}$$
Therefore,
$W_2 = (48 \text{ lb})\sin 37°/\cos 37° =$ **36 lb**.

Quiz answers

1. d	5. d	9. c	13. b	17. c	21. a
2. c	6. c	10. a	14. d	18. c	22. c
3. b	7. a	11. b	15. d	19. b	
4. b	8. b	12. b	16. a	20. b	

MCAT REVIEW PROBLEMS

PASSAGE 1

A student performs a series of experiments to determine the coefficient of static friction and the coefficient of kinetic friction between a large crate and the floor. The magnitude of the force of static friction is always less than or equal to $\mu_s N$, where μ_s denotes the coefficient of static friction, and N denotes the normal force exerted by the floor on the crate:

$$f_s \leq \mu_s N.$$

Static friction exists only when the crate is not sliding across the floor.

The force of kinetic friction is given by

$$f_k = \mu_k N,$$

where μ_k denotes the coefficient of kinetic friction. Kinetic friction exists only when the crate is sliding across the floor.

The crate has mass 100 kg. In this situation, the normal force points upward.

Experiment 1

The student pushes horizontally (rightward) on the crate, and gradually increases the strength of this push force. The crate does not begin to move until the push force reaches 400 N.

Experiment 2

The student applies a constant horizontal (rightward) push force for 1.0 seconds and measures how far the crate moves during that time interval. In each trial, the crate starts at rest, and the student stops pushing after the 1.0-second interval. The following table summarizes the results.

trial	push force (N)	distance (m)
1	500	1.0
2	600	1.5
3	700	2.0

---

1. The coefficient of static friction between the crate and floor is approximately:

A. 0.25 **B.** 0.40 **C.** 2.5 **D.** 4.0

---

2. In experiment 1, when the rightward push force was 50 N, the crate didn't move. *Why* didn't it move?

A. The push force was weaker than the gravitational force on the crate.
B. The push force had the same strength as the gravitational force on the crate.
C. The push force was weaker than the frictional force on the crate.
D. The push force had the same strength as the frictional force on the crate.

---

3. The coefficient of kinetic friction between the crate and the floor is approximately:

A. 0.20 **B.** 0.30 **C.** 0.40 **D.** 0.50

---

4. In trial 3, what is the crate's speed at the moment the student stops pushing it?

A. 1.0 meters per second **B.** 2.0 meters per second
C. 3.0 meters per second **D.** 4.0 meters per second

---

5. For trial 3, which of the following graphs best shows the position of the crate as a function of time? The student first starts pushing the crate at time $t = 0$.

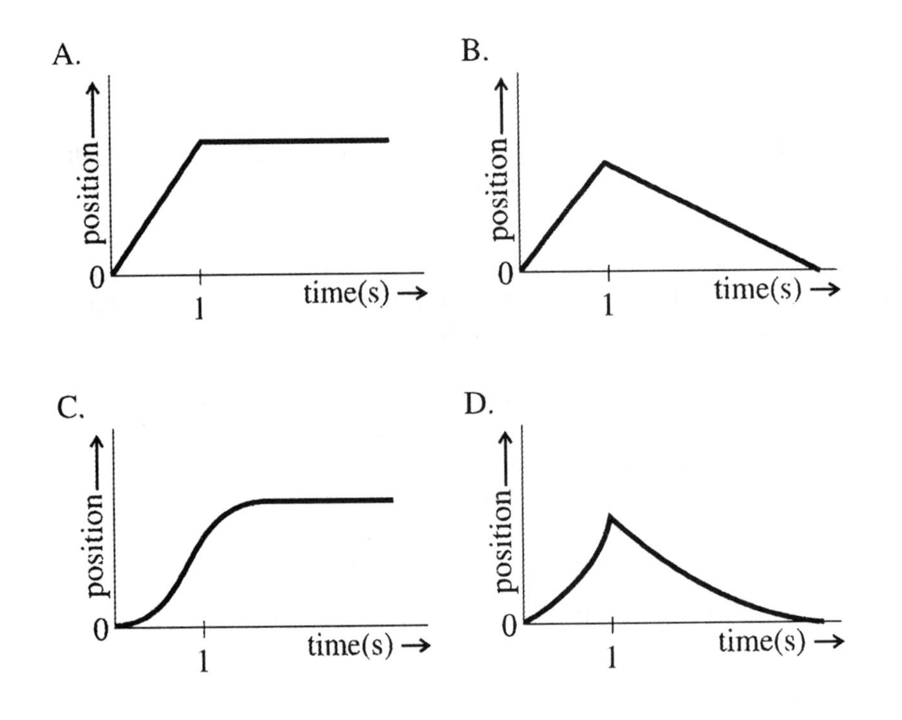

ANSWERS TO MCAT REVIEW PROBLEMS

---

1. **B.** Since f_s is always less than or equal to $\mu_s N$, its maximum possible value is

$$f_{s\,max} = \mu_s N = \mu_s mg.$$

(The normal force must equal the crate's weight, because the vertical forces cancel.)

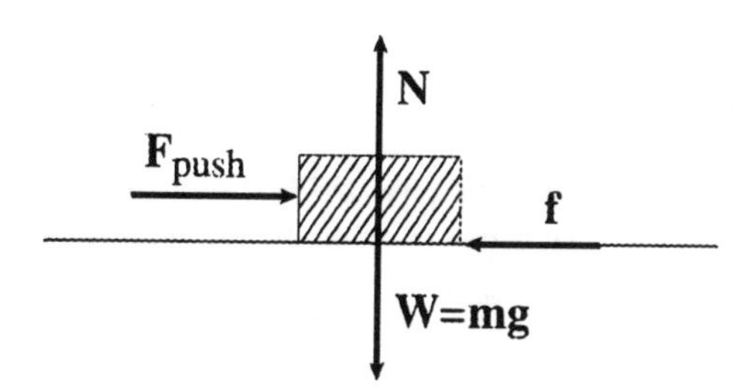

The crate starts to move when the push force barely exceeds $f_{s\,max}$. This happens when $F_{push} \approx 400$ N. So,

$$400\text{N} \approx f_{s\,max} = \mu_s mg = \mu_s(100\text{ kg})(10\text{ m/s}^2) = \mu_s(1000\text{ N}).$$

Therefore, $\mu_s \approx 0.40$.

---

2. **D.** Don't get sucked into choosing C. If f_s were *bigger* than F_{push}, the crate would accelerate leftward, because it would feel a net leftward force.

Many students choose C because they calculate $f_s = \mu_s mg = 400$ N. But that's the *maximum* possible force of static friction. Static friction "adjusts" itself, becoming bigger or smaller as needed in order to cancel the push force. When the push force is only 50 N, static friction reduces itself to 50 N. That's why we write $f_s \leq \mu_s N$ instead of $f_s = \mu_s N$.

---

3. **B.** First, use a kinematic equation to find the crate's acceleration during a particular trial. Then, apply Newton's 2nd law, $F_{net} = ma$. This reasoning works no matter which trial you consider. Here, we'll use trial 1. Since $x = v_0 t + \frac{1}{2}at^2$, and since the crate begins with no velocity, we get

$$x = \frac{1}{2}at^2$$

$$1.0\text{ m} = \frac{1}{2}a(1.0\text{ s})^2,$$

and hence, $a = 2.0$ m/s^2. That's the *horizontal* acceleration.

Since the crate only moves horizontally, the vertical forces cancel, and therefore $N = mg$ (as above). Therefore, the frictional force has magnitude

$$f_k = \mu_k N = \mu_k mg = \mu_k(100\text{ kg})(10\text{ m/s}^2)$$
$$= \mu_k(1000\text{ N}).$$

Newton's 2nd law, applied to the horizontal component of the forces, gives us

$$\begin{aligned} F_{\text{net}} &= ma \\ F_{\text{push}} - f_k &= ma \\ 500\text{ N} - (0.3)(1000\text{ N}) &= (100\text{ kg})(2.0\text{ m/s}^2) \\ &= 200\text{ N}. \end{aligned}$$

Solve for μ_k to get 0.30.

---

4. **D.** Given μ_k, we can use Newton's 2nd law to figure out the crate's acceleration during trial 3. Then we can use $v = v_0 + at$ to find the velocity at time $t = 1.0$ s, the moment the student stops pushing.

From Newton's 2nd law applied to trial 3,

$$\begin{aligned} F_{\text{net}} &= ma \\ F_{\text{push}} - f_k &= ma \\ 700\text{ N} - (0.3)(1000\text{ N}) &= (100\text{ kg})a, \end{aligned}$$

and hence, $a = 4.0\text{ m/s}^2$.

Since the crate speeds up by 4.0 m/s each second, its speed after 1 second is simply

$$v = v_0 + at = 0 + (4.0\text{ m/s}^2)(1.0\text{ s}) = 4.0\text{ m/s}.$$

---

5. **C.** Graph B would correctly show the crate's *velocity* vs. time. The crate speeds up while the student pushes it, and then slows down while sliding freely across the floor. Crucially, after the student stops pushing, the crate does not move backwards, as represented in graph D. It continues moving forward, but at a slower and slower rate. Therefore, after $t = 1$ s, the position vs. time graph continues upward, but with a smaller and smaller slope. When the crate stops, the graph levels off.

PASSAGE 2

A moving company uses the pulley system in figure 1 to lift heavy crates up a ramp. The ramp is coated with rollers that make the crate's motion essentially frictionless. A worker piles cinder blocks onto the plate until the plate moves down, pulling the crate up the ramp. Each cinder block has mass 10 kg. The plate has mass 5 kg. The rope is nearly massless, and the pulley is essentially frictionless. The ramp makes a 30° angle with the ground. The crate has mass 100 kg.

Let W_1 denote the combined weight of the plate and the cinder blocks piled on the plate. Let T denote the tension in the rope. And let W_2 denote the crate's weight.

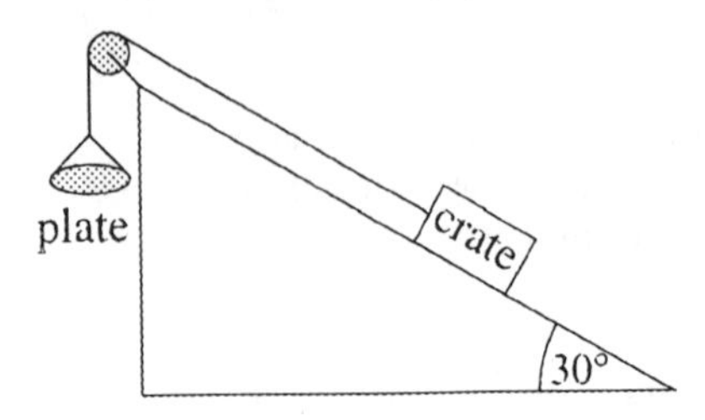

---

6. What is the smallest number of cinder blocks that need to be placed on the plate in order to lift the crate up the ramp?

   A. 3 **B.** 5 **C.** 7 **D.** 10

---

7. Ten cinder blocks are placed on the plate. As a result, the crate accelerates up the ramp. Which of the following is true?

   A. $W_1 = T = W_2 \sin 30°$ **B.** $W_1 = T > W_2 \sin 30°$
   C. $W_1 > T = W_2 \sin 30°$ **D.** $W_1 > T > W_2 \sin 30°$

---

8. The ramp exerts a "normal" force on the crate, directed perpendicular to the ramp's surface. This normal force has magnitude:

   A. W_2 **B.** $W_2 \sin 30°$ **C.** $W_2 \cos 30°$ **D.** $W_2(\sin 30° + \cos 30°)$

---

9. The net force on the crate has magnitude:

   A. $W_1 - W_2 \sin 30°$ **B.** $W_1 - W_2$ **C.** $T - W_2 \sin 30°$ **D.** $T - W_2$

---

10. After the crate is already moving, the cinder blocks suddenly fall off the plate. Which of the following graphs best shows the subsequent velocity of the crate, *after* the cinder blocks have fallen off the plate? (Up-the-ramp is the positive direction.)

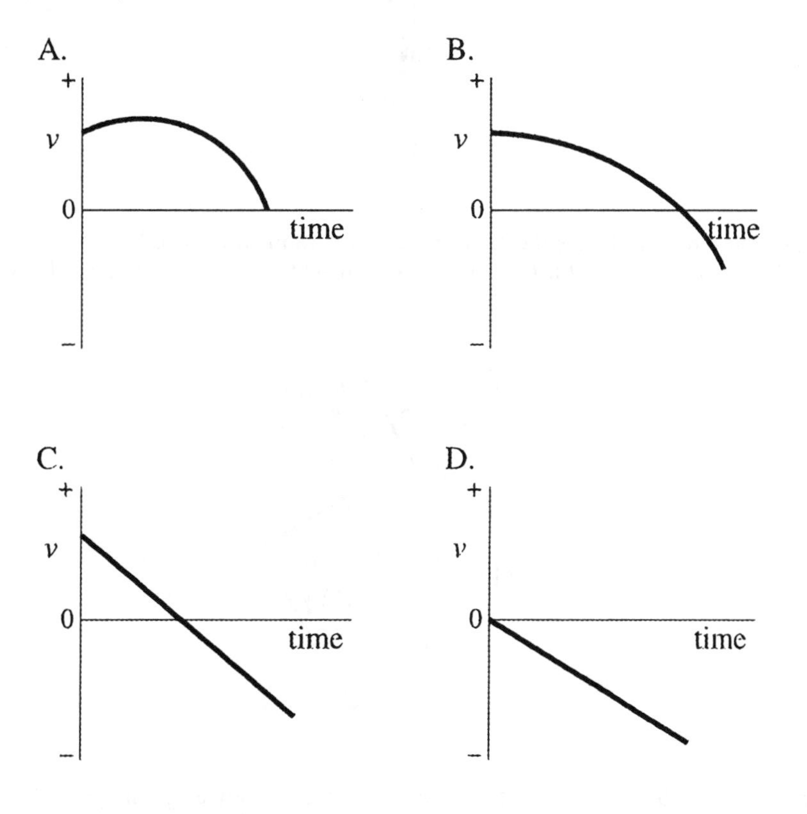

ANSWERS TO MCAT REVIEW PROBLEMS

6. **B.** Intuitively, the downward force on the plate and cinder blocks must overcome the component of gravity pulling the crate down the ramp. In symbols, W_1 must be bigger than $W_2 \sin 30°$. (See the free-body diagrams in the answer to question 7, if this isn't clear.)

For the crate,

$$\begin{aligned} W_2 \sin 30° &= m_{\text{crate}} g \sin 30° \\ &= (100 \text{ kg})(10 \text{ m/s}^2)(0.5) \\ &= 500 \text{ N}. \end{aligned}$$

Therefore, the weight of the plate and cinder blocks must exceed 500 N, in order to pull the crate up the ramp.

Each cinder block weighs

$$m_{\text{block}} g = (10 \text{ kg})(10 \text{ m/s}^2) = 100 \text{ N}.$$

By similar reasoning, the plate weighs 50 N. So, we need 5 blocks, to ensure that the plate and blocks together weigh more than 500 N.

7. **D.** If W_1 equaled T, then the plates and cinder blocks would experience *zero* net force.

As a result, the plate would just hang there, instead of accelerating downward.

Similarly, if T equaled $W_2 \sin 30°$, the crate would feel no net force. It wouldn't accelerate up the ramp.

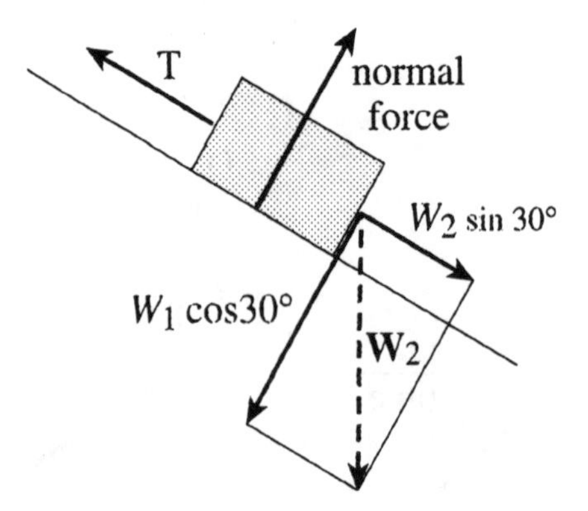

This problem illustrates the importance of drawing *separate* free-body diagrams for each relevant object.

8. **C.** Because the crate never moves perpendicular to the ramp's surface, the force components perpendicular to the ramp must cancel out. So, the normal force must cancel the component of gravity perpendicular to the ramp. Therefore, from the crate's free-body diagram,

$$\text{normal force} = W_2 \cos 30°$$

---

9. **C.** As just noted, the forces perpendicular to the ramp cancel out. The net force on the crate points *along* the ramp's surface. From the free-body diagram, that force is $T - W_2 \sin 30°$.

   Options C and A are not the same answer. As explained in question 2, W_1 must exceed T.

---

10. **C.** Graph A would accurately show the *position* of the crate, since it slides up the ramp while losing speed, and then slides back down the ramp while gaining speed. So, as the block moves forward (positive velocity), the velocity gets smaller and smaller, eventually reaching zero at the "peak." That's why the first segment of the velocity graph stays positive but heads toward zero.

    Next, the block moves backwards while speeding up. So, its negative velocity gets bigger and bigger. Hence, the next segment of the velocity graph becomes "more and more negative." (For a similar example, see MCAT review question 5 from chapter 3.) Both segments of the graph are straight lines, because the forces on the crate stay constant. Therefore, since $F_{net} = ma$, the acceleration—which is the slope of the velocity vs. time graph—also stays constant.

CHAPTER 5
Dynamics of Uniform Circular Motion

PREVIEW

In this chapter you will be introduced to the physics of an object traveling in a circular path. After completion of the chapter, you will be able to find the speed and acceleration of an object traveling in a circle at constant speed along with the period of the motion and the force necessary to keep the object moving in the circle. Applications which will be explored are the motion of a satellite in a circular orbit and artificial gravity.

QUICK REFERENCE

Important Terms

Uniform circular motion
The motion of an object traveling with constant speed in a circular path.

Period of uniform circular motion
The time necessary for an object to complete one revolution around the circle.

Centripetal acceleration
The acceleration that an object must have in order to stay in a circular path. The direction of this acceleration is always toward the center of the circle.

Centripetal force
The net force that must act on an object to keep it in a circular path. This net force is in the direction of the centripetal acceleration, that is, toward the center of the circle.

Equations of Uniform Circular Motion

The **speed** of an object moving in a circular path of radius, r, with period, T, is

$$v = \frac{2\pi r}{T} \tag{5.1}$$

The magnitude of the **centripetal acceleration** of an object moving with uniform circular motion is

$$a_c = \frac{v^2}{r} \tag{5.2}$$

The magnitude of the **centripetal force** acting on an object moving in uniform circular motion is

$$F_c = \frac{mv^2}{r} \tag{5.3}$$

Banked Curves

The following applies to an object moving without friction around a curve of radius, r, and an angle of bank, θ.

$$\tan\theta = \frac{v^2}{gr} \tag{5.4}$$

Satellites in Circular Orbits

The **speed of a satellite** moving in a circular orbit of radius, r, about the earth is

$$v = \sqrt{\frac{GM_E}{r}} \tag{5.5}$$

The **period of a satellite** in circular orbit about the earth is

$$T = \frac{2\pi r^{3/2}}{\sqrt{GM_E}} \tag{5.6}$$

DISCUSSION OF SELECTED SECTIONS

5.1 Uniform Circular Motion

Example 1 *Finding the speed of a planet*
The planet Neptune travels in a nearly circular orbit of radius, $r = 4.5 \times 10^9$ km, about the sun. It takes Neptune 165 y to make a complete trip around the sun. How fast (in km/h) does Neptune travel in its orbit?

Equation (5.1) is

$$v = \frac{2\pi r}{T} = \frac{2(3.14)(4.5 \times 10^9\ \text{km})}{(165\ \text{d})(365\ \text{d}/1\ \text{y})(24\ \text{h}/1\ \text{d})} = 2.0 \times 10^4\ \text{km/h}.$$

Example 2 *Finding the period*
A satellite travels in a circular orbit 3.59×10^4 km above the earth with a speed of 1.11×10^4 km/h. What is the period of the satellite's motion? The radius of the earth is 6.38×10^3 km.

Rearranging equation (5.1) yields

$$T = \frac{2\pi r}{v} = \frac{2(3.14)(3.59 \times 10^4\ \text{km} + 6.38 \times 10^3\ \text{km})}{1.11 \times 10^4\ \text{km/h}} = 23.9\ \text{h}$$

5.2 Centripetal Acceleration

Example 3
Using the data from example 1, find the centripetal acceleration of Neptune in m/s^2.

Equation (5.2) gives

$$a_c = \frac{v^2}{r} = \frac{\left[(2.0 \text{ X } 10^4 \text{ km/h})(10^3 \text{ m/km})(1 \text{ h/3600 s})\right]^2}{(4.5 \text{ X } 10^9 \text{ km})(10^3 \text{ m/1 km})} = 6.9 \text{ X } 10^{-6} \text{ m/s}^2$$

The direction of the centripetal acceleration of Neptune is toward the center of its orbit, that is, toward the sun.

5.3 Centripetal Force

The "centripetal force" is a generic term pertaining to the net force which holds an object in uniform circular motion. Regardless of the identity of the physical force, equation (5.3) must be satisfied if the object is to remain in uniform circular motion. Some possible contributors to the centripetal force are: friction, tension in a rope or string, the gravitational force, the electrical force and the magnetic force. You are familiar with some of these already, and the others will be introduced later in the course.

Example 4
What is the minimum coefficient of static friction necessary to allow a penny to rotate along with a 33 1/3 rpm record (diameter = 0.300 m), when the penny is placed on the outer edge of the record?

The static frictional force acting on the penny is $f_s^{max} = \mu_s F_N = \mu_s mg$. This frictional force plays the role of the centripetal force necessary to hold the penny in its circular path as it revolves with the record so,

$$\mu_s mg = \frac{mv^2}{r}$$

$$\mu_s = \frac{v^2}{rg}$$

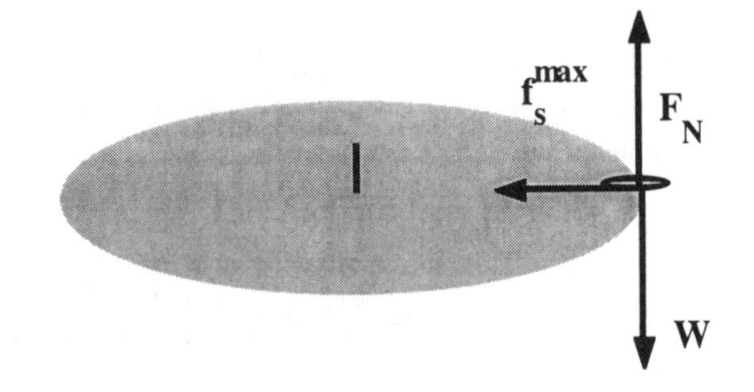

We now need the speed of the penny. Equation (5.1) is $v = 2\pi r/T$, where $T = (60 \text{ s})/(33 \text{ 1/3})$. Therefore,

$$\mu_s = \frac{4\pi^2 r}{gT^2} = \frac{4\pi^2 (0.150 \text{ m})}{(9.80 \text{ m/s}^2)\left[(60 \text{ s})/(33 \text{ 1/3})\right]^2} = 0.187$$

5.5 Satellites in Circular Orbits

Example 5

At what height above the surface of Mars must a satellite be placed to be in a synchronous orbit? The radius of Mars is 6.79×10^6 m, its mass is 6.40×10^{23} kg and its period of rotation is 1.03 d.

Equation (5.6) written for Mars is

$$T = \frac{2\pi r^{3/2}}{\sqrt{GM_M}}$$

Squaring

$$T^2 = \frac{4\pi^2 r^3}{GM_M}$$

Rearranging and taking the cube root

$$r = \sqrt[3]{\frac{GM_M T^2}{4\pi^2}} = \sqrt[3]{\frac{(6.67 \times 10^{-11}\ \text{Nm}^2/\text{kg}^2)(6.40 \times 10^{23}\ \text{kg})\left[(1.03\ \text{d})(8.64 \times 10^4\ \text{s}/1\ \text{d})\right]^2}{4\pi^2}}$$

$$r = 2.05 \times 10^7\ \text{m}$$

The height of the satellite above the surface of Mars is

$$h = 2.05 \times 10^7\ \text{m} - 6.79 \times 10^6\ \text{m} = 1.37 \times 10^7\ \text{m}.$$

PRACTICE PROBLEMS

1. A grinding wheel (radius 7.6 cm) is rotating at 1750 rpm.
   a. What is the speed of a point on the outer edge of the wheel?

   b. What is the centripetal acceleration of the point?

2. An airplane is flying in a horizontal circle of radius 1.5 km with a speed of 450 km/h. What is the magnitude of the centripetal acceleration of the plane?

3. An airplane is flying in a horizontal circle of radius 1.0 km. What must be the speed of the plane if the pilot is to experience a centripetal acceleration three times that of gravity?

4. If the pilot in problem 3 has a mass of 75 kg, what centripetal force acts on him?

5. A trick motorcyclist moves in a horizontal circle inside a large vertical drum which has a radius of 10.0 m. If the coefficient of static friction between the tires and the drum is 0.60, what is the smallest possible speed that the motorcycle and rider can have?

6. A car can barely negotiate a 50.0 m unbanked curve when the coefficient of static friction between the tires and road is 0.80. How much bank would the curve require if the car is to safely go around the curve without relying on friction?

7. A satellite is placed in a circular polar orbit. What must be the height above the surface of the earth of its orbit if it is to pass over the same point on the equator twice per day?

8. A spacecraft orbits Mars (mass = $6.40 \text{ X } 10^{23}$ kg) in a circular orbit of radius $8.01 \text{ X } 10^{5}$ km. What is the period of the spacecraft?

9. How many revolutions per minute must a rotating space station ($r = 1200$ m) turn to provide an artificial gravity of 0.50 g?

10. An airplane pulls out of a dive in a vertical circle of radius 1.0 km traveling with a speed of 550 km/h. How many times greater is the apparent weight of the pilot than his true weight?

11. The moon of a planet is observed to have a nearly circular orbit, $r = 4.00 \times 10^5$ km, and an orbital period of 21.5 days. What is the mass of the planet?

12. A 1.5 kg toy motorcycle is moving on the inside of a vertical circular track ($r = 1.0$ m). It arrives at the top of the track with a speed of 5.0 m/s. What force does the track exert on the motorcycle?

HELPFUL SUGGESTIONS

1. A centripetal force is **not** a new kind of force. The adjective *centripetal* just gives information concerning the *direction* of the force (i.e., inward toward the center of the circle). All the forces that we have studied thus far can act *centripetally*. A centripetal force can be friction, tension in a string, a component of the lift force on an airplane, gravity, etc. The first step in analyzing a problem which deals with circular motion is to identify the centripetal force.

2. In most calculations speeds, angles of bank, etc. will not depend on the mass of the object in uniform circular motion. If you perform an analysis and result depends on the mass of the object, BEWARE.

3. The occupants of a spacecraft in circular orbit about the earth are not really "weightless". Rather they are in free-fall about the earth. For each meter the spacecraft occupants "fall" away from a straight line path, the earth's surface curves away by one meter keeping their height above the surface of the earth constant.

EVERYDAY PHYSICS

1. Many of the results of this chapter can be qualitatively verified by simply paying attention to what happens to you as you round curves in your car. The centripetal force can be sensed as the pressing force that the car door exerts on you as you round a curve. Try taking a curve at different speeds and note the pressing force. Obviously, you should not exceed the posted speed limit for the curve. Also, try taking curves of different radii at the same speed. Again, take note of the pressing force.

2. If you have access to a string-guided model airplane, the magnitude of the centripetal force can be sensed as the pull on the string as the airplane moves in a circle. Try changing the length of the string as the airplane travels with the same speed. If you can adjust the fuel setting (speed) of the airplane, try the same length of string with different speeds.

CHAPTER QUIZ

1. The direction of the centripetal acceleration of an object in uniform circular motion is
   a. in the direction of motion.
   b. radially outward from the center of the circle.
   c. opposite the direction of motion.
   d. radially inward toward the center of the circle.

2. The centripetal force responsible for holding a car in a frictionless banked curve is
   a. the horizontal component of the car's weight.
   b. the vertical component of the car's weight.
   c. the horizontal component of the normal force.
   d. the vertical component of the normal force.

3. A ball on a string is being whirled in a horizontal circle. If the speed is doubled what happens to the tension in the string?
   a. It remains the same.
   b. It doubles.
   c. It quadruples.
   d. It decreases by a factor of two.

4. A 1500 kg car rounds a 85 m unbanked curve with a speed of 28 m/s. If its wheels are on the verge of slipping on the road, what is the coefficient of friction between the tires and the road?
   a. 0.94
   b. 0.87
   c. 0.60
   d. 0.78

5. What angle of bank is necessary for a car to make it around a 105 m curve at a speed of 35.0 m/s without relying on friction?
   a. 50.0°
   b. 38.2°
   c. 14.5°
   d. 23.2°

6. A satellite is orbiting the earth. What one piece of information about the orbit would you need in order to find the period of the satellite?
   a. The speed.
   b. The centripetal force
   c. The radius.
   d. The centripetal acceleration.

7. A boy is swinging a bucket of water, mass m, in a vertical circle at constant speed. When the bucket is at the top of the circle, the boy "feels" no force. What force does the boy "feel" when the bucket is at the bottom of the circle?
   a. 2mg
   b. mg
   c. zero
   d. 3mg

8. Two satellites orbit the earth in circular orbits. One satellite's speed is four times the other's speed. How do their orbital radii compare?
   a. The faster satellite's orbital radius is four times the slower satellite's orbital radius.
   b. The faster satellite's orbital radius is two times the slower satellite's orbital radius.
   c. The faster satellite's orbital radius is one fourth the slower satellite's orbital radius.
   d. The faster satellite's orbital radius is one sixteenth the slower satellite's orbital radius.

9. A roller coaster enters a loop of radius 20.0 m. What is the minimum speed that the roller coaster must have at the top of the circle to just make it without falling?
   a. 196 m/s
   b. 2.0 m/s
   c. 14.0 m/s
   d. 21.0 m/s

10. A banked airplane can fly in a circle. What provides the centripetal force in this case?
    a. The horizontal component of the plane's weight.
    b. The horizontal component of the lift force.
    c. The vertical component of the lift force.
    d. The drag force.

SOLUTIONS AND ANSWERS

Practice Problems

---

1. a. Equation (5.1) gives $v = 2\pi r/T$. The period of the motion is

$$T = (1/1750)(60 \text{ s}) = 3.43 \times 10^{-2} \text{ s}.$$

so

$$v = 2\pi(7.6 \times 10^{-2} \text{ m})/(3.43 \times 10^{-2} \text{ s}) = \mathbf{14 \text{ m/s}}.$$

b. Equation (5.2) gives

$$a_c = v^2/r = (14 \text{ m/s})^2/(7.6 \times 10^{-2} \text{ m}) = \mathbf{2.6 \times 10^3 \text{ m/s}^2}.$$

---

2. Equation (5.2) gives

$$a_c = v^2/r = [(450 \times 10^3 \text{ m/h})(1/3600 \text{ h/s})]^2/(1.5 \times 10^3 \text{ m}) = \mathbf{1.0 \times 10^1 \text{ m/s}^2}.$$

---

3. Equation (5.2) gives $v^2 = a_c r$, but $a_c = 3g$ so

$$v^2 = 3gr = 3(9.80 \text{ m/s}^2)(1.0 \times 10^3 \text{ m})$$

or

$$v = \mathbf{170 \text{ m/s} = 620 \text{ km/h}}.$$

---

4. Equation (5.3) is $F_c = mv^2/r$. Now $v^2 = 3gr$ from problem 3, so

$$F_c = 3mg = 3(75 \text{ kg})(9.80 \text{ m/s}^2) = \mathbf{2200 \text{ N}}.$$

---

5. An application of Newton's second law in the vertical direction gives $F_f = \mu_s F_N = mg$. The centripetal force is provided entirely by the normal force acting on the motorcycle, so $mv^2/r = mg/\mu_s$ or

$$v^2 = rg/\mu_s = (10.0 \text{ m})(9.80 \text{ m/s}^2)/(0.60).$$

Hence,

$$v = \mathbf{13 \text{ m/s}}.$$

---

6. When the car goes around the unbanked curve the centripetal force is provided entirely by friction, so

$$mv^2/r = \mu_s mg$$

or

$$v^2 = \mu_s rg.$$

Using this result in equation (5.4) which applies to a frictionless banked curve, we have

$$\theta = \tan^{-1} \mu_s = \tan^{-1} (0.80) = \mathbf{39°}.$$

---

7. The satellite must have a period of 24 h to pass over the same point on the equator twice in one day. Equation (5.6) gives after some rearrangement

$$r^3 = GM_E T^2/(4\pi^2) = (6.67 \times 10^{-11} \text{ N m}^2/\text{kg}^2)(5.98 \times 10^{24} \text{ kg})(8.64 \times 10^4 \text{ s})^2/(4\pi^2),$$

so $r = 4.23 \times 10^7$ m. The height of the orbit above the earth is

$$H = 4.23 \times 10^7 \text{ m} - 0.64 \times 10^7 \text{ m} = \mathbf{3.59 \times 10^7 \text{ m} = 22\ 300 \text{ mi}}.$$

---

8. Equation (5.6) gives the period of the orbit.

$$T = \frac{2\pi r^{3/2}}{\sqrt{GM_M}} = \frac{2\pi(8.01 \times 10^8 \text{ m})^{3/2}}{\sqrt{(6.67 \times 10^{-11} \text{ N m}^2/\text{kg}^2)(6.40 \times 10^{23} \text{ kg})}} = 2.18 \times 10^7 \text{ s} = \mathbf{6.06 \times 10^3 \text{ h}}.$$

---

9. The centripetal acceleration of the space station is $a_c = v^2/r$ where $v = 2\pi r/T$ so $a_c = 4\pi^2 r/T^2$. We want $a_c = 0.50g$, so

$$T^2 = 4\pi^2 r/(0.50g) = 4\pi^2(1200 \text{ m})/(4.90 \text{ m/s}^2).$$

So,

$$T = 98 \text{ s} = 98/60 \text{ min}.$$

The desired result is

$$1/T = \mathbf{0.61 \text{ rev/min}}.$$

---

10. The apparent weight of the pilot is $W' = F_c + mg = mv^2/r + mg$. Now

$$W'/W = 1 + v^2/(rg) = 1 + [(550 \times 10^3 \text{ m/h})(1/3600 \text{ h/s})]^2/[(1.0 \times 10^3 \text{ m})(9.80 \text{ m/s}^2)] = \mathbf{3.4}.$$

---

11. Equation (5.6) applied to the planet and its moon gives $M = 4\pi^2 r^3/(GT^2)$.
The period of the moon is

$$T = (21.5 \text{ d})(8.64 \times 10^4 \text{ s/d}).$$

Evaluation gives

$$M = \mathbf{1.10 \times 10^{25} \text{ kg}}.$$

---

12. Newton's second law applied to the vertical forces acting on the motorcycle at the top of the track gives

$$F_N + mg = mv^2/r$$

or

$$F_N = m(v^2/r - g) = (1.5 \text{ kg})[(5.0 \text{ m/s})^2/(1.0 \text{ m}) - 9.80 \text{ m/s}^2] = \mathbf{23 \text{ N}}.$$

Quiz answers

1. d	3. c	5. a	7. a	9. c
2. c	4. a	6. c	8. d	10. b

MCAT REVIEW PROBLEMS

Whenever an object undergoes uniform circular motion, it experiences a net force directed towards the center of the circle, of magnitude

$$F_{\text{net}} = \frac{mv^2}{r},$$

where m denotes the object's mass, v denotes its speed, and r denotes the radius of the circle it traces.

As an example, consider a biker riding her bicycle in a tight circle. She must "bank" (tilt) the bicycle at some angle θ, to avoid falling off the seat. As shown in figure 1, the biker experiences two major forces: her weight, and the normal force exerted by the bicycle seat. The normal force pushes perpendicular to the surface of the seat. To excellent approximation, no other forces act on the biker.

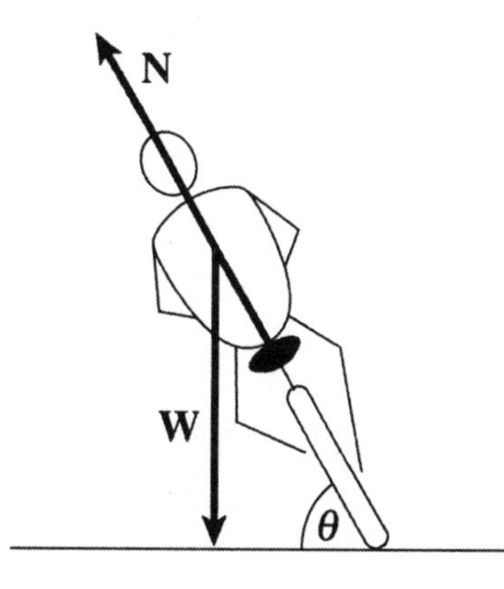

FIGURE 1 :
View of biker from behind

The biker, who weighs $W = 500$ N, rides in a circle at speed 10 m/s, with the bicycle banked at $\theta = 75°$.

---

1. The normal force on the biker is:

   A. (500 N)sin 75° **B.** (500 N)cos 75° **C.** $\frac{500\text{ N}}{\sin 75°}$ **D.** $\frac{500\text{ N}}{\sin 75°}$

---

2. Let N denote the answer to question 1. What net force does the biker experience?

   A. $N\cos 75°$ **B.** $N\cos 75° - W$ **C.** $N\sin 75°$ **D.** $N\sin 75° - W$

---

3. To calculate the radius of the circle traced out by the biker, what additional piece of information do you need?

   A. μ_k, the coefficient of kinetic friction between the bicycle tires and the ground
   B. μ_s, the coefficient of static friction between the bicycle tires and the ground.
   C. g, the acceleration due to gravity near the Earth's surface.
   D. none of the above; you already have enough information.

---

4. If the biker didn't bank the bicycle while going in a circle, she would fall off the seat, partly because:

   A. A force would push her "outward" away from the center of the circle.
   B. A force would push her "inward" towards the center of the circle.
   C. A force would push her backwards.
   D. The net force on her would be zero.

ANSWERS TO MCAT REVIEW PROBLEMS

1. **C.** Because the biker undergoes no vertical motion, the net vertical force on her must vanish. To write an expression for the net vertical force, break the normal force into its vertical and horizontal components:

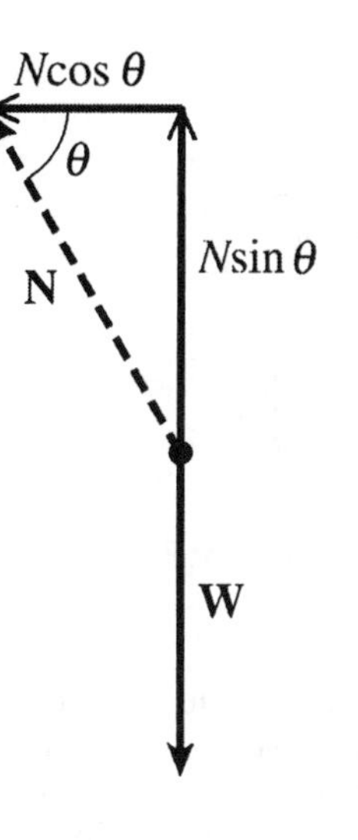

FIGURE 2

Set the net vertical force equal to zero. This gives $N\sin 75° - W = 0$, and hence,

$$N\sin 75° = W.$$

Intuitively, the vertical component of the normal force holds the biker up by "canceling" gravity. Solve for N to get $N = W/\sin 75° = (500\text{ N})/\sin 75°$.

---

2. **A.** As just explained, the vertical forces in figure 2 cancel. Therefore, the net force is simply the horizontal force, $N\cos 75°$.

---

3. **C.** According to the passage, since the biker undergoes uniform circular motion, the net force found in question 2 must equal $\frac{mv^2}{r}$:

$$N\cos 75° = \frac{mv^2}{r}.$$

To solve for r, we need to know N, m, and v. Well, we found N, the normal force, in question 1. And the passage says that $v = 10$ m/s. Therefore, to find r, we just need to know m, the biker's mass. The passage doesn't tell us m. But it *does* specify the biker's weight, $W = 500$ N. Since

$$W = mg,$$

we can easily calculate the biker's mass from her weight, given g.

---

4. **D.** If the bank angle were zero, the biker's free-body diagram would look like this:

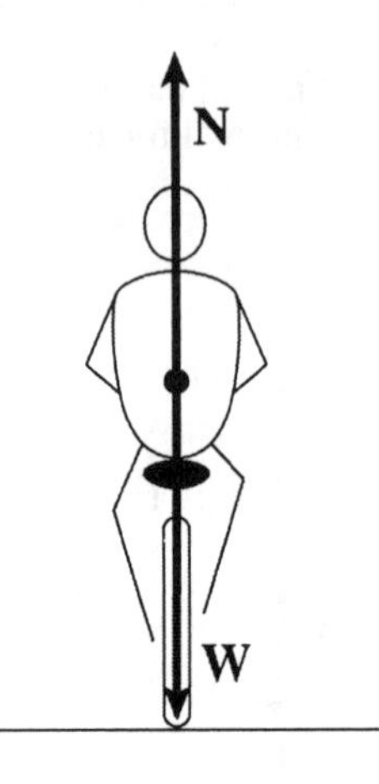

FIGURE 3

Since the vertical forces cancel, N equals W, and the net force vanishes.
Why does the biker fall off the seat when the net force is zero? According to Newton's laws, an object that experiences zero net force travels in a straight line. When the biker feels no net force, she "wants" to keep moving straight. Therefore, when the bicycle veers off along a circular path, it curves out from under the biker, who keeps going straight. As a result, the biker travels "outward" relative to the seat, a shown in figure 4. The dots show the position of the biker and bicycle at 0.1-second intervals.

FIGURE 4

Notice that the biker moves sideways *relative to the bicycle.* That's why she feels thrown outward, even though no outward force acts upon her. Roughly put, the biker doesn't get thrown outward; instead the biker travels straight while the bicycle veers inward!

CHAPTER 6
Work and Energy

PREVIEW

In this chapter you will be introduced to the concept of work and the related concepts of kinetic energy and potential energy. After completion of the chapter you will be able to calculate the work done by constant forces, the kinetic energy of a moving object, the gravitational potential energy of an object, and the power developed by a force.

In addition, you will be able to use the principle of conservation of mechanical energy to solve a variety of problems in a much simpler way than you have done previously.

QUICK REFERENCE

Important Terms

Work
The work done by a CONSTANT force acting on an object is the component of the force along the displacement of the object times the magnitude of the displacement.

Kinetic energy
The energy of an object due to its motion.

Gravitational potential energy
The energy of an object due to its position relative to the earth.

Total mechanical energy
The sum of the kinetic and potential energies of an object.

Conservative force
A force which does work on an object which is independent of the path taken by the object between its starting point and its ending point.

Average power
The work done by a force on an object divided by the time taken to do the work.

Work

The **work done on an object** by a force, **F**, is

$$W = (F \cos \theta)\, s \qquad (6.1)$$

where s is the magnitude of the displacement of the object and θ is the angle between the force and the displacement.

The **work done by gravity** on an object is

$$W_{gravity} = -\, mg\,(h_f - h_0) \qquad (6.4)$$

where m is the mass of the object, h_0 is the initial height of the object, and h_f is the final height of the object.

The **work done by nonconservative forces acting** on an object is

$$W_{nc} = (1/2\, mv_f^2 - 1/2\, mv_0^2) + (mgh_f - mgh_0) \qquad (6.6)$$

$$W_{nc} = (KE_f - KE_0) + (PE_f - PE_0) \qquad (6.7a)$$

$$W_{nc} = \Delta KE + \Delta PE \qquad (6.7b)$$

Energy

The **kinetic energy** of an object of mass, m, and speed, v, is

$$KE = 1/2\ mv^2 \tag{6.2}$$

The **gravitational potential energy** of an object a height, h, above the reference level is

$$PE = mgh \tag{6.5}$$

The **total mechanical energy** of an object is the sum of its kinetic and potential energies.

$$E = KE + PE$$

The work done by nonconservative forces produces a change in the total mechanical energy of the object.

$$W_{nc} = E_f - E_0 \tag{6.8}$$

Power

The average power developed by a force which has done work, W, in time, t, is

$$\overline{P} = \frac{\text{Work}}{\text{Time}} = \frac{W}{t} \tag{6.10}$$

The average power developed by a force, F, moving an object with an average speed, v is

$$\overline{P} = F\,\overline{v} \tag{6.11}$$

Theorems and Principles

Work-Energy Theorem - the total work done by all forces acting on an object is

$$W = KE_f - KE_0 = 1/2\ mv_f^2 - 1/2\ mv_0^2 \tag{6.3}$$

Principle of Conservation of Mechanical Energy - the total mechanical energy of an object remains constant as the object moves, provided that no net work is done by nonconservative forces.

$$E_f = E_0 \tag{6.9}$$

DISCUSSION OF SELECTED SECTIONS

6.1 Work Done by a Constant Force

The concept of work plays an important role in physics since it connects, via the work-energy theorem, Newton's second law to the important scalar quantities of kinetic energy and potential energy.

Although work is a scalar quantity, it comes in two varieties-positive and negative. POSITIVE work indicates that work is done ON the object, whereas NEGATIVE work indicates that the work is done BY the object.

Example 1 *Positive and negative work*
A constant force of 40.0 N is needed to keep a car traveling at constant speed as it moves 5.0 km down the road. How much work is done? Is the work done on or by the car?

The force needed to keep the car moving is in the direction of the car's displacement. Equation (6.1) gives the work.

$$W = (F \cos 0°)\, s = +(40.0 \text{ N})(5.0 \times 10^3 \text{ m}) = +2.0 \times 10^5 \text{ J}$$

The work is POSITIVE so the work is done ON the car.

Example 2
There must be a retarding force acting on the car in example 1 if it is to remain traveling at constant speed. What is this force? How much work does the force do? Is this work done on or by the car?

Newton's second law applied along the direction of motion of the car gives

$$F - F_r = ma = 0$$

or

$$F_r = F = 40.0 \text{ N}$$

The work done by the retarding force is then given by (6.1) to be

$$W_r = (F_r \cos 180°)\, s = -(40.0 \text{ N})(5.0 \times 10^3 \text{ m}) = -2.0 \times 10^5 \text{ J}$$

The work is NEGATIVE so the work is done BY the car.

Example 3
A 2.40×10^2 N force is pulling an 85.0 kg block across a horizontal surface. The force acts at an angle of 20.0° above the surface. The coefficient of kinetic friction is 0.200, and the block moves a distance of 8.00 m. Find (a) the work done by the pulling force, (b) the work done by the kinetic frictional force, (c) the total work done by all the forces.

a. The work done by the pulling force is

$$W_F = (F \cos 20.0°)\, s$$
$$W_F = (2.40 \times 10^2 \text{ N})(\cos 20.0°)(8.00 \text{ m})$$
$$W_F = +1.80 \times 10^3 \text{ J}$$

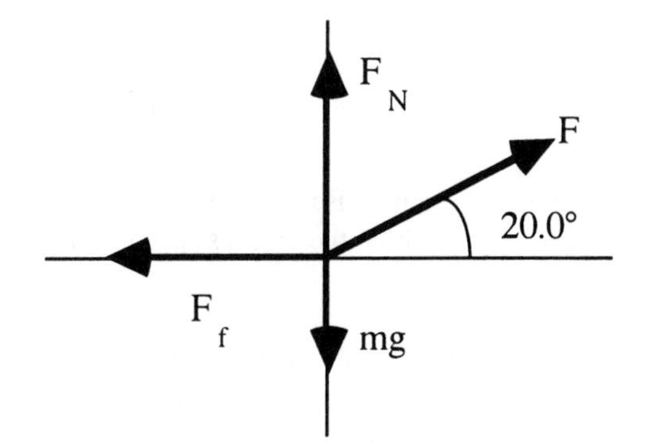

b. The frictional force, $F_f = \mu F_N$, acting on the block is found by applying Newton's second law to the free body diagram.
In the vertical direction

$$F_N + F \sin 20.0° - mg = 0$$

then

$$F_f = \mu(mg - F \sin 20.0°)$$

The work done by this force is

$$W_f = \mu(mg - F \sin 20.0°)(\cos 180°)\ s$$
$$W_f = -\ (0.200)\{(85.0\ kg)(9.80\ m/s^2) - (2.40\ X\ 10^2\ N)(\sin 20.0°)\}(8.00\ m)$$
$$W_f = -\ 1.20\ X\ 10^3\ J$$

c. The total work done by all forces is

$$W = W_F + W_f + W_N + W_g = 1.80\ X\ 10^3\ J - 1.20\ X\ 10^3\ J + 0\ J + 0\ J$$
$$W = 6.0\ X\ 10^2\ J$$

Note that the work due to the gravitational force and the normal force is zero since these forces act at a 90° angle to the motion of the block.

6.2 The Work-Energy Theorem and Kinetic Energy

Net work done on an object always produces a change in the kinetic energy of the object according to the work energy theorem (6.3). The theorem is often very useful in finding the speed of an object if the net work done by the forces acting on it are known or can be calculated.

Example 4 *Using the work energy theorem to find the speed of an object*
If the block in example 3 starts from rest, what is the speed of the block after it has traveled 8.00 m?

The work energy theorem gives

$$W = 1/2\ mv_f^2 - 1/2\ mv_0^2$$
$$W = 1/2\ mv_f^2$$

Solving for v_f gives

$$v_f = \sqrt{\frac{2W}{m}} = \sqrt{\frac{2(6.0\ X\ 10^2\ J)}{85.0\ kg}} = 3.8\ m/s$$

Another common use of the work energy theorem is finding information about the forces acting on an object when information about the object's speed changes is known.

Example 5 *Finding an average force when the speeds are known*
A baseball pitcher can throw a 9.00 ounce ($1.75\ X\ 10^{-2}$ sl) baseball with speed measured by a radar gun of 90.0 miles per hour (132 ft/s). Assuming that the force exerted by the pitcher on the ball acts over a distance of two arm lengths, each 28.0 inches, what is the average force exerted by the pitcher on the ball?

The work energy theorem gives $W = 1/2\ mv_f^2$ if the ball starts at rest. The work done is W = Fs, so

$$F = \frac{mv_f^2}{2s} = \frac{(1.75\ X\ 10^{-2}\ sl)(132\ ft/s)^2}{2(4.67\ ft)} = 32.6\ lb$$

6.3 Gravitational Potential Energy

The gravitational force is one of a class of "special" forces called conservative forces. What makes the gravitational force special is that the work done on an object by gravity depends only on the initial and final positions of the object. The work does NOT depend on how the object got from the initial to the final position.

Example 6 *Work done on an object by gravity*
How much work is done by gravity when a 5.0 kg object moves 1.5 m horizontally and then moves 3.0 m vertically upward? Vertically downward?

The work done by gravity during the horizontal motion is

$$W = (mg \cos 90°)s = 0 \text{ J}$$

The work done by gravity during the vertical motion is

$$W = (mg \cos 180°)s = -(5.0 \text{ kg})(9.80 \text{ m/s}^2)(3.0 \text{ m}) = -150 \text{ J}$$

The work is negative so the work is done BY the object.

If the object is moving vertically downward, the work done during the vertical motion is

$$W = (mg \cos 0°)s = +(5.0 \text{ kg})(9.80 \text{ m/s}^2)(3.0 \text{ m}) = +150 \text{ J}.$$

In this case the work is done ON the object since the work is positive.

As the previous example shows, the magnitude of the work done by gravity on the object when it is going up is the same as the work done by gravity on the object when it is going down. The negative of the work done by gravity is called the potential energy of the object. The work energy theorem then gives rise to equation (6.6).

Example 7 *Using gravitational potential energy*
A 3.00 kg model rocket is launched vertically upward with sufficient initial speed to reach a height of 1.00×10^2 m, even though air resistance (a nonconservative force) performs -8.00×10^2 J of work on the rocket. How high would the rocket have gone if there were no air resistance?

The equation (6.7b) written for the actual motion of the rocket gives

$$\Delta KE = W - \Delta PE$$
$$\Delta KE = W - mgh$$
$$\Delta KE = -8.00 \times 10^2 \text{ J} - (3.00 \text{ kg})(9.80 \text{ m/s}^2)(1.00 \times 10^2 \text{ m})$$
$$\Delta KE = -3740 \text{ J}$$

If the rocket were launched with the same initial kinetic energy and no air resistance acts, then (6.7b) is

$$\Delta KE = -\Delta PE = -mgh$$

or

$$h = -\frac{\Delta KE}{mg} = -\frac{(-370\text{J})}{(3.00\text{kg})(9.80\text{m/s}^2)}$$
$$h = 127 \text{ m}$$

6.5 The Conservation of Mechanical Energy

If no friction or other nonconservative force acts on an object then its total mechanical energy remains constant. This principle is very useful in solving problems involving the motion of an object. It is often easier to use than Newton's laws or the kinematic equations since it involves scalars rather than vectors. Also, it only requires that you have knowledge about the motion at the beginning and end.

Example 8
A truck is descending a winding mountain road. When the truck is 1480 m above sea level and traveling 15 m/s, its brakes fail. What is the maximum possible speed of the truck at the foot of the mountain 550 m above sea level?

The truck would have the maximum possible speed if friction were ignored. In this case, the total mechanical energy of the truck is conserved as it moves down the mountain road.

$$1/2\ mv_f^2 + mgh_f = 1/2\ mv_0^2 + mgh_0$$

Taking the zero of potential energy at sea level we have

$$v_f = \sqrt{v_0^2 + 2g(h_0 - h_f}$$

$$v_f = \sqrt{(15\ \text{m/s})^2 + 2(9.80\ \text{m/s}^2)(1480\ \text{m} - 550\ \text{m})}$$

$$v_f = 140\ \text{m/s}$$

6.7 Power

The average power is the rate at which work is done as given by equation (6.10).

Example 9
A 73 kg sprinter, starting from rest, reaches a speed of 7.0 m/s in 1.8 s with negligible effect due to air resistance. The sprinter then runs the remainder of the race at a steady speed of 7.0 m/s under the influence of a 35 N force due to air resistance. Calculate the average power needed (a) to accelerate the runner and (b) to sustain the steady speed at which most of the race is run.

(a) In order to find the power, we need to find the work needed to accelerate the runner. The work energy theorem gives $W = \Delta KE = \frac{1}{2}\ mv_f^2$ so

$$\bar{P} = \frac{1/2\ mv_f^2}{t} = \frac{(1/2)(73\ \text{kg})(7.0\ \text{m/s})^2}{1.8\ \text{s}}$$

$$\bar{P} = 990\ \text{W}$$

$$\bar{P} = \frac{990\ \text{W}}{(746\ \text{W})/(1\ \text{hp})} = 1.3\ \text{hp}$$

(b) The force that the sprinter must exert to run at constant speed is equal and opposite to the force exerted on him by air resistance. The power needed to run the race at a constant speed is then

$$\bar{P} = Fv = (35\ \text{N})(7.0\ \text{m/s})$$

$$\bar{P} = 250\ \text{W}$$

$$\bar{P} = \frac{250\ \text{W}}{(746\ \text{W})/(1\ \text{hp})} = 0.34\ \text{hp}$$

6.9 Work Done by a Variable Force

The work done by a variable force in moving an object is equal to the area under the (F cos θ) versus s curve.

Example 10

The graph below shows the component (F cos θ) of the net force that acts on a 2.0 kg block as it moves along a flat horizontal surface. Find (a) the net work done on the block, and (b) the final speed of the block if it starts from rest at s = 0.

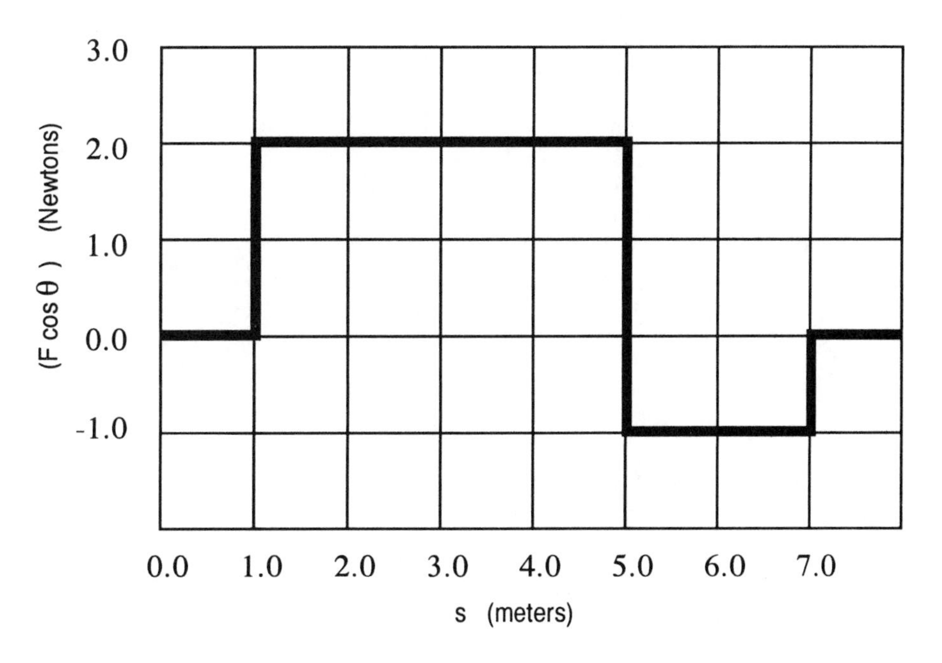

(a) The net work done on the block is equal to the area under the curve.

From 0.0 to 1.0 m, the net force is zero, so no work is done.

From 1.0 m to 5.0 m, there are 8 blocks under the curve. Each block has an area of (1.0 N)(1.0 m) = 1.0 J so

$$W_{1\text{-}5} = 8\ \text{blocks}\left(\frac{1.0\ \text{J}}{1\ \text{block}}\right) = 8.0\ \text{J}$$

From 5.0 to 7.0 m, there are 2 blocks under the curve. Note that the net force is *negative* for this area, so, each block has an area of - 1.0 J.

$$W_{5\text{-}7} = 2 \text{ blocks} \left(\frac{-1.0 \text{ J}}{1 \text{ block}}\right) = -2.0 \text{ J}$$

The net work done on the block is therefore,

$$W_{net} = W_{1\text{-}5} + W_{5\text{-}7} = (+8.0 \text{ J}) + (-2.0 \text{ J}) = 6.0 \text{ J}.$$

(b) The final speed of the block can be found from the Work-energy theorem. Since the block starts from rest, we have $W = \Delta KE = \frac{1}{2} m v_f^2 - 0$ so

$$W = \frac{1}{2} m v_f^2.$$

Solving for v_f gives

$$v_f = \sqrt{\frac{2W}{m}} = \sqrt{\frac{2(6.0 \text{ J})}{2.0 \text{ kg}}} = 2.45 \text{ m/s}.$$

PRACTICE PROBLEMS

1. The driver of a 1500 kg car slams on the brakes locking the wheels. A total retarding force of 1800 N acts to stop the car in a distance of 70.0 m. How much work is done in bringing the car to a halt? Is the work done on or by the car?

2. A 2.0 ton truck descending a 5.0° hill is brought to a stop in 250 ft. The driver applies the brakes so that the wheels lock. (a) If the coefficient of kinetic friction between the truck tires and the road is 0.60, how much work is done by friction in stopping the truck? (b) How much work is done on the truck by gravity?

3. In problem 1, how fast was the car traveling initially?

4. In problem 2, how fast was the truck traveling immediately before the brakes were applied?

5. A 2100 lb airplane requires a power of 100.0 hp at the propeller to climb at an angle of 20.0° to the horizontal at constant speed. How much altitude could it gain in 10.0 minutes if air resistance is ignored?

6. A bicyclist tops the crest of a 20.0 m high hill moving with a speed of 1.0 m/s. Ignoring friction find the speed of the bicycle at the bottom of the hill.

7. A 0.60 kg ball thrown vertically upward with an initial speed of 12.0 m/s rises to a height of 5.1 m. What is the average force exerted by the air on the ball? How high would the ball have gone if no air resistance acted on it?

8. A swing is observed to rise to a maximum height of 1.5 m above its lowest point. How fast is it going at its lowest point? Ignore friction and air resistance.

9. A 1500 kg car accelerates from rest to 75 km/h in 45 s. How much power is supplied by the engine to accelerate the car?

10. A locomotive engine can supply 2.0×10^3 hp to accelerate a train from rest to 20.0 m/s in 9.0 min. Find the mass of the train. (Ignore friction.)

HELPFUL SUGGESTIONS

1. Remember that when the NET work is POSITIVE it is work done on an object and it results in an increase in the kinetic energy of the object. NEGATIVE net work is work done BY the object and results in a decrease in the kinetic energy of the object according to the work energy theorem.

2. When confronted with a problem involving motion, check first to see if the conservation of mechanical energy principle can be used. If not, then check to see if the work energy theorem can be conveniently used. It may sometimes be necessary to use Newton's laws and/or the kinematic equations directly.

3. The zero of gravitational potential energy is totally arbitrary, so you are free to choose it to be anywhere you want. Before attacking a problem involving the conservation of energy spend some time deciding what level you will call the zero level. Choose it to either eliminate a height that is unknown or to make your conservation of energy equation the simplest one.

EVERYDAY PHYSICS

1. Can we really ignore friction and air resistance? Test this question yourself by pulling a swing back some distance, releasing it and observing how close it comes back to it release point. The conservation of energy principle demands that the swing return to its release point if no friction or air resistance acts. It should come back very close to the release point if friction and air resistance can be ignored. Try the test with the swing empty and with someone in it. Is there a difference? Why or why not?

2. When a ball is dropped on a hard floor, there are energy losses due to work performed by frictional forces while the ball is in contact with the floor and by air resistance while the ball is in the air. Try this experiment by dropping a hard ball from a known height onto a concrete floor and measuring the rebound height of the ball. Can you calculate the average resistive force exerted on the ball from the time it was dropped to the time it returned to its maximum rebound height? Use various types of balls and compare the results. If you can find one, a "superball" gives interesting results.

CHAPTER QUIZ

1. Work done on an object is ________, whereas work done by an object is ________.
   a. negative, negative b. positive, negative c. negative, positive d. positive, positive

2. The work in the work energy theorem is
   a. work due to gravity.
   b. work due to friction.
   c. work due to nonconservative forces.
   d. net work.

3. What parts of a force are capable of doing work?
   a. All parts of the force.
   b. Only the component along the direction of the displacement of the object.
   c. Only the component perpendicular to the displacement of the object.

4. A conservative force is one which
   a. does no work in moving an object from one point to another.
   b. does only positive work.
   c. does no net work in moving an object from one point to another and back again.
   d. does only negative work.

5. The gravitational potential energy of an object is ________ the work done by gravity on the object.
   a. unrelated to b. the positive value of c. the negative value of

6. The zero of gravitational potential energy is
   a. always at ground level b. arbitrary c. at sea level d. at the center of the earth

7. Mechanical energy is conserved provided that ________work is done by nonconservative forces .
   a. no net b. no negative c. no positive d. only positive

8. An object's speed increases by a factor of 4. Its kinetic energy
   a. remains unchanged.
   b. increases by a factor of 4.
   c. decreases by a factor of 16.
   d. increases by a factor of 16.

9. In physics power is
   a. the same as force.
   b. another form of kinetic energy.
   c. the rate at which work is done.
   d. has the units of ft-lb.

10. The total mechanical energy of an object is
    a. proportional to the speed squared.
    b. proportional to the mass of the object.
    c. proportional to the height of the object.
    d. all of the above.

SOLUTIONS AND ANSWERS

Practice Problems

---

1. The work done by the retarding force is

$$W = (F \cos 180°)\, s$$
$$W = -\,(1800\ N)(70.0\ m) = \mathbf{-\ 1.3\ X\ 10^5\ J.}$$

This work is negative so the work was actually **done by the car**.

---

2. a. The frictional force acting on the truck is

$$F_f = \mu F_N = \mu mg \cos 5.0°.$$

The work done by friction is then

$$W_f = (F_f \cos 180°)\, s = -\,\mu mgs \cos 5.0°$$
$$W_f = -\,(0.60)(4.0\ X\ 10^3\ lb)(250\ ft) \cos 5.0° = \mathbf{-\ 6.0\ X\ 10^5\ ft\text{-}lb.}$$

b. The work done by gravity is

$$W_{gravity} = (mg \cos 85.0°)\, s$$
$$W_{gravity} = (4.0\ X\ 10^3\ lb)(\cos 85.0°)(250\ ft) = \mathbf{+\ 8.7\ X\ 10^4\ ft\text{-}lb.}$$

---

3. The work energy theorem applied to the car's motion gives

$$W = -\ 1/2\ mv_0^2$$

since the final speed of the car is zero. Solving yields

$$v_0^2 = -\ 2W/m = -\ 2(-\ 1.3\ X\ 10^5\ J)/(1500\ kg)$$

so

$$v_0 = \mathbf{13\ m/s.}$$

---

4. The work energy theorem gives for the truck

$$W = -\ 1/2\ mv_0^2$$

so

$$v_0^2 = -\ 2W/m = -\ 2(-\ 6.0\ X\ 10^5\ ft\text{-}lb + 8.7\ X\ 10^4\ ft\text{-}lb)(32.2\ ft/s^2)/(4.0\ X\ 10^3\ lb)$$

or

$$v_0 = \mathbf{91\ ft/s.}$$

---

5. The work done by the propeller in 10.0 min is

$$W_{nc} = Pt = (100.0\ hp)(550\ ft\text{-}lb/hp)(10.0\ min)(60\ s/min) = 3.3\ X\ 10^7\ ft\text{-}lb.$$

Equation (6.6) written for constant speed motion is

$$W_{nc} = mg(h_f - h_0)$$

then

$$h_f - h_0 = W_{nc\ r}/mg = (3.3 \text{ X } 10^7 \text{ ft-lb})/(2.1 \text{ X } 10^3 \text{ lb}) = \mathbf{1.6 \text{ X } 10^4 \text{ ft.}}$$

---

6. The conservation of energy applied between the top and bottom of the hill with the bottom of the hill chosen as the zero of potential energy gives

$$1/2\ mv_f^2 + mgh_f = 1/2\ mv_0^2$$

or

$$v_0^2 = v_f^2 + 2gh_f = (1.0 \text{ m/s})^2 + 2(9.80 \text{ m/s}^2)(20.0 \text{ m})$$

so

$$v_0 = \mathbf{2.0 \text{ X } 10^1 \text{ m/s}}.$$

---

7. If the initial position of the ball is taken as the zero of potential energy, then equation (6.6) gives

$$W_{nc} = -\ 1/2\ mv_0^2 + mgh_f.$$

The work done by the force of air resistance is

$$W_{nc} = (F \cos 180°)\ h_f.$$

Then

$$F = m(v_0^2/2h_f - g) = (0.60 \text{ kg})\{(12.0 \text{ m/s})^2/(10.2 \text{ m}) - 9.80 \text{ m/s}^2]$$

$$F = \mathbf{2.6 \text{ N}}.$$

In the absence of air resistance the total mechanical energy of the ball is conserved, hence

$$mgh_f = 1/2\ mv_0^2.$$

or

$$h_f = v_0^2/2g = (12.0 \text{ m/s})^2/(19.6 \text{ m/s}^2) = \mathbf{7.35 \text{ m}}.$$

---

8. Let the lowest point of the swing be the level of zero potential energy. The conservation of energy applied between the lowest and highest level of the swing's motion is

$$1/2\ mv_0^2 = mgh_f$$

or

$$v_0^2 = 2gh_f = 2(9.80 \text{ m/s}^2)(1.5 \text{ m})$$

which gives

$$v_0 = \mathbf{5.4 \text{ m/s}}.$$

---

9. The work done on the car by the engine is, according to the work energy theorem,

$$W = 1/2\ mv_f^2 = 1/2(1500 \text{ kg})\{(75 \text{ X } 10^3 \text{ m/h})(1/3600 \text{ h/s})\}^2 = 3.3 \text{ X } 10^5 \text{ J}.$$

The average power supplied by the engine is

$$P = W/t = (3.3 \text{ X } 10^5 \text{ J})/(45 \text{ s}) = \mathbf{7.3 \text{ X } 10^3 \text{ W}}$$

---

10. The work done by the engine on the train is given by the work energy theorem to be

$$W = 1/2\ mv_f^2.$$

Equation (6.10) gives the work in terms of the power so

$$m = 2W/v_f^2 = 2Pt/v_f^2$$

$$m = 2(2.0 \times 10^3 \text{ hp})(746 \text{ W}/1 \text{ hp})(9.0 \text{ min})(60 \text{ s}/1 \text{ min})/(20.0 \text{ m/s})^2 = \mathbf{4.0 \times 10^6\ kg}.$$

Quiz answers

1. b	3. b	5. c	7. a	9. c
2. d	4. c	6. b	8. d	10. b

MCAT REVIEW PROBLEMS

PASSAGE 1

Two physicists, both of mass 50 kg, climb up identical ropes suspended from the ceiling of a gymnasium. The ropes are 15 meters long. Physicist 1 reaches the top twice as quickly as physicist 2 does. After physicist 2 also reaches the top, the they argue about who did more work against gravity:

Physicist 1

"I did more work fighting gravity, because I was overcoming gravity more quickly. Your climb was lazier, and therefore, you did less work."

Physicist 2

"No way. I did more work fighting gravity, because I spent *more time* climbing the rope. Since we both ended up at the same height, but I spent more time getting there, I had to work harder."

---

1. Which physicist, if either, did more work against gravity while climbing from the floor to the ceiling?

   A. Physicist 1 did more work.
   B. Physicist 2 did more work.
   C. Both physicists did the same positive work.
   D. Neither physicist did any work.

---

2. Physicist 1 climbed her rope in 30 seconds. What average power did she exert fighting gravity?

   A. 25 watts
   B. 250 watts
   C. 1500 watts
   D. 15,000 watts

---

3. Physicist 2 started at rest from the floor and ended at rest near the ceiling. Which of the following best expresses the net energy transfer during this process?

   A. Chemical to kinetic
   B. Chemical to potential
   C. Kinetic to chemical
   D. Kinetic to potential

---

4. Physicist 2 now lets go of the rope and falls onto a heavily-padded cushion, safely coming to rest. During this process, the energy transfer is best described as:

   A. Potential to kinetic to chemical
   B. Potential to kinetic to heat
   C. Kinetic to potential to chemical
   D. Kinetic to potential to heat

---

5. When physicist 2 has fallen one third of the way from the ceiling to the floor, her kinetic energy is approximately:

   A. 10 joules
   B. 250 joules
   C. 1000 joules
   D. 2500 joules

ANSWERS TO MCAT REVIEW PROBLEMS

---

1. **C.** Because physicist 1 did the work more quickly, she exerted more *power*. But both physicists changed their gravitational potential energy by the same amount, mgh. So, they both did the same work fighting gravity.

   To reach this conclusion another way, recall that work = $(F\cos\theta)s$, where s denotes the distance covered fighting the force F, and θ denotes the angle between the force and the direction of motion. To climb the rope, both physicists must fight the same gravitational force (mg) over the same distance, $s = h = 15$ m. Put another way, the work done fighting a "conservative" force such as gravity depends only on your starting and ending points, not on your path. Slow versus fast makes no difference. You do the same work either way.

---

2. **B.** Power is the rate at which work gets done. In 30 seconds, physicist 2 does work

$$\begin{aligned}(F\cos\theta)s &= (mg\cos 0°)h \\ &= (50\text{ kg})(10\text{ m/s}^2)(15\text{ m}) \\ &= 7500\text{ J},\end{aligned}$$

   since cos 0° = 1. Therefore, she does work at an average rate of

$$\bar{P} = \frac{\text{work}}{\text{time}} = \frac{7500\text{ W}}{30\text{ s}} = 250\text{ W}.$$

   By contrast, since physicist 1 took twice as long to complete the same work, she exerted only half as much power.

---

3. **B.** At the top of the rope, physicist 2 has potential energy but no kinetic energy. So, the answer must be B or D. In order to do work, the physicist "burns calories," i.e., she uses some of the chemical energy stored in her cells.

   Physicist 2 has kinetic energy while climbing the rope. But her kinetic energy is neither the initial source nor the final form of the energy. While climbing, the physicist must continue to burn calories in order to reach the top; the kinetic energy alone doesn't carry her up.

---

4. **B.** While she falls, the physicist's potential energy gradually converts into kinetic energy; she gets faster and faster. But when she lands, this kinetic energy disappears. It turns mostly into heat that dissipates in the cushion.

---

5. **D.** To solve, you could calculate the physicist's speed after falling 5 meters, and then plug that speed into the kinetic energy formula, $\frac{1}{2}mv^2$. But we can jump to the answer using energy conservation. While falling through $h = 5$ meters (a third of the way to the floor), the physicist loses the following amount of potential energy:

$$\Delta U = mgh = (50\text{ kg})(10\text{ m/s}^2)(5.0\text{ m}) = 2500\text{ J}.$$

   Energy conservation implies that the "lost" energy converts into another form—in this case, kinetic. So, after falling through 5 meters, the physicist has gained 2500 joules of kinetic energy. End of story. We need not calculate her speed.

PASSAGE 2

Any two spherical objects gravitationally attract each other with force $F_{grav} = \frac{GMm}{r^2}$, where M and m are the masses of the two objects, and r is the distance from the center of one object to the center of the other. The potential energy of the system is $U_{grav} = -\frac{GMm}{r}$; the old formula $U = mgh$ does not apply.

Two astronomers are figuring out the speed with which a spherical meteor of mass m_1 will crash into the Earth, which has mass M_E and radius R_E. To make the calculation simpler, the astronomers assume that the meteor starts at rest a distance D from the center of the Earth, where D is several times bigger than R_E. They also neglect air resistance and treat the Earth as fixed. But the astronomers disagree about how to proceed from there.

Astronomer 1:

"Let's invoke Newton's 2nd law and kinematics. First, using $\frac{GM_E m_1}{D^2} = m_1 a$, we can find the meteor's acceleration. Given that acceleration, we can then use those old constant-acceleration kinematic formulas, such as $v^2 = v_0^2 + 2ax$, to find the speed with which the meter hits the Earth."

Astronomer 2:

"Let's use conservation of energy. The meteor starts with potential energy $-\frac{GM_E m_1}{D}$. All of that potential energy converts into kinetic energy as the meteor rushes towards the Earth. So, we can set the magnitude of the initial potential energy equal to the final kinetic energy:

$$\frac{GM_E m_1}{D} = \frac{1}{2} m_1 v^2.$$

Just solve for v to find the speed with which the meteor crashes into the Earth

---

6. Which astronomer will correctly find the speed with which the meteor hits the Earth?

   A. Astronomer 1 only
   B. Astronomer 2 only
   C. Astronomer 1 and astronomer 2
   D. Neither astronomer 1 nor astronomer 2

---

7. Neglecting air resistance, the speed with which the meteor hits the Earth does *not* depend on:

   A. M_E
   B. m_1
   C. R_E
   D. G

---

8. As the meteor gets closer to the Earth, the gravitational force on it:

   A. Increases
   B. Stays constant
   C. Decreases, but stays above zero
   D. Decreases to zero

---

9. Which of the following best describes the transfer of energy as the meteor crashes? (The "crash" begins when the meteor first touches the Earth.)

   A. Potential to kinetic
   B. Potential to heat
   C. Kinetic to potential
   D. Kinetic to heat

ANSWERS TO MCAT REVIEW PROBLEMS

---

6. **D.** The constant-acceleration kinematic formulas from chapter 2 of your textbook apply only when the acceleration is constant. But as the meteor approaches the Earth, the Earth-to-meteor distance (r) decreases.

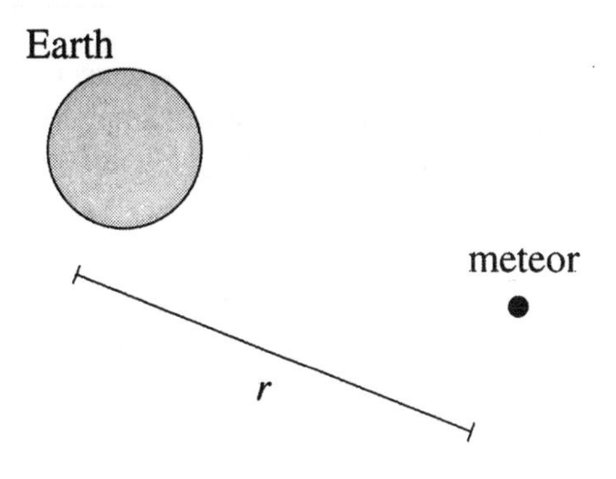

Therefore, $F_{grav} = \frac{GM_E m_1}{r^2}$ increases, because the denominator gets smaller. Since the force increases, so does the acceleration, according to Newton's 2nd law. Since the acceleration changes, the astronomer can't use $v^2 = v_0^2 + 2ax$ or related formulas.

Although energy conservation would work if applied properly, astronomer 2 makes a crucial mistake. She assumes that the "final" potential energy of the meteor at the Earth's surface is zero. This would be true if U equaled mgh. But as the problem tells us, that old formula doesn't apply. We must use $U_{grav} = -\frac{GMm}{r}$ instead. Here, "r" means the distance from the center of the meteor to the *center* of the Earth. And at the surface of the Earth, the meteor is still a distance R_E from the Earth's center.

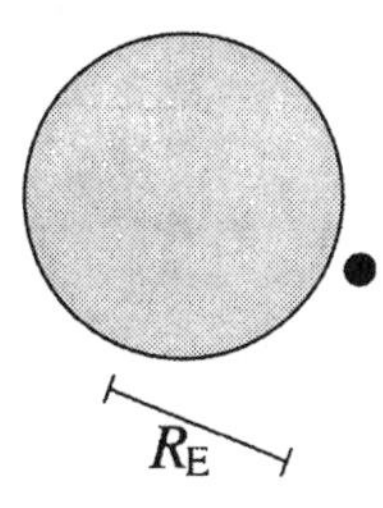

Therefore, according to energy conservation,

$$-\frac{GM_E m_1}{D} = \frac{1}{2} m_1 v^2 + \left(-\frac{GM_E m_1}{R_E}\right).$$

---

7. **B.** In this energy-conservation equation, the m_1's cancel. But when you solve for v, the answer contains M_E, R_E, and G.

Intuitively, whenever gravity alone acts on an object, the object's mass has no effect on the motion. To see why, apply Newton's 2nd law to the meteor:

$$\frac{GM_E m_1}{r^2} = m_1 a.$$

Solve for a to get $\frac{GM_E}{r^2}$. The meteor's acceleration depends on the *Earth's* mass, but not on the meteor's mass.

---

8. **A.** As the Earth-to-meteor distance r gets smaller, the gravitational attraction ($F_{grav} = \frac{GM_E m_1}{r^2}$) gets larger.

---

9. **D.** During the crash, the meteor loses very little potential energy, because it falls through a small distance. But its huge kinetic energy disappears as it comes to rest. The kinetic energy converts mostly into heat. In general, during an inelastic collision, some kinetic energy "dissipates non-mechanical forms.

CHAPTER 7
Impulse and Momentum

PREVIEW

In this chapter you will learn about the concepts of impulse and momentum. In addition, the important principle of conservation of linear momentum is discussed. This concept is then applied to analysis of collisions in one and two dimensions.

QUICK REFERENCE

Important Terms

Impulse (of a force)
The product of the average force and the time interval during which the force acts. Impulse is a vector quantity.

Linear momentum
The mass times velocity of an object. Linear momentum is a vector quantity.

Total linear momentum of a system of objects
The vector sum of the individual momenta of the objects.

Internal forces
Forces which act between the objects of a system.

External forces
Forces which act on the objects of a system which come from the outside.

Inelastic collision
A collision in which the total kinetic energy of the system of objects is NOT constant.

Elastic collision
A collision in which the total kinetic energy of the system is constant.

Equations Involving Impulse and Momentum

If **F** is the average force and Δt is the time over which the force acts, then

$$\textbf{Impulse} = \overline{\mathbf{F}}\,\Delta t \tag{7.1}$$

The **linear momentum** of an object is

$$\mathbf{p} = m\mathbf{v} \tag{7.2}$$

where m is the mass of the object and **v** is the velocity of the object.

Newton's second law written in terms of momentum

$$\overline{\mathbf{F}} = \frac{m\mathbf{v}_f - m\mathbf{v}_0}{\Delta t} \tag{7.3}$$

For a **system of objects** where $\mathbf{P_f}$ is the final total momentum of the system and $\mathbf{P_0}$ is the initial total momentum of the system

$$(\textbf{Sum of external forces})\,\Delta t = \mathbf{P_f} - \mathbf{P_0} \tag{7.6}$$

The final velocities of two objects participating in a **head-on elastic collision** in which ball 2 is initially at rest are:

$$v_{f1} = \left(\frac{m_1 - m_2}{m_1 + m_2}\right) v_{01} \quad (7.8a)$$

$$v_{f2} = \left(\frac{2m_1}{m_1 + m_2}\right) v_{01} \quad (7.8b)$$

The **thrust** of a rocket which ejects gases at a rate, $\Delta m/\Delta t$ and a final velocity, $\mathbf{v_f}$, and an initial velocity, $\mathbf{v_0}$, is

$$\mathbf{T} = -\left(\frac{\Delta m}{\Delta t}\right)(\mathbf{v_f} - \mathbf{v_0}) \quad (7.10)$$

Theorems and Principles

Impulse - Momentum Theorem

$$\overline{\mathbf{F}}\,\Delta t = m\mathbf{v_f} - m\mathbf{v_0} \quad (7.4)$$

The Principle of Conservation of Linear Momentum

If no external forces act on a system of objects, then the total linear momentum of the system remains constant in time. That is,

$$\mathbf{P_0} = \mathbf{P_f} \quad (7.7)$$

DISCUSSION OF SELECTED SECTIONS

7.1 The Impulse - Momentum Theorem

Example 1
A 0.25 kg ball is dropped onto a hard surface. Immediately before the ball hits the surface it is traveling with a speed of 2.5 m/s. As it rebounds it travels with a reduced speed of 2.0 m/s. What is the impulse of the force that the surface exerts on the ball?

Taking up as the positive direction, the impulse-momentum theorem gives

$$\textbf{impulse} = \overline{\mathbf{F}}\,\Delta t = m\mathbf{v_f} - m\mathbf{v_C}$$
$$\textbf{impulse} = (0.25\text{ kg})(2.0\text{ m/s} + 2.5\text{ m/s}) = +1.1\text{ kg m/s}$$

The direction of the impulse is UP.

Example 2
If the ball in example 1 is in contact with the floor for 15 ms, what is the average force exerted on it by the surface?

$$\overline{\mathbf{F}} = \frac{m\mathbf{v_f} - m\mathbf{v_0}}{\Delta t} = \frac{(0.25\text{ kg})(2.0\text{ m/s} + 2.5\text{ m/s})}{15 \times 10^{-3}\text{ s}} = +75\text{ N}$$

The direction of the force on the ball is UP.

7.2 The Principle of Conservation of Linear Momentum

The conservation of linear momentum principle can often be used to analyze the motion of systems of colliding objects. The only criterion for its validity is that no EXTERNAL forces act on the system of objects during the collision. Practically, collisions are often of such short duration that external forces such as gravity and friction have little time to influence the motion and may be neglected. In these cases, the conservation of linear momentum principle may be applied to the motion IMMEDIATELY BEFORE and IMMEDIATELY AFTER the collision.

Example 3
Two identical spaceships perform a docking maneuver in which one spaceship catches up with the other and the two lock together. If the faster spaceship travels with a speed of 9015 m/s and the slower travels with 9005 m/s before the docking maneuver, what is the speed of the spaceships after the docking is complete?

The total momentum of the two spaceship system is constant since no external forces act during the docking maneuver. Of course, internal forces do act during the maneuver to lock the spaceships together, but they cannot affect the overall motion of the system.

$$\mathbf{P_0} = \mathbf{P_f}$$

$$m\mathbf{v_{01}} + m\mathbf{v_{02}} = 2m\mathbf{v_f}$$

Taking the initial direction of travel of the two spaceships as the positive direction, we have

$$\mathbf{v_f} = \frac{\mathbf{v_{01}} + \mathbf{v_{02}}}{2} = \frac{9015 \text{ m/s} + 9005 \text{ m/s}}{2} = +\,9010 \text{ m/s}$$

The spaceships move together with a speed of 9010 m/s in the direction of the initial motion.

Example 4
A 1550 kg cannon, initially at rest, fires a 60.0 kg shell horizontally with a speed of 525 m/s. What is the maximum possible recoil velocity of the cannon?

The maximum possible recoil velocity of the cannon corresponds to no friction acting on the cannon to impede its motion. In this case there are no external forces acting in the directions of motion of the cannon and shell, so the conservation of momentum principle can be applied in the horizontal direction. Note that gravity does not affect the motion of the system since it acts vertically and is balanced by the normal force of the ground.

$$\mathbf{P_f} = \mathbf{P_0}$$

$$m_1\mathbf{v_{f1}} + m_2\mathbf{v_{f2}} = m_1\mathbf{v_{01}} + m_2\mathbf{v_{02}}$$

The initial velocities of both the cannon and shell are zero, that is $\mathbf{v_{01}} = \mathbf{v_{02}} = 0$ m/s. Assigning the initial velocity of the shell to be positive, we have for the final velocity of the cannon

$$\mathbf{v_{f1}} = -\frac{m_2}{m_1}\mathbf{v_{f2}} = -\left(\frac{60.0 \text{ kg}}{1550 \text{ kg}}\right)(+\,525 \text{ m/s}) = -\,20.3 \text{ m/s}$$

The cannon moves in the negative direction opposite to the motion of the shell with a speed of 20.3 m/s.

Example 5
Two 45 kg boys are playing pitch and catch in a 105 kg boat. The boys and boat are initially at rest; one boy stands in the front of the boat and the other in the back of the boat. One of the boys throws the 0.50 kg ball with a

horizontal speed of 26 m/s. What is the speed of the boat immediately after the ball is thrown? What is the speed of the boat after the second boy catches the ball? Ignore friction.

If friction can be ignored then the conservation of linear momentum principle can be applied to the horizontal motion of the system of boat, boys, and ball.

$$\mathbf{P}_f = \mathbf{P}_0$$

$$m_1\mathbf{v}_{f1} + m_2\mathbf{v}_{f2} = m_1\mathbf{v}_{01} + m_2\mathbf{v}_{02}$$

The initial velocity of the boys and boat, $\mathbf{v}_{01}$, is zero as is the initial velocity of the ball, $\mathbf{v}_{02}$. Taking the direction in which the ball is thrown as positive, the final velocity of the boat is

$$\mathbf{v}_{f1} = -\left(\frac{m_2}{m_1}\right)\mathbf{v}_{f2} = -\left[\frac{0.50 \text{ kg}}{2(45 \text{ kg}) + 105 \text{ kg}}\right](+\ 26 \text{ m/s}) = -\ 0.067 \text{ m/s}.$$

7.3 Collisions in One Dimension

When objects collide in the absence of external forces, the total momentum of the system is conserved. The conservation of momentum principle then yields an equation connecting the masses and initial and final velocities of the objects. This equation can then be used to solve for one quantity if all the others are known.

If the collision also happens to be elastic, then the conservation of kinetic energy will provide a second equation containing the masses and initial and final speeds of the objects. The two equations allow you to simultaneously find TWO quantities.

Example 6 *Collision of two objects of equal mass traveling in opposite directions*
Two identical pool balls make a perfectly elastic head-on collision on a frictionless table. The speeds of the balls before the collision are 2.0 m/s and 3.0 m/s. What are the speeds and directions of motion of the balls after the collision?

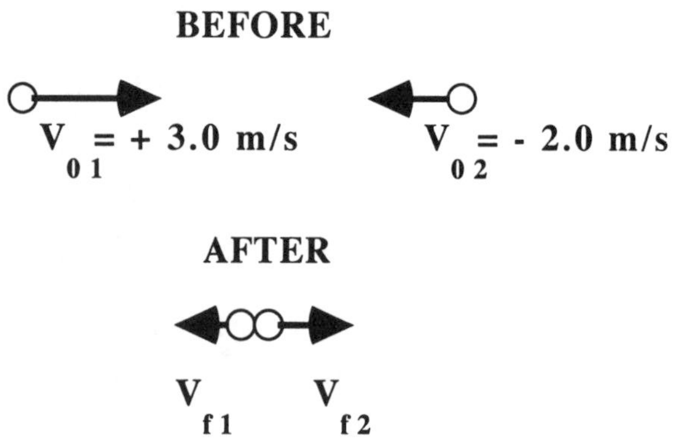

Before the collision, the velocities of the balls are as shown in the figure. The conservation of momentum principle gives

BEFORE

$V_{01} = +\ 3.0$ m/s $\quad V_{02} = -\ 2.0$ m/s

AFTER

$V_{f1} \quad V_{f2}$

$$\mathbf{P}_f = \mathbf{P}_0$$

$$m\mathbf{v}_{f1} + m\mathbf{v}_{f2} = m\mathbf{v}_{01} + m\mathbf{v}_{02}$$

$$\mathbf{v}_{f1} + \mathbf{v}_{f2} = \mathbf{v}_{01} + \mathbf{v}_{02}$$

$$v_{f1} + v_{f2} = +\ 3.0 \text{ m/s} - 2.0 \text{ m/s}$$

$$v_{f1} + v_{f2} = +\ 1.0 \text{ m/s} \qquad (1)$$

The conservation of kinetic energy gives

$$1/2\ mv_{f1}^2 + 1/2\ mv_{f2}^2 = 1/2\ mv_{01}^2 + 1/2\ mv_{02}^2$$

$$v_{f1}^2 + v_{f2}^2 = 13.0\ m^2/s^2 \qquad (2)$$

The equations (1) and (2) may be solved simultaneously for the velocities by solving (1) for v_{f1}, substituting into (2) and solving the resulting quadratic equation for v_{f2}. In this calculation we omit the units for convenience.

Equation (1) gives

$$v_{f1} = 1.0 - v_{f2}$$

Substitution of this into equation (2) yields

$$(1.0 - v_{f2})^2 + v_{f2}^2 = 13.0$$

or

$$2v_{f2}^2 - 2.0v_{f2} - 12.0 = 0$$

The quadratic formula gives

$$v_{f2} = \frac{-(-2.0) \pm \sqrt{(-2.0)^2 - 4(2)(-12.0)}}{4}$$

$$v_{f2} = \frac{2.0 \pm 10.0}{4} = -2.0\ \text{m/s OR} + 3.00\ \text{m/s}$$

The final velocity of ball one is found by substituting these values into equation (1)

$$v_{f1} = 1.0\ \text{m/s} - 3.00\ \text{m/s} = -2.0\ \text{m/s}$$

$$v_{f1} = 1.0\ \text{m/s} - (-2.0\ \text{m/s}) = +3.0\ \text{m/s}$$

The result, $v_{1f} = +3.0$ m/s and $v_{2f} = -2.0$ m/s, indicates that the balls have passed through each other without colliding and can be ruled out as being physically unrealistic. Hence, after the collision ball 1 has reversed direction and is moving to the LEFT with a speed of 2.0 m/s, while ball 2 has also reversed direction and is moving to the RIGHT with a speed of 3.0 m/s. Note that the balls simply exchanged speeds during the collision. This is true for all one dimensional elastic collisions of two objects with EQUAL MASSES.

Example 7 *Collision of two objects with equal mass traveling in the same direction*
Repeat example 6 assuming that both balls are traveling in the same direction with the 3.0 m/s ball overtaking the 2.0 m/s ball.

The analysis in example 6 applies to this problem with the exception that the velocity $v_{02} = +2.0$ m/s so that equation (1) becomes

$$v_{f1} + v_{f2} = 5.0\ \text{m/s} \qquad (1')$$

Following the same procedure except using (1') instead of (1) results in the quadratic equation

$$2v_{f2}^2 - 10.0v_{f2} + 12.0 = 0$$

with the solutions $v_{f2} = +3.0$ m/s and $+2.0$ m/s.

Equation (1') now gives

$$v_{f1} = 5.0 \text{ m/s} - 3.0 \text{ m/s} = +\,2.0 \text{ m/s}$$

and

$$v_{f1} = 5.0 \text{ m/s} - 2.0 \text{ m/s} = +\,3.0 \text{ m/s}$$

The case where $v_{f1} = +\,3.0$ m/s and $v_{f2} = +\,2.0$ m/s corresponds to no collision at all. Again, the balls have exchanged their speeds, but their directions of motion after the collision are the same as before the collision.

Example 8 *Elastic collision of two objects with different masses when one object is initially at rest*
A 1.0 kg ball initially traveling with a speed of 2.5 m/s strikes a stationary 0.50 kg ball. Determine the final velocities of the balls. How would the result differ if the 0.50 kg ball were moving with 2.5 m/s and the 1.0 kg ball were at rest?

Equations (7.8a,b) apply to this motion. Note that these were derived by an application of the conservation of kinetic energy and the conservation of momentum principle.

$$v_{f1} = \left(\frac{m_1 - m_2}{m_1 + m_2}\right) v_{01} = \left(\frac{1.0 \text{ kg} - 0.50 \text{ kg}}{1.0 \text{ kg} + 0.50 \text{ kg}}\right)(2.5 \text{ m/s}) = 0.8 \text{ m/s}$$

$$v_{f2} = \left(\frac{2m_1}{m_1 + m_2}\right) v_{01} = \left(\frac{2(1.0 \text{ kg})}{1.0 \text{ kg} + 0.50 \text{ kg}}\right)(2.5 \text{ m/s}) = 3.3 \text{ m/s}$$

In the second case, where the 0.50 kg ball is moving, it becomes ball 1 and the 1.0 kg ball becomes ball 2.

$$v_{f1} = \left(\frac{0.50 \text{ kg} - 1.0 \text{ kg}}{0.50 \text{ kg} + 1.0 \text{ kg}}\right)(2.5 \text{ m/s}) = -\,0.8 \text{ m/s}$$

$$v_{f2} = \left(\frac{2(0.50 \text{ kg})}{0.50 \text{ kg} + 1.0 \text{ kg}}\right)(2.5 \text{ m/s}) = +\,1.7 \text{ m/s}$$

These results indicate that the direction of the final velocity of the striking ball depends on whether or not it is more massive than the struck ball. If the striking ball is more massive than the struck ball, then it will continue in its original direction of travel. If the striking ball is less massive than the struck ball then it will rebound in a direction opposite its original direction of travel. This is true independently of the initial speed of the striking ball as can be seen from an examination of equation (7.8a).

The conservation of momentum principle may still apply even if kinetic energy is not conserved during the collision. This is true if the frictional forces which take energy from the system are INTERNAL rather than EXTERNAL forces.

Example 9 *A completely inelastic collision*
A 1500 kg automobile traveling with a speed of 55 km/h makes a head-on collision with an 1800 kg automobile traveling with a speed of 35 km/h in the opposite direction. The automobiles lock together during the collision and move together with a final speed, v. Find v.

The collision is completely inelastic since the automobiles lock together. If the friction between the automobile's tires and the road and air resistance can be neglected during the collision, then no net external force acts on the cars and the total linear momentum of the system is conserved.

$$\mathbf{P_f} = \mathbf{P_0}$$

$$m_1\mathbf{v_{f1}} + m_2\mathbf{v_{f2}} = m_1\mathbf{v_{01}} + m_2\mathbf{v_{02}}$$

The final velocities of the automobiles are equal so

$$v = \frac{m_1 v_{01} + m_2 v_{02}}{m_1 + m_2} = \frac{(1500\text{ kg})(55\text{ km/h}) + (1800\text{ kg})(-35\text{ km/h})}{1500\text{ kg} + 1800\text{ kg}} = +5.9\text{ km/h}$$

The direction the velocity of the cars after the collision is the same as the initial direction of the 1500 kg automobile.

7.4 Collisions in Two Dimensions

Example 10
A 1200 kg automobile traveling with a velocity of 65 km/h, 35° north of east makes a completely inelastic collision with an identical automobile traveling with a velocity of 32 km/h, 45° south of west. What is the magnitude and direction of the velocity of the cars after the collision?

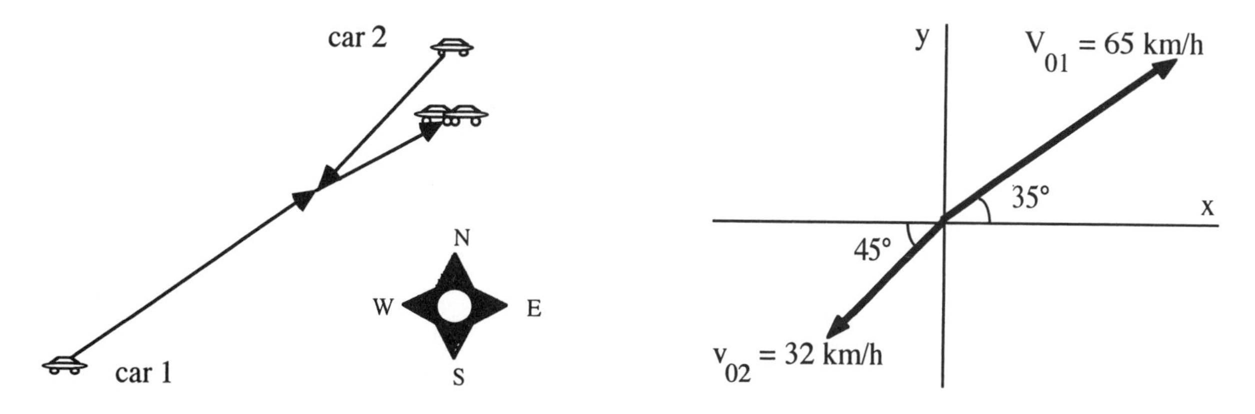

If friction between the tires and the road and air resistance can be ignored, then no net external force acts on the cars and the total VECTOR momentum of the system is constant during the collision.

In the x direction:

$$P_{fx} = P_{0x}$$

$$(m_1 + m_2)v_f \cos\theta = m_1 v_{01} \cos 35° - m_2 v_{02} \cos 45°$$

$$(2400\text{ kg})v_f \cos\theta = (1200\text{ kg})(65\text{ km/h}) \cos 35° - (1200\text{ kg})(32\text{ km/h}) \cos 45°$$

$$v_f \cos\theta = 15.3\text{ km/h} \qquad (1)$$

One extra figure has been retained in the answer.

In the y direction:

$$P_{fy} = P_{0y}$$

$$(m_1 + m_2)v_f \sin\theta = m_1 v_{10} \sin 35° - m_2 v_{02} \sin 45°$$

$$v_f \sin\theta = 7.33 \text{ km/h} \qquad (2)$$

Again, one extra figure has been retained.

Division of (2) by (1) yields

$$\theta = \tan^{-1}(7.33 \text{ km/h}/15.3 \text{ km/h}) = 25.6°$$

Substitution of this angle into (2) gives the final speed of the cars to be v_f = 17 km/h. The angle that the final velocity makes with the x-axis rounded to proper significant figures is $\theta = 26°$.

PRACTICE PROBLEMS

1. A batter hits a 0.14 kg ball which is initially traveling at a speed of 40.0 m/s. The ball loses 25% of its kinetic energy during the impact with the bat. What is the impulse of the force that the bat exerts on the ball?

2. A large ball has a mass which is ten times larger than a small ball. The small ball has speed which is fifteen times larger than the large ball's speed. Compare the momenta of the two balls. Which ball would be harder to stop in a given time period? Why?

3. Two 75 kg campers are in the middle of a creek in a 45 kg boat without a paddle. All they have on board the boat is a quantity of 0.500 kg cans of beans. One of the campers is able to throw a can with a speed of 26.8 m/s. How many cans of beans must the camper throw to give the boat a speed of 1.0 m/s. Ignore friction.

4. A 45 kg boy is standing on a 2.0 kg skateboard which is at rest on the ground. The boy jumps off the front of the skateboard with a horizontal speed of 0.50 m/s. What is the maximum possible speed of the skateboard? Which way does it move?

5. A 35 kg child runs with a speed of 4.5 m/s and "belly flops" onto a 7.0 kg snow sled. What is the speed of the sled and child immediately after the child comes to rest on the sled? Assume that no friction acts between the sled runners and the snow.

6. A 0.50 kg wooden block is dropped from a height of 15.0 m. At what height must a 5.0 g bullet traveling upward strike the block in order to momentarily stop its fall? Assume that the time required for the bullet to come to rest in the block is negligibly short and that the bullet travels with a speed of 320 m/s immediately before striking the block.

7. A 0.60 kg brick is thrown into a stationary 20.0 kg wagon with a velocity of 15.0 m/s directed 20.0° below the horizontal. What is the speed of the wagon and brick after the brick comes to rest in the wagon?

8. A 5.0 kg ball traveling with a speed of 5.0 m/s makes an elastic collision with a 3.0 kg ball which is traveling in the same direction with a speed of 1.5 m/s. What are the velocities of the balls immediately after the collision?

9. A 1500 kg car is traveling east with a speed of 85 km/h when it collides with a 1200 kg car traveling north with a speed of 65 km/h. The cars lock together during the collision. What is the velocity of the cars after the collision?

10. A pool ball moving with a speed of 2.5 m/s makes an elastic head-on collision with an identical ball traveling in the opposite direction with a speed of 5.9 m/s. Find the velocities of the balls after the collision.

HELPFUL SUGGESTIONS

1. Any time that you encounter a problem involving two or more objects which interact in such a way that the interaction can be interpreted as a "collision", check to see if total momentum is conserved in some direction. If so, the conservation of linear momentum principle will give an equation which may be useful.

2. If an external force such as gravity or friction acts on a system, then, strictly speaking, total momentum is NOT conserved. If the time of interaction of the objects is short, however, the total linear momentum of the system is approximately constant during the interaction. Then you can use the principle of conservation of linear momentum to relate velocities, etc. IMMEDIATELY BEFORE the interaction to those IMMEDIATELY AFTER the interaction. A bullet fired into a ballistic pendulum is an excellent example of this.

3. The key to solving collision problems is to generate as many equations containing the unknown quantities as there are unknown quantities. The equations can then be solved simultaneously for the unknowns. The conservation of linear momentum principle will give one equation for each direction in which no external forces act. The conservation of kinetic energy will give one equation if the collision is elastic.

4. The application of the impulse-momentum principle applied to rocket propulsion (equation 7.10) is also applicable to situations where mass is accumulated by an object. See example 11 in this guide.

EVERYDAY PHYSICS

An excellent way to get an intuitive "feel" for the principles introduced in this chapter is to spend some time "colliding" pool balls on a pool table. You will need to remember that this real situation is somewhat more complicated than the ideal situations presented thus far, in that real pool balls roll, are large, are affected by friction and do not make true elastic collisions.

The simplest experiment that you can do is to collide a ball with stationary one. Keeping in mind that the balls have equal masses, do they respond as you would expect? Also, try colliding two balls head-on with both balls moving and see what happens. What do you expect to happen?

CHAPTER QUIZ

1. A 0.5 kg ball traveling with a speed of 2 m/s strikes a wall. If the ball rebounds with the same speed, what is the impulse of the force exerted by the wall on the ball?
   a. 1 kg m/s b. 2 kg m/s c. 4 kg m/s d. zero

2. A 1.5 kg object moving with a speed of 2.5 m/s strikes a wall and rebounds with a speed of 1.5 m/s. The ball is in contact with the wall for 45 ms. What is the magnitude of the average force exerted on the ball by the wall?
   a. 2.7 N b. 0.7 N c. 130 N d. 270 N

3. A 5.0 kg object, initially at rest, explodes breaking into two pieces of mass 1.0 kg and 4.0 kg. The 1.0 kg piece has a speed ________ the speed of the 4.0 kg piece.
   a. five times b. four times c. one-fourth as large as d. one-fifth as large as

4. The total linear momentum of a system of objects is constant if
   a. friction can be ignored.
   b. gravity can be ignored.
   c. the only forces acting are external forces.
   d. the only forces acting are internal forces.

5. Two identical cars collide at an intersection and lock together. Before the collision one car was traveling east with a speed of 42 km/h and the other car was traveling south with the same speed. What is the velocity of the cars immediately after the collision?
   a. 30 km/h, 45° south of east
   b. 30 km/h, The direction can not be found.
   c. 42 km/h, 45° south of east
   d. 42 km/h, The direction can not be found.

6. A pool ball (ball 1) moving with a speed of 5.0 m/s overtakes and collides elastically with an identical pool ball (ball 2) moving with a speed of 1.5 m/s in the same direction. What are the velocities of the balls after the collision?
   a. Ball 1 - 1.5 m/s opposite its original direction. Ball 2 - 5.0 m/s in its original direction.
   b. Ball 1 - 1.5 m/s in its original direction. Ball 2 - 5.0 m/s in its original direction.
   c. Ball 1 - 3.9 m/s opposite its original direction. Ball 2 - 6.1 m/s in its original direction.
   d. Ball 1 - 3.9 m/s in its original direction. Ball 2 - 6.1 m/s opposite its original direction.

7. Which of the following is a true statement concerning a system of objects?
   a. If the total momentum is conserved, then the total energy is definitely conserved.
   b. If the total energy is not conserved, then the total momentum is definitely not conserved.
   c. If the total energy is conserved, the total momentum is definitely conserved.
   d. The total energy and total momentum may be conserved independently of each other.

8. A 1100 kg rocket in space is accelerated by a 5.0 s "burn" from rest to a speed of 15 m/s. The exhaust gases have a speed of 4500 m/s. What is the average rate at which mass is ejected from the rocket engine?
   a. 1.2 kg/s b. 0.98 kg/s c. 0.73 kg/s d. 0.51 kg/s

9. A 1.0 kg object moving with a speed of 1.0 m/s collides head-on with a 4.0 kg object moving in the opposite direction with a speed of 2.0 m/s. After the collision the 1.0 kg object moves with a speed of 7.0 m/s opposite its original direction while the 4.0 kg object is at rest. Which of the following is true concerning the collision?
   a. Momentum is conserved, but energy is not.
   b. Energy is conserved, but momentum is not.
   c. Neither momentum or energy is conserved.
   d. Both energy and momentum are conserved.

10. An object with mass, m_1, makes a head-on elastic collision with a stationary object of mass, m_2. After the collision both objects are moving in the same direction as the initial velocity of mass 1. Which is true?
    a. $m_1 > m_2$ b. $m_2 > m_1$ c. $m_1 = m_2$ d. No relationship exists.

SOLUTIONS AND ANSWERS

Practice Problems

---

1. The impulse of the force is equal to change of momentum of the ball. In order to find the change in momentum of the ball we need to find its final speed.

$$1/2\ mv_f^2 = (0.75)(1/2)mv_0^2,$$

gives the final speed of the ball to be $v_f = \sqrt{(0.75)}\ v_0$. If the direction of the ball's initial velocity is taken as positive, then impulse of the force is

Impulse $= m\mathbf{v_f} - m\mathbf{v_0} = -\ mv_0[\sqrt{(0.75)} + 1)] = -\ (0.14\ kg)(40.0\ m/s)[\sqrt{(0.75)} + 1] =$ **- 1.0 X 10¹ kg m/s.**

The negative sign indicates that the impulse is in a direction opposite the initial direction of travel of the ball.

---

2. The ball with the larger mass has a momentum $p_L = m_L v_L$ and the ball with the smaller mass has a momentum $p_S = m_S v_S$. Now

$$m_L = 10\ m_S \text{ and } v_L = (1/15)v_S$$

so

$$p_L = 2/3\ m_S v_S = 2/3\ p_S.$$

The ball with the larger mass has the smaller momentum. The smaller mass ball will be the hardest to stop since the impulse momentum theorem requires that more force is needed to produce a larger momentum change in the same time.

---

3. If friction is ignored, no external force acts on the boat in the horizontal direction, so the total momentum of the boat plus contents is conserved in that direction. After N cans of beans have been thrown with a horizontal speed of v_c the boat will have a speed of v_B. The conservation of momentum requires

$$(M - Nm_c)v_B - Nm_c v_c = 0$$

or

$$N = \frac{mv_B}{m_c(v_B + v_c)} = \frac{(195\ kg)(1.0\ m/s)}{(0.500\ kg)(1.0\ m/s + 26.8\ m/s)} = \mathbf{14.}$$

---

4. If friction is ignored, no external forces act on the boy and skateboard in the horizontal direction. Total momentum is conserved in that direction. The conservation of momentum principle gives $m_B v_B + m_S v_S = 0$ or

$$v_S = -\ (m_B/m_S)v_B = -\ (45/2.0)(0.50\ m/s) = \mathbf{-\ 11\ m/s}.$$

The negative indicates that the skateboard moves in a direction opposite the boy.

---

5. The child and sled make a completely inelastic collision. The total momentum is conserved in the horizontal direction if friction is ignored. Then

$$(m_c + m_s)v = m_c v_{0c} \text{ or } (42\ kg)v = (35\ kg)(4.5\ m/s) \text{ so } v = \mathbf{3.8\ m/s}.$$

6. The mechanical energy of the block (object 1) is conserved as the block falls so

$$1/2\ m_1v_1^2 + m_1gh = m_1g(15.0\ \text{m})$$

or

$$h = 15.0\ \text{m} - 1/2\ v_1^2/g.$$

If the collision time is short, then the total momentum of the bullet (object 2) and block system will be approximately conserved since gravity cannot change it very much in a short time. Then

$$m_2v_2 - m_1v_1 = 0 \text{ or } v_1 = (m_2/m_1)v_2.$$

Using this result in the conservation of energy expression above yields

$$h = 15.0\ \text{m} - (1/2)(m_2/m_1)^2v_2^2/g = 15.0\ \text{m} - (1/2)(5.0 \times 10^{-3}\ \text{kg}/0.50\ \text{kg})^2(320\ \text{m/s})^2/(9.80\ \text{m/s}^2) = \mathbf{14.5\ m}.$$

---

7. The brick and wagon make a totally inelastic collision. If friction between the wagon and ground is ignored, then no net force acts on the system in the horizontal direction so the total momentum is conserved in that direction.

$$(m_B + m_W)v = m_Bv_B \cos 20.0°$$

or

$$(20.6\ \text{kg})v = (0.60\ \text{kg})(15.0\ \text{m/s}) \cos 20.0°$$

or

$$v = \mathbf{0.41\ m/s}.$$

---

8. Both the conservation of momentum and the conservation of kinetic energy apply. The former gives

$$(5.0\ \text{kg})v_{1f} + (3.0\ \text{kg})v_{2f} = (5.0\ \text{kg})(5.0\ \text{m/s}) + (3.0\ \text{kg})(1.5\ \text{m/s})$$

$$5.0v_{1f} + 3.0v_{2f} = 29.5 \qquad (1)$$

The conservation of kinetic energy gives

$$1/2\ (5.0\ \text{kg})v_{1f}^2 + 1/2\ (3.0\ \text{kg})v_{2f}^2 = 1/2\ (5.0\ \text{kg})(5.0\ \text{m/s})^2 + 1/2\ (3.0\ \text{kg})(1.5\ \text{m/s})^2$$

$$2.5v_{1f}^2 + 1.5v_{2f}^2 = 65.9 \qquad (2)$$

Solving (1) for v_{1f} and substituting the result into (2) and rearranging gives

$$2.4v_{2f}^2 - 17.7v_{2f} + 21.1 = 0$$

The quadratic formula yields solutions of $v_{2f} = +1.5$ m/s and **+ 5.9 m/s.**
Substituting these values in to (1) results in $v_{1f} = +\ 5.0$ m/s and **+ 2.4 m/s.**
The first of these values correspond to no collision at all so they are rejected.

---

9. The cars make a completely inelastic collision in which the total momentum is conserved if friction between the tires and the road is ignored. In the north-south (y) direction

$$m_1 v_{01y} = (m_1 + m_2) v_{fy}$$

or

$$v_{fy} = (1200/2700)(65\ \text{km/h}) = 29\ \text{km/h}$$

and in the east-west (x) direction

$$m_2 v_{02x} = (m_1 + m_2) v_{fx}.$$

or

$$v_{fx} = (1500/2700)(85\ \text{km/h}) = 47\ \text{km/h}.$$

The speed of the cars after the collision is found from the Pythagorean theorem,

$$v_f^2 = v_{fx}^2 + v_{fy}^2,$$

to be

$$\mathbf{v_f = 55\ km/h}.$$

The direction that the velocity vector makes with east is

$$\theta = \tan^{-1}(29/47) = \mathbf{32° \text{ north of east.}}$$

---

10. The balls have the same mass, so we immediately know that **they will exchange speeds**. The directions of travel must be determined. Before the collision, the total momentum of the system is

$$P_0 = m(v_{01} + v_{02}) = m[+2.5\ \text{m/s} + (-5.9\ \text{m/s})] = -(3.4\ \text{m/s})m$$

which is in the direction of motion of the 5.9 m/s ball, say to the left. After the collision, the total momentum must be the same in magnitude and still to the left.

$$P_f = m(v_{f1} + v_{f2}) = -(3.4\ \text{m/s})m \text{ so } v_{f1} + v_{f2} = -3.4\ \text{m/s}$$

where the magnitude of v_{f1} is 5.9 m/s and the magnitude of v_{f2} is 2.5 m/s. The only way for these to be true is for $v_{f1} = -5.9$ m/s and $v_{f2} = +2.5$ m/s. That is, **the balls also exchange directions of motion.**

Quiz answers

1. b	3. b	5. a	7. d	9. a
2. c	4. d	6. b	8. c	10. a

MCAT REVIEW PROBLEMS

A wrecking ball attached to the end of a long, essentially massless chain is released from rest, from a height $H = 5$ meters off the ground, as shown in figure 1. At the lowest point of its swing, it collides with a piece of modern sculpture. Since the sculpture is made of wet clay, it sticks to the wrecking ball.

The wrecking ball has mass 200 kg. The sculpture has mass 100 kg.

"Mechanical energy" is the sum of potential and kinetic energy.

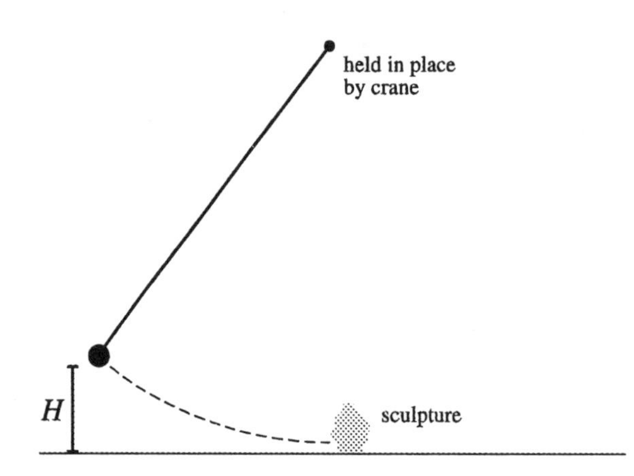

FIGURE 1

1. As the ball swings down from its starting point, its acceleration along the direction of motion:

   A. Stays constant at about 10 m/s
   B. Stays constant at some other value
   C. Increases
   D. Decreases

2. Neglect air resistance. As the ball swings down from its starting point, but before it collides with the sculpture, the ball's:

   A. Momentum stays the same, and mechanical energy stays the same.
   B. Momentum stays the same, but mechanical energy changes.
   C. Momentum changes, but mechanical energy stays the same.
   D. Momentum changes, and mechanical energy changes.

3. During the collision between the ball and the sculpture, the mechanical energy of the whole system:

   A. Increases
   B. Stays the same
   C. Decreases
   D. We cannot determine the answer from the given information.

4. Immediately before colliding with the sculpture, the ball's speed was 10 m/s. Immediately after the collision, the ball's speed is most nearly

A. 3 m/s **B.** 5 m/s **C.** 7 m/s **D.** 8 m/s

---

5. The ball weighs W = 2000 N. Immediately before the ball collided with the sculpture, the tension in the chain was:

A. zero **B.** positive, but less than 2000 N
C. 2000 N **D.** greater than 2000 N

ANSWERS TO MCAT REVIEW PROBLEMS

---

1. **D.** The ball keeps speeding up, but at a slower and slower rate of change. Near the top of its swing, the ball is "pulled" by a large component of gravity—$mg\sin\theta_1$, in figure 2. By contrast, near the bottom of its swing, the ball is pulled by a smaller force, $mg\sin\theta_2$. The decreasing force produces a decreasing acceleration. It's as if the ball were sliding down a ramp that got less and less steep near the bottom. Again, this decreasing acceleration does *not* mean the ball slows down. The ball speeds up until smashing into the sculpture. But the *rate* at which it speeds up gets smaller and smaller.

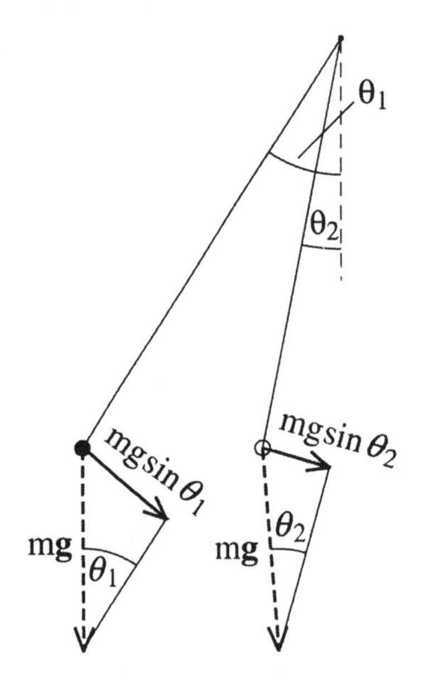

FIGURE 2

---

2. **C.** Since $\mathbf{p} = m\mathbf{v}$, the momentum changes as the ball speeds up. In general, whenever a system feels a net external force (such as gravity or some other outside "push"), momentum isn't conserved.
   By contrast, up until the collision, no energy dissipates away as heat or any other non-mechanical form of energy (since we're neglecting air resistance). As the ball swings down, potential energy converts into kinetic energy. But the sum of potential and kinetic energy stays constant.

---

3. **C.** During the collision, some of the ball's kinetic energy turns into heat. In other words, some mechanical energy dissipates into non-mechanical forms of energy. This *always* happens during an inelastic collision. By contrast, momentum is conserved during *any* collision, elastic or inelastic.

---

4. **C.** Let v_{before} and v_{after} denote the ball's speed immediately before and immediately after the collision. According to momentum conservation,

$$M_{\text{ball}}v_{\text{before}} = (M_{\text{ball}} + m_{\text{sculpture}})v_{\text{after}},$$

since the ball and sculpture stick together as one big mass. Solve for v_{after} to get

$$\begin{aligned} v_{after} &= \frac{M_{ball}}{M_{ball} + m_{sculpture}} v_{before} \\ &= \frac{200 \text{ kg}}{200 \text{ kg} + 100 \text{ kg}} (10 \text{ m/s}) \\ &= \frac{2}{3} (10 \text{ m/s}) \\ &\approx 7 \text{ m/s}. \end{aligned}$$

A common error is to "conserve" kinetic energy. As explained in question 3, however, mechanical energy isn't conserved during an inelastic collision. Some of it converts into heat:

$$\frac{1}{2} M_{ball} v_{before}^2 = \frac{1}{2} (M_{ball} + m_{sculpture}) v_{after}^2 + \text{Heat}.$$

By contrast, elastic collisions generate no heat.

---

5. **D.** If the ball were hanging still, then the tension would equal the ball's weight. The net force on the ball would be zero. But since the ball swings in a circle, it experiences a "centripetal" acceleration, i.e., an acceleration towards the center of the circle:

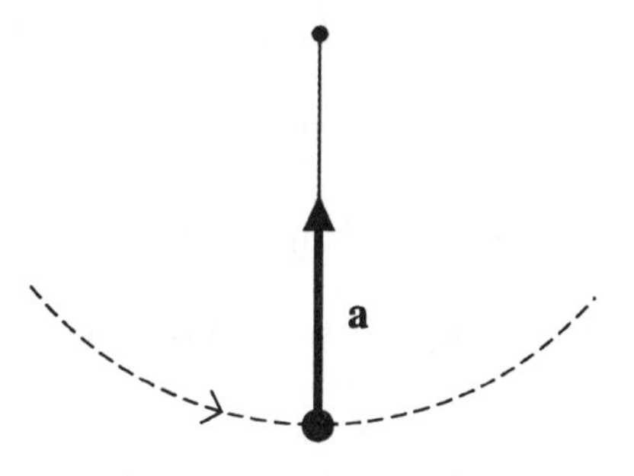

Therefore, according to Newton's 2nd law, the ball must feel a net force towards the center of the circle. In other words, the tension must exceed the ball's weight. If the net force were zero, the ball would be sitting still or moving in a straight line, not moving along a circular arc.

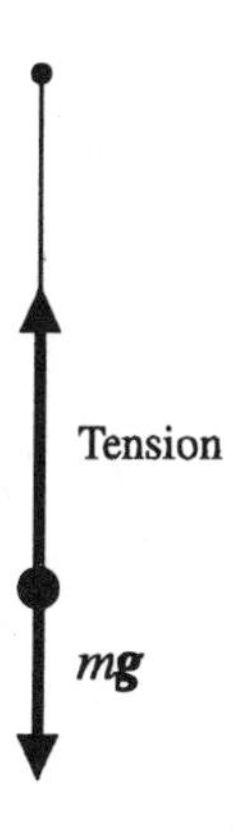

CHAPTER 8
Rotational Kinematics

PREVIEW

In this chapter we will discuss rotational motion. Basic quantities such as angular displacement, angular velocity and angular acceleration are introduced along with the kinematic equations of rotation. Centripetal and tangential accelerations are also studied in detail.

QUICK REFERENCE

Important Terms

Angular displacement
The angle through which a rigid object rotates about a fixed axis. By convention, this quantity is positive if the rotation is counterclockwise, negative if it is clockwise. Units are radians (rad).

Radian
An SI unit of angular displacement defined as the circular arc length s traveled by a point on a rotating body divided by the radial distance r of the point from the axis: $\theta = s/r$. Also, 2π radians = 1 revolution = 360°.

Average angular velocity
The angular displacement of an object divided by the elapsed time. Units are (rad/s).

Instantaneous angular velocity
The angular velocity of an object at a given instant of time.

Average angular acceleration
The change in angular velocity divided by the elapsed time. Units are (rad/s^2).

Tangential velocity
The linear velocity of a point on a rotating body. It represents how fast the point is moving along the arc of the circle. Units are (m/s).

Tangential acceleration
The linear acceleration of a point on a rotating body, measured in the tangential direction. Units are (m/s^2).

Uniform circular motion
The motion of an object moving at constant tangential speed in a circular path.

Angular Quantities

Angular Displacement - the angle θ swept out when a rigid body rotates about a fixed axis. If s is the arc length swept out and r the radius, as shown in the diagram, we have

$$\theta \text{ (radians)} = \frac{\text{arc length}}{\text{radius}} = \frac{s}{r} \qquad (8.1)$$

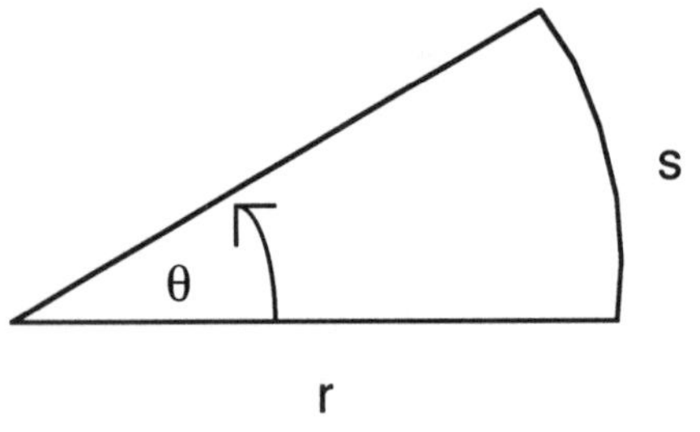

s

θ

r

The angle θ can be measured in revolutions, degrees, or radians. The number of radians that corresponds to one revolution or 360° is

$$\theta = \frac{s}{r} = \frac{2\pi r}{r} = 2\pi \text{ rad}$$

Average Angular Velocity - the angular displacement divided by the elapsed time. That is

$$\overline{\omega} = \frac{\theta - \theta_0}{t - t_0} = \frac{\Delta\theta}{\Delta t} \qquad (8.2)$$

Instantaneous Angular Velocity - the angular velocity at a given instant of time. That is

$$\omega = \lim_{\Delta t \to 0} \overline{\omega} = \lim_{\Delta t \to 0} \frac{\Delta\theta}{\Delta t} \qquad (8.3)$$

Average Angular Acceleration - the change in angular velocity divided by the elapsed time. That is

$$\overline{\alpha} = \frac{\omega - \omega_0}{t - t_0} = \frac{\Delta\omega}{\Delta t} \qquad (8.4)$$

Equations of Rotational Kinematics

If the **angular acceleration is constant** (uniform), the following equations are valid for rotational motion:

$$\omega = \omega_0 + \alpha t \qquad (8.4) \qquad\qquad \theta = \omega_0 t + \frac{1}{2}\alpha t^2 \qquad (8.7)$$

$$\theta = \frac{1}{2}(\omega_0 + \omega)t \qquad (8.6) \qquad\qquad \omega^2 = \omega_0^2 + 2\,\alpha\theta \qquad (8.8)$$

Note the similarity of these equations to the equations of kinematics for linear motion found in Chapter 2.

Angular and Tangential Variables and Centripetal Acceleration

Tangential velocity is given by: $\quad v_T = r\omega \qquad (8.9)$

Tangential acceleration is given by: $\quad a_T = r\alpha \qquad (8.10)$

Centripetal acceleration is:

$$a_c = \frac{v_T^2}{r} = \frac{(r\omega)^2}{r} = r\omega^2 \qquad (8.11)$$

In **nonuniform circular motion**, the centripetal and tangential accelerations are perpendicular to one another. The total acceleration, a, and the angle between a and a_c, ϕ, can then be expressed as

$$a = \sqrt{a_c^2 + a_T^2} \quad \text{while} \quad \tan\phi = \frac{a_T}{a_c}$$

Rolling Motion

A point on an object rolling without slipping has a speed

$$v = v_T = r \qquad (8.12)$$

and an acceleration

$$a = a_T = r \qquad (8.13)$$

DISCUSSION OF SELECTED SECTIONS

8.1 Rotational Motion and Angular Displacement

Example 1

Find the angle subtended by the moon as seen from earth. The moon's diameter is 3.48 x 10^6 m and its distance from earth is 3.85 x 10^8 m. Express your answer in radians and degrees.

Using equation (8.1) with s = 3.48 x 10^6 m and r = 3.85 x 10^8 m, we have

$$\theta = \frac{s}{r} = \frac{3.48 \times 10^6 \text{ m}}{3.85 \times 10^8 \text{ m}} = 9.04 \times 10^{-3} \text{ rad}$$

$$9.04 \times 10^{-3} \text{ rad} \left(\frac{360°}{2\pi \text{ rad}} \right) = 0.52°.$$

8.2 Angular Velocity and Angular Acceleration

The following examples make use of the definitions of angular velocity and angular acceleration.

Example 2

Find the average angular velocity of the earth as it rotates on its axis. Express your answer in rad/s.

We know that the earth makes one revolution per day. We must convert "revolutions" to the unit of angular displacement, "radians", and convert "days" to "seconds". Using equation (8.2) we can write:

$$\bar{\omega} = \frac{\Delta\theta}{\Delta t} = \left(\frac{1 \text{ rev}}{\text{day}} \right) \left(\frac{2\pi \text{ rad}}{\text{rev}} \right) \left(\frac{1 \text{ day}}{24 \text{ h}} \right) \left(\frac{1 \text{ h}}{3600 \text{ s}} \right) = 7.27 \times 10^{-5} \text{ rad/s.}$$

Example 3

A phonograph turntable changes from 33.3 rev/min to 78.0 rev/min in 3.00 s. What is the average angular acceleration? Express your answer in rad/s^2.

Using equation (8.4):

$$\omega_0 = \left(33.3\,\frac{\text{rev}}{\text{min}}\right)\left(\frac{2\pi\ \text{rad}}{1\ \text{rev}}\right)\left(\frac{1\ \text{min}}{60\ \text{s}}\right) = 3.49\ \text{rad/s}$$

$$\omega = \left(78.0\,\frac{\text{rev}}{\text{min}}\right)\left(\frac{2\pi\ \text{rad}}{1\ \text{rev}}\right)\left(\frac{1\ \text{min}}{60\ \text{s}}\right) = 8.17\ \text{rad/s}$$

$$\alpha = \frac{\Delta\omega}{\Delta t} = \frac{\omega - \omega_0}{\Delta t} = \frac{8.17\ \text{rad/s} - 3.49\ \text{rad/s}}{3.00\ \text{s}} = 1.56\ \text{rad/s}^2.$$

8.3 The Equations of Rotational Kinematics

Equations (8.4), (8.6), (8.7), and (8.8) are the equations for rotational kinematics for constant angular acceleration. Note their similarity to the equations of kinematics for linear motion. In fact, the equations are identical if the following linear quantities are replaced by their rotational analogs:

Quantity	*Linear Motion*	*Rotational Motion*
Displacement	x	θ
Initial velocity	v_0	ω_0
Final velocity	v	ω
Acceleration	a	α
Time	t	t

Example 4
If it takes 20.0 s for the shaft of a motor to start from rest and attain its full speed of 1800 rev/min, what is the magnitude of the average angular acceleration of the shaft? What is the angular displacement of the shaft?

$$\omega = \omega_0 + \alpha t \qquad (8.4)$$

$$\omega_0 = 0$$

$$\omega = \left(1800\,\frac{\text{rev}}{\text{min}}\right)\left(\frac{1\ \text{min}}{60\ \text{s}}\right)\left(\frac{2\pi\ \text{rad}}{1\ \text{rev}}\right) = 188\ \text{rad/s}$$

$$t = 20.0\ \text{s}$$

$$\alpha = \frac{188\ \text{rad/s}}{20.0\ \text{s}} = 9.40\ \text{rad/s}^2$$

$$\theta = \omega_0 t + \frac{1}{2}\alpha t^2 = \frac{1}{2}(9.40\ \text{rad/s}^2)(20.0\ \text{s})^2 = 1880\ \text{rad}$$

Example 5
A grinding wheel has an initial angular velocity of 215 rad/s when the driving motor is shut off. After 10.0 s, its angular velocity is 125 rad/s. What is the angular acceleration of the wheel? What is the angular displacement during the ten second interval? Assuming the angular acceleration remains constant, find the total number of revolutions the wheel turns through in coming to rest.

$$\omega = \omega_0 + \alpha t \Rightarrow \alpha = \frac{\omega - \omega_0}{t} = \frac{125 \text{ rad/s} - 215 \text{ rad/s}}{10.0 \text{ s}} = -9.0 \text{ rad/s}^2$$

$$\theta = \omega_0 t + \frac{1}{2}\alpha t^2 = (215 \text{ rad/s})(10.0 \text{ s}) + \frac{1}{2}(-9.0 \text{ rad/s}^2)(10.0 \text{ s})^2 = 1.70 \times 10^3 \text{ rad}$$

$$\omega^2 = \omega_0^2 + 2\alpha\theta' \Rightarrow \theta' = \frac{\omega^2 - \omega_0^2}{2\alpha} = \frac{-(215 \text{ rad/s})^2}{2(-9.0 \text{ rad/s}^2)} = 2600 \text{ rad}\left(\frac{1 \text{ rev}}{2\pi \text{ rad}}\right) = 410 \text{ rev.}$$

8.4 Angular Variables and Tangential Variables

Consider two people standing on a moving merry-go-round; one person near the center and one near the outer edge of the merry-go-round. Both people experience the same angular speed ω, since the merry-go-round is a rigid body rotating about a fixed axis. However, the person on the outer edge is moving faster, since he is sweeping out a larger arc length in a given amount of time. The arc length traveled by each person, Δs, is given by $\Delta s = r\Delta\theta$, so that the linear speed is

$$v = \frac{\Delta s}{\Delta t} = r\frac{\Delta\theta}{\Delta t} = r\,\omega.$$

Therefore, the greater the value of r, the greater the linear or **tangential** speed. A similar expression exists for the tangential acceleration. The relationships between v_T and ω, and a_T and α are given by

$$v_T = r\omega \qquad (8.9) \qquad\qquad a_T = r\alpha \qquad (8.10)$$

A word of caution: The expression "$s = r\theta$" is valid only when the angle θ is expressed in radians. Any expression or equation derived from "$s = r\theta$" has the same restrictions. Therefore, in using equations (8.9) and (8.10), all angular quantities (ω and α) must be expressed in radians; that is ω must be expressed in rad/s and α must be expressed in rad/s^2

Example 6
A large disk is rotating with an angular speed of 225 rad/s. Find the tangential speed at a distance of 2.0 m from the center of the disk. If the disk uniformly speeds up to 450 rad/s in 25 s, what is the tangential acceleration of this point?

Using equation (8.9) we have:

$$v_T = r\omega = (2.0 \text{ m})(225 \text{ rad/s}) = 450 \text{ m/s}.$$

In order to determine the tangential acceleration, we need to find the angular acceleration.

Using equation (8.4) we have:

$$\alpha = (\omega - \omega_0)/t = (450 \text{ rad/s} - 225 \text{ rad/s})/(25 \text{ s}) = 9.0 \text{ rad/s}^2.$$

So the tangential acceleration is:

$$a_T = r\alpha = (2.0\ \text{m})(9.0\ \text{rad/s}^2) = 18\ \text{m/s}^2.$$

8.5 Centripetal Acceleration and Tangential Acceleration

Suppose you tie a rock to the end of a string and spin it in a circle. The string holding the rock is exerting a force inward toward your hand. This force is called a **centripetal force**. Since the rock is spinning in a circle, and thus constantly changing direction, it is being accelerated by this centripetal force. This acceleration is directed INWARD, in the same direction as the force, and is known as the **centripetal acceleration**, a_c. The magnitude of this acceleration depends on how fast you spin the rock and how long the string is. That is,

$$a_c = \frac{v_T^2}{r} = \frac{(r\omega)^2}{r} = r\omega^2 \qquad (8.11)$$

Example 7
The moon revolves around the earth, making a complete revolution in 27.3 days. Assuming that the moon's orbit is a circle of radius 3.85×10^8 m, what is the magnitude of the acceleration of the moon toward the earth?

The moon's angular velocity is:

$$\omega = \Delta\theta/\Delta t = (2\pi\ \text{rad}/27.3\ \text{d})(1\ \text{d}/24\ \text{h})(1\ \text{h}/3600\ \text{s})$$
$$\omega = 2.66 \times 10^{-6}\ \text{rad/s}$$

So that the centripetal acceleration is:

$$a_c = r\omega^2 = (3.85 \times 10^8\ \text{m})(2.66 \times 10^{-6}\ \text{rad/s})^2$$
$$a_c = 2.72 \times 10^{-3}\ \text{m/s}^2.$$

8.6 Rolling Motion

As an automobile moves along the street, its tires roll without slipping. When no slippage occurs, we refer to the motion as rolling motion. The following equations are valid under these conditions:

$$v = v_T = r\omega \quad (8.12) \qquad\qquad a = a_T = r\alpha \quad (8.13)$$

Example 8
A wheel rolls without slipping for a linear distance of 628 m. If the diameter of the wheel is 1.00 m, how many revolutions does the wheel make in this time?

Using equation (8.1) we have:

$$\theta = s/r = (628\ \text{m})/(0.500\ \text{m})$$
$$\theta = 1260\ \text{rad}(1\ \text{rev}/2\pi\ \text{rad})$$
$$\theta = 200\ \text{rev}.$$

Example 9
A car accelerates uniformly from rest and reaches a velocity of 60.0 mi/h in 7.0 s. If the wheels have a diameter of 24 inches, (a) how many revolutions do the tires make? (b) What is the angular acceleration of the wheels?

(a) The final speed of the car is v = (60.0 mi/h)(5280 ft/1 mi)(1 h/3600 s) = 88 ft/s. This corresponds to the tangential velocity of a point on the edge of the tire. The final angular velocity a tire is therefore, using equation (8.12),

$$\omega = v/r = (88 \text{ ft/s})/(1.0 \text{ ft}) = 88 \text{ rad/s}.$$

Using equation (8.6),

$$\theta = (1/2)(\omega_0 + \omega)t$$

$$\theta = (1/2)(88 \text{ rad/s})(7.0 \text{ s})$$

$$\theta = 310 \text{ rad}(1 \text{ rev}/2\pi \text{ rad})$$

$$\theta = 49 \text{ rev}.$$

(b) Using equation (8.4),

$$\omega = \omega_0 + \alpha t$$

which gives

$$\alpha = \omega/t = (88 \text{ rad/s})/(7.0 \text{ s}) = 13 \text{ rad/s}^2.$$

Note that this last result can also be obtained using equations (2.4) and (8.13), i.e.,

$$v = v_0 + at$$

which gives

$$a = v/t = (88 \text{ ft/s})/(7.0 \text{ s}) = 13 \text{ ft/s}^2.$$

then

$$\alpha = a/r = (13 \text{ ft/s}^2)/(1 \text{ ft}) = 13 \text{ rad/s}^2.$$

PRACTICE PROBLEMS

1. An electric fan is set on its HIGH setting. After the LOW push button is depressed, the angular speed of the fan blades decreases to a value of 8.00×10^2 rev/min in 1.75 s. The deceleration is 42.0 rad/s^2. Determine the initial angular speed of the blades in rev/min.

2. An electric grinder uses a grinding wheel of 12-cm radius. It takes the electric motor 3.0 s to reach its rated speed of 1500 rev/min starting from rest. Find (a) the angular acceleration, (b) the total angular displacement, and (c) the tangential velocity at the rim of the grinding wheel at the rated speed.

3. The high speed drill of a dentist is driven by an air turbine and achieves a rotational speed of 450 000 rev/min. The radius of the cutting burr is 1.0 mm. (a) Find the linear speed of the outer edge of the cutting burr. (b) Determine the acceleration of the outer edge and express this as a multiple of the gravitational acceleration, g.

4. A small bug sits on a turntable, 10.0 cm from the center. What is the centripetal acceleration of the bug when the turntable rotates at 45 rev/min? What is the bug's tangential velocity?

5. A space station is constructed in the shape of a doughnut whose outer radius is 480 m and whose inner radius is 410 m. At what rate should the space station rotate so that its occupants will experience an acceleration of magnitude g when they are located at the outer diameter of the station? By what fraction would a person's weight change if they stood at the inner diameter of the station?

6. An automobile traveling 30.0 m/s has wheels of 0.750 m diameter. (a) What is the angular speed of the wheels about the axle? (b) If the car is brought to a stop uniformly in 10.0 s, what is the angular acceleration? (c) How many times do the wheels turn during this braking period?

7. Wheel A has a radius $r_A = 75$ cm and is coupled to wheel B, radius $r_B = 25$ cm, by belt C. Wheel A accelerates from rest at an angular acceleration of 2.0 rad/s^2. How long does it take for wheel B to reach a rotational speed of 1200 rad/s assuming that the belt does not slip?

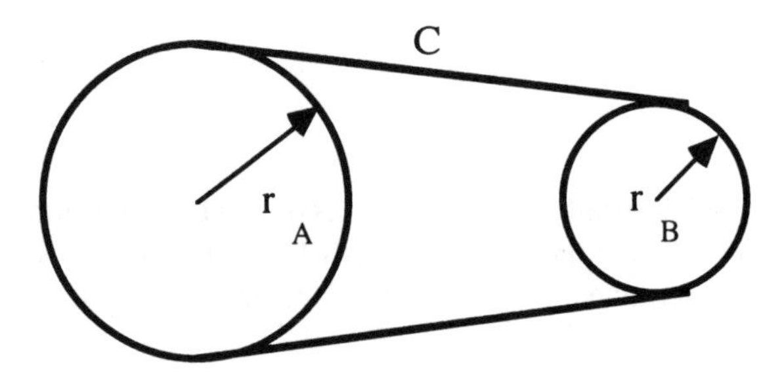

8. One end of a string is attached to the rim of a wheel 0.50 ft in radius. A small object hangs from the string as shown in the diagram below. If the wheel is given an initial angular velocity of 9.0 rad/s "into the page" and the constant angular acceleration of the wheel is 3.0 rad/s^2 "out of the page" (a) how many turns does the wheel make before coming momentarily to rest? (b) What length of string is then wrapped around the wheel? (c) At what time will the string again be entirely unwrapped from the wheel?

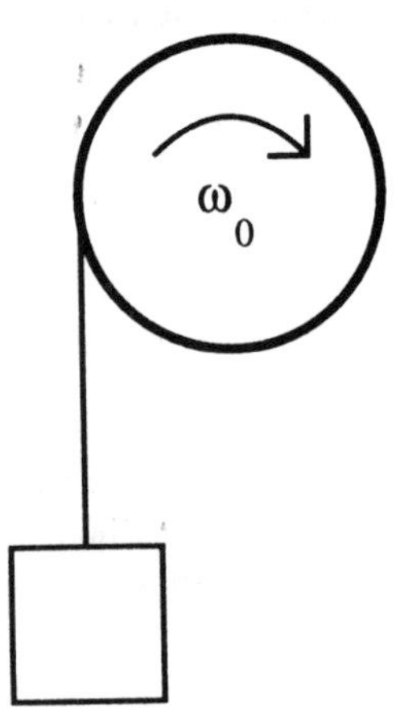

HELPFUL SUGGESTIONS

1. Rotational motion has many analogies with translational motion. Remember that the kinematic equations of rotational motion can be obtained from the kinematic equations of linear motion by simply making the substitutions shown in Table 8.1. That is, the linear quantities such as displacement, s, velocity, v, and acceleration, a, need to simply be replaced by angular displacement, θ, angular velocity, ω, and angular acceleration, α, respectively.

2. When expressing rotational quantities, the preferred angular unit is the radian. Angles measured in revolutions or degrees usually need to be converted to radians. Remember that 2π radians = 1 revolution = 360°. Also remember that when using $s = \theta r$, $\Delta s = r\Delta\theta$, text equations (8.9), (8.10), (8.11), or (8.12) the angular unit must be the radian.

3. If an object is moving at a constant speed in a circle, there must be a force acting on the object (otherwise it would move in a straight line). Therefore, there is also an acceleration present in this situation, the centripetal acceleration. Also keep in mind that this acceleration MUST be directed INWARD towards the center of the circle.

4. Tangential acceleration and centripetal acceleration represent two DIFFERENT changes in motion. Centripetal acceleration is always directed inward along the radius vector, while tangential acceleration is always perpendicular to the radius vector. Centripetal acceleration involves only a change in **direction** while tangential acceleration occurs ONLY when the object's **speed** is changing.

EVERYDAY PHYSICS

1. Tie an object to a piece of string and spin it in a horizontal circle. If you suddenly let go of the string (make sure your roommate isn't nearby) which way does the object fly off? Does it go straight out from the center of rotation or does it fly off tangent to the circle?

2. There are many examples of centripetal acceleration in nature. Consider, for example, a planet orbiting around the sun, or the swirling motion of a hurricane. What are the forces producing the accelerations in each situation? Can you think of some other examples of this kind of motion?

3. Amusement parks often have a "centripetal force" ride that goes by various names (the roundup is one) but operates on the principles discussed. In one version of this ride you stand with your back to the wall of a large drum that spins on a vertical axis. When the drum is rotating fast enough, the floor drops out from beneath your feet and you are pinned to the wall. The normal force is the centripetal force as well as the force determining the static friction between you and the wall.

CHAPTER QUIZ

Questions 1 - 4 are based on the following:

A fly and a mosquito are sitting on a phonograph record which is turning at a constant rate. The fly is near the outer edge of the record and the mosquito is near the center. Answer the following questions.

1. How do their angular velocities compare?
   a. The fly has the greater angular velocity.
   b. The mosquito has the greater angular velocity.
   c. Both the fly and the mosquito have the same nonzero angular velocity.
   d. The angular velocity for both is zero.

2. How do their translational velocities compare?
   a. The fly has the greater translational velocity.
   b. The mosquito has the greater translational velocity.
   c. Both the fly and mosquito have the same nonzero translational velocity.
   d. The translational velocity for both is zero.

3. How do their angular accelerations compare?
   a. The fly has the greater angular acceleration.
   b. The mosquito has the greater angular acceleration.
   c. Both the fly and the mosquito have the same nonzero angular acceleration.
   d. The angular acceleration for both is zero.

4. How do their centripetal accelerations compare?
   a. The fly has the greater centripetal acceleration.
   b. The mosquito has the greater centripetal acceleration.
   c. Both the fly and the mosquito have the same nonzero centripetal acceleration.
   d. The centripetal acceleration for both is zero.

5. A flywheel requires 3.00 s to rotate through 234 rad. Its angular velocity at the end of this time is 108 rad/s. What is the flywheel's angular acceleration (assumed constant)?
   a. 20 rad/s^2 b. 48 rad/s^2 c. 72 rad/s^2 d. 108 rad/s^2

6. A wheel of radius 2 m is spinning with an angular velocity of 4 rad/s. What is the translational velocity of a point on the wheel's rim?
   a. 2 m/s b. 4 m/s c. 8 m/s d. 32 m/s

7. A wheel of radius 2.0 m is spinning with an angular velocity of 4.0 rad/s. What is the centripetal acceleration of a point on the wheel's rim?
   a. 2.0 m/s^2 b. 4.0 m/s^2 c. 8.0 m/s^2 d. 32 m/s^2

8. A wheel has an angular acceleration of 8 rad/s^2. If the wheel starts from rest, through what angle does it turn through in 5 s?
   a. 40 rad b. 100 rad c. 200 rad d. 300 rad

9. An object has a centripetal acceleration of 30 m/s^2 and a tangential acceleration of 40 m/s^2. What is the total acceleration of the object?
   a. 1000 m/s^2 b. 70 m/s^2 c. 50 m/s^2 d. 10 m/s^2

SOLUTIONS AND ANSWERS

Practice Problems

---

1. $\omega = (8.00 \times 10^2 \text{ rev/min})(1 \text{ min}/60 \text{ s})(2\pi \text{ rad}/1 \text{ rev})$

$\omega = 83.8 \text{ rad/s},$

Using equation (8.4),

$\omega_0 = \omega - \alpha t = (83.8 \text{ rad/s}) - (-42.0 \text{ rad/s}^2)(1.75 \text{ s})$

$\omega_0 = 157 \text{ rad/s}(1 \text{ rev}/2\pi \text{ rad})(60 \text{ s}/1 \text{ min})$

$\omega_0 = \mathbf{1500 \text{ rev/min.}}$

---

2. $\omega = 1500 \text{ rev/min} = 160 \text{ rad/s}, \omega_0 = 0, t = 3.0 \text{ s}, r = 0.12 \text{ s}.$

(a) Using equation (8.4) we have,

$$\alpha = (\omega - \omega_0)/t = (160 \text{ rad/s})/(3.0 \text{ s}) = \mathbf{53 \text{ rad/s}^2}.$$

(b) Using equation (8.7) we have,

$$\theta = \omega_0 t + (1/2)\alpha t^2 = (1/2)(53 \text{ rad/s}^2)(3.0 \text{ s})^2 = \mathbf{240 \text{ rad}}.$$

(c) Using equation (8.9) we have,

$$v_T = r\omega = (0.12 \text{ m})(160 \text{ rad/s}) = \mathbf{19 \text{ m/s}}.$$

---

3. $\omega = (4.5 \times 10^5 \text{ rev/min})(1 \text{ min}/60 \text{ s})(2\pi \text{ rad}/1 \text{ rev}) = 4.7 \times 10^4 \text{ rad/s}, r = 1.0 \times 10^{-3} \text{ m}.$

(a) Using equation (8.9) we have,

$$v_T = r\omega = (1.0 \times 10^{-3} \text{ m})(4.7 \times 10^4 \text{ rad/s}) = \mathbf{47 \text{ m/s}}.$$

(b) Using equation (8.11) we have

$$a_c = r\omega^2 = (1.0 \times 10^{-3} \text{ m})(4.7 \times 10^4 \text{ rad/s})^2 = 2.2 \times 10^6 \text{ m/s}^2.$$

To obtain the acceleration in terms of gravity, divide a_c by 9.8 m/s^2 to get

$$a_c = \mathbf{2.3 \times 10^5 \text{ g}}.$$

---

4. $\omega = (45 \text{ rev/min})(2\pi \text{ rad}/1 \text{ rev})(1 \text{ min}/60 \text{ s}) = 4.7 \text{ rad/s}, r = 0.10 \text{ m}.$

Using equation (8.11) we can find the centripetal acceleration,

$$a_c = r\omega^2 = (0.10 \text{ m})(4.7 \text{ rad/s})^2 = \mathbf{2.2 \text{ m/s}^2}.$$

The tangential velocity can be obtained using equation (8.9),

$$v_T = r\omega = (0.10 \text{ m})(4.7 \text{ rad/s}) = \mathbf{0.47 \text{ m/s}}.$$

---

5. The centripetal acceleration is used to simulate gravity. Using equation (8.11): $a_c = r\omega^2 = 9.8\ m/s^2$ we can solve for ω, i.e,

$$\omega = \sqrt{(a_c/r)} = \sqrt{(9.8\ m/s^2/480\ m)} = 0.14\ rad/s = \mathbf{1.3\ rev/min.}$$

At r = 410 m,

$$a_c = r\omega^2 = (410\ m)(0.14\ rad/s)^2 = 8.0\ m/s^2.$$

The fractional weight is then

$$(8.0\ m/s^2)/(9.8\ m/s^2) = \mathbf{0.82}.$$

---

6. (a) Using equation (8.12) we have,

$$\omega = v/r = (30.0\ m/s)/(0.375\ m) = \mathbf{8.00 \times 10^1\ rad/s}.$$

(b) Using $\omega_0 = 80.0$ rad/s, t = 10.0 s, $\omega = 0$ and equation (8.4) we have,

$$\alpha = (\omega - \omega_0)/t = \mathbf{-\ 8.00\ rad/s^2}.$$

(c) Using equation (8.6),

$$\theta = (1/2)(\omega_0 + \omega)t = (1/2)(80.0\ rad/s)(10.0\ s) = 4.00 \times 10^2\ rad = \mathbf{63.7\ rev.}$$

---

7. In this situation, the tangential acceleration at the rim of each wheel must be the same. Using equation (8.10)

$$a_T = r_A\alpha_A = r_B\alpha_B.$$

Therefore,

$$\alpha_B = \alpha_A(r_A/r_B) = (2.0\ rad/s^2)(75\ cm/25\ cm) = 6.0\ rad/s^2.$$

We can now use equation (8.4) with $\omega_0 = 0$ and $\omega = 1200$ rad/s, to find

$$t = (\omega - \omega_0)/\alpha = (1200\ rad/s)/(6.0\ rad/s^2) = \mathbf{2.0 \times 10^2\ s.}$$

---

8. (a) Use $\omega_0 = 9.0$ rad/s, $\alpha = -3.0\ rad/s^2$, $\omega = 0$ and equation (8.8): $\omega^2 = \omega_0^2 + 2\alpha\theta$. Solving for θ we get,

$$\theta = -\omega_0^2/2\alpha = -(9.0\ rad/s)^2/[2(-3.0\ rad/s^2) = 13.5\ rad(1\ rev/2\pi\ rad) = \mathbf{2.1\ rev.}$$

(b) Using equation (8.1),

$$s = r\theta = (0.50\ ft)(13.5\ rad) = \mathbf{6.8\ ft.}$$

(c) Using $\theta = -\ 13.5$ rad, $\omega_0 = 0$, and $\alpha = -\ 3.0\ rad/s^2$, equation (8.7) gives for t,

$$t = \sqrt{(2\theta/\alpha)} = \mathbf{3.0\ s.}$$

Quiz answers

1. c
2. a
3. d
4. a
5. a
6. c
7. d
8. b
9. c

MCAT REVIEW PROBLEMS

Angular speed, ω, is the rate at which an object rotates. Although usually measured in radians per second, ω can also be expressed in degrees per second. For instance, consider a merry-go-round. Figure 1 shows its angular speed as a function of time, t. In this graph, a positive value of ω indicates counterclockwise rotation.

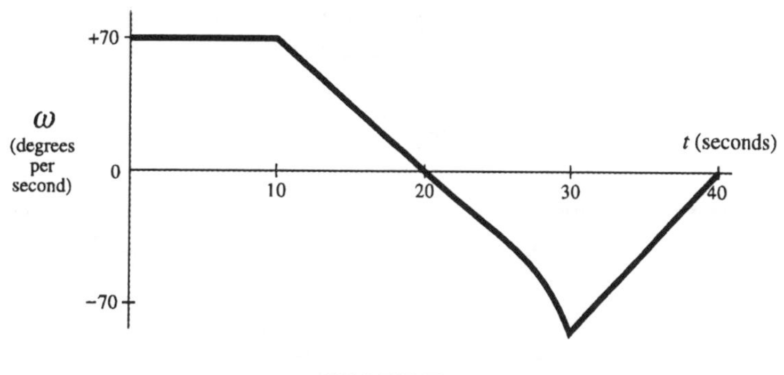

FIGURE 1

A "complete revolution" occurs when the merry-go-round turns through one full circle.

1. Between $t = 0$ and $t = 10$ s, about how many revolutions does the merry-go-round complete?

   A. 1 **B.** 2 **C.** 3 **D.** 4

2. Between $t = 10$ s and $t = 20$ s, the merry-go-round:

   A. rotates clockwise, at a constant rate.
   B. rotates clockwise, and slows down.
   C. rotates counterclockwise, at a constant rate.
   D. rotates counterclockwise, and slows down.

3. Between $t = 30$ s and $t = 40$ s, the merry-go-round:

   A. rotates clockwise, at a constant rate.
   B. rotates clockwise, and slows down.
   C. rotates counterclockwise, at a constant rate.
   D. rotates counterclockwise, and slows down.

4. Consider two children riding on the merry-go-round. Child 1 sits near the edge, as shown in figure 2. Child 2 sits closer to the center.

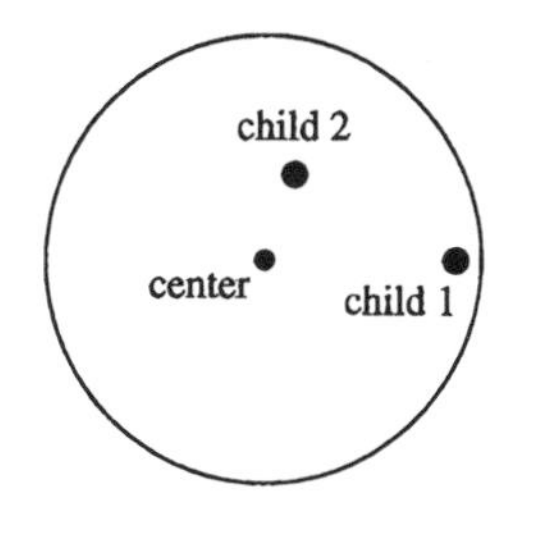

FIGURE 2
Top view of merry-go-round

Let v_1 and v_2 denote the linear speed of child 1 and child 2, respectively. Which of the following is true?

A. $v_1 > v_2$
B. $v_1 = v_2$
C. $v_1 < v_2$
D. We cannot determine which is true, without more information.

---

5. The scalar angular acceleration is the rate of change of the angular speed with time. Which graph best represents the merry-go-round's scalar angular acceleration between $t = 0$ and $t = 20$ s?

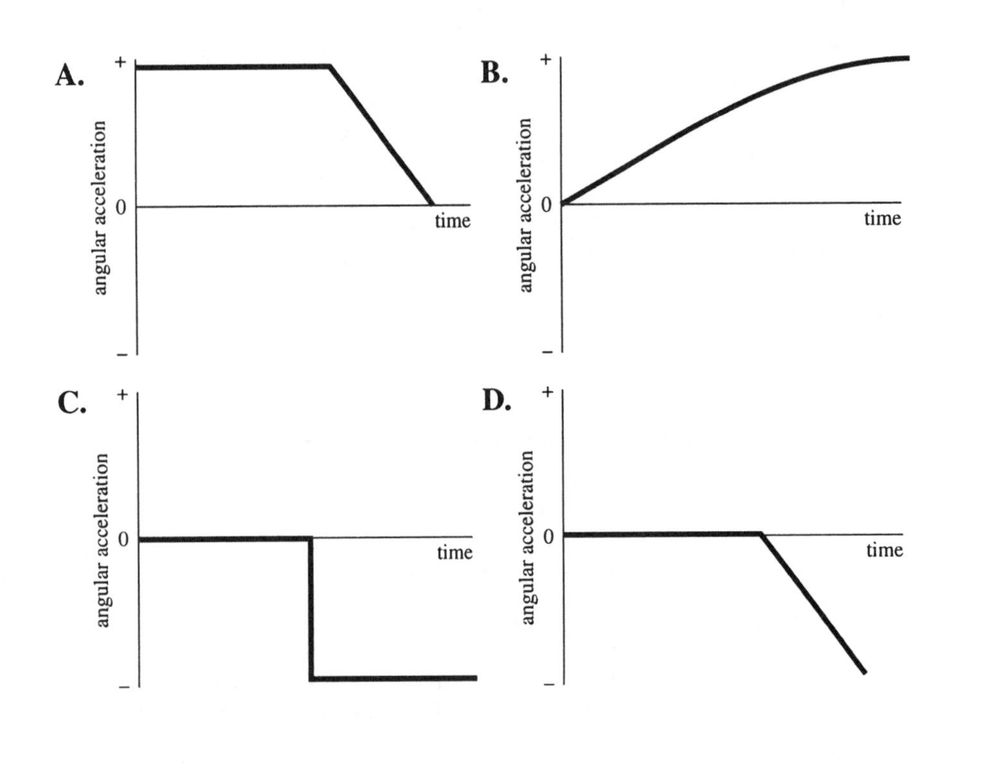

ANSWERS TO MCAT REVIEW PROBLEMS

---

1. **B.** Between $t = 0$ and $t = 10$ s, the merry-go-round rotates at a constant rate of 70 degrees per second. Therefore, it spins through

$$70 \frac{\text{degrees}}{\text{second}} \times (10 \text{ seconds}) = 700 \text{ degrees}.$$

Since 1 revolution = 360 degrees, 2 revolutions correspond to 720°. So, the merry-go-round completes slightly less than 2 revolutions.

---

2. **D.** Don't get distracted by the negative slope, which indicates a negative angular *acceleration.* Between t = 10 s and $t = 20$ s, ω remains positive. In other words, the rotation remains counterclockwise. But ω gets smaller and smaller, indicating that the merry-go-round slows down.

You've seen this before, in velocity vs. time graphs. When a velocity graph "comes down" while remaining positive, it doesn't mean the object travels backwards. It means the object continues traveling forward, but at a slower and slower rate—like a car when the driver hits the brakes.

---

3. **B.** Between $t = 30$ s and $t = 40$ s, ω is negative. So, the merry-go-round spins "backwards," i.e., clockwise. But the magnitude (absolute value) of ω decreases, until reaching zero. For instance, at $t = 30$ s, the merry-go-round spins clockwise at 100 degrees per second. At $t = 35$ s, it spins clockwise at only 50 degrees per second. As time passes, the merry-go-round continues to slow down, until stopping at $t = 40$ s.

---

4. **A.** Although both children share the same *angular* speed, child 1 has more linear speed. To see why, consider one complete revolution of the merry-go-round. During that time, both children sweep through 360°. So, they both cover the same *angle* per time. But during that revolution, child 1 traces out a bigger circle. He covers more *distance* per time than child 2 does. And linear speed *is* distance per time.

---

5. **C.** Solving this problem is just like figuring out an acceleration graph from a velocity graph. Between $t = 0$ and $t = 10$ s, the merry-go-round spins at a steady rate. Since ω neither increases nor decreases, the rate of change of ω is zero.

By contrast, between $t = 10$ s and $t = 20$ s, the positive angular speed *decreases.* Roughly speaking, the merry-go-round *decelerates*. So, the angular acceleration is negative. Furthermore, since the merry-go-round slows down at a steady rate, the negative acceleration is constant.

CHAPTER 9
Rotational Dynamics

PREVIEW

In this chapter you will learn about the dynamics of rotational motion. The topics include Newton's second law for rotations, moment of inertia, rotational work, rotational kinetic energy, and angular momentum.

QUICK REFERENCE

Important Terms

Axis of rotation
A line of points about which an object rotates.

Line of action
A line drawn colinear with a force.

Lever arm
The shortest distance from a rotation axis to a line of action of a force.

Torque
A vector quantity whose magnitude is the magnitude of the force times the lever arm and whose direction is positive if it tends to cause a counterclockwise rotation and is negative if it tends to cause a clockwise rotation.

Translational motion
The motion of an object in which all points of the object move in parallel paths.

Rotational motion
The motion of a rigid object in which the points of the object move in circles about some axis.

Equilibrium
The state of an object when it has no translational or rotational acceleration. It follows that no net torque and no net force act on an object in equilibrium.

Center of gravity (of a rigid body)
Is the point at which the weight of the body can be assumed to act when calculating the torque due to the weight of the body.

Moment of inertia of an object
Is a rotational quantity which plays the same role as mass in linear motion. The net torque acting on an object is proportional to the angular acceleration of the object. The constant of proportionality is the moment of inertia.

Rotational work
The work done by a torque in rotating an object through an angle.

Rotational kinetic energy
The total kinetic energy of all parts of a rigid object due to their rotational motion about an axis. It is equal to one-half the moment of inertia of the object times the angular speed of the object squared.

Angular momentum (of a rigid object)
Is the product of the moment of inertia of the object and its angular velocity.

Equations

The magnitude of the **torque** about an axis is

$$\tau = F\ell \qquad (9.1)$$

where F is the magnitude of the force and ℓ is the lever arm.

A rigid body is in **equilibrium** when

$$\Sigma F_x = 0 \qquad \Sigma F_y = 0 \qquad (4.9a \text{ and } 4.9b)$$

$$\Sigma\tau = 0 \qquad (9.2)$$

The coordinate of the **center of gravity** of a collection of objects which lie along a line is

$$X_{cg} = \frac{W_1X_1 + W_2X_2 + \ldots}{W_1 + W_2 + \ldots} \qquad (9.3)$$

Newton's second law for rotations

for a single mass $\tau = (mr^2)\alpha$ (9.4)

for a rigid collection of masses $\Sigma\tau = (\Sigma mr^2)\alpha$ (9.5)

$$\Sigma\tau = I\alpha \qquad (9.7)$$

Moment of inertia of a rigid body composed of many masses

$$I = \Sigma mr^2 \qquad (9.6)$$

Rotational Work

$$W_R = \tau\theta \qquad (9.8)$$

Rotational Kinetic Energy of a rigid body

$$\text{Rotational KE} = 1/2\ I\omega^2 \qquad (9.9)$$

Angular Momentum of a rigid body

$$L = I\omega \qquad (9.10)$$

Principles

The **principle of conservation of angular momentum** - the total angular momentum of a system remains constant if the net external torque acting on the system is zero.

DISCUSSION OF SELECTED SECTIONS

9.1 The Effects of Forces and Torques on the Motion of Rigid Objects

Torque is a vector quantity and as such can point in any direction. Many rotations caused by torques can, however, be described as either clockwise (CW) or counterclockwise (CCW) as viewed by someone looking along the axis of rotation. This is why it is convenient to classify torques as positive if they tend to produce a counterclockwise rotation or negative if they tend to produce a clockwise rotation of the object.

In order to calculate a torque due to a force using (9.1) you will need to know the magnitude of the force and the lever arm. The lever arm can usually be found from the geometry of the problem.

Example 1
A person pushes on the edge of a 0.90 m wide door with a horizontal force of 5.0 N acting at an angle of 35° to the plane of the door. What is the torque of the force about the door hinge?

The lever arm is

$$\ell = W \sin 35°$$

so the torque is

$$\tau = F\ell$$

$$\tau = FW \sin 35°$$

$$\tau = (5.0 \text{ N})(0.90 \text{ m}) \sin 35°$$

$$\tau = 2.6 \text{ N m}.$$

The torque is POSITIVE since the door would rotate in a counterclockwise sense if it could.

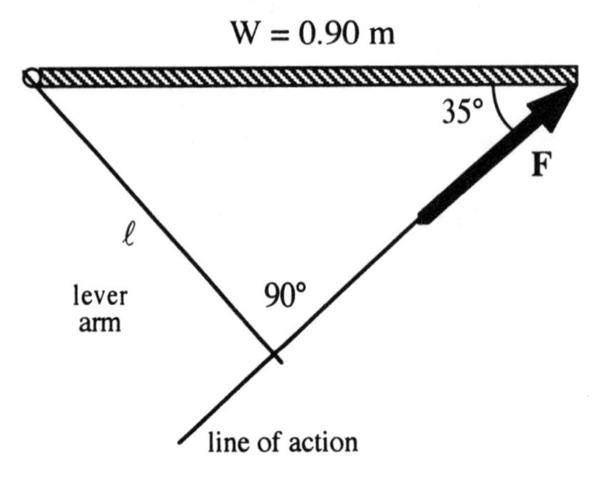

Note that the same result could have been obtained if we had found the component of the force perpendicular to the door and multiplied it by the distance from the hinge to the point where the force acts. That is,

$$\tau = (F \sin 35°)W$$

Sometimes this alternative way of calculating torques is more convenient than finding the lever arm.

9.2 Rigid Objects in Equilibrium

In solving problems involving rigid objects in equilibrium, it is usually necessary to generate a number of equations to be solved simultaneously. These equations can be derived by an application of Newton's second law for translation and rotation. That is, $\Sigma F_x = 0$, $\Sigma F_y = 0$ and $\Sigma\tau = 0$. The force equations are valid only in perpendicular directions and yield two equations at most. The torque equation, however, is valid about ANY axis and will yield at least one useful equation.

Often a judicious choice of rotation axis will result in a simple torque equation which will allow you to solve for one unknown immediately. As a general rule, try to write torque equations about axes which will eliminate torques due to any forces which are not of interest.

Example 2
A clothesline pole requires a guy wire to keep it in from being pulled over by the tension in the line. If the guy wire makes an angle of 45° to the horizontal and is attached 1.5 m above the ground, and the tension in the line is 550 N and it is attached 2.0 m above the ground, how much tension is required in the guy wire?

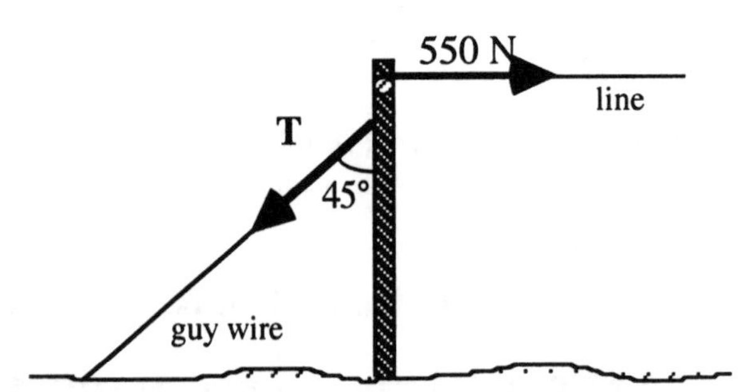

At first glance, this problem appears to require only a simple application of Newton's second law in the horizontal direction. However, the ground exerts forces on the pole which are completely unknown. A choice of axis through the point where the pole contacts the ground will eliminate these forces from the torque equation and will allow us to solve for the desired tension in the guy wire.

As shown by the free-body diagram, the lever arm of the tension in the line is simply 2.0 m and the lever arm of the torque due to the tension in the guy wire, T, is $\ell = (1.5 \text{ m}) \sin 45°$.

The torque equation is then

$$\Sigma\tau = 0$$

$$\Sigma\tau = +T(1.5\text{ m})\sin 45° - (550\text{ N})(2.0\text{ m}) = 0$$

$$T = 1.0 \times 10^3\text{ N}$$

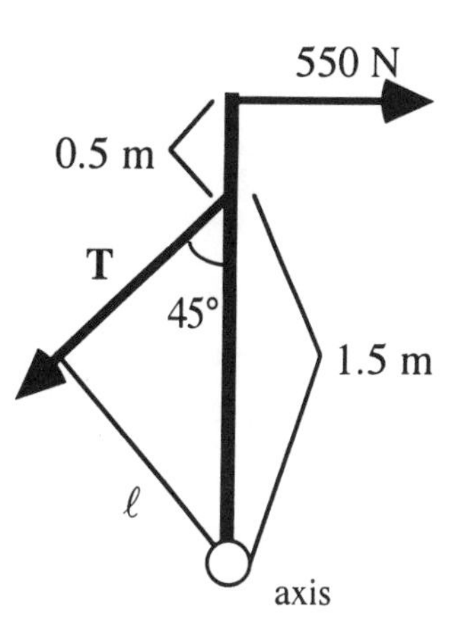

Example 3

Find the forces exerted by the ground on the pole in the previous example, if the pole has a weight of 120 N.

Newton's second law applied to the horizontal forces acting on the pole gives

$$F_h + 550\text{ N} - (1.0 \times 10^3\text{ N})\sin 45° = 0$$

$$F_h = 160\text{ N}$$

A similar application in the vertical direction gives

$$F_v - 120\text{ N} - (1.0 \times 10^3\text{ N})\cos 45° = 0$$

$$F_v = 830\text{ N}$$

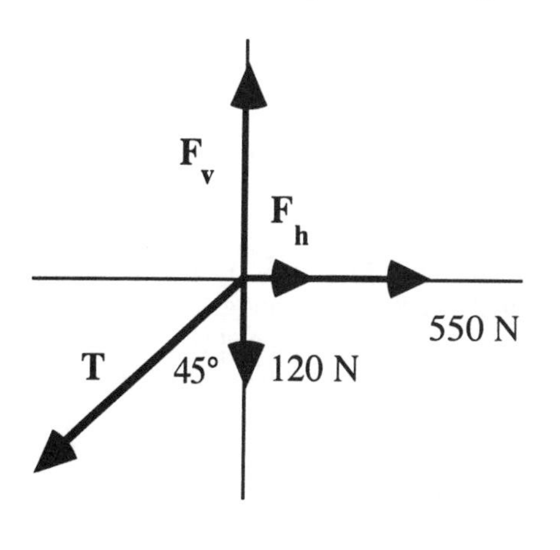

Example 4

A scaffold is constructed by placing two supports located a distance of 12 ft apart under a 16 ft board. Two painters stand on the board; one weighs 150 lb and stands 4.0 ft from one end of the board. Where must the other 180 lb painter stand so that each support carries the same weight? Neglect the weight of the board.

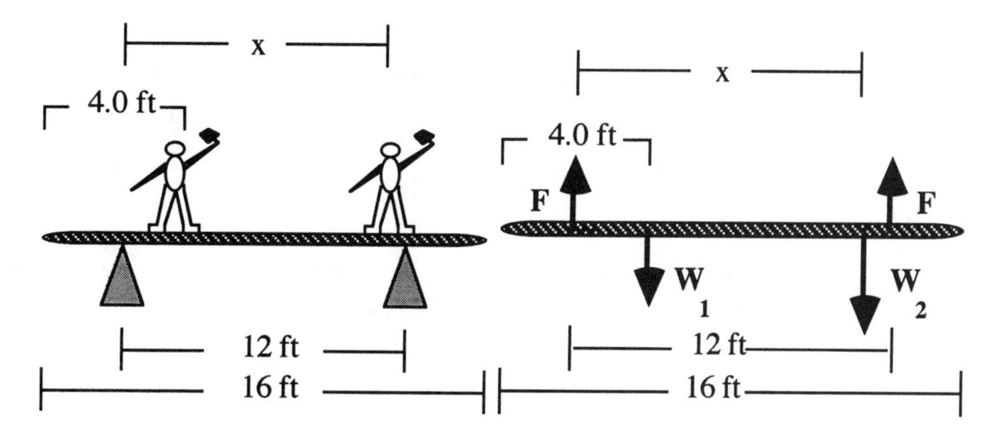

A torque equation written for an axis through the contact point between the board and the left support gives

$$(12\text{ ft})F - (2.0\text{ ft})W_1 - xW_2 = 0$$

$$x = (12\text{ ft})(F/W_2) - (2.0\text{ ft})(W_1/W_2) \qquad (1)$$

Newton's second law applied in the vertical direction gives

$$2F - W_1 - W_2 = 0$$

$$F = 1/2\,(W_1 + W_2) \qquad (2)$$

Substitution of (2) into (1) yields

$$x = (4.0\text{ ft})(W_1/W_2) + 6.0\text{ ft}$$

$$x = (4.0\text{ ft})(150/180) + 6.0\text{ ft}$$

$$x = 9.3\text{ ft}$$

The 180 lb man should stand 9.3 ft to the right of the left support.

Example 5

A 3.0 m flag pole with a mass of 2.0 kg is attached to a wall by a rigid support and makes an angle of 20.0° to the horizontal as shown. A 12 kg flag is hung so that its weight acts at a point 2.5 m from the support. In order for the flag pole to be in equilibrium, an additional force must be supplied by a slight bending of the pole. If the magnitude of this force, assumed to be perpendicular to the pole, can be at most 175 N, where is it effectively located?

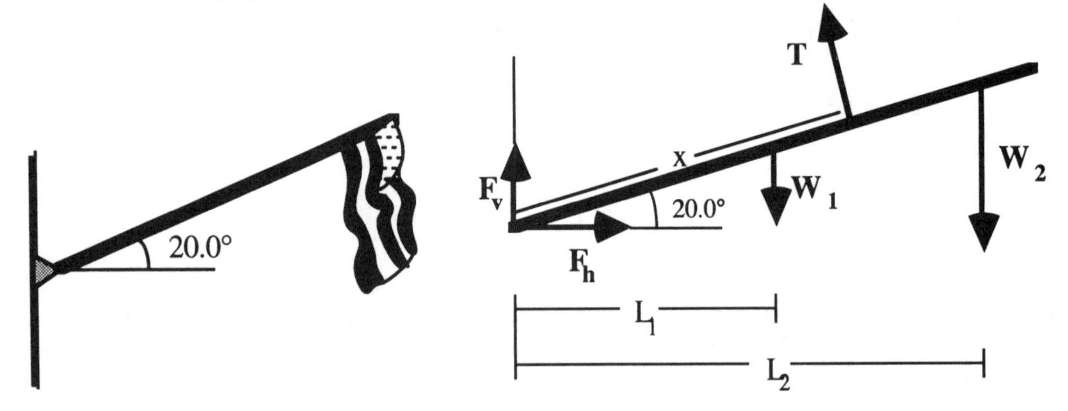

The forces exerted by the wall on the flag pole, $\mathbf{F_v}$ and $\mathbf{F_h}$, are unknown and unwanted. A good place to start is to write a torque equation about an axis through the point of contact of the pole with the wall so that these forces are eliminated.

$$Tx - W_1L_1 - W_2L_2 = 0$$

$$x = \frac{W_1L_1 + W_2L_2}{T} = \frac{(2.0\text{ kg})(9.8\text{ m/s}^2)(1.5\text{ m})\cos 20.0^\circ + (12\text{ kg})(9.8\text{ m/s}^2)(2.5\text{ m})\cos 20.0^\circ}{175\text{ N}}$$

$$x = 1.7\text{ m}$$

9.3 Center of Gravity

The center of gravity of an object is a point where the weight of an object can be assumed to act. The position of the center of gravity of a collection of masses arranged along a line is given by equation (9.3). The center of gravity of a group of masses lying in a plane can be found by applying (9.3) in each of the two directions, x and y.

Example 6

The center of gravity of an empty 1200 kg airplane is located 2.0 m behind the pilot's seat. When loaded the airplane has a 75 kg pilot and a 55 kg passenger in a seat beside the pilot's seat and 95 kg of luggage located 1.0 m behind the pilot's seat. Where along the midline of the airplane is the center of gravity of the loaded airplane?

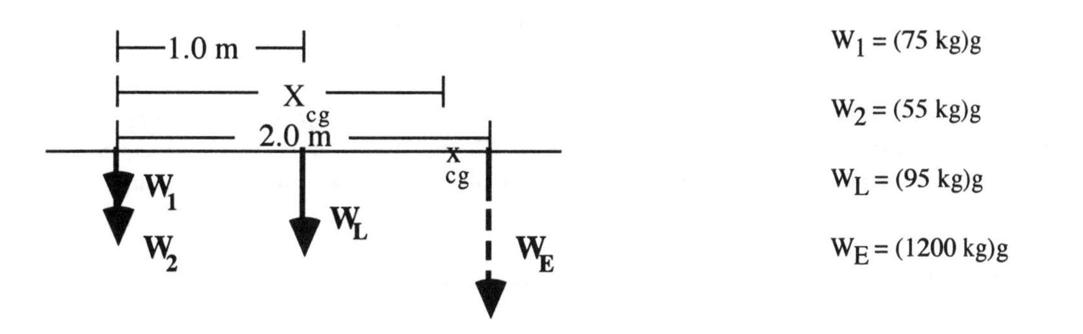

$W_1 = (75\text{ kg})g$

$W_2 = (55\text{ kg})g$

$W_L = (95\text{ kg})g$

$W_E = (1200\text{ kg})g$

Let the pilot's seat be the origin of the coordinate system. Equation (9.3) gives

$$X_{cg} = \frac{(75\text{ kg})(0\text{ m}) + (55\text{ kg})(0\text{ m}) + (95\text{ kg})(1.0\text{ m}) + (1200\text{ kg})(2.0\text{ m})}{75\text{ kg} + 55\text{ kg} + 95\text{ kg} + 1200\text{ kg}}$$

$$X_{cg} = 1.8\text{ m}$$

Note that the common factor, g, appearing in the numerator and denominator was canceled. The center of gravity of the loaded airplane is 1.8 m behind the pilot's seat.

9.4 Newton's Second Law for Rotational Motion About a Fixed Axis

The application of Newton's second law for rotations is very similar to the procedures which you have already learned for using Newton's second law for linear motion. The main difference is that the moment of inertia of the system in question must usually be determined either from a table, such as table 9.1 in your text, or from equation (9.6).

Example 7

Three masses are rigidly located in the x-y plane as shown in the figure. What is the moment of inertia of the system about an axis through the origin and perpendicular to the x-y plane? What is the moment of inertia about the x-axis?

$$I = \Sigma mr^2 = m_1r_1^2 + m_2r_2^2 + m_3r_3^2$$

For an axis through the origin the distances from the axis can be found by using the Pythagorean theorem.

$$I = (2.0\ \text{kg})(13\ \text{m}^2) + (1.5\ \text{kg})(17\ \text{m}^2) + (1.0\ \text{kg})(5.0\ \text{m}^2)$$

$$I = 57\ \text{kg m}^2$$

The moment of inertia about the x-axis is

$$I = (2.0\ \text{kg})(4.0\ \text{m}^2) + (1.5\ \text{kg})(1.0\ \text{m}^2) + (1.0\ \text{kg})(4.0\ \text{m}^2)$$

$$I = 14\ \text{kg m}^2$$

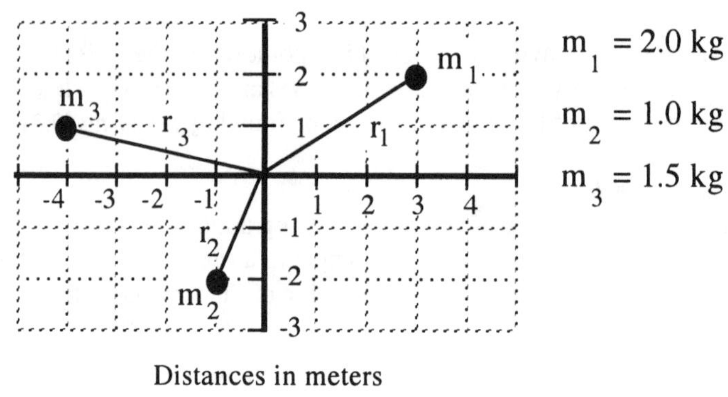

Distances in meters

Example 8

A mechanic is "spin balancing" an automobile tire, radius R = 0.36 m, by placing it in a machine which applies a torque to accelerate the tire to a speed equivalent to traveling 95 km/h. The tire has a mass of 32 kg and has a moment of inertia equivalent to a hoop of radius 0.31 m. How much torque must the machine apply to the tire to bring it up to speed in 15 s?

Newton's second law for rotations requires that $\Sigma\tau = I\alpha$. The angular acceleration, α, can be found from rotational kinematics.

$$\alpha = \frac{\omega - \omega_0}{t} = \frac{\omega}{t}$$

A point on the rim of the tire mounted on a car traveling at a speed, v, will move about the axle with a speed v. The angular velocity of the wheel is then $\omega = v/R$ so

$$\alpha = \frac{v}{Rt}$$

The moment of inertia of the equivalent hoop is $I = Mr^2$, so the torque needed to accelerate the tire is

$$\Sigma\tau = (Mr^2)\left(\frac{v}{Rt}\right) = \frac{Mr^2 v}{Rt}$$

$$\Sigma\tau = \frac{(32\ \text{kg})(0.31\ \text{m})^2(95\ \text{km/h})(10^3\ \text{m/1 km})(1\ \text{h/3600 s})}{(0.36\ \text{m})(15\ \text{s})} = 15\ \text{N m}$$

9.5 Rotational Work and Energy

Example 9

It has been suggested that an automobile can be powered for short trips by removing energy from a spinning flywheel which has been "charged" overnight by a small electric motor. How much energy can be stored in a solid cylindrical flywheel of radius 0.5 m and mass 50 kg rotating 10 000 rpm? If the car normally requires an average of 50 hp, how long could it run if it can extract 50 % of the energy from the flywheel?

The kinetic energy of rotation of the flywheel is $KE_r = 1/2\ I\omega^2$ where $I = 1/2\ MR^2$ so

$$KE_r = 1/4\ MR^2\omega^2$$

$$KE_r = (1/4)(50\text{ kg})(0.5\text{ m})^2[(1 \text{ X } 10^4\text{ rev/min})(1\text{ min/60 s})(2\pi\text{ rev}^{-1})]^2$$

$$KE_r = 3 \text{ X } 10^8\text{ J}$$

The car has 0.5 KE_r energy available for doing work, so it can run a time t = W/P = 0.5 KE_r/P. Hence,

$$t = \frac{(0.5)(3 \text{ X } 10^8\text{ J})}{(50\text{ hp})(746\text{ W/1 hp})} = 4 \text{ X } 10^3\text{ s} = 1\text{ h}.$$

9.6 Angular Momentum

The angular momentum of a system of objects is conserved if no net external torque acts on the system.

Example 10
A centrifugal clutch consists of an outer cylindrical shell mounted on a shaft and an independent inner mechanism mounted on a shaft colinear with the shell's. As the angular speed of the inner mechanism increases, it expands ultimately engaging the outer shell. After a short time the two pieces rotate with the same angular speed. Assume that the outer shell has a moment of inertia one-half that of the inner mechanism and is initially at rest while the inner mechanism is rotating with an angular speed of 150 rpm. What is the final angular speed of the clutch after the parts engage? What fraction of the kinetic energy of the clutch remains after it is engaged?

No external torques act on the clutch so angular momentum is conserved.

$$L_f = L_0$$

$$I_1\omega + I_2\omega = I_1\omega_1$$

$$I_1\omega + 1/2\, I_1\omega = I_1\omega_1$$

$$\omega = 2/3\, \omega_1 = (2/3)(150\text{ rpm}) = 1.0 \text{ X } 10^2\text{ rpm}$$

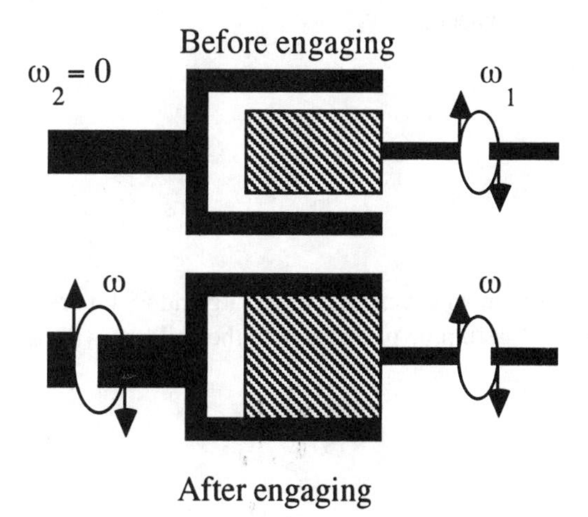

The rotational kinetic energy before the parts engage is $KE_0 = 1/2\, I_1\omega_1^2$ and after they engage it is $KE_f = 1/2\,(I_1 + I_2)\omega^2$. The fraction remaining is

$$\frac{KE_f}{KE_0} = \frac{(I_1 + I_2)\omega^2}{I_1\omega_1^2}$$

Using $I_2 = 1/2\, I_1$ and $\omega = 2/3\, \omega_1$ yields

$$\frac{KE_f}{KE_0} = \frac{2}{3}$$

PRACTICE PROBLEMS

1. The crank handle on an ice cream freezer is 0.20 m from the shaft. At some point in making ice cream a boy must exert a force of 50 N to turn the crank. How much torque does the boy apply about an axis through the shaft?

2. A bolt on an automobile engine is required to be "torqued" to 95 ft lb. A mechanic wants to use a wrench to torque the bolt to this specification without having to exert more than 55 lb of force. How long should the wrench handle be?

3. A 1500 kg car is driven onto a pair of ramps so that its front wheels have been raised 0.30 m above the rear wheels. The wheel base of the car is 2.4 m and its center of gravity is located at the midpoint between the front and rear wheels. How much of the car's weight is supported by the front wheels? How much weight is supported by the rear wheels?

4. Two children whose masses are 35 kg and 55 kg are on a playground "teeter totter" which is 3.0 m long. Where must the fulcrum be placed so that the children will balance?

5. An airplane lands with a speed of 150 km/h. Before landing the airplane's wheels are not rotating. The ground exerts a constant normal force of 12 000 N on each wheel. The coefficient of kinetic friction between the runway and a tire is 0.60. A wheel has a moment of inertia of 2.9 kg m^2 and a radius of 0.15 m. How long is the skid mark that the tire leaves on the runway?

6. A 55 kg ice skater is traveling with a speed of 2.5 m/s in a circle of radius 4.0 m. The skater is held in the circle by a rope held by a friend standing at the center of the circle. If the friend slowly pulls in 1.5 m of the rope, what is the resulting speed of the skater? Neglect friction between the ice and the skater's skates.

7. Compare the initial and final kinetic energies of the skater in practice problem 6. Explain why they are not equal.

8. A bowling ball encounters a 2.50 ft vertical rise on the way back to the ball rack. Neglect frictional losses and assume that the mass of the ball is distributed uniformly. If the linear speed of the ball is 11.5 ft/s at the bottom of the rise, find the linear speed of the ball at the top.

9. A rifleman shoots a 35 g bullet at a 1.0 m high target which is a uniform 0.60 m X 0.60 board with a mass of 9.0 kg hinged at the bottom. If the bullet enters the target at the top with a speed of 250 m/s, what is the speed of the top of the target when the target is horizontal?

HELPFUL SUGGESTIONS

1. An alternative method of finding torques is to draw a line from the rotation axis to the point where the force acts, find the component of the force perpendicular to this line and multiply by the length of the line. The result is entirely equivalent to finding the lever arm and multiplying by the magnitude of the force. Learn to use the method that is most convenient for a particular situation.

2. If a body is in equilibrium, $\Sigma\tau = 0$ MUST be true for ANY axis of rotation. Give some thought to which quantities are known, which are asked for, and which are unknown and unwanted. Choose the axis so that the torque equation will contain quantities that are known or asked for and will not contain unwanted quantities.

3. The center of gravity of an object is determined by the distribution of the mass of the object and may not be located within the object at all. A quick estimate of the location of the center of gravity of a complicated object can be made by breaking the object into simpler parts, estimating the center of gravity of each, treating each part as a point mass located at its center of gravity, and finding the center of gravity of the point masses.

4. Each physical quantity, equation and law that you learned in studying linear kinematics and dynamics has an analog in rotational motion. Many techniques that you learned previously can be used to treat rotational problems.

5. An object can have both rotational and translational kinetic energy. If no non conservative forces act, then the total mechanical energy must be conserved. The conservation of energy equation now has an extra term on each side due to the rotational kinetic energy. For the case of rolling objects this kinetic energy of rotation is related to the linear kinetic energy of the object through $v = r\omega$.

EVERYDAY PHYSICS

1. The torque necessary to just start a door rotating is a fixed quantity determined by the friction in the hinges. Investigate the relationship between applied force and torque by pushing on a door with your index finger at various distances from the hinge. Is there a place where your push will not start the door moving at all? What happens when you push at an angle other than 90° to the door?

2. Determine the location of the center of gravity of various objects by hanging them by a string as discussed in the text. Also, try finding the center of gravity by balancing the object on your index finger. If the object balances, the center of gravity must be on a vertical line through your finger.

3. A cafeteria trick popular among physics students is to mesh the tines of two forks together to form a "V". Place a wooden match stick through the tines and try to place the match stick on the rim of a glass so that the forks balance. Where is the center of gravity of the forks?

4. A rotating object not constrained to rotate about a particular axis by a shaft, bearing, etc, will rotate about an axis through its center of gravity. Test this by locating the center of gravity of an object such as a stick, wrench or hammer, and mark it with a piece of white tape. Then throw the object in the air so that is rotating. How does the tape move?

5. You can perform a favorite lecture demonstration yourself if you have a rotating chair or stool with good bearings. Sit on the stool while holding some weight in your hands and your arms outstretched to your sides. Have a friend give you a gentle push to start you rotating. Then pull your arms in slowly to decrease your moment of inertia. According to the conservation of angular momentum principle you should increase your rotation rate. BE CAREFUL. If you rotate too fast, then you may become dizzy. Have your friend watch closely to catch you if you should begin to fall off the stool. Also, it is best to wait for a few moments after stopping the rotation before getting up.

CHAPTER QUIZ

1. A person pushes on a door 0.50 m from the hinge with a force of 25 N at an angle of 35° to the door. What is the magnitude of the torque that the person applies about the hinge?
   a. 10.2 Nm b. 7.2 Nm c. 13 Nm d. 8.8 Nm

2. Which of the following has the largest moment of inertia?
   a. A solid cylinder of mass, M, and radius, R, about the cylinder axis.
   b. A hollow cylinder of mass, M, and radius, R, about the cylinder axis.
   c. A solid sphere of mass, M, and radius, R, about an axis through its center.
   d. A hollow sphere of mass, M, and radius, R, about an axis through its center.

3. A 5.0 kg object and a 3.0 kg object are separated by a distance of 1.0 m. Where is the center of gravity of the system?
   a. Between the objects 3/5 m from the 5.0 kg object.
   b. Between the objects 5/8 m from the 5.0 kg object.
   c. 3/8 m from the 2.0 kg object and 11/8 m from the 5.0 kg object.
   d. Between the objects and 3/8 m from the 5.0 kg object.

4. A 75 kg man uses a lever with arms of 0.5 m and 3.0 m. What is the most massive object that he can lift with the lever?
   a. 75 kg b. 450 kg c. 260 kg d. 190 kg

5. A solid cylinder starts from rest and rolls without slipping down a ramp. What is the ratio of its rotational kinetic energy to its translational kinetic energy?
   a. 1:2 b. 1:4 c. 1:3 d. 2:1

6. A turntable with a moment of inertia, $I = 0.050$ kg m^2, requires 30 s to stop after the motor is shut off when playing at 33 1/3 rpm. What is the average torque exerted on the turntable in stopping it?
   a. 0.06 N m b. 0.12 N m c. 0.03 N m d. 0.006 N m

7. A spinning ice skater stretches out her arms. Her moment of inertia _______ and her angular speed _______.
   a. increases, decreases
   b. increases, increases
   c. decreases, decreases
   d. decreases, increases

8. The earth is actually in an elliptical orbit about the sun. As it moves closer to the sun its linear speed increases. What happens to its angular momentum and its angular kinetic energy?
   a. Its angular momentum increases while its angular kinetic energy decreases.
   b. Its angular momentum remains constant, while its angular kinetic energy increases.
   c. Its angular momentum decreases as does its angular kinetic energy.
   d. Its angular momentum and its angular kinetic energy remains constant.

9. The center of gravity and the center of mass of an object coincide
   a. if the object is uniform.
   b. in all circumstances.
   c. if acceleration of gravity does not vary over the object.
   d. only if both a and b are true.

SOLUTIONS AND ANSWERS

Practice Problems

---

1. Assuming that the boy always exerts the force perpendicular to the crank, the moment arm of the torque is just 0.2 m. Then

$$\tau = F\ell = (50.0\ \text{N})(0.20\ \text{m}) = \mathbf{10.0\ N\ m}.$$

---

2. The mechanic will exert the smallest force if he pushes perpendicular to and at the end of the wrench handle. The lever arm for the torque is then the length of the wrench,

$$\ell = \tau/F = 95\ \text{ft lb}/55\ \text{lb} = \mathbf{1.7\ ft}.$$

---

3. The ramp makes an angle of $\theta = \tan^{-1}(0.30/2.4) = 7.1°$ with the horizontal. The normal force exerted by the ramp on the front wheels is perpendicular to the ramp, while the weight of the car is 7.1° from the perpendicular to the ramp. Taking torques about an axis through the contact point between the rear wheels and the ground gives

$$F_f\,(2.4\ \text{m}) - (1500\ \text{kg})(9.80\ \text{m/s}^2)(1/2)(2.4\ \text{m})\cos 7.1° = 0$$

$$F_f = \mathbf{7300\ N}.$$

The rest of the weight of the car is supported by the rear wheels so

$$F_r = (1500\ \text{kg})(9.80\ \text{m/s}^2) - 7300\ \text{N} = \mathbf{7400\ N}.$$

---

4. Let x be the distance from the 35 kg child to the fulcrum. The torque equation for an axis through the fulcrum is

$$(35\ \text{kg})x - (55\ \text{kg})(3.0\ \text{m} - x) = 0.$$

Solving for x gives

$$x = \mathbf{1.8\ m}.$$

---

5. The wheel is accelerated by the torque provided by friction, $\tau = \mu F_N R = I\alpha$. The wheel will skid on the runway until its angular speed is $\omega = v/R$ where v is the speed of the airplane and R is the radius of the wheel. The wheel skids for a time, $t = \omega/\alpha$. During this time it has moved a distance, $x = vt = v\omega/\alpha$ which is the length of the skid mark. Now

$$x = (v^2/R^2)(I/\mu F_N) = \mathbf{31\ m}.$$

---

6. If friction is ignored, the angular momentum of the ice skater is conserved since no external torque acts on him. $I_f\omega_f = I_0\omega_0$. The skater's initial and final paths are circular so $\omega_f = v_f/R_f$ and $\omega_0 = v_0/R_0$. The initial and final moments of inertia are $I_f = mR_f^2$ and $I_0 = mR_0^2$. Using these in the conservation of angular momentum equation results in

$$v_f = (R_0/R_f)v_0 = (4.0/1.5)(2.5\ \text{m/s}) = \mathbf{6.7\ m/s}.$$

---

7. The initial kinetic energy of the skater is $KE_0 = 1/2\ mv_0^2$ and the final kinetic energy of the skater is

$$KE_f = 1/2\ mv_f^2 \text{ so } KE_f/KE_0 = (v_f/v_0)^2 = (6.7/2.5)^2 = \mathbf{7.2}.$$

The final kinetic energy of the skater is larger than the initial kinetic energy because the skater's friend did work in pulling the skater into a smaller circle.

---

8. The total mechanical energy of the ball is conserved.

$$1/2\ mv_f^2 + 1/2\ I\omega_f^2 + mgh = 1/2\ mv_0^2 + 1/2\ I\omega_0^2$$

If the ball rolls without slipping, then $\omega_0 = v_0/R$ and $\omega_f = v_f/R$. The moment of inertia of the solid ball is $I = 2/5\ mR^2$. Using these in the conservation of energy equation and simplifying yields

$$v_f^2 = v_0^2 - (10/7)\ gh = (11.5\ ft/s)^2 - (10/7)(32.2\ ft/s^2)(2.5\ ft)$$

so

$$v_f = \mathbf{4.15\ ft/s.}$$

---

9. Total angular momentum is conserved during the time required for the bullet to imbed in the target. $(I + ms^2)\omega_0 = msv$ where s is the length of a side of the target, m is the mass of the bullet, v is the initial speed of the bullet, ω_0 is the resulting angular speed of the bullet and target system, and I is the moment of inertia of the target. The moment of inertia of the target is $I = 1/3\ Ms^2$ which is much larger than ms^2, so the initial angular speed is approximately $\omega_0 = 3(m/M)(v/s)$. After the bullet imbeds in the target, the total mechanical energy of the system is conserved.

$$1/2\ (I + ms^2)\omega_f^2 = 1/2\ (I + ms^2)\omega_0^2 + Mg(1/2\ s)$$

where ω_f is the angular speed of the target as it passed through the horizontal and M is the mass of the target. The mass of the bullet has been neglected since it is much less than the mass of the target. The final angular speed of the target is given by

$$\omega_f^2 = \omega_0^2 + Mgs/I = 9(m/M)^2(v/s)^2 + 3g/s$$

so that

$$\omega_f = 8.5\ rad/s$$

Now the speed of the top of the target is

$$v_f = s\omega_f = \mathbf{5.1\ m/s.}$$

Quiz answers

1. b
2. b
3. d
4. b
5. a
6. d
7. a
8. b
9. c

MCAT REVIEW PROBLEMS

In figure 1, the winch is mounted on an axle, and the 6-sided nut is welded to the winch. By turning the nut with a wrench, a person can rotate the winch. For instance, turning the nut clockwise lifts the block off the ground, because more and more rope gets wrapped around the winch.

Three students agree that using a longer wrench makes it easier to turn the winch. But they disagree about why. All three students are talking about the case where the winch is used, over a 10-second time interval, to lift the block one meter off the ground.

Student 1

By using a longer wrench, the person decreases the average *force* he must exert on the wrench, in order to lift the block one meter in ten seconds.

Student 2

Using a longer wrench reduces the *work* done by the person as he uses the winch to lift the block one meter in ten seconds.

Student 3

Using a longer wrench reduces the *power* that the person must exert to lift the block one meter in ten seconds.

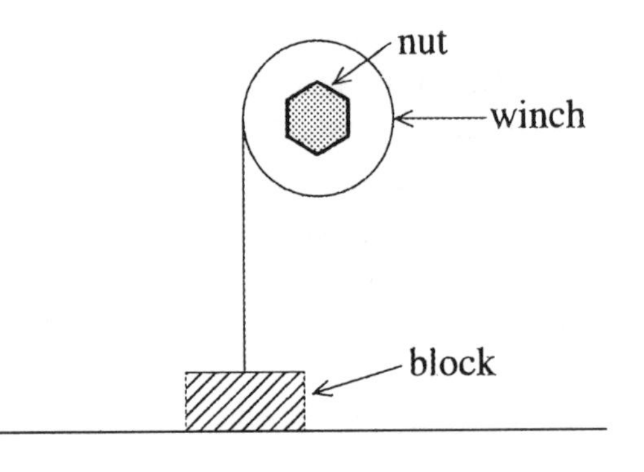

FIGURE 1

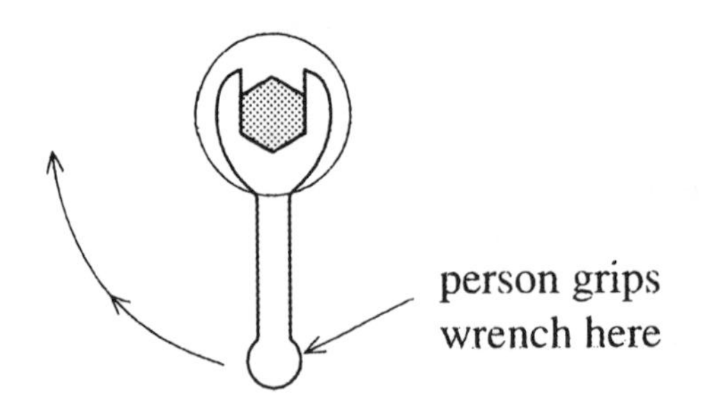

FIGURE 2
Wrench turns winch clockwise

1. Student 1 is:

   A. Correct, because the torque that the wrench must exert to lift the block doesn't depend on the wrench's length.
   B. Correct, because using a longer wrench decreases the torque it must exert on the winch.
   C. Incorrect, because the torque that the wrench must exert to lift the block doesn't depend on the wrench's length.
   D. Incorrect, because using a longer wrench decreases the torque it must exert on the winch.

---

2. Which of the following is true about students 2 and 3?

   A. Students 2 and 3 are both correct.
   B. Student 2 is correct, but student 3 is incorrect.
   C. Student 3 is correct, but student 2 is incorrect.
   D. Students 2 and 3 are both incorrect.

---

3. If several wrenches all apply the *same* torque to a nut, which graph best expresses the relationship between the force the person must apply to the wrench, and the length of the wrench?

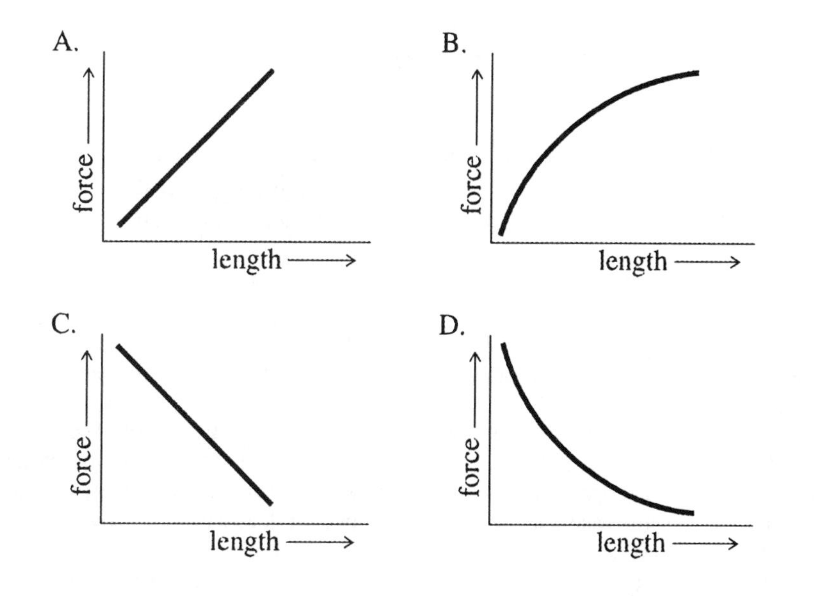

ANSWERS TO MCAT REVIEW PROBLEMS

---

1. **A.** While rising, the block pulls down on the rope, which therefore exerts a torque on the winch. This torque "tries" to spin the winch counterclockwise. In order to keep the block rising, the wrench's torque on the winch—and hence, the person's torque on the wrench—must counterbalance the rope's torque on the winch. For instance, if the rope exerts a 4 N·m torque on the winch, then the person must exert at least a 4 N·m torque on the wrench. This is true no matter what wrench the person chooses. Using a longer wrench makes it easier for the person to create this torque, but it doesn't change the size of the torque he must create.

   Let us review why using a longer wrench makes it easier to apply a fixed torque. Recall the relationship between force and torque:

$$\tau = Fl,$$

where l denotes the lever arm, i.e., the distance from the force's "line of action" to the axis of rotation. If the person pushes perpendicular to the wrench, as drawn in figure 3, then l is simply the wrench's length.

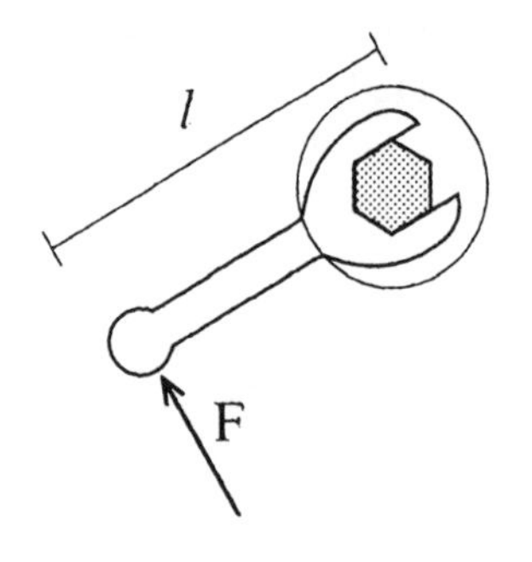

FIGURE 3

Therefore, since $\tau = Fl$, lengthening the wrench (increasing l) decreases the force needed to generate a fixed torque. For instance, suppose the person must produce a 4 N·m torque. If the wrench is .1 m long, then he must apply a 40 N force. But if the wrench is .2 m long, he needs only a 20 N force.

---

2. **D.** Using two different arguments, we can see why the work performed by the person does not depend on the wrench's length.

   Potential energy. Here, all of the person's work gets "devoted" to lifting the block. In other words, every joule of work done by the person gets transferred to the block, in the form of potential energy. As you've seen, the block's potential energy rises by mgh, where m denotes the block's mass and h denotes the height through which it rises, in this case one meter. So, the person does work mgh on the block. This is true no matter which wrench he uses, for reasons explained more intuitively below. . .

   Force and distance. When a force F pushes something a distance s along its direction of motion, the force does work

$$W = Fs.$$

For the person, F denotes the force exerted by his hand on the end of the wrench, while s denotes the distance through which his hand moves. As shown in question 1, using a longer wrench reduces F. But with a longer wrench, the person's hand traces a *bigger* circle while turning the winch all the way around. See figure 4.

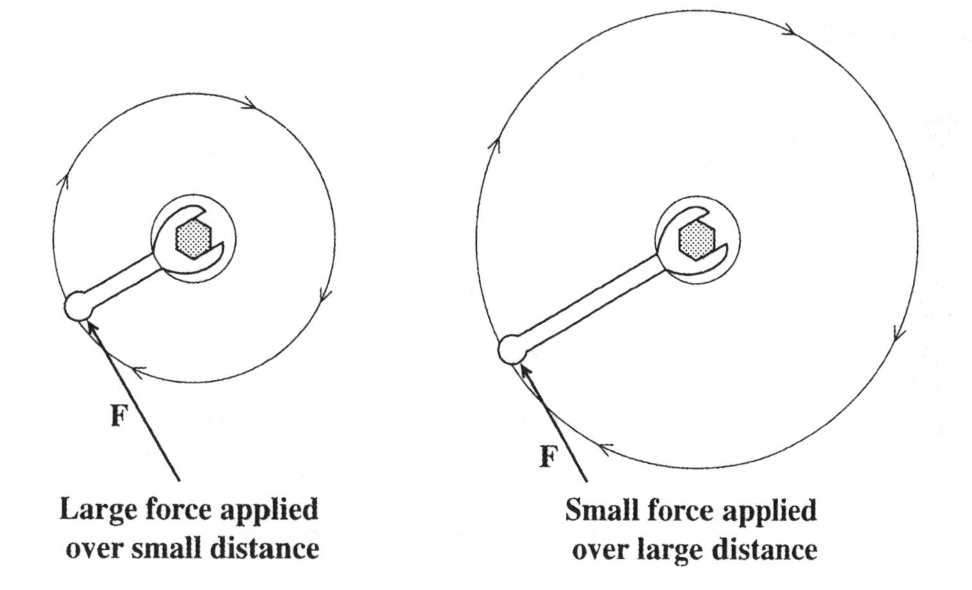

FIGURE 4

When using a shorter wrench, the person exerts a larger force over a smaller distance. When using a longer wrench, he exerts a smaller force over a larger distance. In both cases, the force *times* the distance comes out the same. For instance, doubling the wrench's length cuts the force in half, but doubles the circumference of the circle around which the person moves his hand. Furthermore, to lift the block one meter, the person must turn the winch through a certain fixed number of rotations, no matter which wrench he uses. Therefore, doubling the wrench's length leaves the work ($W = Fs$) unchanged.

What about power? It's work per time. In this scenario, the block rises one meter in ten seconds, no matter which wrench the person uses. So, the choice of wrench has no effect on the rate at which work gets done, i.e., the power.

---

3. **D.** As explained in question 1, for a given torque, F and l are inversely proportional. Mathematically, $\tau = Fl$, and hence,

$$F = \tau / l.$$

Graph D, but not graph C, expresses inverse proportionality. In graph D, doubling l cuts F in half. Tripling l cuts F in third. And so on. By contrast, graph C represents a linear decrease. It corresponds to an equation such as $F = F_0 - cl$, where c denotes the slope of the line. This equation differs from $F = \tau / l$.

CHAPTER 10
Elasticity and Simple Harmonic Motion

PREVIEW

In this chapter you will study the elastic properties of materials and the kinds of deformations a material can experience when subjected to outside forces. A special case, the spring, will be examined in detail. You will also study simple harmonic motion, the oscillatory motion that occurs when the restoring force of an ideal spring acts on an object.

QUICK REFERENCE

Important Terms

Elasticity
The tendency for a solid object to return to its original shape after it has been stretched, compressed, or otherwise distorted by outside forces.

Young's Modulus
The proportionality constant which relates the fractional change in length of a material to the force per unit area applied to the material.

Shear Modulus
The proportionality constant which relates the shear deformation per unit length of a material to the force per unit area applied to the material.

Bulk Modulus
The proportionality constant which relates the fractional change in volume of a material to the pressure applied to the material.

Stress
The magnitude of the force per unit area required to cause an elastic deformation.

Strain
The fractional change in length or volume that results from an applied force.

Hooke's Law
The relationship which states that stress is directly proportional to strain.

Ideal Spring
A spring which behaves according to the Hooke's Law relationship $F = kx$ and which has no internal friction.

Restoring Force
A force which tends to restore an object to its original state. For example, the force which tends to pull a stretched spring back to its unstretched or "equilibrium" position.

Simple Harmonic Motion
The oscillatory motion that results when an elastic material is subjected to the restoring force of an ideal spring.

Period
The time required to complete one oscillatory cycle of motion.

Frequency
The number of cycles of oscillatory motion that occurs in one second.

Amplitude
The maximum distance that an oscillating object moves away from its equilibrium position.

Equations

The equation for **stretching or compression** is:

$$F = Y\left(\frac{\Delta L}{L_0}\right)A \qquad (10.1)$$

The equation for **shear deformation** is:

$$F = S\left(\frac{\Delta X}{L_0}\right)A \qquad (10.2)$$

The **fractional volume change** is given by:

$$\Delta P = -B\left(\frac{\Delta V}{V_0}\right) \qquad (10.4)$$

Hooke's law restoring force for an ideal spring:

$$F = -kx \qquad (10.6)$$

The **displacement** of an object moving with **simple harmonic motion** is:

$$x = A \cos \omega t \qquad (10.7)$$

The **velocity** of an object moving with **simple harmonic motion** is:

$$v = -A\omega \sin \omega t \qquad (10.11)$$

The **acceleration** of an object moving with **simple harmonic motion** is:

$$a = -A\omega^2 \cos \omega t \qquad (10.13)$$

DISCUSSION OF SELECTED SECTIONS

10.1 Elastic Deformation

Refer to the following figure in which an object of length L_0 and cross-sectional area A is subjected to a force, **F**, which pulls (or pushes) along the length of the object (i.e., perpendicular to A). The object will stretch (or compress) by an amount ΔL as shown in the figure.

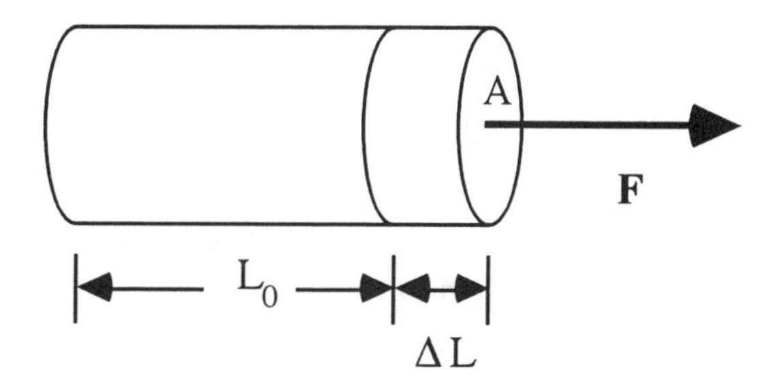

If the amount of stretching is small compared to the original length of the object, the elastic behavior of the material is maintained. We must incorporate a proportionality constant (which depends on the type of material being stretched) which relates the fractional change in length ($\Delta L/L_0$) to the force per unit area (F/A) being applied, i.e.,

$$F = Y\left(\frac{\Delta L}{L_0}\right)A \qquad (10.1)$$

Here, the proportionality constant, Y, is known as **Young's Modulus**. Solving equation (10.1) for Y shows that Young's modulus has units of force per unit area (N/m^2), the exact value depends on the nature of the material.

Example 1

A 25.0 kg mass is suspended from an aluminum wire of length 2.00 m and diameter 1.00 mm. By how much does the wire stretch?

The cross-sectional area of the wire is

$$A = \pi r^2 = \pi(5.00 \times 10^{-4}\ m)^2 = 7.85 \times 10^{-7}\ m^2.$$

The force being applied is the weight of the 25.0 kg mass, i.e.,

$$F = mg = (25.0\ kg)(9.80\ m/s^2) = 245\ N.$$

Using Table 10.1 we look up the value of Young's modulus for aluminum. We find $Y = 6.9 \times 10^{10}\ N/m^2$. The amount the wire stretches is therefore,

$$\Delta L = \frac{FL_0}{YA} = \frac{(245\ N)(2.00\ m)}{(6.9 \times 10^{10}\ N/m^2)(7.85 \times 10^{-7}\ m^2)} = 9.0 \times 10^{-3}\ m = 9.0\ mm.$$

It is possible to deform a solid object in a way other than stretching or compressing it. Suppose a force is applied parallel to the cross-sectional area of the object, as shown in the figure below.

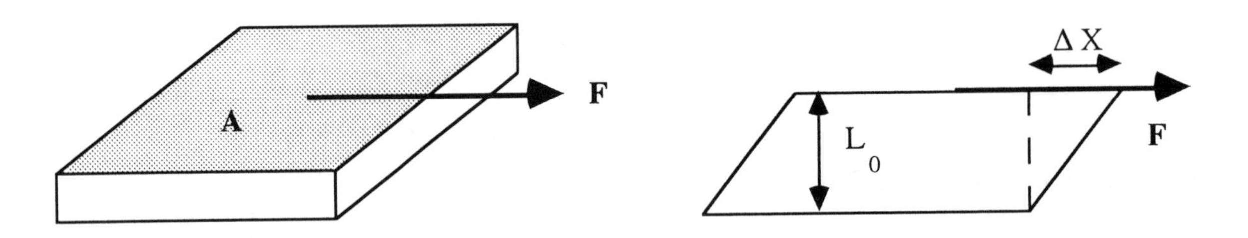

The resulting deformation in the material is called a **shear deformation**. The magnitude of the force **F** needed to produce an amount of shear ΔX for an object of cross-sectional area A and thickness L_0 is given by

$$F = S\left(\frac{\Delta X}{L_0}\right)A \qquad (10.2)$$

The constant of proportionality, S, is known as the **shear modulus**. Like Young's modulus, it has units of force per unit area (N/m^2), and depends on the nature of the material.

Example 2

A lead block of length 0.50 m, width 0.30 m, and height 0.080 m is subjected to a shearing force of 2.0×10^6 N directed perpendicular to its height. The shear modulus for lead (from Table 10.2) is $S = 5.4 \times 10^9$ N/m^2. Find the amount of shear, ΔX.

Using equation (10.2) with A = length x width = (0.50 m)(0.30 m) = 0.15 m^2 and L_0 = height = 0.080 m, we have

$$\Delta X = \frac{FL_0}{SA} = \frac{(2.0 \times 10^6 \text{ N})(0.080 \text{ m})}{(5.4 \times 10^9 \text{ N/m}^2)(0.15 \text{ m}^2)} = 2.0 \times 10^{-4} \text{ m.}$$

When a compressive force is applied to all three dimensions of a solid, the length of each dimension decreases, leading to a decrease in volume for the object. The forces acting in such situations are applied perpendicular to every available surface. Because of this, we speak of the perpendicular force per unit area, rather than the total force involved. The perpendicular force per unit area is called the **pressure.** When such a pressure is applied to a volume V, the volume changes by an amount ΔV, according to the relation

$$\Delta P = -B\left(\frac{\Delta V}{V_0}\right) \qquad (10.4)$$

The proportionality constant, B, is known as the **bulk modulus of the material**. The minus sign appears in equation (10.3) because an increase in pressure (ΔP positive) always results in a decrease in volume (ΔV negative). B is always a positive quantity.

Example 3

A brass sphere of radius 1.50 m is subjected to a pressure increase of 3.0×10^6 N/m^2. What change is volume results?

The volume of the sphere is

$$V = (4/3)\pi r^3 = (4/3)\pi(1.50 \text{ m})^3 = 14.1 \text{ m}^3.$$

The bulk modulus for brass (obtained from Table 10.3) is $B = 6.7 \times 10^{10}$ N/m^2. The change in volume is therefore

$$\Delta V = \frac{-V_0 \Delta P}{B} = \frac{-(14.1 \text{ m}^3)(3.0 \times 10^6 \text{ N/m}^2)}{6.7 \times 10^{10} \text{ N/m}^2} = -6.3 \times 10^{-4} \text{ m}^3$$

10.2 Stress, Strain, and Hooke's Law

The force per unit area required to produce any elastic deformation is referred to as the **stress**. The fractional change in a quantity ($\Delta L/L_0$, $\Delta X/L_0$, or $\Delta V/V_0$) that results when a stress is applied is referred to as the **strain**. In each situation described in section 10.2, the stress was found to be directly proportional to the strain, the constants of proportionality being either Young's modulus, the shear modulus, or the bulk modulus. The relationship between stress and strain is known as **Hooke's Law**. The SI unit of stress is the **pascal** (Pa) where 1 Pa = 1 N/m^2, while the strain is a dimensionless quantity. Materials obey Hooke's law up to a certain limit. If too much stress is applied to a material, it becomes unable to "spring back" to its original size. We would say that the material has exceeded its "elastic limit". Beyond this limit the material can no longer obey Hooke's law.

Example 4

A steel cable is able to withstand a stress of 8.0×10^7 Pa up to its elastic limit. If Young's modulus for steel is $Y = 2.0 \times 10^{11}$ N/m^2, what is the strain?

Hooke's law states that the stress is proportional to the strain. In this case, since the wire is being stretched, the constant of proportionality is Young's modulus. We have, therefore,

$$Y = \frac{\text{Stress}}{\text{Strain}} \quad \Rightarrow \quad \text{Strain} = \frac{\text{Stress}}{Y} = \frac{8.0 \times 10^{7}\ \text{Pa}}{2.0 \times 10^{11}\ \text{N/m}^2} = 4.0 \times 10^{-4}.$$

10.3 The Ideal Spring and Simple Harmonic Motion

Consider the following figure in which an unstretched spring is pulled on by a force **F**. In the process, the spring will be stretched by an amount x from its unstrained or equilibrium position (x = 0).

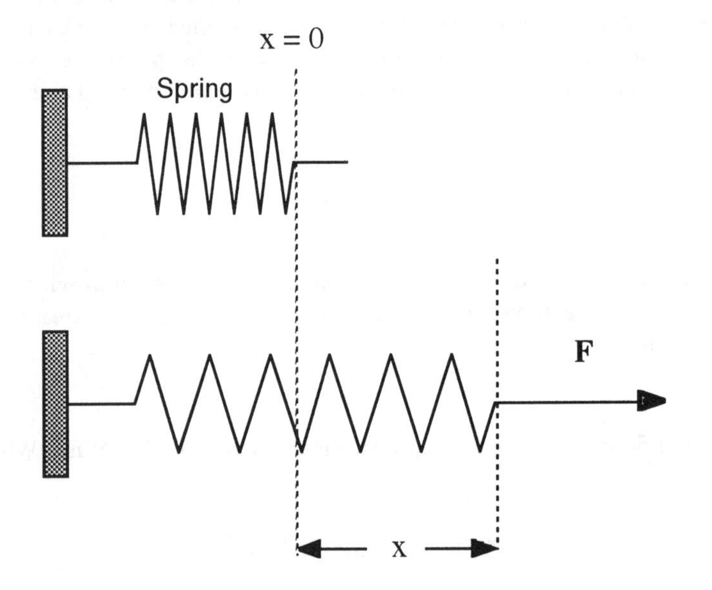

For relatively small deformations, the force required to stretch (or compress) the spring obeys the following equation:

$$F = kx \qquad (10.5)$$

The constant k is called the **spring constant** which is a measure of the stiffness of the spring (units for k are N/m). The larger the value of k, the harder it is to stretch the spring. Any spring that obeys equation (10.5), which is a form of Hooke's law and has no internal friction, is said to be an **ideal spring**.

To stretch or compress a spring, a force must be applied to it. As a consequence of Newton's third law, the spring must exert an oppositely-directed force of equal magnitude. The force of the spring is a "restoring force" in the sense that it tries to pull or push the spring back to its equilibrium (x = 0) position. This restoring force can thus be written

$$F = -kx \qquad (10.6)$$

The minus sign means that the restoring force always points in a direction opposite to the direction of deformation.

Example 5
A spring has a spring constant of 155 N/m. What is the magnitude of the force needed (a) to stretch the spring by 5.00 cm from its unstretched length and (b) to compress the spring by the same amount.

(a) Using equation (10.5) we have:

$$F = kx = (155\ \text{N/m})(5.00 \times 10^{-2}\ \text{m}) = 7.75\ \text{N}.$$

(b) The magnitude of the force required to compress the spring by the same amount (5.00 cm) is EXACTLY the same as that found in part (a), 7.75 N. That is, the same force is needed for stretching or compressing the spring by equal amounts. The difference lies in the direction of the force.

Imagine that we now attach one end of a spring to a wall, and the other end to a block which rests on a horizontal frictionless surface, as shown in the figure below. If we were to pull the block to the right, stretching the spring, and then let go, what would happen? The restoring force, given by equation (10.6), would act to pull the block back to the left. The block would begin to move to the left. However, because of the block's momentum it would not stop when it returned to its equilibrium position. It would continue to move until the spring compressed enough to stop the block. The restoring force of the spring would now push the block back to the right, and again it would begin to move (this time to the right). As it passed through its equilibrium position it would keep moving until stopped by the force of the once again stretched spring. The process would then repeat, over and over again. This is referred to as **Simple Harmonic Motion**.

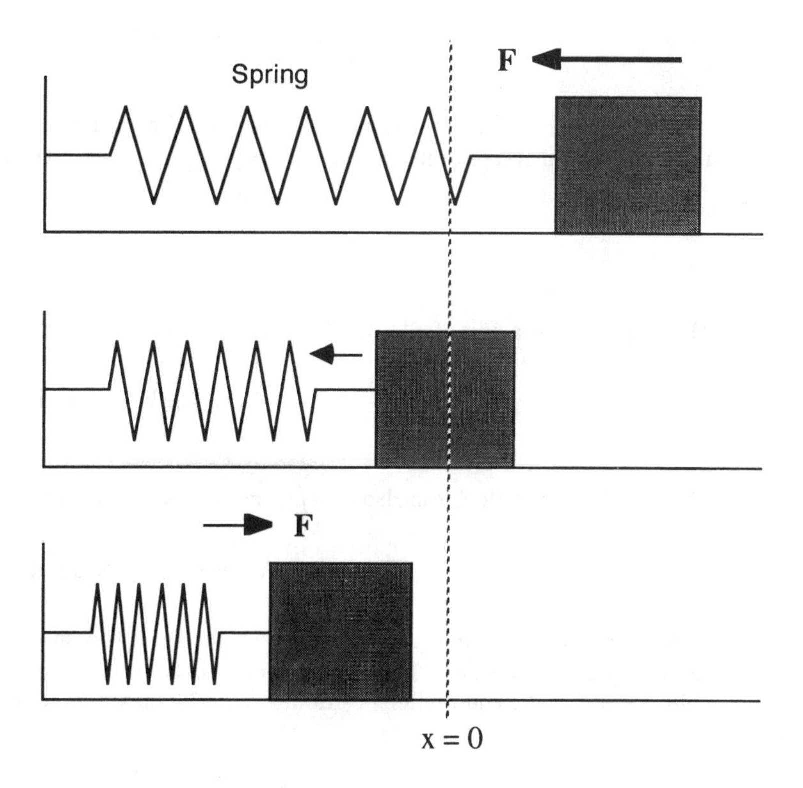

For an ideal spring, and on a frictionless surface, the block will continue to oscillate back and forth forever. We can describe this oscillatory motion by first defining a number of important quantities that are basic to any type of repetitive motion. This follows in the next section.

10.4 Simple Harmonic Motion and the Reference Circle

If we were to plot the motion depicted in the above figure, that is, the displacement of the block x as a function of time t, we would obtain the following graph. Note that a displacement of the block to the right of the equilibrium position is positive (+) and a displacement to the left of x = 0 is negative (-).

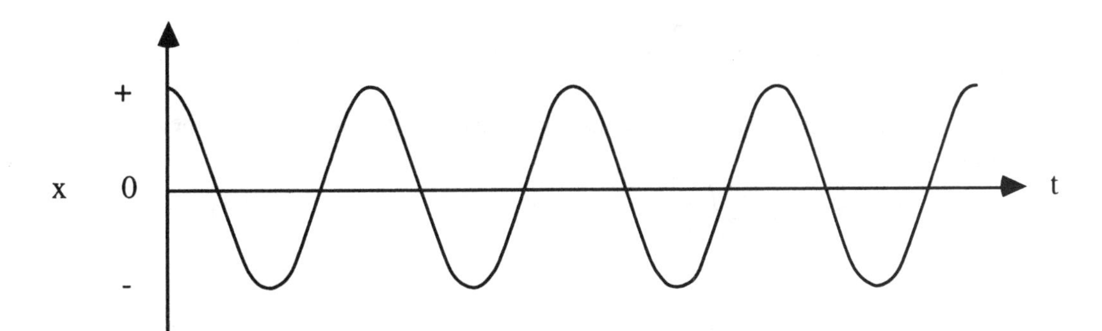

The graph above shows that the displacement is represented by a **cosine** function which can be expressed as

$$x = A \cos \omega t \qquad (10.7)$$

In equation (10.7), A is called the **amplitude** of the motion, that is, the magnitude of the maximum x-displacement of the block from its equilibrium position. The quantity, ω is the **angular frequency** where

$$\omega = \frac{2\pi}{T} \qquad (10.8)$$

and T is the **period**, the time required for the block to complete one cycle of its motion. A related quantity is the **frequency**, which represents the number of cycles of motion completed each second. We have

$$f = \frac{1}{T} \qquad (10.9)$$

The frequency has a unit of s^{-1} which is referred to as one **hertz** (Hz). Finally, the angular frequency is written

$$\omega = \frac{2\pi}{T} = 2\pi f \qquad (10.10)$$

The velocity of the block can also be represented by a sinusoidally varying function. We can write

$$v = - A\omega \sin \theta = - A\omega \sin \omega t \qquad (10.11)$$

Note the negative sign in equation (10.11). This is needed because the block begins to move to the left initially (i.e., in the negative x direction). The block's velocity is zero at each end point of its motion and it reaches its maximum value (+ or -) at the equilibrium position, x = 0. The maximum of equation (10.11) occurs when $\sin \omega t = \pm 1$. So,

$$v_{max} = A\omega \quad (\omega \text{ in rad/s}) \qquad (10.12)$$

In simple harmonic motion, the velocity is NOT constant, therefore, there IS acceleration. Furthermore, since the force varies as a function of x (i.e., the more you stretch the spring, the harder it becomes to keep it stretched), the acceleration is not constant either. The acceleration of the block can be written as

$$a = - A\omega^2 \cos \theta = - A\omega^2 \cos \omega t \qquad (10.13)$$

The acceleration is zero at x = 0 (since F = - kx = ma = 0 because x = 0) and reaches its maximum value at the endpoints of its motion (i.e., when x = A). The expression for a_{max} is obtained by setting $\cos \omega t = \pm 1$ in the above equation (10.13). We therefore have,

$$a_{max} = A\omega^2 \ (\omega \text{ in rad/s}) \qquad (10.14)$$

Finally, it can be shown that the angular frequency is related to the mass of the block, m, and the stiffness of the spring, k, by the equation

$$\omega = \sqrt{\frac{k}{m}} \qquad (10.15)$$

Example 6

A spring whose spring constant is 196 N/m is attached to a block of mass 5.00 kg which sits on a horizontal, frictionless surface. The block is pulled 20.0 cm to the side and released. Find (a) the frequency, (b) the period, (c) the amplitude, (d) the maximum velocity, and (e) the maximum acceleration of the block; (f) write expressions for the displacement, velocity, and acceleration of the block as a function of time.

(a) Using equation (10.15),

$$\omega = \sqrt{\frac{k}{m}} = \sqrt{\frac{196\ \text{N/m}}{5.00\ \text{kg}}} = 6.26\ \text{rad/s}$$

Using equation (10.10),

$$f = \frac{\omega}{2\pi} = \frac{6.26\ \text{rad/s}}{2\pi} = 0.996\ \text{Hz}$$

(b) Using equation (10.9),

$$T = \frac{1}{f} = \frac{1}{0.996\ \text{Hz}} = 1.00\ \text{s}$$

(c) The amplitude is,

$$A = 20.0\ \text{cm} = 0.200\ \text{m}$$

(d) Using equation (10.12),

$$v_{max} = A\omega = (0.200\ \text{m})(6.26\ \text{rad/s}) = 1.25\ \text{m/s}$$

(e) Using equation (10.14),

$$a_{max} = A\omega^2 = (0.200\ \text{m})(6.26\ \text{rad/s})^2 = 7.84\ \text{m/s}^2$$

(f) Using equations (10.7), (10.11), and (10.13), and the values of A and ω found above, we can write

$$x = A\cos\omega t = (0.200)[\cos(6.26t)]\ \text{m}$$

$$v = -A\omega\sin\omega t = -(0.200\ \text{m})(6.26\ \text{rad/s})[\sin(6.26t)]$$

$$= -(1.25)[\sin(6.26t)]\ \text{m/s}$$

$$a = -A\omega^2\cos\omega t = -(0.200\ \text{m})(6.26\ \text{rad/s})^2\ [\cos(6.26\ t)]$$

$$= -(7.84)[\cos(6.26t)]\ \text{m/s}^2.$$

10.5 Energy and Simple Harmonic Motion

When an object is attached to a spring, the spring force can do work $W_{elastic}$ on the object. If in addition, nonconservative forces do work W_{nc} on the object, the work-energy theorem discussed in Chapter 6 states that this work will result in a change in energy for the system. This can be stated mathematically as

$$W_{nc} + W_{elastic} = \Delta KE + \Delta PE_{grav}$$

The work required to stretch a spring from its initial position x_0 to its final position x_f, is given by

$$W_{elastic} = 1/2\ kx_0^2 - 1/2\ kx_f^2 \qquad (10.16)$$

The work done to stretch the spring therefore gives rise to a change in the **elastic potential energy** of the spring. In general, if a spring is stretched (or compressed) by an amount x, we can write

$$PE_{elastic} = 1/2\ kx^2 \qquad (10.17)$$

In the special case where only the forces of the spring and gravity are present, the total mechanical energy of the system, E, can be written

$$E = KE + PE_{grav} + PE_{elastic}$$

Example 7
A spring (spring constant 150 N/m) is attached to a 1.5 kg block which rests on a horizontal frictionless surface. If the block is set in motion and oscillates with an amplitude of 15 cm, what is the velocity of the block as it passes through its equilibrium position?

The conservation of energy principle states that, in the absence of friction or air resistance, the total mechanical energy of the system is conserved. That is

$$E_f = E_0$$

$$KE_f + \frac{1}{2}kx_f^2 + mgh_f = KE_0 + \frac{1}{2}kx_0^2 + mgh_0$$

Since the spring is horizontal, there is no change in PE_{grav} since $h_f = h_0$. Take the initial position to be when the spring is fully stretched, and the final position to be the equilibrium position. We can make the following substitutions: $x_0 = A = 0.15$ m, $x_f = 0$, $v_0 = 0$. We are therefore left with

$$\frac{1}{2}mv_f^2 = \frac{1}{2}kx_0^2$$

and we can solve for the final velocity,

$$v_f = x_0\sqrt{\frac{k}{m}} = (0.15\ \text{m})\sqrt{\frac{150\ \text{N/m}}{1.5\ \text{kg}}} = 1.5\ \text{m/s}.$$

10.6 The Pendulum

A **simple pendulum** consists of a particle of mass m, attached to a frictionless pivot by a thin support wire of length L, as shown in the figure.

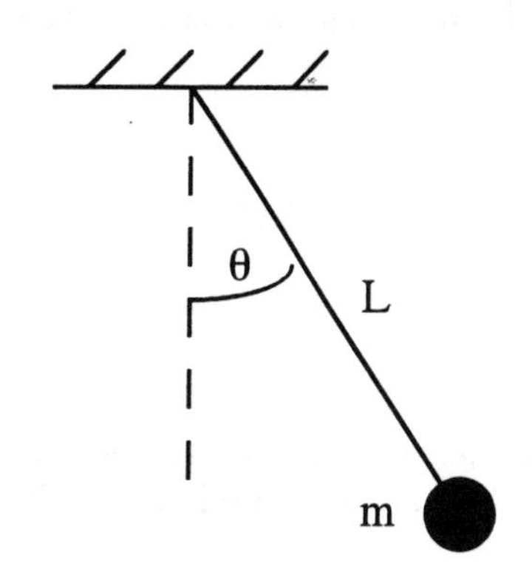

If the mass is pulled aside by a small angle, θ and released, the pendulum will swing back and forth in simple harmonic motion. It can be shown that the frequency of this motion is given by equation (10.20)

$$f = \frac{1}{2\pi}\sqrt{\frac{g}{L}}$$

Note that the frequency does NOT depend on the mass of the particle suspended from the wire! Only the length of the wire and the value for the gravitational acceleration, g, affect the pendulum's motion.

Example 8
A pendulum of length 4.0 m is pulled aside and released. Find (a) the frequency and period of the pendulum on earth, and (b) the frequency and period for the same pendulum on the moon ($g' = 1.57\ m/s^2$).

(a) Using equation (10.20),

$$f = \frac{1}{2\pi}\sqrt{\frac{g}{L}} = \frac{1}{2\pi}\sqrt{\frac{9.80\ m/s^2}{4.0\ m}} = 0.25\ Hz$$

The period is,

$$T = 1/f = 1/(0.25\ Hz) = 4.0\ s.$$

(b) If we take this pendulum to the moon,

$$f' = \frac{1}{2\pi}\sqrt{\frac{g'}{L}} = \frac{1}{2\pi}\sqrt{\frac{1.57\ m/s^2}{4.0\ m}} = 0.10\ Hz$$

$$T' = 1/f' = 1/(0.10\ Hz) = 10\ s.$$

PRACTICE PROBLEMS

1. A steel wire can withstand a tensile stress of 6.0×10^8 N/m^2. If such a wire is used to support a mass of 750 kg, what is the minimum diameter wire that can be used?

2. A copper sphere is subjected to a pressure of 2.0×10^{10} Pa. What is the resulting fractional change in the volume of the sphere? What is the fractional change in the sphere's radius?

3. A rectangular block of tungsten is 5.00 cm high and has a cross-sectional area of 144 cm^2. A force of 3.6×10^7 N is applied to the block. (a) If the force pushes perpendicular to the block's area, by how much will the block compress? (b) If the force is applied parallel to the surface (i.e., a shearing force), what will the amount of shear deformation be?

4. A 2.0 kg block hung from a vertical spring causes it to stretch by 20.0 cm. If the 2.0 kg block is replaced by a 0.50 kg mass, and the spring is stretched and released, what are the frequency and period of the oscillations?

5. A small object of mass 0.20 kg is undergoing simple harmonic motion of period 0.80s and amplitude 0.50 m. (a) What is the maximum force acting on the object? (b) What is the maximum velocity of the object? (c) If the oscillations are produced by a spring, what is the force constant of the spring?

6. A 16-lb object hangs from a spring. It undergoes simple harmonic motion in a vertical direction with an amplitude of 6.0 in and a frequency of 2.0 Hz. Find (a) the period, (b) the force constant of the spring, (c) the maximum speed of the object, (d) the maximum acceleration of the object, (e) the speed when the object is at half its maximum displacement.

7. A 1200 kg car oscillates vertically on its suspension springs with a period of 0.60 s when empty. If six persons, each of mass 80.0 kg, get into the car, how far down will the supporting springs be depressed?

8. A block of mass 0.10 kg is placed on top of a block of mass 0.20 kg. The coefficient of static friction between the blocks is 0.20. The lower block is now moved back and forth horizontally in simple harmonic oscillations with an amplitude of 6.0 cm. Keeping the amplitude constant, what is the highest frequency for which the upper block will not slip relative to the lower block?

9. A pendulum clock keeps perfect time at a location where the acceleration due to gravity is exactly 9.80 m/s^2. When the clock is moved to a higher altitude, it loses 80.0 s per day. Find the value of g at this new location.

10. A spring (spring constant of 2500 N/m) is fixed vertically to the floor. A 10.0 kg ball is placed on the spring, pushed down 0.50 m, and released. How high does the ball fly above its release position?

HELPFUL SUGGESTIONS

1. Hooke's law applies only up to the proportionality limit of a material. If a material is stretched beyond its elastic limit, Young's modulus, the shear modulus, and the bulk modulus are no longer constants..

2. Remember that the strain ($\Delta L/L_0$, $\Delta V/V_0$, etc.) is a dimensionless quantity. The variables making up these fractional quantities (for example ΔL and L_0) must be expressed in the same units.

3. Note that stress, pressure, Young's modulus, shear modulus, and bulk modulus ALL have the same units. These are N/m^2 or Pa.

4. For simple harmonic motion, displacements and amplitudes are measured from the equilibrium position. The displacement is a vector so + and - values are used to indicate direction in one dimension.

5. For a simple harmonic oscillator, the velocity is zero when the displacement is a maximum (positive or negative). The velocity is greatest at the equilibrium position. The acceleration, on the other hand, is a maximum at the end points (i.e., at maximum displacement, $x = A$) but is zero at the equilibrium position.

6. The force causing simple harmonic motion is NOT constant (therefore, neither is the acceleration). The kinematic equations of motion discussed in Chapter 2 do not apply.

7. The frequency of simple harmonic motion is independent of the amplitude of the motion, and depends only on the mass and spring constant. Likewise, for small oscillations, the frequency of a simple pendulum doesn't depend on the amplitude (for small amplitudes), only on gravity and the length of the pendulum (and is independent of the mass).

8. Due to conservation of energy, the kinetic energy of a simple harmonic oscillator at the equilibrium position equals its potential energy at maximum displacement, i.e.,

$$\frac{1}{2}kA^2 = \frac{1}{2}mv_{max}^2$$

EVERYDAY PHYSICS

1. It's easy to experiment with a pendulum. Tie a small object to a string and let it swing back and forth. You can change the period by adjusting the length of the string. You can try objects of different masses and show that the period is mass independent. Vary the amplitude (but don't make it too large) and demonstrate that this too has no effect on the period of oscillation. The same types of experiments can be performed with a spring/mass system. Use a slinky suspended from the ceiling with light masses taped on.

2. A good demonstration of shear deformation is to place a book on a rough surface. Place the palm of your hand on top of the book, and push parallel to the surface, as if you were sliding your hand across. The top surface of the book is shifted relative to the fixed bottom surface. Experiment with a thicker book; what happens?

3. There are many examples of oscillatory motion in everyday life. The shock absorbers and springs in your car exhibit **damped harmonic motion** (fortunately), the diaphragm of the loudspeaker in your stereo system vibrates with simple harmonic motion to produce sound, and some clocks keep time using mechanisms based on the vibration of a pendulum. Can you think of some other examples?

CHAPTER QUIZ

1. Equal weights are hung from two steel wires. One wire has twice the diameter of the other. Compare the amount of stretch in each wire.
   a. The smaller diameter wire stretches twice as much.
   b. The smaller diameter wire stretches four times as much.
   c. The larger diameter wire stretches more.
   d. Both wires stretch by the same amount.

2. An aluminum sphere sinks to the bottom of a deep lake. Which statement is true about the sphere?
   a. The sphere's density increases slightly.
   b. The sphere's density increases greatly.
   c. The sphere's density decreases slightly.
   d. The sphere's density remains unchanged.

3. A 1.0×10^3 N force is applied to a wire of cross-sectional area 1.0×10^{-5} m^2. The wire stretches by 0.10% of its total length. What is Young's modulus for the wire?
   a. 1.0×10^5 Pa b. 1.0×10^8 Pa c. 1.0×10^{10} Pa d. 1.0×10^{11} Pa

4. A 3 kg object is hung from a spring whose spring constant is 300 N/m. When set in motion, the frequency of the mass-spring system is
   a. 10 Hz b. 2π Hz c. $5/\pi$ Hz d. 100 Hz

Questions 5-7 are based on the following: A system executes simple harmonic motion of the form $x = 0.300 \cos 7.50t$ where x is in meters, t is in seconds, and 7.50t is in radians.

5. What is the amplitude of the motion?
   a. 0.300 m b. 7.50 m c. 2.25 m d. 25.0 m

6. What is the maximum velocity of the system?
   a. 0.300 m/s b. 7.50 m/s c. 2.25 m/s d. 16.9 m/s

7. What is the maximum acceleration of the system?
   a. 0.300 m/s^2 b. 7.50 m/s^2 c. 2.25 m/s^2 d. 16.9 m/s^2

8. What must the approximate length of a pendulum be on the earth ($g = 32$ ft/s^2) to have a period of 1 second?
   a. 0.25 ft b. 0.81 ft c. 1.2 ft d. 32 ft

9. A mass is attached to a vertical spring that bobs up and down between points A and B. Where is the mass located when its kinetic energy is a minimum?
   a. At either A or B
   b. Midway between A and B
   c. One fourth of the way between A and B
   d. None of the above

10. A spring (spring constant = 35 N/m) is stretched by 4.0 m. How much energy is stored in the spring?
    a. 140 J b. 280 J c. 560 J d. 1120 J

11. A 5.6 kg object is attached to the spring in problem 10. The mass oscillates with an amplitude of 4.0 m. What is the maximum speed of the object?
    a. 7 m/s b. 10 m/s c. 50 m/s d. 100 m/s

SOLUTIONS AND ANSWERS

Practice Problems

---

1. We know that stress = F/A where $A = (1/4)\pi d^2$ and d = diameter. Solving for d we obtain

$$d^2 = 4F/[\pi(\text{stress})] = 4(750\text{ kg})(9.80\text{ m/s}^2)/[\pi(6.0 \times 10^8\text{ N/m}^2)]$$

so

$$d = \mathbf{3.9 \times 10^{-3}\ m}.$$

---

2. Using equation (10.4),

$$\Delta P = -B(\Delta V/V_0).$$

Solving for $(\Delta V/V_0)$ and substituting for B for copper (Table 10.3), we find

$$(\Delta V/V_0) = \Delta P/B = (2.0 \times 10^{10}\text{ Pa})/(1.3 \times 10^{11}\text{ Pa}) = \mathbf{1.5 \times 10^{-1}}.$$

The fractional change in radius is

$$(\Delta r/r_0) = (1/3)(\Delta V/V_0) = \mathbf{5.0 \times 10^{-2}}.$$

---

3. (a) Equation (10.1),

$$\Delta L = FL_0/YA = (3.6 \times 10^7\text{ N})(5.00 \times 10^{-2}\text{ m})/[(3.6 \times 10^{11}\text{ Pa})(144 \times 10^{-4}\text{ m}^2)]$$
$$\Delta L = \mathbf{0.35\ mm}$$

(b) Equation (10.2),

$$\Delta X = FL_0/SA = (3.6 \times 10^7\text{ N})(5.00 \times 10^{-2}\text{ m})/[(1.5 \times 10^{11}\text{ Pa})(144 \times 10^{-4}\text{ m}^2)]$$
$$\Delta X = \mathbf{0.83\ mm}$$

---

4. Find the spring constant using equation (10.5), F = kx which gives

$$k = F/x = (2.0\text{ kg})(9.8\text{ m/s}^2)/(0.20\text{ m}) = 98\text{ N/m}.$$

Now use equation (10.15),

$$\omega = \sqrt{(k/m)} = \sqrt{(98\text{ N/m})/(0.50\text{ kg})} = 14\text{ rad/s}$$

to find

$$f = \omega/2\pi = \mathbf{2.2\ Hz}.$$

Therefore, from equation (10.9), the period is

$$T = 1/f = \mathbf{0.45\ s}.$$

---

5. (a) The maximum acceleration, equation (10.14), is

$$a_{max} = A\omega^2 = A(2\pi/T)^2$$
$$a_{max} = (0.50\text{ m})[2\pi/(0.80\text{ s})]^2$$
$$a_{max} = 31\text{ m/s}^2.$$

The maximum force is obtained from

$$F_{max} = ma_{max} = (0.20\ \text{kg})(31\ \text{m/s}^2) = \mathbf{6.2\ N}.$$

(b) Using equation (10.12),

$$v_{max} = A\omega = A(2\pi/T) = (0.50\ \text{m})[2\pi/(0.80\ \text{s})] = \mathbf{3.9\ m/s}.$$

(c) Using $F = kx$

$$k = F/x = (6.2\ \text{N})/(0.50\ \text{m}) = \mathbf{12\ N/m}.$$

---

6. (a) Equation (10.9),

$$T = 1/f = 1/(2.0\ \text{Hz}) = \mathbf{0.50\ s},$$

(b) Equation (10.10),

$$\omega = 2\pi f = \sqrt{(k/m)},$$

solving for k,

$$k = 4\pi^2 f^2 m = 4\pi^2(2.0\ \text{Hz})^2(0.5\ \text{slugs}) = \mathbf{79\ lb/ft},$$

(c)

$$v_{max} = A\omega = A(2\pi f) = (0.5\ \text{ft})(2\pi)(2.0\ \text{Hz}) = \mathbf{6.3\ ft/s},$$

(d)

$$a_{max} = A\omega^2 = (0.5\ \text{ft})[2\pi(2.0\ \text{Hz})]^2 = \mathbf{79\ ft/s^2},$$

(e) Using conservation of energy,

$$(1/2)kA^2 = (1/2)k(A/2)^2 + (1/2)mv^2.$$

Thus,

$$v^2 = (3/4)kA^2/m = (3/4)(79\ \text{lb/ft})(0.5\ \text{ft})^2/(0.5\ \text{sl}),$$

or

$$v = \mathbf{5.4\ ft/s}.$$

---

7. Find the spring constant using equation (10.15),

$$k = m\omega^2 = (1200\ \text{kg})[2\pi/(0.6\ \text{s})]^2 = 1.3 \times 10^5\ \text{N/m}.$$

Now use $F = kx$ to find

$$x = F/k = 6(80.0\ \text{kg})(9.80\ \text{m/s}^2)/(1.3 \times 10^5\ \text{N/m}) = \mathbf{3.6\ cm}.$$

---

8. The force of static friction for the top block (m_1) is

$$f_s = \mu m_1 g = (0.20)(0.10\ \text{kg})(9.8\ \text{m/s}^2) = 0.20\ \text{N}.$$

The maximum acceleration the top block can sustain without slipping is

$$a_{max} = f_s/m = (0.20\ \text{N})/(0.10\ \text{kg}) = 2.0\ \text{m/s}^2$$

Using equation (10.14),

$$a_{max} = A\omega^2 = A(2\pi f)^2$$

or

$$f^2 = a_{max}/4\pi^2 A = (2.0\ \text{m/s}^2)/[4\pi^2(0.06\ \text{m})],$$

then

$$f = \mathbf{0.92\ Hz}.$$

---

9. If the clock loses 80.0 s per day, it must be swinging slower (T greater). It loses (80.0s/86 400 s) = 0.926 ms each second. Therefore, the new period is 1.000926 s. If the original period is

$$T = 2\pi\sqrt{(L/g)}$$

and the new period is

$$T' = 2\pi\sqrt{(L/g')}$$

then

$$T'/T = \sqrt{(g/g')}$$

gives

$$g' = g\,(T/T')^2 = (9.80\ m/s^2)[(1.000\ s)/(1.001\ s)]^2 = \mathbf{9.78\ m/s^2}.$$

---

10. Using conservation of energy,

$$E_A = (1/2)kx^2 = E_B = mgh.$$

We have,

$$h = kx^2/2mg = (2500\ N/m)(0.5\ m)^2/[2(10.0\ kg)(9.80\ m/s^2)] = \mathbf{3.2\ m}.$$

Quiz answers

1. b	3. d	5. a	7. d	9. a	11. b
2. a	4. c	6. c	8. b	10. b	

MCAT REVIEW PROBLEMS

A student measures g, the acceleration due to gravity near the Earth's surface, by performing an experiment using a simple pendulum. She intends to use the formula $T = 2\pi \sqrt{L/g}$.

Experiment 1

The pendulum consists of a small "bob," of mass $m = 0.50$ kg, attached to the end of a very light string of length $L = 2.0$ m. The string is initially hanging straight down. The student then displaces the pendulum bob horizontally by about 4 cm, and releases it from rest. She observes how many times the pendulum oscillates back and forth in one minute.

Experiment 2

Same as experiment 1, except the student displaces the pendulum bob by 8 cm instead of 4 cm.

Experiment 3

Same as experiment 1, except the student uses a bob of mass 0.25 kg instead of 0.50 kg.

In experiment 1, the pendulum completed 20 oscillations in one minute.

---

1. In experiment 1, what is the frequency of the pendulum, in hertz (cycles per second)?

   A. 0.05 **B.** 0.33 **C.** 3.0 **D.** 20

---

2. In experiment 2, about how many oscillations will the pendulum complete in one minute?

   A. 10 **B.** 14 **C.** 20 **D.** 28

---

3. In experiment 3, the number of oscillations completed by the pendulum in one minute is:

   A. Less than 20
   B. About 20
   C. More than 20
   D. We cannot determine the answer from the information provided.

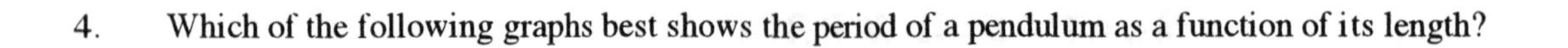

4. Which of the following graphs best shows the period of a pendulum as a function of its length?

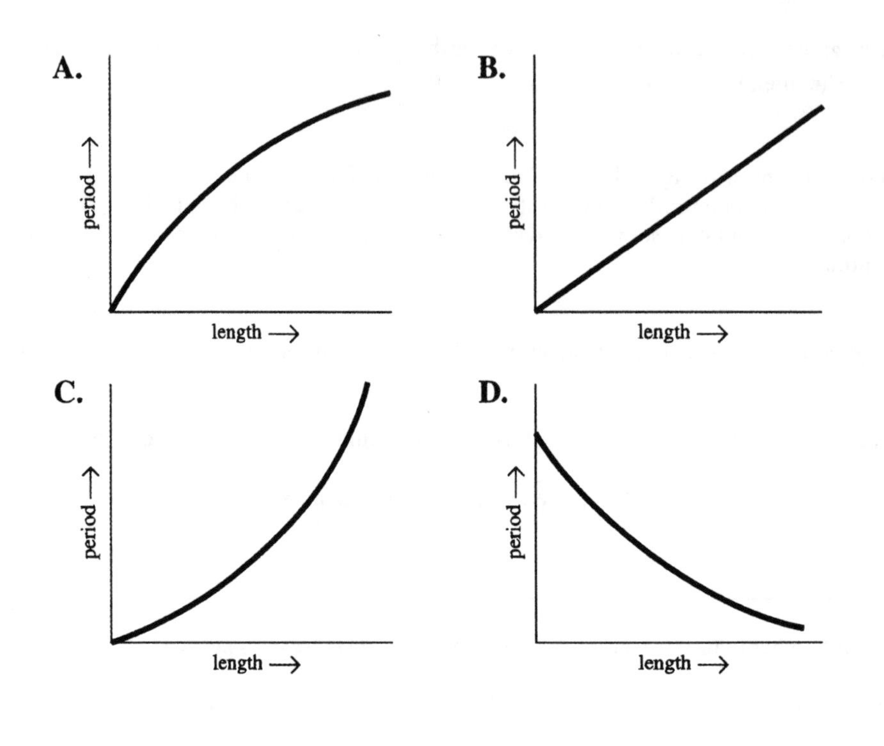

---

5. The pendulum bob detaches from the string right when it reaches the highest point of its arc (the endpoint of the swing). Which of the following diagrams best represents the subsequent trajectory of the pendulum bob?

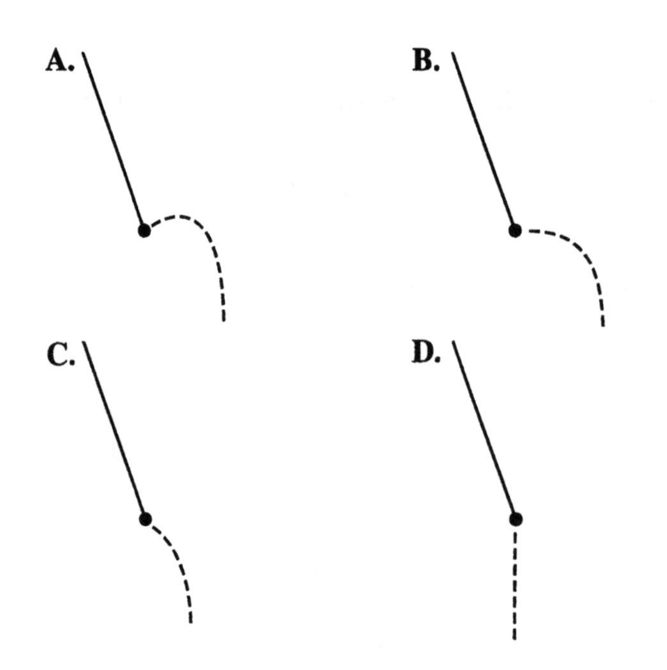

ANSWERS TO MCAT REVIEW PROBLEMS

---

1. **B.** Frequency (in hertz) is the number of oscillations per second. Don't confuse this with the period, which is the number of seconds per oscillation. Since 20 oscillations require 60 seconds, the *period* is 3.0 seconds. So, during each second, the pendulum completes only a third of an oscillation. In other words, the frequency is 1/3 oscillation per second.

---

2. **C.** The period of a simple harmonic oscillator does *not* depend on the amplitude, for small-amplitude oscillations. You might think the period increases, because the pendulum has more distance to cover. But it covers that distance at a higher average speed. The period comes out the same. Try it!
   Mathematically speaking, since $T = 2\pi\sqrt{L/g}$, the period depends only on L and g, not on amplitude.

---

3. **B.** Again, we can invoke the fact that T depends only on L and g. Intuitively, changing the mass makes no difference for the same reason that a bowling ball and a pencil, when simultaneously dropped from the same height, reach the ground at the same time. Gravity accelerates the pendulum bob at the same rate, no matter what mass it has.

---

4. **A.** From the formula $T = 2\pi\sqrt{L/g}$, T is proportional to the square root of L. So, you must quadruple L in order to double T. Graph A captures the fact that T increases "slowly" with L. By contrast, according to graph B, doubling L would double T. According to graph C, doubling L would *more* than double T. And according to graph D, doubling L would *decrease* T.

---

5. **D.** Many students pick graph A, thinking that the pendulum "remembers" its swinging motion when it becomes a projectile. But at the endpoint of its arc, the pendulum is *motionless*. Therefore, when the bob detaches at that point, it has zero velocity. In other words, the bob's initial velocity when it becomes a projectile is zero. For this reason, the bob behaves as if it were dropped from rest. It *was* "dropped" from rest. That's why the bob falls straight down. By contrast, if the bob detached from the string at any point other than the endpoint of the swing, then the bob would have some sideways motion.]

CHAPTER 11
Fluids

PREVIEW

In this chapter you will be introduced to the properties of fluids. You will learn to describe the inertial properties and the weight of a fluid in terms of its mass density and weight density. Also you will describe the forces acting on a fluid and exerted by a fluid in terms of pressure.

Pascal's principle reveals how pressures applied to fluids are transmitted to other portions of the fluid. Archimedes' principle describes how and why objects float in fluids. Using these principles, you will be able to understand such things as how hydraulic jacks work and how submarines can sometimes float on and other times dive below the surface of the ocean.

In this chapter you will also learn about flowing fluids including the equation of continuity, Bernoulli's equation, and viscous flow. Numerous applications are discussed, such as fluid flow through pipes, streamline air flow around an airplane wing, what makes a spinning baseball curve, and many others.

QUICK REFERENCE

Important Terms

Fluid
A material which has the ability to flow. Usually fluids are liquids and gases.

Mass density
The mass of a volume of a substance divided by the volume.

Specify gravity
Is the mass density of the material compared to the mass density of water at 4 °C.

Pressure on a surface
The perpendicular component of the force acting on the surface divided by the area of the surface.

Absolute pressure
The amount by which a pressure exceeds the pressure of a true vacuum.

Gauge pressure
The amount by which a pressure exceeds atmospheric pressure.

Buoyant force
The net force exerted by a fluid on an object which is partially or wholly immersed in the fluid. The direction of the buoyant force is up.

Steady fluid flow
Movement in which the velocity of the fluid particles at any point in the fluid is constant with time. Every particle that passes through a particular point in the fluid has the same velocity.

Incompressible fluid
One in which the density of the liquid remains constant as the pressure changes.

Viscous fluid
A fluid that does not flow readily (like honey) due to the viscosity of the fluid.

Ideal fluid
An incompressible, nonviscous fluid.

Streamline
Lines representing the trajectory of the fluid particles in a steadily flowing fluid.

Mass flow rate
The mass of fluid which flows past a given point per unit time. Units are (kg/s).

Equation of continuity
The mass of fluid that enters one end of a pipe must leave at the other end. In other words, the mass flow rate must remain constant between any two positions along a tube of flowing fluid.

Equations

The **mass density** of a substance

$$\rho = \frac{m}{V} \tag{11.1}$$

The **specific gravity** of a substance

$$\text{Specific gravity} = \frac{\text{Density of substance}}{1.000 \text{ X } 10^3 \text{ kg/m}^3} \tag{11.2}$$

The **pressure** exerted by a force acting perpendicular to a surface of area A

$$P = \frac{F}{A} \tag{11.3}$$

The **pressure within a fluid** at a depth, h, when an external pressure, P_1, is applied to the fluid

$$P_2 = P_1 + \rho g h \tag{11.4}$$

The force transmitted by a fluid from one piston to another piston

$$F_2 = F_1 \left(\frac{A_2}{A_1}\right) \tag{11.5}$$

The **equation of continuity** is:

$$\rho_1 A_1 v_1 = \rho_2 A_2 v_2 \tag{11.8}$$

For an **incompressible fluid**:

$$A_1 v_1 = A_2 v_2 \tag{11.9}$$

Bernoulli's equation is:

$$P_1 + \frac{1}{2}\rho v_1^2 + \rho g y_1 = P_2 + \frac{1}{2}\rho v_2^2 + \rho g y_2 \tag{11.11}$$

The force needed to move a layer of **viscous fluid** is:

$$F = \frac{\eta A v}{y} \tag{11.13}$$

Poiseuille's law can be written as:

$$Q = \frac{\pi R^4 (P_2 - P_1)}{8 \eta L} \tag{11.14}$$

Principles

Pascal's principle - any change in the pressure applied to a completely enclosed fluid is transmitted undiminished to all parts of the fluid and the walls of the container.

Archimedes' principle - a fluid will exert an upward buoyant force on an object wholly or partially immersed in it equal to the weight of the fluid displaced by the object.

Important Numbers

Mass density of water at 4 °C	1.00×10^3 kg/m^3 or 1.00 g/cm^3
Mass density of air at 0 °C	1.29 kg/m^3
Atmospheric pressure at sea level	1.013×10^5 Pa or 14.70 lb/in^2.

DISCUSSION OF SELECTED SECTIONS

11.1 Mass Density

Example 1
What is the volume of 1.00 kg of air?

Equation (11.1) gives the volume of air

$$V = \frac{m}{\rho} = \frac{1.00 \text{ kg}}{1.29 \text{ kg/m}^3} = 0.775 \text{ m}^3$$

This is roughly one cubic yard.

Example 2
An object weighs 450 N and has a volume of 0.083 m^3. Use the density data in Table 11.1, to determine the material from which the object is made.

Equation (11.1) gives the density of the material to be

$$\rho = \frac{M}{V} = \frac{\dfrac{450 \text{ N}}{9.8 \text{ m/s}^2}}{0.083 \text{ m}^3} = 550 \text{ kg/m}^3$$

Comparing this density with the values in Table 11.1 reveals that the object is made of PINE.

11.2 Pressure

Example 3
What is the downward force on the top of your head due to atmospheric pressure. Assume that the top of your head is a flat circle with a diameter of 7.00 inches.

$$F = P_{atm}A = (14.70 \text{ lb/in}^2)\pi(3.50 \text{ in})^2 = 566 \text{ lb}.$$

11.3 Pressure and Depth in a Static Fluid

A static fluid generates a pressure within itself due to its own weight. This pressure varies with the depth below the surface of the fluid of the point in question.

Example 4
What is the pressure due to the sea water at a depth of 500.0 m below the surface of the ocean? The density of sea water is 1025 kg/m^3.

The pressure due to the weight of the sea water is

$$P_2 - P_1 = \rho gh = (1025\ kg/m^3)(9.80\ m/s^2)(500.0\ m) = 5.02 \times 10^6\ Pa$$

This is about 730 lb/in^2!

Example 5
A small boy stands with his finger plugging a hole in a dike. The hole has a diameter of 12 mm and is located 3.4 m below the surface level of the water behind the dike. How much force must the boy exert to hold back the water?

The pressure acting on the boy's finger due to the water is ρgh. The force exerted by the water on the boy, hence the force exerted by the boy is then

$$F = \rho ghA$$
$$F = (1.00 \times 10^3\ kg/m^3)(9.80\ m/s^2)(3.4\ m)\pi(6.0 \times 10^{-3}\ m)^2 = 3.8\ N$$

11.4 Pressure Gauges

Example 6
A gas orifice in a furnace is designed to work with a gauge pressure of 3.0 in of water as measured by an open tube manometer. What is this pressure in lb/in^2? Water has a weight density of 62.4 lb/ft^3.

The gauge pressure of the gas is the absolute gas pressure in the left arm of the tube, P_2, minus the atmospheric pressure in the right arm of tube.

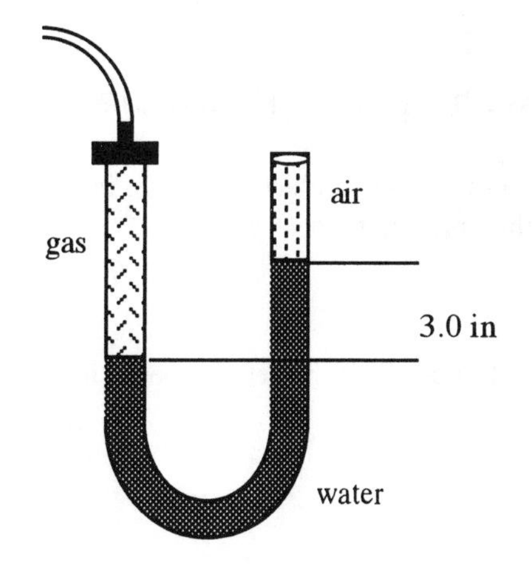

$$\text{gauge pressure} = P_2 - P_{atm} = \rho gh$$

$$\text{gauge pressure} = (62.4\ lb/ft^3)(0.25\ ft)$$
$$= 15.6\ lb/ft^2$$

$$\text{gauge pressure} = (15.6\ lb/ft^2)(1\ ft^2/144\ in^2)$$
$$= 0.11\ lb/in^2$$

11.5 Pascal's Principle

Pascal's principle lies at the heart of many useful devices such as hydraulic jacks, hydraulic brake systems, etc. In any situation where a force is transmitted by a fluid, Pascal's principle is involved.

Example 7

The handle of a hydraulic jack is 15 cm long and is pivoted 2.5 cm from the input piston which has a radius of 0.60 cm. The output piston has a radius of 1.2 cm. What weight could be lifted by the jack if the person pushing on the handle is to exert no more than 110 N of force?

The weight that can be lifted is equal of the magnitude of the force exerted by the output piston. The output force is related to the input force by

$$F_2 = F_1\left(\frac{A_2}{A_1}\right) = F_1\left(\frac{R_2}{R_1}\right)^2$$

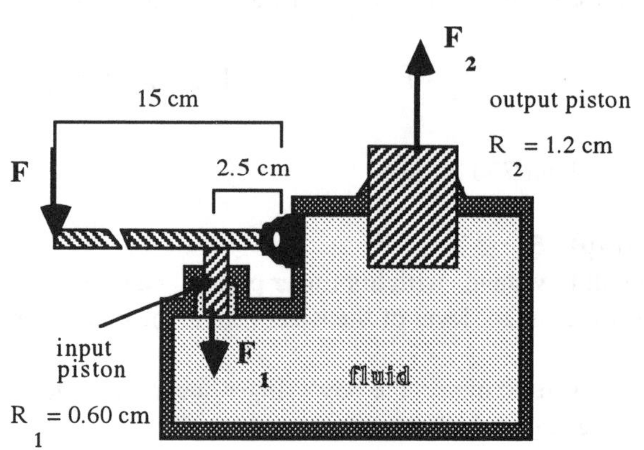

We need the force, F_1, on the input piston. Taking torques on the handle about an axis through the pivot gives

$$F(15\text{ cm}) - F_1'(2.5\text{ cm}) = 0$$

$$F_1' = 6.0\,F = 660\text{ N}.$$

The force on the input piston due to the handle is equal in magnitude to the force on the handle exerted by the input piston , F_1'.

The force exerted by the output piston is then

$$F_2 = (660\text{ N})\left(\frac{1.2\text{ cm}}{0.60\text{ cm}}\right)^2 = 2600\text{ N}$$

11.6 Archimedes' Principle

The buoyant force on an object in a fluid is equal to the weight of the fluid displaced by the object. The net upward force on the object is then

$$F = B - W = \rho_f g V_f - \rho_o g V_o.$$

If the object floats, then the net force on it is zero and

$$\rho_o = \rho_f\left(\frac{V_f}{V_o}\right)$$

If the object floats partially submerged, then it displaces less than it own volume of fluid. This implies that the density of the object is less than the density of the fluid. However, if the object floats submerged in the fluid, then it displaces exactly its own volume of fluid and the densities of the object and fluid must be equal.

In the case where the object sinks, it displaces its own volume of fluid and the net force is directed downward and is negative. Now the density of the object must be greater than the density of the fluid.

Example 8
Archie is playing with a small boat in the bathtub. The boat has a mass of $M = 0.22$ kg and a volume of $V = 0.0011\ m^3$. Archie has several lead weights each with a mass of $m = 0.028$ kg. How many lead weights could Archie place in the boat before it sinks?

In order for the boat to float just submerged, its density must be equal to the density of the water. The average density of the boat with N lead weights in it is $(M + Nm)/V$, so

$$N = \frac{\rho_w V - M}{m} = \frac{(1.00 \times 10^3\ kg/m^3)(0.0011\ m^3) - 0.22\ kg}{0.028\ kg} = 31$$

He can place **31 lead weights** in the boat before it sinks. The thirty-second weight will sink the boat.

Example 9
An empty barge 25 m long and 4 .0 m wide floats with the waterline at the 1.0 m mark. That is, 1.0 m of its hull is submerged. Gravel is added to the barge until it floats with the waterline at the 3.5 m mark. What is the weight of the gravel? Assume that the barge is a rectangular solid.

Since the empty barge is floating, its weight must be equal to the buoyant force exerted by the water.

$$W_E = \rho_w g V_E = \rho_w g A h_E$$

A similar expression must hold for the loaded barge.

$$W_E + W_L = \rho_w g A h_L$$

The weight of the load of gravel is then

$$W_L = \rho_w g A(h_L - h_E)$$
$$W_L = (1.00 \times 10^3\ kg/m^3)(9.80\ m/s^2)(25\ m \times 4.0\ m)(3.5\ m - 1.0\ m)$$
$$W_L = 2.5 \times 10^6\ N$$

Example 10
An 80.0 kg person weighs 30.0 N when weighed in water. If body fat has a density close to that of water and the lean portion of the body has an average density of $1.060 \times 10^3\ kg/m^3$, what is the person's percentage of body fat?

The percentage of body fat is the mass of fat divided by the total mass of the person. Also, the mass of fat, $\rho_f V_f$, and lean, $\rho_L V_L$, must add to give 80.0 kg. So

$$\rho_L V_L + \rho_f V_f = 80.0\ kg$$

$$1.060\ V_L + V_f = 8.00 \times 10^{-2} \qquad (1)$$

Archimedes' principle applied to the person weighed in water gives

$$\rho_w g V_L + \rho_w g V_f = (80.0\ kg)g - 30.0\ N$$

$$V_L + V_f = 7.69 \times 10^{-2} \qquad (2)$$

Solving (2) for V_L, substituting into (1) and rearranging yields $V_f = 0.025\ m^3$. The mass of fat in the person is then $(1.00 \times 10^3\ kg/m^3)(0.025\ m^3) = 25$ kg. This represents $25/80.0 = 0.31$ or **31 %** of the person's weight.

11.8 The Equation of Continuity

In the figure below, a mass of fluid flows along a tube. The fluid has a density ρ, and travels with a velocity v_2, through a segment of the tube whose cross-sectional area is A_2. The tube then tapers off to cross-sectional area A_1, and the fluid, whose density is now ρ, now travels with velocity v_1.

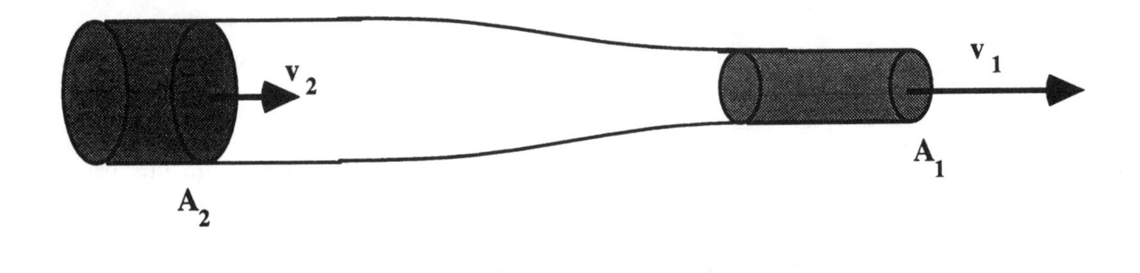

At position 1 we have: $$\text{Mass Flow Rate} = \rho_1 A_1 v_1 \tag{11.7b}$$

At position 2 we have: $$\text{Mass Flow Rate} = \rho_2 A_2 v_2 \tag{11.7a}$$

The **equation of continuity** states that the mass flow rate must remain constant, that is,

$$\rho_1 A_1 v_1 = \rho_2 A_2 v_2 \tag{11.8}$$

If the fluid is incompressible, i.e., $\rho_1 = \rho_2$, equation (11.8) can be written as

$$A_1 v_1 = A_2 v_2 \tag{11.9}$$

where $Q = Av$ is the **volume flow rate**. Note: Units are $Q = m^3/s$, $\rho = kg/m^3$, $A = m^2$, $v = m/s$.

Example 11
Water is flowing through a 3.0 mm diameter pipe at a speed of 5.0 m/s. The pipe has a constriction of 1.5 mm diameter. What is the velocity of the water through the constriction?

We assume the flow is incompressible, so equation (11.9) is valid. Noting that $A = \pi r^2$ where r is the radius of the pipe and $r = d/2$ we can write

$$A_1 v_1 = (\pi r_1^2) v_1 = A_2 v_2 = (\pi r_2^2) v_2, \text{ and solving for } v_2 \text{ yields,}$$

$$v_2 = v_1 \left(\frac{r_1}{r_2}\right)^2 = (5.0 \text{ m/s}) \left(\frac{1.5 \times 10^{-3} \text{ m}}{7.5 \times 10^{-4} \text{ m}}\right)^2 = 20 \text{ m/s}.$$

Example 12
Oil is flowing with a speed of 4.00 ft/s through a 2.00-ft diameter pipeline. How many gallons of oil flow through the pipeline in one day? (1 gal = 0.134 ft^3)

The volume flow rate is: $Q = Av = \pi r^2 v = \pi (1.00 \text{ ft})^2 (4.00 \text{ ft/s}) = 12.6 \text{ ft}^3/\text{s}.$

Converting units gives:

$$Q = \left(12.6 \frac{\text{ft}^3}{\text{s}}\right)\left(\frac{1 \text{ gal}}{0.134 \text{ ft}^3}\right)\left(\frac{3600 \text{ s}}{1 \text{ h}}\right)\left(\frac{24 \text{ h}}{1 \text{ day}}\right) = 8.12 \times 10^6 \text{ gal/day}$$

11.9 Bernoulli's Equation

For any two points (1 and 2) in the steady, irrotational flow of an incompressible, nonviscous fluid, the pressure P, the fluid speed v, and the elevation y, are related by

$$P_1 + \frac{1}{2}\rho v_1^2 + \rho g y_1 = P_2 + \frac{1}{2}\rho v_2^2 + \rho g y_2 \tag{11.11}$$

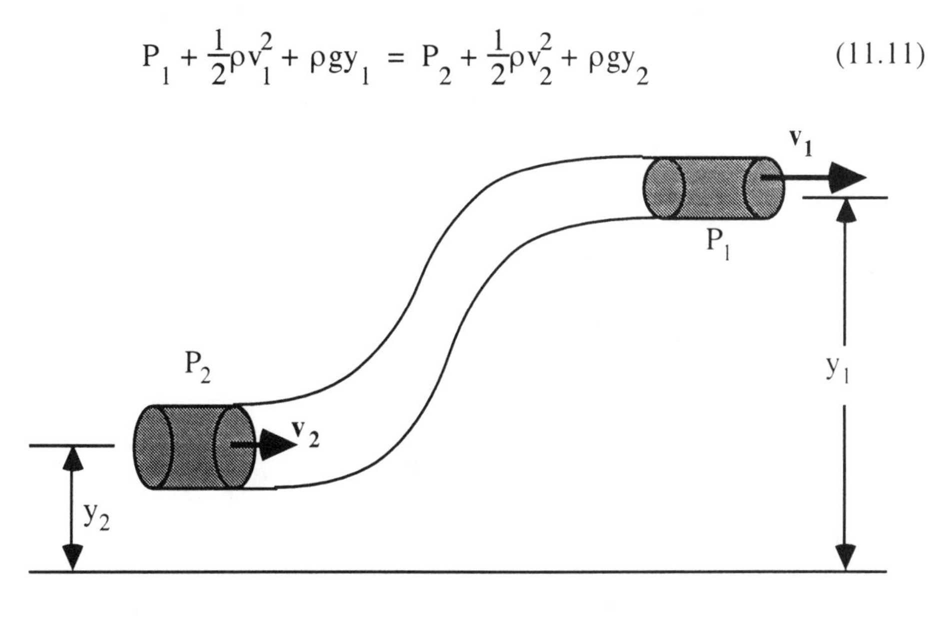

Example 13
In the diagram above, let the difference in pressure be $\Delta P = P_2 - P_1 = 5.0 \times 10^4$ Pa, take $\Delta y = y_1 - y_2 = 2.0$ m, and let $v_2 = 4.0$ m/s. What is the speed v_1 of water flowing through this pipe? (For water, $\rho = 1.0 \times 10^3$ kg/m^3)

Using equation (11.11) we have,

$$\frac{1}{2}\rho v_1^2 = (P_2 - P_1) + \rho g(y_2 - y_1) + \frac{1}{2}\rho v_2^2, \text{ and solving for } v_1$$

$$v_1 = \sqrt{\frac{2\Delta P}{\rho} - 2g\Delta y + v_2^2} = \sqrt{\frac{2(5.0 \times 10^4 \text{ Pa})}{1.0 \times 10^3 \text{ kg/m}^3} - 2(9.8 \text{ m/s}^2)(2.0 \text{ m}) + (4.0 \text{ m/s})^2}$$

$$= 8.8 \text{ m/s}.$$

11.10 Applications of Bernoulli's Equation

For a fluid moving through a horizontal pipe ($y_1 = y_2$), equation (11.11) becomes

$$P_1 + \frac{1}{2}\rho v_1^2 = P_2 + \frac{1}{2}\rho v_2^2 \tag{11.12}$$

Example 14
Water is flowing at a speed of 0.500 m/s through a 4.00-cm diameter hose. The hose is horizontal. (a) What is the mass flow rate of the water? (b) At what speed does the water pass through a nozzle whose effective diameter is 0.600 cm? (c) What must be the absolute pressure of the water entering the hose if the pressure at the nozzle is atmospheric pressure?

(a) Using equation (11.7) the mass flow rate is,

$$MFR = \rho A v = (1.00 \times 10^3 \text{ kg/m}^3)\pi(2.00 \times 10^{-2} \text{ m})^2(0.500 \text{ m/s}) = 0.628 \text{ kg/s}.$$

(b) Since the water is incompressible, use equation (11.9), $A_1v_1 = A_2v_2$, and therefore,

$$v_1 = v_2 (A_2/A_1) = v_2 (r_2/r_1)^2 = (0.500 \text{ m/s})[(2.00 \text{ cm})/(0.300 \text{ cm})]^2 = 22.2 \text{ m/s}.$$

(c) Using equation (11.12) we can write,

$$P_2 = P_1 + \frac{1}{2}\rho(v_1^2 - v_2^2)$$

$$= (1.01 \times 10^5 \text{ Pa}) + \frac{1}{2}(1.00 \times 10^3 \text{ kg/m}^3)[(22.2 \text{ m/s})^2 - (0.500 \text{ m/s})^2]$$

$$= 3.47 \times 10^5 \text{ Pa}.$$

11.11 Viscous Flow

In all real fluids, the different fluid layers do NOT all have the same velocity. The fluid at the center of a pipe moves fastest while the fluid near the walls of the pipe moves slowest. In fact, the fluid directly in contact with the wall does not move at all because it is held tightly by intermolecular forces. Such forces give rise to the **viscosity** of the fluid. Examples of fluids that are highly viscous are honey or tar. Fluids such as water, and especially oil, have low viscosities.

The tangential force **F** required to move a fluid layer at a constant speed v, when the layer has an area A and is located a perpendicular distance y from an immobile surface, is given by

$$F = \frac{\eta A v}{y} \qquad (11.13)$$

Where η is known as the **coefficient of viscosity** or simply the viscosity. The SI unit for viscosity is the poise (P) where 1 P = 0.1 Pa s.

In many cases of interest, it is necessary to know the volume flow rate in a viscous fluid. Consider two points (1 and 2) separated by a distance L in a pipe of constant radius R. If the pressure difference between these two points is $P_2 - P_1$, and η is the viscosity of the fluid in the pipe, the volume flow rate Q is given by **Poiseuille's law**, which is

$$Q = \frac{\pi R^4 (P_2 - P_1)}{8\eta L} \qquad (11.14)$$

Example 15
Glycerin flows through a tube of 1.0 cm radius. If the ends of the pipe are 5.0 m apart and a pressure difference of 3.0×10^4 Pa is maintained between the ends, find the volume flow rate.

Using equation (11.14), and the value of η for glycerin found in the table we have

$$Q = \frac{\pi R^4 (P_2 - P_1)}{8\eta L} = \frac{\pi(0.010 \text{ m})^4(3.0 \times 10^4 \text{ Pa})}{8(1.5 \text{ Pa s})(5.0 \text{ m})} = 1.6 \times 10^{-5} \text{ m}^3\text{/s}.$$

PRACTICE PROBLEMS

1. The tires of a 1600 kg automobile are inflated to a pressure of 25 lb/in^2. Assuming that each tire supports 1/4 the weight of the car, calculate the area of contact between a tire and the road.

2. A research diving vessel is shaped like a sphere of radius 3.5 m. If the inside of the vessel is maintained at atmospheric pressure, what is the net force trying to crush the vessel when it is at a depth of 1500 m below the surface of the ocean?

3. What is the weight of the water in a waterbed which is 2.4 m long, 1.8 m wide and 0.23 m deep?

4. A 80.0 kg person weighs 40.0 N in fresh water and finds it difficult to stay afloat. In the Great Salt Lake the density of the water is increased due to dissolved salts and the person finds that he can float just submerged with no effort at all. What is the density of the water in the Great Salt Lake?

5. A "U" tube of volume, V, and uniform cross-section is initially filled with water. Then 1/2 V of mercury is poured into one arm of the tube and pushes the water through the tube. How much water spills out of the other arm of the tube?

6. Water is being siphoned from a fish tank. The procedure begins with filling a hose of diameter 1.0 cm with water and then placing one end of the hose in the tank and the other end 1.0 m below the surface of the water, but outside the tank. A net force then acts to push the water from the tank through the tube. Where does this force originate? What is its magnitude?

7. A garden hose having an internal diameter of 2.0 cm is connected to a lawn sprinkler that consists of an enclosure with 36 holes, each 0.10 cm in diameter. If the water in the hose has a speed of 0.75 m/s, at what speed does it leave the sprinkler holes?

8. A venturi meter has a cross section of 40.0 cm^2 at its entry and exit ports and a cross section of 25.0 cm^2 in its constricted region. Water enters at 3.5 m/s. (a) What is the speed of the water through the constricted region? (b) What is the difference in pressure between the two regions?

9. Water is moving with a speed of 8.0 m/s through a pipe of 4.0 cm diameter. The water gradually descends 15 m as the pipe diameter increases to 6.0 cm. (a) What is the speed of flow at the lower level? (b) If the pressure at the upper level is 2.0×10^5 Pa, what is the pressure at the lower level?

10. A stream of water (initially horizontal) flows from a small hole in the side of a tank. The stream is 2.5 m from the side of the tank after falling 0.70 m. How far below the surface of the water in the tank is the hole?

11. A straight horizontal pipe with a diameter of 1.0 cm and a length of 50.0 m carries oil with a density of $\rho = 930$ kg/m^3 and a viscosity of 0.12 Pa s. The discharge rate is 0.80 kg/s at atmospheric pressure. Find the gauge pressure at the pipe input.

12. During a heavy rain, a 3.0 m x 4.6 m family room is flooded to a depth of 15 cm. To remove the water ($\eta = 1.00 \times 10^{-3}$ Pa-s), a pump is used that does the job in two hours. The water flows through a horizontal pipe of radius 0.64 cm and length 6.7 m. What gauge pressure does the pump produce?

HELPFUL SUGGESTIONS

1. Keep in mind that the pressure generated by a force acting on a surface area depends ONLY on the component of force perpendicular to the surface. The component of the force parallel to the surface produces no pressure. Also, a pressure acting on a surface area produces only a force perpendicular to the surface.

2. Pressures cited in everyday life are almost always gauge pressures. That is, the pressure above atmospheric pressure that would be measured by a pressure gauge exposed to the atmosphere. An example is the pressure in your car's tires, say, 28.0 lb/in^2. This is a gauge pressure. The corresponding absolute pressure is
28.0 lb/in^2 + 14.7 lb/in^2 = 42.7 lb/in^2.

3. To find the pressure due to the weight of a fluid, one only has to find the vertical distance from the surface of the fluid to the point in question, that is, the depth. The pressure is the same at all points within the fluid that are at the same depth. This depth is so important in finding pressures due to fluid weight that engineers give it a special name, the "pressure head". It is the large pressure head of a tall water tower that is responsible for providing the pressure in a city's water system.

4. Bernoulli's equation is only valid under a strict set of assumptions. The equation is valid only for an incompressible, nonviscous, steady flow of fluid with no turbulence.

5. Be careful to be consistent with units when applying the continuity equation or Bernoulli's equation. SI units should be used as a rule; so that liters (or gallons) should be converted to m^3,pressure in atmospheres should be converted to Pa (N/m^2), and so on. Do not mix units.

6. If a pipe is an open pipe, and exposed to the atmosphere, the pressure at that point is 1 atm = 1.0 x 10^5 Pa.

7. Bernoulli's equation relates the conditions of pressure, velocity, and height at two different points in the fluid. When choosing these points in applying the equation, be sure that there are no obstructions between the two points. That is, the two points must lie along the same **streamline**.

EVERYDAY PHYSICS

1. The next time you put air in your tires, pay particular attention to the "contact patch" of a tire as you add air. Can you explain what you see?

2. You can estimate the volume of a helium filled balloon by attaching small fishing weights to the string of the balloon until the balloon neither rises or sinks and then applying Archimedes' principle.

3. Take a glass jar filled to the brim with water and place an oversized lid on it. Turn the jar and lid upside down while holding the lid. Remove your hand from the lid. What happens? Why? It is best to do this experiment over a sink!

4. Airplane wings are an excellent example of Bernoulli's effect. The bottom side of the wing is relatively flat, while the top side is curved. The air must travel a greater distance over the top of the wing and hence, moves faster over the top side. According to equation (11.12), the higher velocity means a lower pressure on the top side. Thus, the greater pressure on the bottom of the wing produces a net force which "lifts" the airplane.

5. The curving of a thrown baseball can be explained in a way similar to that in #4. The spin of the ball produces pressure differences which displace the ball on a curved path.

6. Another illustration of Bernoulli's effect is to take a strip of paper, hold one end between your lips and blow across the top. As long as you blow, the strip will remain in an essentially horizontal position. When you stop, the paper relaxes and dangles below your chin. Explain.

CHAPTER QUIZ

1. The inside of a spaceship is maintained at 75 % of atmospheric pressure while the outside is a perfect vacuum. What is the total force on a 1.0 m X 1.0 m hatch in the wall of the spacecraft?
   a. 5.1×10^4 N b. 2.5×10^4 N c. 3.8×10^4 N d. 7.6×10^4 N

2. An object floats in water half submerged. What is the density of the object?
   a. 500 kg/m^3 b. 2000 kg/m^3 c. 750 kg/m^3 d. It is impossible to know.

3. A diver descends to a depth of 25 m in fresh water. What is total pressure on the diver?
   a. 2.5×10^5 N/m^2 b. 3.5×10^5 N/m^2 c. 1.5×10^5 N/m^2 d. 2.5×10^4 N/m^2

4. A diver descends in the ocean to a depth of 50 m. How does the buoyant force on the diver change as he descends?
   a. It increases linearly with the depth.
   b. It decreases linearly with the depth.
   c. It increases quadratically with the depth.
   d. It doesn't change.

5. A helium filled balloon with a total mass of 580 kg rises with an acceleration of 0.50 m/s^2. What is the volume of the balloon? The density of air is 1.29 kg/m^3.
   a. 4900 m^3 b. 6500 m^3 c. 5400 m^3 d. 470 m^3

6. A hydraulic lift has an input piston with a diameter of 1.0 cm and an output piston with a diameter of 25 cm. How much force acts on the input piston if the output piston supports a 1500 kg car?
   a. 6.0 N b. 59 N c. 0.24 N d. 2.4 N

7. The radius of a pipe carrying water increases by a factor of two. As a result, the speed of the water in the pipe
   a. increases by a factor of two.
   b. decreases by a factor of two.
   c. increases by a factor of four.
   d. decreases by a factor of four.

8. Water flows downward through a pipe of constant diameter. If the water falls through a height of 10.0 m, what is the change in pressure between the two heights? (The density of water is 1.0×10^3 kg/m^3.)
   a. 9.8 Pa b. 980 Pa c. 9800 Pa d. 98 000 Pa

9. Blood flows through a blood vessel at 0.10 m/s. The vessel has a cross-sectional area of 1.0×10^{-4} m^2. The volume flow rate of the blood is
   a. 1.0×10^{-5} kg/s b. 1.0×10^{-5} m^3/s c. 1.0×10^{-3} m^3/s d. 1.0×10^{-2} m^3/s

10. Water flows with a volume flow rate of 2.5 m^3/s. What is the approximate speed of the water where the pipe diameter is 0.5 m?
    a. 2.5 m/s b. 5.0 m/s c. 13 m/s d. 25 m/s

11. A tank holds water which empties out through a hole located 10.0 meters below the surface of the water. How fast is the water moving when it leaves the tank?
    a. 10 m/s b. 14 m/s c. 25 m/s d. 98 m/s

12. The Sears tower is 512 m high. What must be the minimum water pressure in a pipe at ground level in order to get water up to a restaurant on the top floor?
    a. 1.0×10^5 Pa b. 5.0×10^5 Pa c. 6.0×10^5 Pa d. 5.0×10^6 Pa

SOLUTIONS AND ANSWERS

Practice Problems

---

1. Each tire pushes against the ground with a force F = PA which is equal and opposite to the force exerted by the ground on the tire. The total force exerted by the ground on the tire is then 4F which must equal the weight of the car, 4F = 4PA = mg. Solving for A gives A = 1/4 mg/P. Now

$$P = (25\text{lb/in}^2/14.7\ \text{lb/in}^2)(1.01 \text{ X } 10^5\ \text{Pa}) = 1.7 \text{ X } 10^5\ \text{Pa}\ ,$$

so that,

$$A = (1/4)(1600\ \text{kg})(9.8\ \text{m/s}^2)/(1.7 \text{ X } 10^5\ \text{Pa}) = \mathbf{0.023\ m^2}.$$

---

2. The net pressure on the vessel is

$$P = \rho gh = (1025\ \text{kg/m}^3)(9.8\ \text{m/s}^2)(1500\ \text{m}) = 1.5 \text{ X } 10^7\ \text{Pa}.$$

The net force on the spherical surface of the vessel is then

$$F = PA = P(4\pi R^2) = (1.5 \text{ X } 10^7\ \text{Pa})(4\pi)(3.5\ \text{m})^2 = \mathbf{2.3 \text{ X } 10^9\ N}.$$

---

3. The weight of the water is

$$W = mg = \rho gV = (1000\ \text{kg/m}^3)(9.8\ \text{m/s}^2)(2.4\ \text{m})(1.8\ \text{m})(0.23\ \text{m}) = \mathbf{9700\ N}.$$

---

4. In fresh water mg - ρ_wgV = 40.0 N. In the salt water mg - ρ_sgV = 0. The first equation gives the volume of the person to be

$$V = [\ (80.0\ \text{kg})(9.8\ \text{m/s}^2) - 40.0\ \text{N}]/[(1000\ \text{kg/m}^3)(9.8\ \text{m/s}^2)] = 0.076\ \text{m}^3.$$

The second equation then gives the density of the salt water to be

$$\rho_s = m/V = (80.0\ \text{kg})/(0.076\ \text{m}^3) = \mathbf{1050\ kg/m^3}.$$

---

5. Let h_w and h_m be the heights of the water column and mercury columns, respectively, measured from the mercury-water interface. If the column is not to move, the weight of the mercury above the interface must equal the weight of the water above the interface.

$$\rho_w g h_w A = \rho_m g h_m A.$$

The length of the column of water lost is $L_w = L - h_w$ where L is the total length of the tube. Additionally, $2h_w - h_m = 1/2\ L$. Elimination of h_w and h_m from the above equations results in

$$L_w = L - 1/2\ L\ \{\rho_m/\rho_w\}/\{2\rho_m/\rho_w - 1\} = L - (1/2\ L)\ \{13.6\}/\{27.2 - 1\} = 0.74\ L.$$

Hence, the volume of water lost is $V_w = \mathbf{0.74\ V}$.

---

6. The force originates from the weight of the water in the portion of the siphoning tube that lies below the surface of the water in the tank. If h is the height of the water level in the tank above the end of the tube, then the net pressure at the end of the tube is

$$p = p_{atm} + \rho gh - p_{atm} = \rho gh.$$

This pressure generates a force,

$$F = pA = \rho ghA = (1000\ kg/m^3)(9.8\ m/s^2)(1.0\ m)(3.14)(0.05 \times 10^{-2}\ m)^2 = \mathbf{7.7 \times 10^{-3}\ N}$$

to push the water from the tube.

---

7. Use $A_1v_1 = A_2v_2$, where

$$A_2 = \pi r_2^2 = \pi(1.0 \times 10^{-2}\ m)^2 = 3.14 \times 10^{-4}\ m^2,\ v_2 = 0.75\ m/s,$$
$$A_1 = 36(\pi r_1^2) = 36\pi(5.0 \times 10^{-4}\ m)^2 = 2.83 \times 10^{-5}\ m^2.$$

Solving for v_1 we obtain,

$$v_1 = v_2(A_2/A_1) = (0.75\ m/s)(3.14 \times 10^{-4}\ m^2)/(2.83 \times 10^{-5}\ m^2) = \mathbf{8.3\ m/s.}$$

---

8. (a) Equation (11.9), $A_1v_1 = A_2v_2$ so that

$$v_1 = v_2(A_2/A_1) = (3.5\ m/s)(40\ cm^2/25\ cm^2) = \mathbf{5.6\ m/s.}$$

(b) Equation (11.12),

$$P_1 + (1/2)\rho v_1^2 = P_2 + (1/2)\rho v_2^2,$$

solving for the change in pressure,

$$\Delta P = (P_2 - P_1) = (1/2)\rho(v_1^2 - v_2^2) = (1/2)(1.0 \times 10^3\ kg/m^3)[(5.6\ m/s)^2 - (3.5\ m/s)^2] = \mathbf{9600\ N/m^2.}$$

---

9. (a) Equation (11.9),

$$v_1 = v_1(A_2/A_1) = (8.0\ m/s)(4.0\ cm/6.0\ cm)^2 = \mathbf{3.6\ m/s.}$$

(b) Solving equation (11.11) for P_1 yields, $P_1 = P_2 + (1/2)\rho(v_2^2 - v_1^2) + \rho gh$, substituting,

$$P_1 = 2.0 \times 10^5\ Pa + (1/2)(1.0 \times 10^3\ kg/m^3)[(8.0\ m/s)^2 - (3.6\ m/s)^2]$$
$$+ (1.0 \times 10^3\ kg/m^3)(9.8\ m/s^2)(15\ m) = \mathbf{3.7 \times 10^5\ Pa.}$$

---

10. Find the speed of the water as it leaves the hole. Using $v_{0y} = 0$ we can write

$$t^2 = 2y/a = 2(0.7\ m)/(9.8\ m/s^2)$$

$$t = 0.38\ s.$$

Using $a_x = 0$ gives $x = v_{0x}t$ so that

$$v_{0x} = x/t = (2.5\ m)/(0.38\ s) = 6.6\ m/s.$$

Using the result $v = \sqrt{2gh}$, we obtain

$$h = v^2/2g = (6.6\ m/s)^2/2(9.8\ m/s^2) = \mathbf{2.2\ m.}$$

---

11. The flow rate is

$$Q = (0.80 \text{ kg/s})/(930 \text{ kg/m}^3) = 8.6 \times 10^{-4} \text{ m}^3/\text{s}.$$

Now use equation (11.14) and solve for ΔP,

$$\Delta P = P_2 - P_1 = 8\eta LQ/\pi R^4 = 8(0.12 \text{ Pa s})(50.0 \text{ m})(8.6 \times 10^{-4} \text{ m}^3/\text{s})/\pi(5.0 \times 10^{-3} \text{ m})^4 = \mathbf{2.1 \times 10^7 \text{ Pa}}.$$

---

12. Find the volume flow rate,

$$Q = V/t = (3.0 \text{ m} \times 4.6 \text{ m} \times 0.15 \text{ m})/(2 \text{ h}) = (2.07 \text{ m}^3)/(7200 \text{ s}) = 2.9 \times 10^{-4} \text{ m}^3/\text{s}.$$

Using equation (11.14), we can solve for the gauge pressure, $P_g = P_2 - P_1$, i.e.,

$$P_g = 8\eta LQ/\pi R^4 = 8(1.00 \times 10^{-3} \text{ Pa s})(6.7 \text{ m})(2.9 \times 10^{-4} \text{ m}^3/\text{s})/\pi(0.64 \times 10^{-2} \text{ m})^4 = \mathbf{2.9 \times 10^3 \text{ Pa}}.$$

Quiz answers

1. d	4. d	7. d	10. c
2. a	5. d	8. d	11. b
3. b	6. d	9. b	12. d

MCAT REVIEW PROBLEMS

Bernoulli's law and the continuity equation can help us understand some aspects of the circulatory system. According to Bernoulli's law,

$$p_1 + \frac{1}{2}\rho v_1^2 + \rho g h_1 = p_2 + \frac{1}{2}\rho v_2^2 + \rho g h_2,$$

where p refers to the pressure, v refers to the speed, and h refers to the height of the fluid. The subscripts "1" and "2" refer to the fluid at two different points in the system.

According to the continuity equation, if a piece of the "pipeline" gets thicker, gets thinner, or branches, the volume of fluid per second passing two imaginary cross-sectional "slices" must be the same. For instance, in both cases drawn in figure 1, the volume of fluid per second passing slice 1 must equal the volume of fluid per second passing slice 2.

In case 1, the pipeline doesn't branch. In such cases, for essentially incompressible fluids, the continuity equation can be written

$$A_1 v_1 = A_2 v_2,$$

where A denotes the cross-sectional area of the "pipe."

FIGURE 1

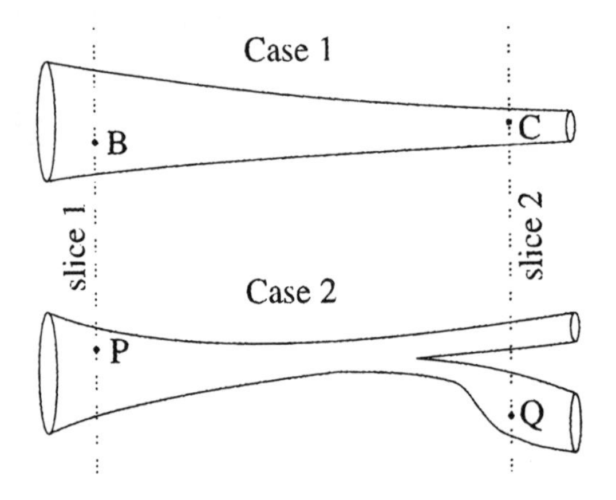

---

1. Consider case 1. The blood at point B flows at 10 m/s. The blood at point C flows at a speed:

   A. greater than 10 m/s **B.** equal to 10 m/s
   C. less than 10 m/s **D.** Impossible to determine from the information provided

---

2. Consider case 2. The blood at point P flows at speed 10 m/s. The blood at point Q flows at a speed:

   A. greater than 10 m/s. **B.** equal to 10 m/s.
   C. less than 10 m/s. **D.** Impossible to determine from the information provided

---

3. Assume that points B and C are at the same height. Let p_B and p_C denote the pressure at points B and C, respectively. Which of the following is true?

   A. $p_B > p_C$ **B.** $p_B = p_C$
   C. $p_B < p_C$ **D.** We cannot determine the relationship between p_B and p_C.

---

4. Suppose a blood vessel gets thinner, but doesn't "branch." Case 1 is an example. Which of the following graphs best shows the relationship between the cross-sectional area of the blood vessel and the speed of the blood flowing through it?

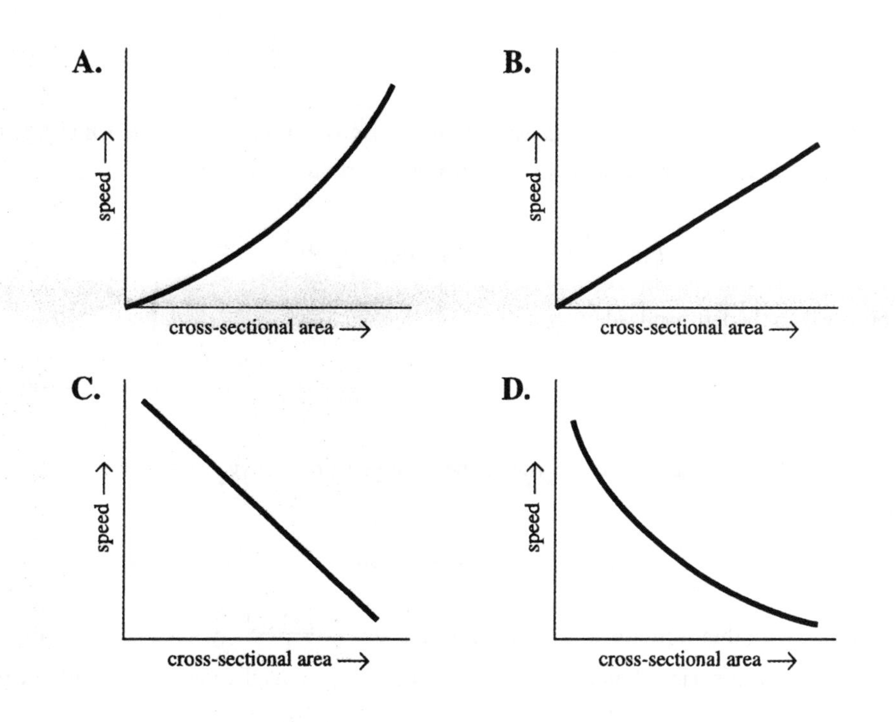

ANSWERS TO MCAT REVIEW PROBLEMS

---

1. **A.** The continuity equation tells us that $A_B v_B = A_C v_C$. And the blood vessel is thinner at point C. In other words, A_C is less than A_B. Therefore, v_C must be *greater* than v_B, to ensure that the two sides of the continuity equation are in fact equal.

---

2. **D.** Because the blood vessel "branches," the equation $A_1 v_1 = A_2 v_2$ doesn't apply. We can't set $A_P v_P$ equal to $A_Q v_Q$. Therefore, without more information, we can't relate v_P to v_Q.

---

3. **A.** As found in question 1, v_B is less than v_C. Since those two points share the same height, Bernoulli's law implies that p_B must be greater than p_C. Otherwise, the right-hand side of

$$p_B + \frac{1}{2}\rho v_B^2 + \rho g h_B = p_C + \frac{1}{2}\rho v_C^2 + \rho g h_C$$

would be bigger than the left-hand side.

---

4. **D.** According to the continuity equation for unbranched pipelines, $A_1 v_1 = A_2 v_2 = A_3 v_3$, and so on. In other words,

$$A v = \text{constant}.$$

By playing around with this formula, or by rewriting it as $v = \frac{\text{constant}}{A}$, you can confirm that v is inversely proportional to A. For instance, doubling the cross-sectional area cuts the velocity in half. Tripling A cuts v in third. Quadrupling A cuts v in fourth.

Graph C does *not* express this relationship. It shows v decreasing *linearly* with A. The following two tables demonstrate the difference between an inverse proportionality and a linear decrease. Make sure you develop an intuitive feel for this distinction.

TABLE 1: INVERSE PROPORTIONALITY

$(v = \frac{1}{A})$

A (mm^2)	v (m/s)
1	1.00
2	0.50
3	0.33
4	0.25

TABLE 2: LINEAR DECREASE
$(v = -\frac{1}{5}A + \frac{6}{5})$

A (mm^2)	v (m/s)
1	1.00
2	0.80
3	0.60
4	0.40

CHAPTER 12
Temperature and Heat

PREVIEW

In this chapter you will become acquainted with the concept of temperature and the three most common temperature measurement scales: Celsius, Fahrenheit and Kelvin. You will learn to convert a temperature reading in one scale to any other. You will also be introduced to various methods of measuring temperatures and will study in detail one of the most common methods: the thermal expansion of materials.

Additionally, you will study heat including the concepts of specific heat, latent heat of phase change, phase diagrams, and humidity. This chapter will serve as an introduction to later chapters involving the kinetic theory of gases and thermodynamics.

QUICK REFERENCE

Important Terms

Fahrenheit scale
The common temperature scale which assigns 32 ° to the freezing point of water and 212 ° to the boiling point of water.

Celsius scale
A scientific temperature scale which assigns 0 ° to the freezing point of water and 100 ° to the boiling point of water.

Kelvin (absolute) scale
A scientific temperature scale which has the same size degree as the Celsius scale, but assigns the lowest possible temperature as 0 °. A degree is called a Kelvin (K) and is the SI unit for temperature.

Absolute zero
The zero point on the Kelvin temperature scale. It is the lowest possible temperature that can be attained by cooling an object.

Thermal expansion
The change in physical size (length, area or volume) of a substance when its temperature changes. For most substances, the physical size increases with an increase in temperature and decreases with a decrease in temperature.

Thermal stress
A stress which develops within an object when it attempts to expand or contract in response to a temperature change, but cannot due to being held rigidly in place.

Internal Energy
The energy associated with individual molecules in a gas, liquid, or solid. This energy may take the form of translational or rotational kinetic energy, vibrational energy, or potential energy.

Heat
The energy that flows from a higher temperature object to a lower temperature object because of the difference in temperatures. The SI unit of heat is the joule (J). Another unit for heat is the calorie (cal) or kilocalorie (kcal).

Specific heat capacity
The heat Q per unit mass per degree change in temperature that must be supplied or removed to change the temperature of a substance. The SI unit for specific heat is J/(kg C°).

Latent heat
The amount of heat energy per kilogram that must be added or removed when a substance changes from one phase to another (i.e., from solid to liquid - heat of fusion or from liquid to gas - heat of vaporization). The SI unit of latent heat is J/kg.

Equilibrium vapor pressure
The pressure of the vapor phase of a substance that is in equilibrium with the liquid or solid phase. Vapor pressure depends only on temperature.

Phase diagram
A plot of pressure vs. temperature for a given substance showing the various phases possible for that particular substance.

Temperature Scale Conversion

$$T_C = \frac{5}{9}(T_F - 32.0)$$

$$T_F = \frac{9}{5}T_C + 32.0$$

$$T = T_C + 273.15 \qquad (12.1)$$

Thermal Expansion

Linear expansion

$$\Delta L = \alpha L_0 \Delta T \qquad (12.2)$$

Volume Expansion

$$\Delta V = \beta V_0 \Delta T \qquad (12.3)$$

DISCUSSION OF SELECTED SECTIONS

12.1 Common Temperature Scales

Example 1
At what temperature will the reading on the Fahrenheit scale be numerically equal to that on the Kelvin scale?

The relationship between the two scales is given by $T_F = (9/5)\ T_C + 32.0$. Setting $T_F = T_C$ and rearranging gives

$$T_F - (9/5)\ T_F = 32.0$$

$$T_F = -\ (5/4)(32.0) = -\ 40.0$$

Example 2
The surface temperature of the sun is approximately 5700 °C. What is this temperature on the Fahrenheit scale?

$$T_F = (9/5)\ T_C + 32.0$$
$$T_F = (9/5)(5700) + 32.0 = 1.0 \text{ X } 10^4\ °F$$

Example 3
Most bank thermometers alternately display the temperature in Fahrenheit and Celsius. If a bank thermometer displays 25 °F, what will it display for the Celsius temperature?

The relationship between Fahrenheit and Celsius temperatures is

$$T_C = (5/9)(T_F - 32.0)$$
$$T_C = (5/9)(25 - 32.0) = -4\ °C$$

12.2 The Kelvin Temperature Scale

The only difference between the Kelvin and Celsius scales is in the choice of the zero point. The zero point for the Celsius scale was chosen to be the freezing point of water since water is abundant and its freezing point is easily measurable. The zero of the Kelvin scale is chosen to be the lowest possible temperature, that is, absolute zero. Absolute zero corresponds to - 273.15 °C. To convert between the two scales it is only necessary to adjust for this difference in zero points by adding or subtracting 273.15.

If you should need to convert from Fahrenheit to Kelvin, simply convert the Fahrenheit temperature to Celsius and then add 273.15 to get the Kelvin temperature.

Example 4
Many people keep their rooms at a temperature of 65 °F. What is this temperature on the Kelvin scale?

First convert the temperature to Celsius.

$$T_C = (5/9)(T_F - 32.0) = (5/9)(65 - 32.0) = 18\ °C$$

Now add 273.15 to get the temperature on the Kelvin scale.

$$T = 18 + 273.15 = 291\ K$$

12.4 Linear Thermal Expansion

Example 5
How much will the length a 1.0 km section of concrete highway change if the temperature varies from - 15 °C in the winter to 41 °C in the summer?

Using equation 12.2 and the value of the coefficient of linear expansion for concrete from Table 12.2, we have

$$\Delta L = \alpha LT = (12 \times 10^{-6} /C°)(1.0 \times 10^3\ m)[(41\ °C - (-15\ °C)]$$
$$\Delta L = 0.67\ m$$

Example 6
Two 12 ft sections of aluminum siding are placed end to end on the outside wall of a house. How large a gap should be left between the pieces to prevent buckling if the temperature can change by 55 C°?

Each piece can expand an amount

$$\Delta L = \alpha L\Delta T = (23 \times 10^{-6} /°C)(12\ ft)(55\ °C) = 0.015\ ft = 0.18\ in.$$

If each piece can expand in both directions, then each will expand 0.09 in. toward the other. The gap should be at least 2(0.09 in.) = 0.18 in. to keep the pieces from contacting each other and buckling.

Example 7
A glass window pane (1.100 m X 0.920 m) is fitted inside an aluminum frame with a clearance of 0.05 cm all around on a day when the temperature is 39 °C. How much must the temperature drop for the frame to begin to exert a stress on the window pane?

The changes of the largest dimensions of the pane and frame are $\Delta L_g = \alpha_g L_g \Delta T$ and $\Delta L_a = \alpha_a L_a \Delta T$. The pieces will contact when $\Delta L_a - \Delta L_g = 0.10$ cm. The necessary change in temperature is then

$$\Delta T = \frac{0.10 \text{ X } 10^{-2} \text{ m}}{\alpha_a L_a - \alpha_g L_g} = \frac{0.10 \text{ X } 10^{-2} \text{ m}}{(23 \text{ X } 10^{-6}/\text{C}°)(1.101 \text{ m}) - (8.5 \text{ X } 10^{-6}/\text{C}°)(1.100 \text{ m})}$$

$$\Delta T = 63 \text{ C}°$$

Similarly, for the smallest dimensions,

$$\Delta T = \frac{0.10 \text{ X } 10^{-2} \text{ m}}{(23 \text{ X } 10^{-6}/\text{C}°)(0.921 \text{ m}) - (8.5 \text{ X } 10^{-6}/\text{C}°)(0.920 \text{ m})} = 75 \text{ C}°$$

The window pane and frame will contact along the large dimension first when the temperature drops by 63 C°.

12.5 Volume Thermal Expansion

Example 8
A diesel powered automobile has its steel tank filled to the top with 22.0 gal of diesel fuel from an underground storage tank. The fuel in the storage tank is initially at 55 °F (13 °C) and warms up to 95 °F (35 °C) as it sits in the car's tank. How much fuel spills out of the car's tank? The coefficient of volume expansion of diesel fuel is $9.5 \text{ X } 10^{-4}$ /C°.

The interior volume of the car's tank expands as if it were made of steel. It will increase by $\Delta V_t = \beta_s V \Delta T$. The fuel will increase its volume by an amount $\Delta V_f = \beta_f V \Delta T$. The amount of fuel which will spill out of the tank, ΔV, is then

$$\Delta V = \Delta V_f - \Delta V_t = (\beta_f - \beta_t) V \Delta T$$
$$\Delta V = (9.5 \text{ X } 10^{-4}/\text{C}° - 36 \text{ X } 10^{-6}/\text{C}°)(22 \text{ gal})(22 \text{ C}°)$$
$$\Delta V = 0.44 \text{ gal}$$

12.7 Heat and Temperature Change: Specific Heat Capacity

In order to raise the temperature of a substance by an amount ΔT, a certain amount of heat, Q, is required. The greater the mass m of the substance, the greater the amount of heat required to raise it by ΔT. The relationship between the heat required, the mass, and the change in temperature is

$$Q = cm\Delta T \qquad (12.4)$$

where the constant of proportionality, c, is referred to as the **specific heat capacity** or simply the specific heat. The specific heat has units of J/(kg C°). Table 12.3 shows the specific heat capacities for various substances. For example, water has a specific heat of 4186 J/(kg C°). In other words, it would take 4186 J of heat to raise the temperature of 1 kilogram of water by 1 °C.

Example 9
How much heat is required to raise 5.0 kg of water from 25 °C to 55° C?

Using equation (12.4) we have:

$$Q = cm\Delta T = [4186 \text{ J/(kg C°)}](5.0 \text{ kg})(55 \text{ °C} - 25 \text{ °C}) = 6.3 \times 10^5 \text{ J}.$$

An common unit for measuring heat is the **kilocalorie** (kcal), defined as the amount of heat energy required to raise the temperature of 1 kilogram of water by 1 °C. From the above discussion it is obvious that 4186 J = 1 kcal. This conversion factor is known as the **mechanical equivalent of heat**.

Example 10
If 15 kcal of heat are added to 5.0 kg of silver, by how much will its temperature rise?

Using equation (12.4) and the specific heat of silver found in Table 12.2, we can solve for ΔT. We have

$$\Delta T = \frac{Q}{cm} = \frac{(15 \text{ kcal})(4186 \text{ J/1 kcal})}{[235 \text{ J/(kg C°)}](5.0 \text{ kg})} = 53 \text{ C°}$$

Specific heat capacity can be measured using the technique of **calorimetry**. A calorimeter is an insulated container much like a thermos for hot coffee or iced tea. An ideal calorimeter would prevent any heat from leaking in or out. However, heat can flow between the materials inside the calorimeter which have different temperatures. Cooler materials gain heat while hotter materials lose heat until a common temperature is reached. That is, until **thermal equilibrium** is achieved. In the process of reaching thermal equilibrium, the total amount of heat gained equals the total amount of heat lost.

Example 11
An aluminum cup having a mass of 250.0 g is filled with 50.0 g of water. The initial temperature of the cup and water is 25.0 °C. A 75.0-g piece of iron initially at 350.0 °C is dropped into the water. What is the final equilibrium temperature of the system assuming that no heat is lost to the outside environment?

Using equation (12.4) and the fact that the heat lost by the iron equals the heat gained by the aluminum + water,

$$\text{Heat Gained} = \text{Heat Lost}$$

$$(cm\Delta T)_{aluminum} + (cm\Delta T)_{water} = (cm\Delta T)_{iron}$$

$$[9.00 \times 10^2 \text{ J/(kg C°)}](0.250 \text{ kg})(T_f - 25.0 \text{ °C}) + [4186 \text{ J/(kg C°)}](0.0500 \text{ kg})(T_f - 25.0 \text{ °C})$$
$$= [452 \text{ J/(kg C°)}](0.0750 \text{ kg})(350.0 \text{ °C} - T_f)$$

$$T_f = 48.5 \text{ °C}$$

12.8 Heat and Phase Change: Latent Heat

There are three phases of matter that exist in nature; solid, liquid, and vapor (gas). Nearly all substances can exist in all three phases, depending on the conditions of temperature and pressure that prevail. For example, consider water at atmospheric pressure. For low temperatures (below 0 °C) water exists as a solid, in the form of ice. At intermediate temperatures (between 0 °C and 100 °C) water is a liquid, and at high temperatures (above 100 °C) it is a gas (water vapor).

We know that water can change phases, either from solid to liquid (or vice versa) or from liquid to gas. The simple act of ice melting represents a phase change.

It is important to note that phase changes always occur at a single temperature, with the addition or subtraction of heat. For example, in order to melt 1 kg of ice, the ice must be at 0 °C and we need to add 33.5 x 10^4 J of heat to get 1 kg of liquid water which will still be at 0 °C.

The amount of heat per kilogram required to produce a change of phase for a substance is known as the **latent heat** of that substance. The **latent heat of fusion, L_f**, refers to a phase change between the liquid and solid phases, and the **latent heat of vaporization, L_v**, refers to a change between liquid and vapor phases.

Example 12

How much heat is required to convert 250 g of ice at 0 °C to steam at 100 °C?

The problem includes three distinct parts. (1) The heat required to melt the ice, (2) the heat required to raise the temperature of the water from 0 °C to 100 °C, and (3) the heat required to convert the hot water to steam at 100 °C. Referring to Table 12.5 for the values of the latent heats of fusion and vaporization we can write

$$Q_{total} = Q_{fusion} + Q_{temp.\ change} + Q_{vaporization}$$

$$Q_{total} = mL_f + cm\Delta T + mL_v$$

$$Q_{total} = (0.25\ \text{kg})(33.5 \times 10^4\ \text{J/kg}) + [4186\ \text{J/(kg C°)}](0.25\ \text{kg})(100\ \text{C°}) + (0.25\ \text{kg})(22.6 \times 10^5\ \text{J/kg})$$

$$Q_{total} = 7.5 \times 10^5\ \text{J.}$$

12.9 Equilibrium Between Phases of Matter

Under certain conditions of temperature and pressure, a substance can exist in equilibrium in more than one phase at the same time. The pressure at which vapor coexists in equilibrium with liquid is called the **equilibrium vapor pressure**. For example, consider the following **vaporization curve** for water.

We can see that, for water, at a pressure of 1.01 x 10^5 Pa, the corresponding vaporization temperature is 100 °C. Thus, liquid water and vapor can exist in equilibrium at this pressure and temperature. Water, therefore, boils at 100 °C at atmospheric pressure. However, if we were to change the pressure, the corresponding temperature needed for equilibrium would change. For example, as evident from the curve, if the pressure were about 2 x 10^5 Pa, the corresponding temperature required for equilibrium would be about 125 °C. So water would boil at 125 °C if it were subjected to two atmospheres of pressure.

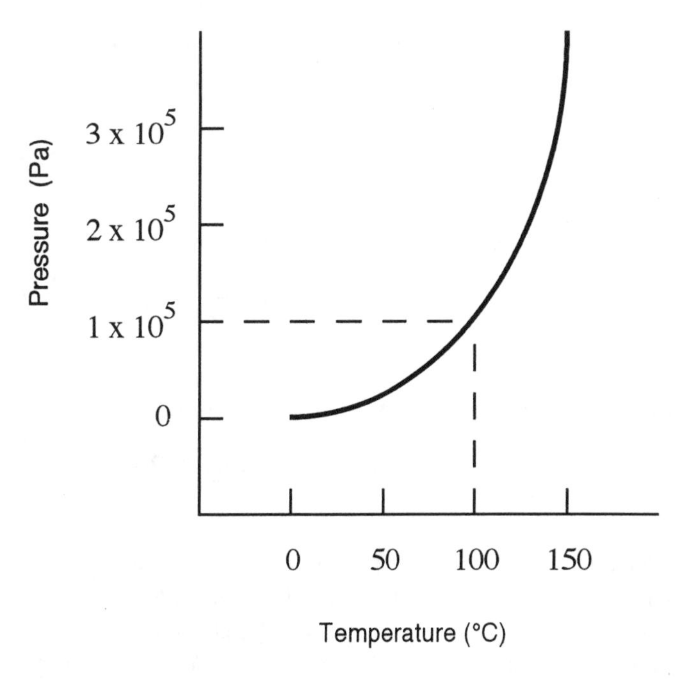

12.10 Humidity

The partial pressure of water vapor is the part of the total atmospheric pressure that is due to water vapor in the air. For example, out of the total 1.01×10^5 Pa of pressure in our atmosphere, 4.0×10^3 Pa may be due to water vapor.

The **relative humidity** is the partial pressure of water vapor at the existing temperature divided by the equilibrium vapor pressure of water at the existing temperature. That is

$$\text{Percent Relative Humidity} = \frac{\text{Partial pressure of water vapor}}{\text{Equil. Vapor Pressure at Given Temperature}} \times 100 \qquad (12.5)$$

Example 13

On a certain day the partial pressure of water vapor is 1.0×10^3 Pa. Using the vaporization curve below, determine the relative humidity if the temperature is (a) 16 °C, and (b) 26 °C.

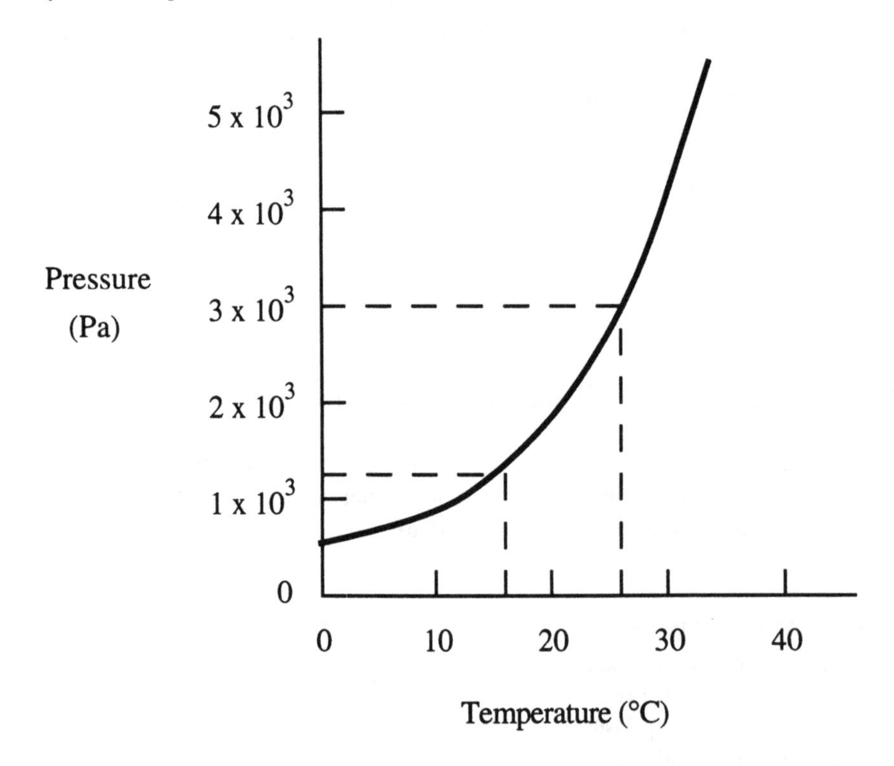

(a) At 16 °C we see that the equilibrium vapor pressure is 1.3×10^3 Pa. Therefore, using equatión (12.5), we have

$$\text{Relative humidity at 16 °C} = \frac{1.0 \times 10^3 \text{ Pa}}{1.3 \times 10^3 \text{ Pa}} \times 100 = 77\%$$

(b) At 26 °C the equilibrium vapor pressure is 3.0×10^3 Pa. Therefore,

$$\text{Relative humidity at 26 °C} = \frac{1.0 \times 10^3 \text{ Pa}}{3.0 \times 10^3 \text{ Pa}} \times 100 = 33\%$$

When the partial pressure of water vapor equals the equilibrium vapor pressure of water at a given temperature, equation (12.5) tells us that the relative humidity is 100%. In this case, the air is said to be saturated. The temperature corresponding to this condition is known as the **dew point**. The dew point is therefore the temperature at which the water vapor in the air would condense in the form of liquid drops (dew or fog). In example 13 it can be seen that the dew point (the temperature corresponding to a vapor pressure of 1.0×10^3 Pa) is about 13 °C.

PRACTICE PROBLEMS

1. At what common temperature are the Fahrenheit and Kelvin temperatures the same?

2. A 1.5 m steel wire is heated from room temperature (22 °C) to 150 °C. How much does its length change?

3. What is the minimum number of 1/2 inch expansion joints that must be cut per mile in a concrete highway to prevent buckling if the temperature can range from - 15 °F to 105 °F?

4. A brass collar (inside radius = 2.500 cm) is to be fitted on a steel shaft (radius 2.505 cm). The collar and shaft are initially at 22.0 °C. To what temperature must the collar and shaft be heated?

5. A bolt hole in the cast iron block of an automobile engine has a diameter of 1.27 cm when the engine is at a cool temperature of 25.0 °C. What is the diameter of the hole when the engine block is at a temperature of 98.0 °C?

6. What is the temperature on the Fahrenheit temperature scale that corresponds to absolute zero?

7. An aluminum airplane has a wingspan of 9.50 m on the ground where the temperature is 25 °C. What is its wingspan when it is at an altitude where the temperature is - 35 °C?

8. A one quart glass canning jar is filled with veggies and liquid at room temperature (72 °F) and heated to 212 °F so that liquid has overflowed the jar, and cooled back to room temperature. Assume that the veggies and liquid expand like water and find the volume of the hot veggies and liquid which have spilled out of the jar.

9. Show that the change in area of a rectangular sheet of material with a linear coefficient of expansion, α, is approximately $\Delta A = 2\alpha A \Delta T$.

10. An automotive cooling system holds 15 quarts of pure water at 10.0 °C. Use Figure 12.19 in the text to determine how much the volume of the water has changed if the temperature drops to 4.0 °C.

11. A 150.0-g glass beaker contains 750 g of water initially at 25.0 °C. A 250.0-g piece of copper is taken out of a hot oven and dropped into the water. After equilibrium is reached between the copper, glass, and water, the final temperature is 32.0 °C. What was the temperature of the oven?

12. Calculate the specific heat of a metal from the following data. A container made of metal has a mass of 1.50 kg and contains 6.00 kg of water. A 0.900 kg piece of the same metal initially at a temperature of 275 °C is dropped into the water. The container and water are initially at a temperature of 17.0 °C and the final equilibrium temperature of the system is 26.0 °C.

13. A well-insulated container holds 1.5 liters of water at 35 °C. How much ice (initially at 0 °C) must be added to this water in order to produce water at 12 °C once equilibrium has been achieved? Neglect the heat capacity of the container.

14. A 50.0-g aluminum calorimeter contains water at a temperature of 26 °C. When 75-g of lead at a temperature of 85 °C is dropped into the calorimeter, the final equilibrium temperature is 28 °C. How much water was in the calorimeter can?

15. Ice at -10.0 °C and water vapor at 130 °C are brought together at atmospheric pressure in a perfectly insulated container whose heat capacity can be neglected. After thermal equilibrium is reached, the liquid phase at 50.0 °C is present. Ignoring the equilibrium vapor pressure at 50.0 °C, find the ratio of the mass of the steam to that of the ice.

16. Using the vaporization curve that accompanies this problem, find the partial pressure of water vapor on a day when the weather forecast gives the relative humidity at 60% and the temperature at 30 °C.

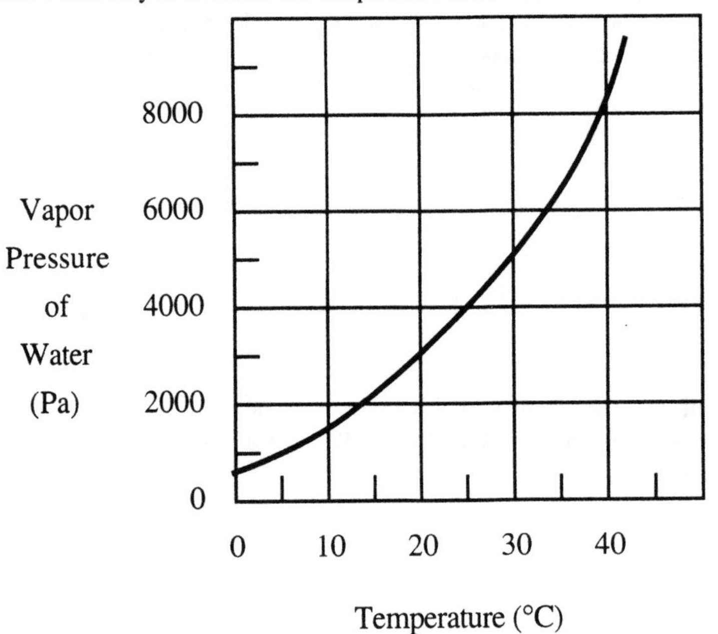

17. Using the vaporization curve which accompanies problem 16, determine the dew point when the temperature is 40.0 °C and the relative humidity is 37.5%.

18. The outdoor temperature is 15 °C and the relative humidity is 75%. The partial pressure of water vapor inside a house is the same as outdoors, but the temperature in the house is 25 °C. Using the vaporization curve for water vapor that accompanies problem 16, determine the relative humidity in the house.

HELPFUL SUGGESTIONS

1. Keep in mind that any hole or void in an object expands or contracts as if it were made of the same material as the object itself.

2. When an object attempts to expand or contract due to a temperature change but is restrained, a stress develops. The results of Chapter 10 can be applied to the situation. This is a good example of how the study of physics is a building process and requires that you keep in mind the concepts and methods that you previously learned.

3. Heat is energy that flows between objects because the objects are at unequal temperatures. Remember that heat is a form of energy and is, therefore, measured in joules (J) or calories (cal). It is tempting to say that the temperature of a substance tells how hot or cold it is, but this is really not correct. Temperature will be more formally defined in Chapter 14 in terms of the average kinetic energy of the individual molecules in a substance.

4. Since the temperature of an object can be raised by adding heat to it or by doing work on it, there is an equivalence between heat and mechanical work. One calorie of heat is equivalent to 4.186 J of energy.

5. The equation, $Q = cm\Delta T$, is valid for heat calculations only if no phase changes occur within the temperature range ΔT. If a change of phase occurs, $Q = mL$ must be used instead. Remember that phase changes occur at a particular temperature and require addition or removal of heat even though the temperature is fixed.

6. When ice is melted into water (or when steam condenses to liquid water), be sure to add the mass of the converted ice (or steam) to the original water mass in calculations involving a subsequent temperature change.

7. The basis for the solution of calorimetry problems is that in a closed environment, hot objects lose the same amount of energy that cold objects gain. In other words, $Q_{gained} = Q_{lost}$. This does NOT mean that the hot and cold objects must undergo the same temperature change, however. Substances with different heat capacities will undergo different temperature changes even if they exchange equal amounts of heat.

8. The temperature at which a substance undergoes a phase change varies with pressure. For example, water vaporizes at 100 °C ONLY at atmospheric pressure. If the we change the pressure, the temperature for vaporization will change as well. We see from the **phase diagram** for water that at about one-half of atmospheric pressure, water boils at 83 °C, rather than 100 °C.

EVERYDAY PHYSICS

1. Almost everyone has watched the rise of a column of liquid in a mercury or alcohol thermometer as it is heated by holding the fingers against the bulb. Removal of the fingers results in an almost immediate drop of the column. Can you explain why the mercury column in a medical thermometer DOES NOT drop after the thermometer is removed from your mouth? HINT: Is the thermometer constructed differently?

2. A good example of thermal expansion is when hot water is used to loosen a metal lid on a glass jar. Why does this work?

3. An inexpensive bimetallic (dial) thermometer can be purchased at most department stores. Obtain one, take it apart, and investigate. Why is the bimetallic strip coiled?

4. Why is water used to cool automobile engines, nuclear reactors, and so on? Notice from Table 12.3 that water has the highest specific heat capacity of any of the substances listed. It is therefore able to carry heat away (or store heat) better than these other materials. Along these same lines, explain why your steak gets cold while your soup remains hot at a restaurant.

5. Why does the temperature in St. Louis (which is inland) vary so much more than the temperature in New York (which is on the coast)?

6. A bowl of water at room temperature is placed in a vacuum jar. The pressure in the jar is slowly reduced so that the temperature for vaporization is lowered. Eventually, we can lower the pressure so much that the water will boil at 0 °C. Continued cooling by evaporation results in the formation of ice over the surface of the boiling water. Boiling and freezing are taking place at the same time! If we spray some drops of coffee into the vacuum chamber they too will boil until they freeze. The water molecules in the coffee will evaporate into the vacuum until little crystals of coffee solids remain. This is how "freeze-dried coffee" is produced.

CHAPTER QUIZ

1. In order to use a substance to make a thermometer the substance MUST__________with a temperature change.
   a. expand b. contract c. change linearly d. change

2. 110 °F corresponds to ______°C.
   a. 43 b. 140 c. 79 d. 316

3. A metal wire changes its length by 0.23 % when its temperature changes by exactly 100 C°. What is the coefficient of linear expansion of the metal?
   a. 4.3×10^{-5} b. 23×10^{-6} c. 43 d. 2.3×10^{-4}

4. A hole is cut in an aluminum sheet and the sheet is subsequently heated. The sheet will ________ while the hole________.
   a. expand, contracts b. contract, contracts c. contract, expands d. expand, expands

5. Brass has $\alpha = 19 \times 10^{-6}\ (C°)^{-1}$ and steel has $\alpha = 12 \times 10^{-6}\ (C°)^{-1}$. It is possible to put a small collar on a larger shaft by heating both the collar and shaft if the shaft is________and the collar is ________.
   a. brass, steel b. steel,brass c. steel, steel d. brass, brass

6. For most solids, the coefficient of volume expansion is ________ the coefficient of linear expansion.
   a. unrelated to b. proportional to c. twice d. three times

7. The volume of a given amount of water ________ as the temperature decreases from 4 °C to 0 °C.
   a. decreases b. increases c. remains constant

8. The temperatures of equal volumes of two liquids are changed by the same amount. One liquid is observed to increase its volume twice as much as the other. The coefficient of volume expansion of the liquid which expands the most is _______ the coefficient of volume expansion of the other.
   a. one-half b. twice c. one-fourth d. four times

9. Liquid hydrogen boils at 17 K. What is this temperature in degrees Celsius?
   a. 290 b. 63 c. - 120 d. - 256

10. The gas in a constant gas thermometer cooled to absolute zero would have
    a. no volume. b. no pressure. c. a zero temperature on all scales.

11. Equal masses of iron and water are initially in thermal equilibrium (note: water has a higher specific heat capacity than iron). The same amount of heat is added to each. Which statement is correct?
    a. They will remain in thermal equilibrium.
    b. They are no longer in thermal equilibrium; the iron is warmer.
    c. They are no longer in thermal equilibrium; the water is warmer.
    d. It is impossible to say which is warmer without knowing the actual amount of heat that was added.

12. When a solid melts,
    a. the temperature of the substance increases.
    b. the temperature of the substance decreases.
    c. heat leaves the substance.
    d. heat enters the substance.

13. A hot piece of silver and a cold piece of lead are thermally insulated from the environment. The silver and lead are placed in thermal contact (Note: the specific heat of silver is higher than that of lead). Which object will experience the greater temperature change?
    a. the silver.
    b. the lead.
    c. neither, both will experience the same temperature change.
    d. it is impossible to tell without knowing the masses.

14. How many kilocalories of heat are required to heat 750 g of water from 35 °C to 55 °C.
    a. 15
    b. 1500
    c. 1.5×10^4
    d. 6.3×10^4

15. Which of the following does **not** determine the amount of internal energy an object has?
    a. temperature
    b. amount of material
    c. type of material
    d. shape of the object

16. How much heat is required to convert 250 g of ice at 0.0 °C to liquid water at 0.0 °C?
    a. 8.4×10^4 J
    b. 8.4×10^4 cal
    c. 1100 J
    d. None, since the temperature is constant

17. A chunk of melting ice at atmospheric pressure is added to an insulated container of cold water (T = 0 °C). What happens in the container?
    a. the ice continues to melt until equilibrium is reached.
    b. the water cools down until equilibrium is reached.
    c. some of the water freezes and the chunk of ice gets bigger.
    d. none of the above.

18. A chunk of ice at T = - 20 °C is added to an insulated container of cold water (T = 0 °C). What happens in the container?
    a. the ice melts until equilibrium is reached.
    b. the water cools down until equilibrium is reached.
    c. some of the water freezes and the chunk of ice gets bigger.
    d. none of the above.

19. On a day when the partial pressure of water vapor remains constant, what happens as the temperature rises?
    a. the relative humidity increases.
    b. the relative humidity decreases.
    c. the relative humidity remains constant.
    d. the air will eventually become saturated.

SOLUTIONS AND ANSWERS

Practice Problems

---

1. We know

$$T = T_c + 273.15$$

and

$$T_c = 5/9\,(T_F - 32.0)$$

so

$$T = 5/9\,(T_F - 32.0) + 273.15. \text{ If } T_F = T,$$

then

$$(4/9)T = (5/9)(-32.0) + 273.15$$

which gives

$$T = \mathbf{574.6}\text{ K or °F.}$$

---

2. Equation (12.2) applied to the steel wire gives

$$\Delta L = \alpha L \Delta T = (12 \text{ X } 10^{-6}\,/\text{C°})(1.5\text{ m})(150\text{ °C} - 22\text{ °C}) = \mathbf{2.3 \text{ X } 10^{-3}\text{ m}}$$

---

3. Equation (12.2) gives the expansion of one mile of concrete to be

$$\Delta L = \alpha L \Delta T = (12 \text{ X } 10^{-6}\,/\text{C°})(5280\text{ ft}/\,1\text{ mi})(12\text{ in}/1\text{ ft})(5/9)(120)\text{C°} = 51\text{ in.}$$

The number of 1/2 in expansion joints needed to just compensate for this increase in length is

$$(51\text{ in})/(1/2\text{ in}) = \mathbf{102}$$

---

4. Equation (12.2) gives for the expansion of the radius of the collar and for the expansion of the radius of the shaft

$$\Delta r_c = \alpha_b r_c\,(T - 22\text{ °C})$$

and

$$\Delta r_s = \alpha_s r_s\,(T - 22\text{ °C}).$$

We want

$$r_c + \Delta r_c = r_s + \Delta r_s$$

for the collar to just fit the shaft.

$$T = 22\text{ °C} + (r_s - r_c)/[{}_b r_c + {}_s r_s]$$
$$T = 22\text{ °C} + (2.505\text{ cm} - 2.500\text{ cm})/[(19 \text{ X } 10^{-6}\,/\text{C°})(2.500\text{ cm}) - (12 \text{ X } 10^{-6}\,/\text{C°})(2.505\text{ cm})$$
$$T = \mathbf{309\text{ °C}}$$

---

5. The bolt hole expands as if it were made of cast iron so its diameter changes by an amount

$$\Delta d = \alpha d \Delta T = (12 \text{ X } 10^{-6}\,/\text{C°})(1.27\text{ cm})(73\text{ C°}) = 1.1 \text{ X } 10^{-3}\text{ cm.}$$

Now

$$d' = d + \Delta d = 1.27\text{ cm} + 1.1 \text{ X } 10^{-3}\text{ cm} = \mathbf{1.27\text{ cm.}}$$

---

6. We know

$$T_F = 9/5\ T_c + 32.0$$

and

$$T_c = T - 273.15;\ \text{so for } T = 0$$

we have

$$T_F = 9/5\,(-273.15) + 32.0 = \mathbf{-459.7\ °F}$$

---

7. The change in the wingspan of the airplane is given by equation (12.2) to be

$$\Delta L = \alpha L \Delta T = (23 \times 10^{-6}/C°)(9.50\ m)(-60\ C°) = -0.013\ m$$

The new wingspan is then

$$9.50\ m - 0.013\ m = \mathbf{9.49\ m}.$$

---

8. The change in volume of the contents of the jar is

$$\Delta V_c = \beta_w V \Delta T$$

and the change in the interior volume of the jar is

$$\Delta V_j = = \beta_g V \Delta T.$$

The amount of contents which spill from the jar is then

$$\Delta V = \Delta V_c - \Delta V_g = (\beta_w - \beta_g)V\Delta T$$
$$\Delta V = (207 \times 10^{-6}/C° - 8.5 \times 10^{-6}/C°)(1.00\ qt)(5/9)(140)C° = \mathbf{0.015\ qt}.$$

---

9. Let the rectangular sheet have a length, a, and a width, b. Each dimension will expand by an amount given by equation (12.2).

$$\Delta a = \alpha a \Delta T \text{ and } \Delta b = \alpha b \Delta T.$$

The expanded area of the rectangle is

$$(a + \Delta a)(b + \Delta b) = ab + a\Delta b + b\Delta a + \Delta a\Delta b = ab + 2\alpha ab\Delta T + \alpha^2 ab(\Delta T)^2.$$

The change in the area of the rectangle is then

$$\Delta A = 2\alpha ab\Delta T + \alpha^2 ab(\Delta T)^2.$$

The last term is very small compared to the first term since it contains α^2, and can be neglected so

$$\Delta A \approx 2\alpha A\Delta T$$

where A = ab, the original area of the rectangle.

---

10. The mass of the water in the radiator does not change so the initial and final densities and volumes are related by $\rho_f V_f = \rho_i V_i$. The change in the volume of the water is

$$\Delta V = V_f - V_i = (\rho_i/\rho_f - 1)V_i = (999.7/999.973 - 1)(15\ qt) = \mathbf{-4.1 \times 10^{-3}\ qt}.$$

The negative sign indicates a decrease in the volume.
By the way, this answer corresponds to about 0.13 ounces.

---

11. We know that

$$Q_{wat} + Q_{glass} = Q_{Cu}$$

or

$$(cm\Delta T)_{wat} + (cm\Delta T)_{glass} = (cm\Delta T)_{Cu}.$$

Substitution yields,

$$[4186 \text{ J/(kg C°)}](0.750 \text{ kg})(7.0 \text{ C°}) + [840 \text{ J/(kg C°)}](0.150 \text{ kg})(7.0 \text{ C°}) = [387 \text{ J/(kg C°)}](0.250 \text{ kg})\Delta T.$$

Solving for ΔT we obtain, $\Delta T = 236$ C°. Therefore,

$$T_i = T_f + \Delta T = 32.0 \text{ °C} + 236 \text{ C°} = \mathbf{268 \text{ °C}}.$$

---

12. Thermal equilibrium implies that

$$Q_{met.\ cont.} + Q_{wat} = Q_{metal}$$

or

$$(cm\Delta T)_{cont} + (cm\Delta T)_{wat} = (cm\Delta T)_{met}.$$

$$(1.50 \text{ kg})(9.0 \text{ C°})c_{met} + [4186 \text{ J/(kg C°)}](6.00 \text{ kg})(9.0 \text{ C°}) = (0.900 \text{ kg})(249 \text{ C°})c_{met}.$$

Solving for the specific heat of the metal we obtain,

$$c_{met} = \mathbf{1070 \text{ J/(kg C°)}}.$$

---

13. The ice melts and then warms up to 12 °C. We have,

$$Q_{wat} = Q_{ice} + Q_{ice\ wat}.$$

Therefore, we can write,

$$(cm\Delta T)_{wat} = m_{ice}L_f + (cm\Delta T)_{ice\ wat}.$$

Using the fact that 1.5 liters of water equals 1.5 kg we have,

$$[4186 \text{ J/(kg C°)}](1.5 \text{ kg})(23 \text{ C°}) = (33.5 \times 10^4 \text{ J/kg})m_{ice} + [4186 \text{ J/(kg C°)}](12 \text{ C°})m_{ice}.$$

Solving for the mass of the ice we obtain,

$$m_{ice} = \mathbf{0.37 \text{ kg}}.$$

---

14. We know that

$$\Delta Q_{alum} + \Delta Q_{wat} = \Delta Q_{lead}$$
$$(cm\Delta T)_{alum} + (cm\Delta T)_{wat} = (cm\Delta T)_{lead}.$$

Substitution yields,

$$[9.00 \times 10^2 \text{ J/(kg C°)}](0.050 \text{ kg})(2.0 \text{ C°}) + [4186 \text{ J/(kg C°)}](2.0 \text{ C°})m_{wat} = [128 \text{ J/(kg C°)}](0.075 \text{ kg})(57 \text{ C°}).$$

Solving for the mass of the water we obtain,

$$m_{wat} = \mathbf{55 \text{ g}}.$$

---

15. The ice and water vapor exchange equal amounts of heat so $Q_{ice} = Q_{steam}$. Each experiences a phase change so

$$(cm\Delta T)_{ice} + m_{ice}L_f + (cm\Delta T)_{wat} = (cm\Delta T)_{steam} + m_{steam}L_v + (cm\Delta T)_{wat}.$$

Substitution yields,

$$[2.00 \times 10^3 \text{ J/(kg C°)}](10.0 \text{ C°})m_{ice} + (33.5 \times 10^4 \text{ J/kg})m_{ice} + [4186 \text{ J/(kg C°)}](50.0 \text{ C°})m_{ice} =$$
$$[2020 \text{ J/(kg C°)}](30.0 \text{ C°})m_{steam} + (22.6 \times 10^5 \text{ J/kg})m_{steam} + [4186 \text{ J/(kg C°)}](50.0 \text{ C°})m_{steam}.$$

Solving for the ratio of the mass of the steam to the mass of the ice,

$$m_{steam}/m_{ice} = \mathbf{0.223}.$$

---

16. Using equation (12.5) we have, R.H. = (P.P./E.V.P.) x 100. From the curve we find that the equilibrium vapor pressure (E.V.P.) at 30 °C is 5000 Pa. The relative humidity (R.H.) is 60% so that the partial pressure of water vapor is

P.P. = (R.H.)(E.V.P.)/100 = (60)(5000 Pa)/100 = **3000 Pa**.

---

17. At 40.0 °C the equilibrium vapor pressure (from the graph) is 8000 Pa. For a relative humidity of 37.5%, the partial pressure of water vapor is

P.P. = (R.H.)(E.V.P.)/100 = (0.375)(8000 Pa) = 3000 Pa.

Therefore, the dew point (saturation) occurs at a temperature when the E.V.P. is 3000 Pa. From the graph we see that T = **20 °C**.

---

18. Find the partial pressure of water vapor outside,

P.P. = (R.H.)(E.V.P. at 15 °C)/100 = (75%)(2000 Pa)/100 = 1500 Pa.

Inside, the E.V.P. at 25 °C is 4000 Pa. Therefore,

(R.H. at 25 °C) = [(P.P.)/(E.V.P. at 25 °C)] x 100.

Thus,

R.H. at 25 °C = [(1500 Pa)/(4000 Pa)] x 100 = **38%**.

Quiz answers

1. d	5. b	9. d	13. d	17. d
2. a	6. d	10. b	14. a	18. c
3. b	7. b	11. b	15. d	19. b
4. d	8. b	12. d	16. a	

MCAT REVIEW PROBLEMS

A student wants to find the specific heat of a dark oil. She pours 100 grams of the oil into a glass beaker of negligible mass, and then submerges a small 50-watt light bulb in the oil. (A watt is a joule per second.) Before the student turns on the light bulb, the oil is 20° C, room temperature. The student then seals the top of the beaker and turns on the light bulb, at time $t = 0$. She can see some light coming from the bulb, though it looks much dimmer than usual. A small magnetic stirrer ensures that the oil has a uniform temperature throughout the beaker, at all times. A thermometer keeps track of the oil's temperature. See figure 1.

After the light bulb has been on for 40 seconds, the student switches it off. The oil's temperature at that moment is 30° C. The student *immediately* unseals the top of the beaker and removes the light bulb.

When the top of the beaker is sealed, assume that negligible heat is absorbed by the beaker or escapes into the environment.

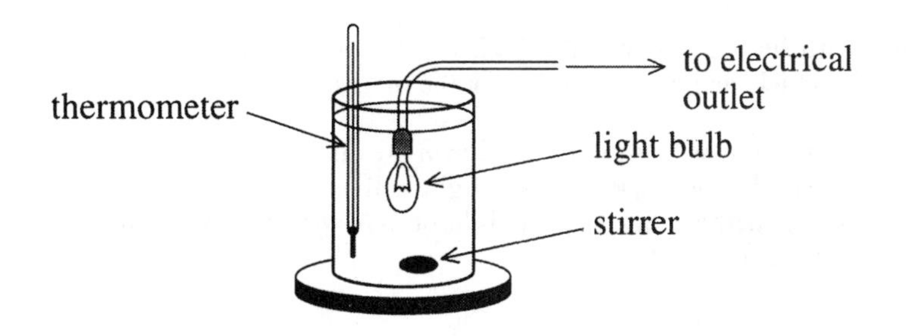

FIGURE 1

1. Which of these graphs best shows the temperature of the oil as a function of time?

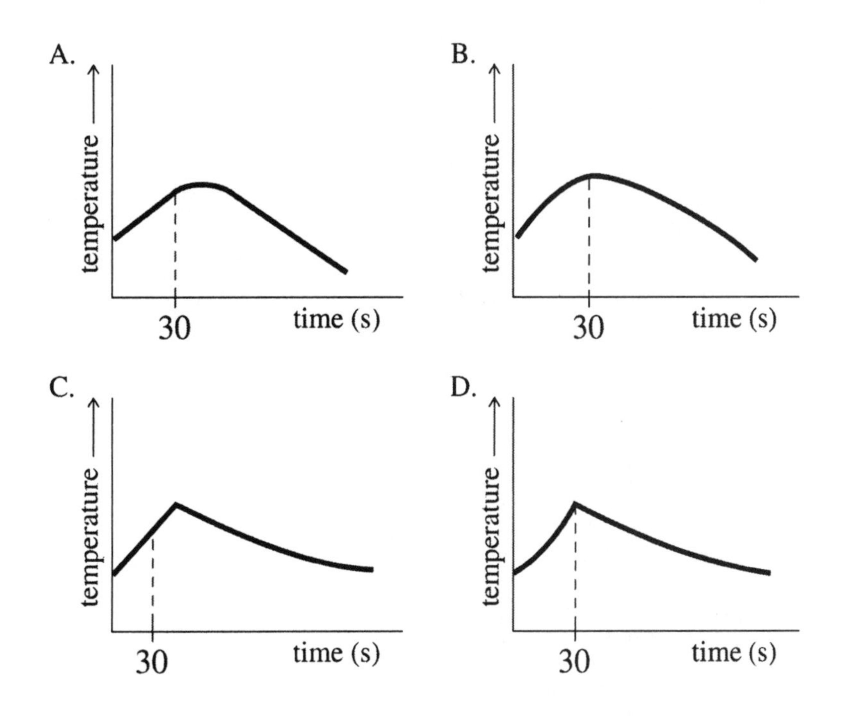

2. If the oil absorbed almost all the energy emitted by the light bulb, the specific heat of the oil, in joules per gram per Celsius degree, is approximately

   A. 0.05 **B.** 0.5 **C.** 0.7 **D.** 2

---

3. Starting again from room temperature, the experiment is repeated. But this time, the student uses a 100-watt light bulb, and leaves it on for 60 seconds. The temperature of the oil after those 60 seconds will be about:

   A. 40° C **B.** 50° C **C.** 60° C **D.** 90° C

---

4. Even if essentially no heat flows into the beaker or the environment while the light bulb is on, this experiment still might yield an inaccurate value for the oil's specific heat. Why?

   A. The stirring action of the stirrer cools down the oil.
   B. When immersed in liquid, a "50-watt" light bulb emits much less than 50 watts of power.
   C. Not all of the energy emitted by the light bulb is absorbed by the oil.
   D. Oil doesn't conduct electricity.

ANSWERS TO MCAT REVIEW PROBLEMS

---

1. **C.** The light bulb emits energy at a steady rate, and (by assumption) the oil absorbs almost all of this energy. So, the molecular kinetic energy of the oil increases steadily. Temperature (in kelvins) is proportional to molecular kinetic energy. So, while the light bulb is on, the temperature increases at a steady rate. (By the way, one Celsius degree equals one kelvin.)

   After the student unseals the beaker and removes the bulb, the oil *immediately* starts to cool off. You might think that some kind of "momentum" carries the oil to a higher temperature before it starts cooling. But since no new energy gets added to the oil after $t = 40$ s, the molecules can't gain any more kinetic energy. In other words, the temperature can't rise. If the temperature keeps rising briefly in a real-life lab, it's because pockets of imperfectly stirred hot oil don't reach the thermometer until a few seconds after the bulb is switched off.

---

2. **D.** If you forget the relevant equation, figure this out using units. Here, the specific heat, also known as specific heat capacity, has units of joules per gram per Celsius degree. So, "specific heat" must mean the number of joules of energy required to heat up one gram of the oil by 1 C°. Well, since the light bulb emits 50 joules per second for 40 seconds, it gives the oil approximately

$$Q = (50 \text{ joules/s})(40 \text{ s}) = 2000 \text{ joules}$$

   of energy. This energy raises the temperature of 100 grams of oil by 10 C°. So, the energy needed to heat up one gram by 1 C° must be

$$\text{specific heat} = \frac{2000 \text{ joules}}{(100 \text{ g})(10 \text{ C°})} = 2 \frac{\text{J}}{\text{g} \cdot \text{C°}}.$$

   Intuitively, if 2000 joules are required to heat up 100 grams of oil, then only 20 joules are needed to heat up 1 gram of oil. And since those 20 joules would heat up the gram of oil by 10 C°, only 2 joules are required to heat up the gram of oil by 1 C°.

   The formula $Q = mc\Delta T$ captures these intuitions. We solved for the specific heat capacity c, given Q = 2000 J, m = 100 g, and ΔT = 10 C°. (Since the oil starts at 20° C and ends up at 30° C, the temperature changes by only 10 C°.)

---

3. **B.** You can solve using ratios, even if you didn't find the specific heat in question 2. Since the student doubles the light bulb's power, energy gets added to the oil twice as quickly as before. And since the bulb stays on for 60 s instead of 40 s, energy gets added for 1.5 times as long. Therefore, as compared to the original experiment, the oil absorbs

   (2 times as many joules per second)
   × (1.5 times as many seconds)
   = 3 times as many joules.

   Consequently, the oil heats up three times as much as it did before. Before, it started at 20° C and gained 10 C°. So now, it starts at 20° C and gains 30 C°.

---

4. **C.** According to the passage, the student sees light coming from the bulb. So, some of the energy emitted by the bulb reaches her eyes, in the form of light. This energy was not absorbed by the oil in the form of heat. Because we don't know how much energy "escapes" from the oil in this way, we don't know how much heat the oil actually absorbs. In other words, we don't know what Q to use in the specific heat equation.

   Now we'll review why the other choices don't work. Stirring the oil cools it down only if heat can "escape" to the surrounding air or beaker. But we're assuming the sealed beaker permits no heat to flow out of the oil.

The light bulb looks dimmer in oil *not* because the bulb emits less energy per second, but only because the oil absorbs most of that energy.

Finally, it's true that oil doesn't conduct electricity. As a result, the bulb doesn't "short out." But this fact causes no inaccuracy in the experiment. Indeed, it makes the experiment possible.

CHAPTER 13
The Transfer of Heat

PREVIEW

In this chapter you will be introduced to the mechanisms of heat transfer - convection, conduction, and radiation. After completion of the chapter you will be able to explain how heat is transferred by convection and calculate the rate at which heat is transferred by conduction through various materials. In addition, you will be able to use the Stefan - Boltzmann law to calculate the rate at which energy is radiated from or absorbed by an object.

QUICK REFERENCE

Important Terms

Convection
A process in which heat energy is transferred by the flow of a fluid.

Convection Current
The flow of a fluid when heat is transferred by convection.

Natural convection
A convection process in which the fluid flows due to buoyant forces produced because the heated fluid is less dense than the surrounding cooler fluid.

Forced convection
A convection process in which an external device, such as a fan, is used to produce the fluid flow.

Conduction
A process by which heat is transferred through a material without a bulk movement of the material.

Thermal conductors
Materials which conduct heat well.

Thermal insulators
Materials which conduct heat poorly.

Radiation
A process by which energy is transferred by electromagnetic waves.

Blackbody
An idealized perfect absorber and perfect emitter of radiation.

Equations

The **heat conducted** through a bar is

$$Q = \frac{kA\Delta Tt}{L} \tag{13.1}$$

The Stefan - Boltzmann law gives the **radiant energy emitted or absorbed** by an object.

$$Q = e\sigma T^4 At \tag{13.2}$$

DISCUSSION OF SELECTED SECTIONS

13.2 Conduction

The ability of a substance to transfer heat by conduction is measured by its **thermal conductivity**, k, which appears in equation (13.1). A glance at the values of k given in Table 13.1 of the text, shows that metals have the highest thermal conductivities and are good conductors of heat, while gases generally have the lowest thermal conductivities and are poor conductors of heat. This is why most commercial insulating materials depend on trapped gases, such as air, for their insulating properties.

Example 1
Fiber glass insulation is composed of small glass fibers tangled together so that air is trapped in the spaces between them. Show that the insulating value of fiber glass insulation is due to the trapped air rather than the glass by comparing the heat which will flow through a 1.0 m^2 sheet of 3 1/2 inch thick glass in 1.0 minute and a similar sheet of air. Assume that the temperature difference maintained across the sheets is 25 C°.

The heat conducted through the glass sheet is

$$Q_g = \frac{k_g A\, \Delta T\, t}{L} = \frac{(0.80 \text{ J/s m C°})(1.0 \text{ m}^2)(25 \text{ C°})(60 \text{ s})}{8.9 \text{ X } 10^{-2} \text{ m}} = 1.3 \text{ X } 10^4 \text{ J}$$

and the heat conducted through the equivalent sheet of air is

$$Q_a = \frac{k_a A\, \Delta T\, t}{L} = \frac{(0.0256 \text{ J/s m C°})(1.0 \text{ m}^2)(25 \text{ C°})(60 \text{ s})}{8.9 \text{ X } 10^{-2} \text{ m}} = 430 \text{ J}$$

The glass conducts about thirty times as much heat as the air.

Example 2
Two identically shaped bars of different materials are placed end to end. A temperature difference, ΔT, is maintained across the bars. Find an expression for the temperature difference across the first bar. Assume that no heat is lost through the sides of the bars.

If no heat is lost through the sides of the bars, then the heat transferred through each bar in a time, t, is Q. Equation 13.1 gives for the first bar

$$Q = \frac{k_1 A\, \Delta T_1\, t}{L}$$

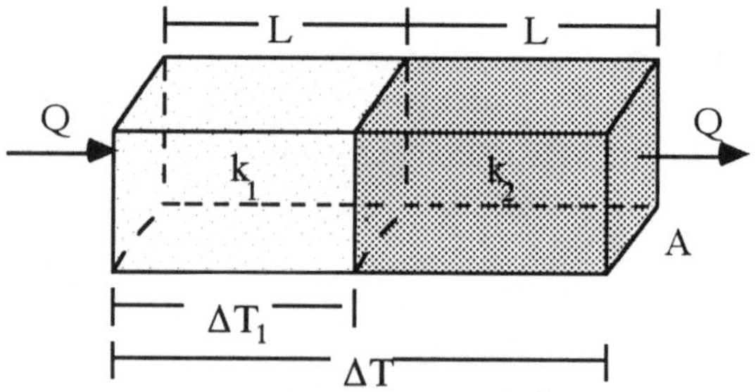

and for the second bar

$$Q = \frac{k_2 A\, (\Delta T - \Delta T_1)\, t}{L}$$

Equating and rearranging yields

$$\Delta T_1 = \left(\frac{k_2}{k_1 + k_2}\right) \Delta T$$

In the above example we can define an equivalent thermal conductivity of the two bars if we use the value that we found for ΔT_1 in the first equation.

$$Q = \frac{k_1 A\, \Delta T\, t}{L} = \left(\frac{k_1 k_2}{k_1 + k_2}\right)\left(\frac{A\, \Delta T\, t}{L}\right)$$

The two bars conduct heat energy as if they were a single bar of length, 2L, and thermal conductivity

$$k = \frac{2k_1 k_2}{k_1 + k_2}$$

Example 3

In example 2 the first bar is brass and the second bar is aluminum. The temperature of the left end is 95 °C and the temperature of the right end is 55 °C. What is the temperature at the junction between the bars? What is the equivalent conductivity of the bars?

Table 13.1 gives the thermal conductivity of brass, $k_1 = 110$ J/s m C°, and the conductivity of aluminum, $k_2 = 240$ J/s m C°. The temperature difference across the brass bar is then

$$\Delta T_1 = \left(\frac{k_2}{k_1 + k_2}\right) \Delta T = \left(\frac{240}{110 + 240}\right)(95\ ^\circ C - 55\ ^\circ C)$$

$$\Delta T_1 = 27\ ^\circ C$$

The units of the conductivities have been suppressed for clarity.

The temperature at the junction between the bars is T = 95 °C - 27 °C = 68 °C.
The equivalent thermal conductivity of the bars is

$$k = \frac{2k_1 k_2}{k_1 + k_2} = \frac{2(110)(240)}{110 + 240}\ \text{J/(s m C°)} = 150\ \text{J/(s m C°)}$$

Example 4

Two bars of different conductivity are oriented as shown in the drawing. The upper bar transfers heat energy, Q_1, while the second bar transfers, Q_2, in the same time, t. Find the total heat transferred in time, t, in terms of the temperature difference, ΔT, the length of the bars, L, and the cross-sectional area of the bars, A. Assume that no heat flows through the sides of the bars.

The heat flowing through the upper bar is

$$Q_1 = \frac{k_1 A\, \Delta T\, t}{L}$$

and the heat flowing through the lower bar is

$$Q_2 = \frac{k_2 A\, \Delta T\, t}{L}$$

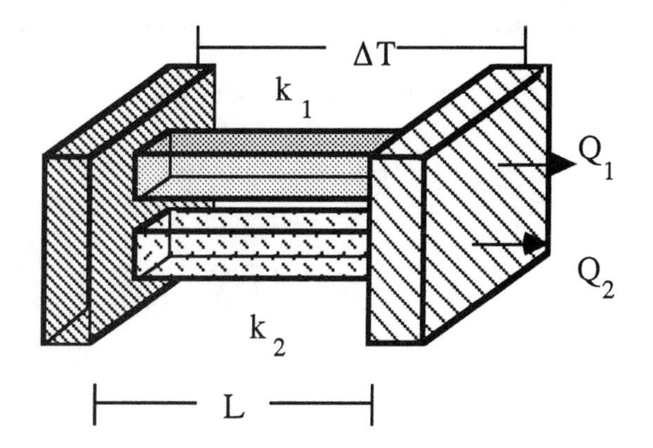

The total heat is $Q = Q_1 + Q_2$ so

$$Q = (k_1 + k_2)\left(\frac{A\,\Delta T\,t}{L}\right)$$

As before, the two bars conduct heat as if they were a single bar. This time the length of the single bar is L and the "equivalent" conductivity is $k = k_1 + k_2$.

13.3 Radiation

Both emission and absorption of radiation by an object are governed by the Stefan- Boltzmann law, equation (13.2). The temperature, T, in the equation is the Kelvin temperature of the object in the case of emission and it is the Kelvin temperature of the object's surroundings in the case of absorption.

In order to use equation (13.2), we need to know the emissivity, e, of the object in question. It ranges from e =1 for a perfect blackbody to e = 0 for a perfect reflector of radiation. The emissivity of real objects varies with the type of radiation (visible, infrared, etc) and is usually determined by experiment.

Example 5

The sun shines directly on one side of a flat black panel on a spacecraft which has an area of 5.0 m^2 and an emissivity of e = 0.95. The spacecraft is located 1.50 X 10^{11} m from the sun. How much power does the panel absorb from the sun if the sun is considered to be an ideal blackbody with a temperature of 5700 K? Assuming that the panel can only lose energy through radiation, what is the equilibrium temperature of the panel?

The total power emitted by the sun is given by equation 13.2 with T = 5700 K, the surface temperature of the sun. The radius of the sun is R_s = 6.96 X 10^8 m.

$$Q_s/t = \sigma T_s^4\,(4\pi R_s^2)$$

This energy spreads out over the surface of a sphere of radius, r, so that the amount reaching the panel is

$$\frac{Q}{t} = \sigma T_s^4\left(\frac{4\pi R_s^2}{4\pi r^2}\right)A$$

$$\frac{Q}{t} = (5.67 \text{ X } 10^{-8}\text{ J/s m}^2\text{ K}^4)(5700\text{ K})^4\left(\frac{6.96 \text{ X } 10^8}{1.50 \text{ X } 10^{11}}\right)^2(5.0\text{ m}^2) = 6400\text{ watts}$$

The panel is in equilibrium if its temperature, T_p, remains constant. This means that the panel is losing as much power as it is gaining. It gains energy only through the one sunlit side, but it can radiate energy from both sides.

$$\text{power lost} = \text{power gained}$$

$$e\sigma T_p^4 A = 6400\text{ watts}$$

$$T_p = \sqrt[4]{\frac{6400\text{ watts}}{(0.95)(5.67 \text{ X } 10^{-8}\text{ J/sm}^2\text{K}^4)(10.0\text{ m}^2)}} = 330\text{ K}$$

PRACTICE PROBLEMS

1. How much heat per square meter is conducted in one hour through a 0.64 cm thick window pane when the inside temperature is 22 °C and the outside temperature is 0 °C?

2. How much heat per square meter is conducted in one hour through two 0.64 cm thick window panes separated by a 0.64 cm stagnant air gap when the inside temperature is 22 °C and the outside temperature is 0 °C?

3. Two cooking pots are identical in every respect except that one has a 0.64 cm thick copper bottom and the other has an aluminum bottom. How thick should the aluminum bottom be in order for it to conduct the same amount of heat from a stove burner in the same time as the copper bottom?

4. A star radiates as if it were a blackbody with a temperature of 3000 K. If it radiates the same power as the sun, what is the ratio of its radius to the radius of the sun? Assume the sun has a blackbody temperature of 6000 K.

5. An object of emissivity, 0.5, radiates a certain amount of power when it is at a temperature of 650 °C. If the object were a perfect blackbody, what temperature would be required for it to radiate the same power?

6. The top of an electric stove burner has an emissivity of 0.800 and an area of 6.00×10^{-3} m^2. The stove is in a kitchen where the ambient temperature is 27° C and is heated so that it glows cherry red (T = 727 °C). What is the net power lost by the top of the burner?

7. How much heat is lost in one hour through a 15 cm X 3.7 m X 6.1 m concrete floor if the inside temperature is 22.0 °C and the ground temperature is 13.0 °C?

8. Compare the heat lost through a brick wall with the heat lost through the same wall with 1/2 inch of styrofoam insulation. Bricks are 4.0 inches thick and have a thermal conductivity of 1.0 J/(s m C°). Assume that the inside and outside temperatures are the same for both walls.

9. A quartz radiant heater has two cylindrical tubes 40.0 cm long and 1.0 cm in radius. The heater is designed to produce 1300 watts of radiant power. What would be the temperature of the tubes if the emissivity of the tubes is 0.60?

EVERYDAY PHYSICS

1. Each winter people are caught in blizzards in which the wind speeds are high and the air temperature is low. Common sense dictates that these people seek an enclosed shelter to avoid cooling of their bodies due to the wind. An unheated shelter affords protection against cooling by the wind, but provides little protection against convective cooling by the air inside the shelter. In some circumstances it may make sense to simply tunnel into a snow bank. Can you explain why a snow bank could be used as protection?

2. A candle can be extinguished by grasping the burning wick between the thumb and forefinger. The method is painless if it is done quickly enough. Why? A margin of safety is provided if the thumb and forefinger are first wetted with saliva since the heat from the candle flame will produce a thin layer of steam which acts both as an insulator and carries heat away before it can reach the fingers.

3. If your coffee is too hot to drink, then its cooling can be aided by simply placing a metal spoon in the cup. The metal conducts heat from the coffee to the handle where is removed by convective currents in the air. Why not use a plastic spoon?

4. Many people confuse whether an object feels "hot" or "cold" to the touch with the actual temperature of the object. Actually your fingers are sensitive to the rate at which heat is transferred to or from them rather than the temperature directly. A metal object with a high thermal conductivity will "feel" colder than an insulator such as wood at the same temperature since heat flows more readily into the metal than the wood. Test this yourself by touching objects of different thermal conductivities which are in the same room and have the same temperature.

5. The door handles on wood-burning stoves are usually chromed metal, metal painted black, or wood. The wooden handle allows you to open the door bare handed with the least discomfort. This is because the wood is a much poorer conductor of heat than is metal. Of the metal handles, the one painted black will be at a lower temperature than the shiny chrome handle since it radiates away more energy due to its higher emissivity. Hence, the black handle would cause less discomfort than the chrome handle.

CHAPTER QUIZ

1. Which of the following have the highest thermal conductivities?
   a. liquids b. gases c. metals d. solids other than metals

2. The wind direction near the ocean is often determined by convection caused by the land warming faster than the water during the day and cooling faster than the ocean during the night. What wind direction would you expect during the day? During the night?
   a. day - toward the shore, night - toward the shore
   b. day - away from the shore, night - away from the shore
   c. day - toward the shore, night - away from the shore
   d. day - away from the shore, night - toward the shore

3. A glider pilot usually takes advantage of "thermals" to gain altitude and stay aloft longer than possible otherwise. If you are a glider pilot and need altitude on a sunny day, which of the following areas would you steer toward?
   a. a lake b. a forest c. a grassy meadow d. a plowed field

4. Identical objects of four different materials are heated to the same high temperature. You are required to quickly move them from one table to another using your bare hand. Which object is the least likely to burn your hand?
   a. aluminum b. brass c. glass d. concrete

5. A 1.0 m X 1.5 m window made from a single pane of glass conducts heat energy at the rate of 2100 watts when the inside temperature is 22 °C and the outside temperature is 0 °C. How thick is the glass pane?
   a. 1.3 cm b. 2.5 cm c. 0.63 cm d. 0.95 cm

6. The sun transfers energy to the earth mainly by ________ while the energy is distributed over the earth primarily by ________.
   a. conduction, convection
   b. radiation, conduction
   c. convection, radiation
   d. radiation, convection

7. The temperature of an electric stove burner is raised from 300 K to 600 K. How much more energy does it radiate per second at the higher temperature than at the lower temperature?
   a. two times b. four times c. eight times d. sixteen times

8. A black object absorbs 84 % of the radiation which falls on it and remains at a constant temperature. How much of this energy does it re-emit as radiation if this is the only method by which it can lose energy.
   a. 100 % b. 84 % c. 42 % d. 21 %

9. How much energy is emitted per second by a perfect blackbody sphere with a radius of 5.0 cm and a temperature of 650 °C?
   a. 600 J/s b. 1300 J/s c. 7.2×10^4 J/s d. 57 J/s

10. Two spacecraft are identical in every respect except that one is black and the other is silver in color. They are side by side and oriented in the same way with respect to the sun. Which spacecraft has the higher surface temperature on the sunlit side? If the two craft are then shaded from the sun by a third craft, which cools quicker?
    a. black, black b. black, silver c. silver, silver d. silver, black

SOLUTIONS AND ANSWERS

Practice Problems

---

1. Equation (13.1) applies.

$$Q/A = kt\Delta T/L = (0.80\ J/(s\ m\ C°))(3600\ s)(22\ °C)/(0.64\ X\ 10^{-2}\ m) = \mathbf{9.9\ X\ 10^6\ J}$$

---

2. Let T_i be the indoor temperature, T_1 be the temperature at the first glass-air interface, T_2 be the temperature at the second air-glass interface, and T_o be the outdoor temperature. Each layer conducts the same quantity of heat in the same time. Equation (13.1) gives for the inside pane

$$Q/A = k_g t(T_i - T_1)/L \qquad (1)$$

for the air gap

$$Q/A = k_a t(T_1 - T_2)/L \qquad (2)$$

and for the outside pane

$$Q/A = k_g t(T_2 - T_0)/L \qquad (3)$$

Equating (1) and (3) and rearranging gives

$$T_i - T_1 = T_0 - T_2 \qquad (4)$$

Similarly (2) and (3) yields

$$k_a T_1 - k_a T_2 = k_g T_2 \qquad (5)$$

Solving (4) for T_1, using the result in (5) and rearranging gives

$$T_1 = (k_g + k_a)T_i/(2k_a + k_g)$$

T_i = 22 °C, k_a = 0.0256 J/(s m C°) and k_g = 0.80 J/(s m C°) so T_1 = 21 °C. This value used in (1) gives

$$Q/A = (0.80\ J/(s\ m\ C°))(3600\ s)(1\ C°)/(0.64\ X\ 10^{-2}\ m) = \mathbf{4.5\ X\ 10^5\ J}$$

---

3. Equation (13.1) applied to the bottom of the aluminum pot is $Q_a = k_a\ tA\Delta T/L_a$. A similar expression for the bottom of the copper pot is $Q_c = k_c\ tADT/L_c$. Equating and solving for L_a gives

$$L_a = (k_a/k_c)L_c = (240\ J/(s\ m\ C°)/390\ J/(s\ m\ C°))(0.64\ cm) = \mathbf{0.39\ cm.}$$

---

4. Equation (13.2) applied to the star is $Q/t = \sigma T^4 A$ and to the sun is $Q/t = \sigma T_s^4 A_s$. Equating and solving gives

$$A/A_s = (T_s/T)^4 \text{ or } (R/R_s) = (T_s/T)^2 = \mathbf{4}.$$

---

5. Equation (13.2) written for the actual object is $Q/t = (0.5)\sigma T^4 A$ and for the equivalent blackbody is $Q/t = \sigma T_b^4 A$. Equating and solving for T_b yields

$$T_b = (0.5)^{1/4}\ T = (0.5)^{1/4}\ (650\ °C + 273) = \mathbf{776\ K\ (or\ 503\ °C)}$$

---

6. The burner loses, due to radiation, energy at a rate of

$$(Q/t)_{lost} = e\sigma T_b{}^4 A = (0.800)(5.67 \times 10^{-8}\ J/s\ m^2\ K^4)(1.00 \times 10^3\ K)^4(6.00 \times 10^{-3}\ m^2) = 272\ W.$$

The burner gains energy from the surroundings at a rate

$$(Q/t)_{gain} = e\sigma T^4 A = (0.800)(5.67 \times 10^{-8}\ J/s\ m^2\ K^4)(3.00 \times 10^2\ K)^4(6.00 \times 10^{-3}\ m^2) = 2.20\ W$$

The net power lost by the burner is $(Q/t)_{lost} - (Q/t)_{gain} = \mathbf{2.70 \times 10^2\ W}$.

---

7. The heat lost through the floor is given by equation (13.1).

$$Q = kAt\Delta T/L = (1.1\ J/s\ m\ C°)(3.7\ m)(6.1\ m)(3600\ s)(9.0\ C°)/(15 \times 10^{-2}\ m) = \mathbf{5.4 \times 10^6\ J}.$$

---

8. The heat lost through the brick wall in time, t, is $Q_b = k_b At\Delta T/L_b$. If ΔT_1 is the temperature difference between the inside and the brick-styrofoam interface and ΔT_2 is the temperature across the styrofoam. The heat lost through the wall is

$$Q_s = k_b At\Delta T_1/L_b = k_s At\Delta T_2/L_s. \text{ Then } k_b\Delta T_1/L_b = k_s\Delta T_2/L_s.$$

We also know that $\Delta T = \Delta T_1 + \Delta T_2$. Elimination of ΔT_1 from these yields

$$\Delta T_2 = (k_b/L_b)\Delta T/[k_b/L_b + k_s/L_s]$$

so that

$$Q_s = (k_s/L_s)(k_b/L_b)At\Delta T/[k_b/L_b + k_s/L_s].$$

Suppressing units, we find that

$$Q_s/Q_b = k_s/L_s/[k_b/L_b + k_s/L_s] = (0.010/0.50)/[(1.0/4.0) + (0.010/0.50)] = 0.074.$$

The heat loss through the insulated wall is on **0.074 times** the loss through the brick wall.

---

9. Equation (13.2) gives

$$T^4 = (Q/t)/(e\sigma A) = (1300\ W)/[(0.60)(5.67 \times 10^{-8}\ J/s\ m^2\ K^4)(40.0 \times 10^{-2}\ m)(2\pi)(1.0 \times 10^{-2}\ m)(2)]$$

Thus,

$$T = \mathbf{930\ K}$$

Quiz answers

1. c	3.d	5. a	7. d	9. b
2. c	4.c	6. d	8. a	10. a

MCAT REVIEW PROBLEMS

A thermos bottle works as follows. A hollow chamber surrounds most of the liquid, as shown in figure 1. In a good thermos bottle, the chamber contains almost no air. As a result, the heat flow between the liquid inside the thermos bottle and the surrounding environment is greatly reduced.

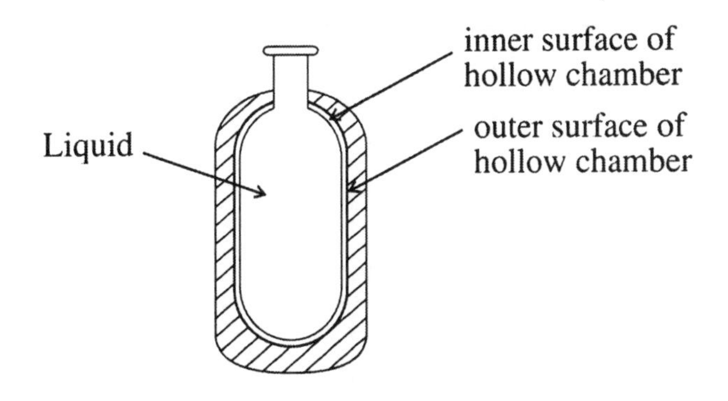

FIGURE 1

While hiking on a hot day, two chemistry students broke their thermos bottle. This sparked an argument about how best to keep their cold cans of juice from getting warm.

Chemistry student 1
We should wrap the cans in thick wool socks—the more socks, the better.

Chemistry student 2
No, it's better to make sure lots of air blows over the cans, so that the "breeze" keeps them cool. We should keep the cans unwrapped, and expose them to a breeze, perhaps by keeping the backpack partially open. (Just make sure no direct sunlight shines on the cans!)

---

1. Even if the chamber of a thermos bottle contained a perfect vacuum, hot coffee in the thermos bottle would still cool off (though extremely slowly), due primarily to

   A. Conduction **B.** Convection **C.** Radiation **D.** None of the above

---

2. A thermos bottle "works" primarily by reducing the heat flow due to

   A. Conduction **B.** Convection **C.** Radiation **D.** None of the above

---

3. Chemistry student 1 wants to wrap the cold cans in thick wool socks. As compared to leaving the cans unwrapped, this plan will:

   A. Make the cans heat up *more* quickly than they otherwise would.
   B. Make no difference in how quickly the cans heat up.
   C. Make the cans heat up *less* quickly than they otherwise would.
   D. Prevent the cans from heating up at all.

---

4. Chemistry student 2 wants to keep the backpack open, so that air flows over the cans. This plan will:

A. Make the cans heat up *more* quickly than they would if the backpack were closed.
B. Make no difference in how quickly the cans heat up.
C. Make the cans heat up *less* quickly than they would if the backpack were closed.
D. Prevent the cans from heating up at all.

ANSWERS TO MCAT REVIEW PROBLEMS

---

1. **C.** A perfect vacuum would eliminate conduction, as explained in answer 2 below. Convection couldn't play a major role either, because it involves the mass movement of fluid. But here, the thermos "traps" the coffee inside.

   All matter emits blackbody electromagnetic radiation. The hotter the material, the more radiation it gives off. Electromagnetic radiation can propagate through a vacuum, as proven by the fact that sunlight reaches the Earth. Here's the point: Since the coffee is hotter than the air outside the thermos, the coffee and the inner surface of the chamber emit more blackbody radiation than they absorb from the air. This net outflow of radiation flows through the hollow chamber, from the inner surface to the outer surface. In this way, the coffee slowly loses energy; it cools off.

---

2. **A.** We should review how conduction works. In figure 1, suppose the coffee is hot, and the hollow chamber is filled with air at atmospheric pressure. The chamber and the air inside it start off cool. So, the coffee molecules initially have more kinetic energy than the "chamber molecules" do. But when those speedy coffee molecules bump into the inner surface of the chamber, they make the chamber molecules "jiggle" more rapidly. During those collisions, the coffee molecules tend to give some of their kinetic energy to the chamber molecules. Consequently, the chamber molecules speed up, and the coffee molecules slow down. In other words, the chamber gets hotter, and the coffee gets cooler. That's what it means for heat to flow by "conduction" from the coffee to the inner surface of the chamber. So, conduction can occur only between substances (or parts of substances) that are *touching*.

   This process of heat conduction continues from the inner surface of the chamber to the air inside the chamber. When the air molecules bump into that surface, the high-kinetic-energy chamber molecules make those air molecules speed up. In other words, the chamber molecules give some of their kinetic energy to the air molecules. Consequently, the air in the chamber gets warmer. In the same way, the air inside the chamber heats up the outer surface of the chamber. And so on, until the heat has been conducted all the way to the air outside the thermos bottle.

   By removing most of the air from the hollow chamber, we reduce the rate at which air molecules bump into the inner and outer surfaces of the chamber. This reduces the rate at which heat conducts from the inner to the outer surface of that chamber. As a result, the coffee stays hot longer.

---

3. **C.** Wool socks conduct heat poorly. For instance, they keep your feet warm by reducing the rate of heat flow from your feet to the cold floor. For this same reason, wool socks reduce the rate at which heat flows from the air into the cans. Remember, the cans warm up *because* they absorb heat from the surrounding air. By slowing down this process, the socks keep the cans cold.

   Some students choose A, reasoning that wool tends to make things warm. But wool doesn't *make* things warm. It merely *keeps* things warm—or cold—by slowing down the heat transfer between that thing and its environment.

---

4. **A.** If the cans were warmer than the surrounding air, then blowing air on the cans would cool them off, by increasing the rate of heat flow. Here's why. Conduction would cause a thin layer of hot air to form right next to the hot can. Blowing air on the can replaces that hot layer with a "fresh" layer of cool air, allowing the conduction process to proceed more efficiently.

   Similar reasoning shows that blowing hot air on a cool can causes the can to heat up more quickly than it otherwise would. As the higher-kinetic-energy air molecules bump into the lower-kinetic-energy can molecules, the air molecules give some of their energy to the can. As a result, the can heats up a little, and a thin layer of cooler air forms right next to the can. When hot air blows on the can, it replaces that cool layer with a "fresh" layer of hot air, allowing the conduction process to proceed more efficiently. For this reason, exposing the cool cans to a hot breeze makes the cans heat up more quickly than they otherwise would.

   If you chose C, it's probably because you associate breezes with cooling. Well, breezes almost always cool *people* off, because we generate our own heat, and the breeze increases the rate of heat flow from our bodies to the air. But if the air were hotter than we are, and if we didn't produce our own heat, then a breeze would make us warmer. We'd be just like the cold cans on a warm day.

CHAPTER 14
The Ideal Gas Law and Kinetic Theory

PREVIEW

In this chapter you will gain an understanding of heat and temperature at the molecular level. The concept of an ideal gas is introduced and the kinetic theory of gases is explored. The ideal gas law is employed to study how a gas behaves under different conditions of pressure, temperature, and volume. The internal energy of a gas is defined and the process of diffusion is discussed.

QUICK REFERENCE

Important Terms

Atomic mass
The mass of an atom of one element as compared to an atom of another. The reference element is chosen to be carbon-12, whose atomic mass is defined to be exactly twelve.

Mole
The mass (in grams) of a substance which contains **Avogadro's number** N_A of atoms (or molecules). One mole of any substance has a mass in grams that is equal to the atomic mass of the constituent particles. For example, 12 g of carbon-12 contains $N_A = 6.022 \times 10^{23}$ atoms of carbon-12.

Ideal gas law
Relates the pressure, volume, and temperature of an ideal gas. It states that PV = nRT where n is the number of moles present and R is the universal gas constant.

Kinetic theory of gases
The theory which states that the temperature of an ideal gas is defined by the average kinetic energy of the particles making up the gas.

Maxwell speed distribution
The distribution of particle speeds in an ideal gas at a given temperature.

Internal energy
The sum of the various kinds of energy that the atoms or molecules of a substance possess. This includes translational and rotational kinetic energy, vibrational energy, and potential energy.

Diffusion
The process whereby solute molecules move from a region of higher concentration to a region of lower concentration.

Equations

The **ideal gas law** can be written:

$$PV = nRT \qquad (14.1)$$

An **alternative way** to write the ideal gas law is:

$$PV = NkT \qquad (14.2)$$

Boyle's law can be stated as:

$$P_iV_i = P_fV_f \qquad (14.3)$$

Charles' law is:

$$\frac{V_i}{T_i} = \frac{V_f}{T_f} \qquad (14.4)$$

The **average translational kinetic energy** of the particles in a gas is given by:

$$\overline{KE} = \frac{1}{2}mv_{rms}^2 = \frac{3}{2}kT \qquad (14.6)$$

The **internal energy** of an **ideal monatomic gas** is:

$$U = \frac{3}{2}nRT \qquad (14.7)$$

Fick's law of diffusion is:

$$m = \frac{DA\,\Delta Ct}{L} \qquad (14.8)$$

DISCUSSION OF SELECTED SECTIONS

14.1 Molecular Mass, the Mole, and Avogadro's Number

The relative masses of the atoms of different elements can be expressed in terms of their **atomic masses**. To set up the scale of atomic masses, carbon-12 is chosen as the reference element. An **atomic mass unit** is then defined as one-twelfth the mass of the carbon-12 atom. For example, the atomic mass of say, oxygen, is 15.9994 atomic mass units (u). It is approximately 16/12 the mass of the carbon-12 atom.

The **molecular mass** of a molecule is the sum of the atomic masses of its constituent atoms. So, for example, a water molecule (H_2O) has a molecular mass which is the sum of the masses of two hydrogen atoms and one oxygen atom, i.e., 2(1.00794 u) + 15.9994 u = 18.0153 u.

Experiment has shown that 12 grams of carbon-12 contain 6.022×10^{23} atoms. This number is referred to as **Avogadro's number**, N_A. A **gram-mole** (or simply a mole) is the amount of a substance that contains Avogadro's number of atoms (or molecules). One mole of any substance has a mass in grams that is equal to the atomic or molecular mass of its constituent particles. Thus, one mole of neon has a mass of 20.179 g.

Example 1
The chemical formula of ethyl alcohol is C_2H_5OH. (a) Determine the mass of one mole of ethyl alcohol. (b) How many molecules are contained in 175 g of ethyl alcohol?

(a) Ethyl alcohol has 2 carbon atoms, 6 hydrogen atoms, and one oxygen atom. Using Figure 14.1 we see that the atomic mass of ethyl alcohol is,

$$2(12.011\text{ u}) + 6(1.00794\text{ u}) + 15.9994\text{ u} = 46.069\text{ u}.$$

Therefore, one mole of ethyl alcohol has a mass of 46.069 grams.
(b) In 175 g of ethyl alcohol, there are (175 g)/(46.069 g/mol) = 3.80 mol. Since one mole contains Avogadro's number of molecules, $N_A = 6.022 \times 10^{23}$ molecules/mol, we therefore have

$$(3.80\text{ mol})(6.022 \times 10^{23}\text{ molecules/mol}) = 2.29 \times 10^{24}\text{ molecules}.$$

14.2 The Ideal Gas Law

The ideal gas law expresses the relationship between the Kelvin temperature, the pressure, the volume, and the number of moles of an ideal gas. An ideal gas is an idealized model of real gases and is a useful concept in studying the properties of gases in general. The ideal gas law can be stated as

$$PV = nRT \tag{14.1}$$

Where P is the absolute pressure (Pa), V the volume (m^3), T the temperature (K), n the number of moles, and R is the **universal gas constant** which has the value R = 8.31 J/(mol K).

An alternative way to write equation (14.1) is in terms of the total number of particles in the gas, N, rather than in terms of the number of moles, n. We have

$$PV = NkT \tag{14.2}$$

Where k is known as **Boltzmann's constant**, k = 1.38×10^{-23} J/K.

Example 2

A cylinder containing carbon dioxide gas (CO_2) has a volume of 0.050 m^3. At 25.0 °C the gas has an absolute pressure of 2.00×10^6 Pa. (a) What is the mass of carbon dioxide in the cylinder? (b) How many molecules of CO_2 are in the cylinder?

(a) First find the number of moles of CO_2 present using equation (14.1).

$$n = \frac{PV}{RT} = \frac{(2.00 \times 10^6\ \text{Pa})(0.050\ \text{m}^3)}{[8.31\ \text{J/(mol K)}](298\ \text{K})} = 40.4\ \text{mol}$$

The atomic mass of CO_2 is 2(15.9994 u) + 12.011 u = 44.010 u. The mass of CO_2 in the cylinder is therefore,

$$(44.010\ \text{g/mol})(40.4\ \text{mol}) = 1.78 \times 10^3\ \text{g} = 1.78\ \text{kg}.$$

(b) The number of CO_2 molecules in the cylinder can be obtained using equation (14.2).

$$N = \frac{PV}{kT} = \frac{(2.00 \times 10^6\ \text{Pa})(0.050\ \text{m}^3)}{(1.38 \times 10^{-23}\ \text{J/K})(298\ \text{K})} = 2.43 \times 10^{25}\ \text{molecules.}$$

Note that this value of N can also be obtained using Avogadro's number, that is,

$$N = nN_A = (40.4\ \text{mol})(6.022 \times 10^{23}\ \text{mol}^{-1}) = 2.43 \times 10^{25}\ \text{molecules.}$$

Historically, the ideal gas law developed as a result of the work of several investigators. One important law relates the conditions of pressure and volume when the temperature and mass remain fixed. **Boyle's law** states

$$P_iV_i = P_fV_f \tag{14.3}$$

Where the subscripts i and f refer to the initial and final states of the gas, respectively.

Example 3

The pressure in a tank of nitrogen is 1.75×10^7 Pa at 20.0 °C. If the nitrogen would occupy 30.0 m^3 at 1 atm and 20.0 °C, find the volume of the tank.

If we take the initial pressure to be 1 atm = 1.01×10^5 Pa, the initial volume to be 30.0 m^3, and the final pressure to be 1.75×10^7 Pa, equation (14.3) yields for the final volume

$$V_f = \frac{P_i V_i}{P_f} = \frac{(1.01 \times 10^5 \text{ Pa})(30.0 \text{ m}^3)}{1.75 \times 10^7 \text{ Pa}} = 0.173 \text{ m}^3.$$

If a situation arises in which the gas pressure and mass remain constant, then **Charles' law** is valid, that is,

$$\frac{V_i}{T_i} = \frac{V_f}{T_f} \qquad (14.4)$$

Example 4

A tire is inflated to a gauge pressure of 2.0 atm. (Gauge pressure is the absolute pressure minus 1 atm). As the car is driven, the temperature of the tire increases from 20.0 °C to 45.0 °C, and simultaneously, the volume of the tire increases by 5.0%. What is the gauge pressure of the tire at the higher temperature?

Since the amount of gas in the tire is constant, equations (14.3) and (14.4) can be combined to yield

$$\frac{P_i V_i}{T_i} = \frac{P_f V_f}{T_f}$$

Using $P_i = 3.0$ atm, $T_i = 273 + 20.0 = 293$ K, $T_f = 273 + 45 = 318$ K, and $(V_f/V_i) = 1.05$ we can solve for P_f.

$$P_f = \frac{P_i V_i T_f}{T_i V_f} = \frac{(3.0 \text{ atm})(318 \text{ K})}{(293 \text{ K})(1.05)} = 3.1 \text{ atm}$$

Therefore, the final gauge pressure is $P_{gauge} = P_f - 1.0$ atm = 2.1 atm.

14.3 Kinetic Theory of Gases

A container filled with a gas contains a large number of particles which are in constant, random motion, colliding with each other and with the walls of the container. At a given temperature, the atoms and molecules are all moving at different speeds. However, it is possible to determine the most-probable particle speed for the gas. As the temperature of the gas rises, the most probable particle speed increases, as shown in the following figure.

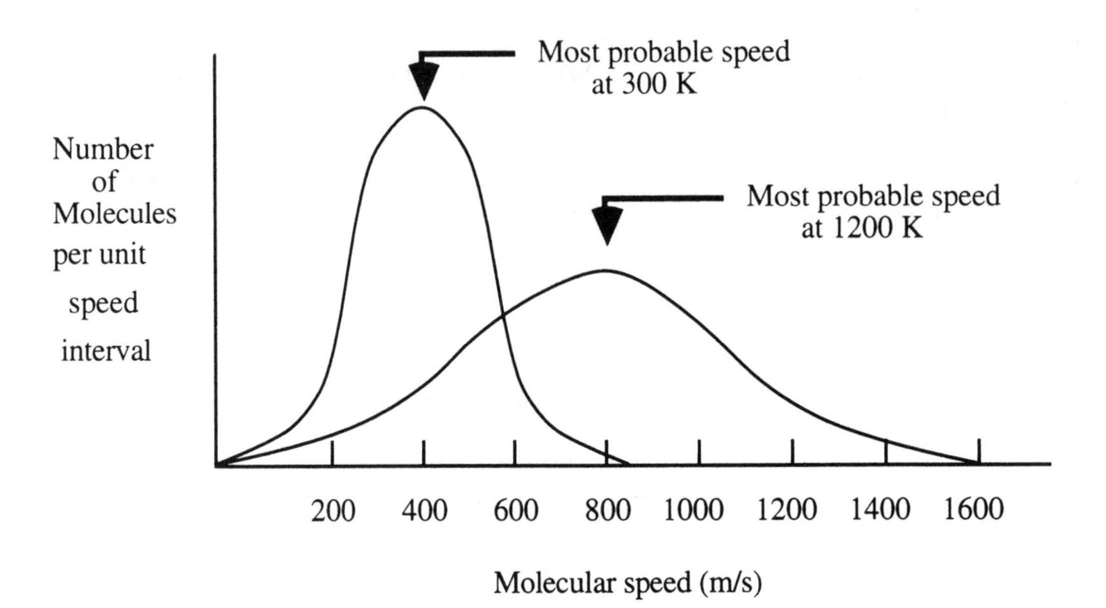

The above plots are called Maxwell distribution curves for particle speeds in a gas. Note that there is actually a wide range of velocities for each temperature, but that particles are more likely to travel at certain speeds (the most probable speed) than at others. It can be shown that the average translational kinetic energy of the particles in a gas is proportional to the temperature of the gas. That is

$$\overline{KE} = \frac{1}{2} mv_{rms}^2 = \frac{3}{2} kT \qquad (14.6)$$

Where m is the mass of an individual molecule, k is Boltzmann's constant, T is the Kelvin temperature, and v_{rms} is the **root-mean-square speed** of the particles (which represents a type of average speed). If the gas can be treated as an ideal monatomic gas, the internal energy U is defined as the total kinetic energy of the N particles that make up the gas. That is

$$U = N(\frac{1}{2} mv_{rms}^2) = N(\frac{3}{2} kT) = \frac{3}{2} nRT \qquad (14.7)$$

Example 5 *Escape of hydrogen gas from the atmosphere*
Hydrogen molecules in the upper atmosphere (where the temperature is about 1500 K) have an appreciable chance of escaping the earth's gravitational attraction. The escape speed for hydrogen gas is about 10.7 km/s. Calculate the rms speed of the H_2 molecules at this altitude and comment on the presence of hydrogen gas in the atmosphere.

Use equation (14.6) to find the rms speed at 1500 K. Note that the mass of a hydrogen molecule can be found from its molecular mass and Avogadro's number, i.e.,

$$m = \frac{\text{molecular mass}}{N_A} = \frac{2(1.00794 \text{ g/mol})}{6.022 \times 10^{23} \text{ mol}^{-1}} = 3.35 \times 10^{-24} \text{ g} = 3.35 \times 10^{-27} \text{ kg}.$$

Solving equation (14.6) for the rms speed therefore yields,

$$v_{rms} = \sqrt{\frac{3kT}{m}} = \sqrt{\frac{3(1.38 \times 10^{-23}\ \text{J/K})(1500\ \text{K})}{(3.35 \times 10^{-27}\ \text{kg})}} = 4300\ \text{m/s}.$$

Although this rms speed is only about 0.40 of the escape speed (10 700 m/s), some molecules will have speeds many times the rms value and therefore will have sufficient energy to escape. In the 4.6 billion years since the formation of the earth, there has been ample time for essentially all of the hydrogen and helium to escape the earth's gravitational attraction and become lost in space.

14.4 Diffusion

Consider a container of water in which we drop a small amount of ink. The ink drop is initially very concentrated, but slowly disperses into the surrounding water. The process by which the ink molecules move from regions of high concentration to regions of low concentration is called **diffusion**. The ink is called the **solute** and diffuses through the water (the solvent).

The mass, m, of solute that diffuses in a time, t, through a solvent contained in a channel of length, L, and cross-sectional area, A, is

$$m = \frac{DA\,\Delta Ct}{L} \qquad (14.8)$$

where ΔC is the concentration difference between the ends of the channel and D is the diffusion constant. The SI unit for the diffusion constant is m^2/s. Equation (14.8) is known as **Fick's law of diffusion**.

Example 6

The diffusion constant of ethanol in water is $12.4 \times 10^{-10}\ m^2/s$. A cylinder has a cross-sectional area of $4.00\ cm^2$ and a length of 2.00 cm. A difference in ethanol concentration of $1.50\ kg/m^3$ is maintained between the ends of the cylinder. In one hour, what is the mass of the ethanol that diffuses through the cylinder?

Using $A = 4.00\ cm^2 = 4.00 \times 10^{-4}\ m^2$, and $L = 2.00\ cm = 2.00 \times 10^{-2}\ m$ in equation (14.8) we obtain,

$$m = \frac{DA\,\Delta Ct}{L} = \frac{(12.4 \times 10^{-10}\ m^2/s)(4.00 \times 10^{-4}\ m^2)(1.50\ kg/m^3)(3600\ s)}{2.00 \times 10^{-2}\ m} = 1.34 \times 10^{-7}\ kg.$$

PRACTICE PROBLEMS

1. How many molecules are there in 30.0 g of $C_9H_8O_4$?

2. Sodium bicarbonate, $NaHCO_3$, is one of the ingredients in baking powder. How many moles are present in a 21.0 g sample of sodium bicarbonate?

3. A 20.0 liter container of oxygen gas has a gauge pressure of 8.00 atm at 25 °C. (a) How many moles of O_2 are present in the container? (b) How many O_2 molecules are present?

4. A bubble of air, 0.010 m^3 in volume, is formed at the bottom of a lake which is 30.0 m deep and where the temperature is 8.0 °C. The bubble rises to the surface, where the water temperature is 26 °C and where the pressure is atmospheric pressure. What is the volume of the bubble just as it reaches the surface?

5. Consider an ideal gas of density ρ and pressure P. Show that the root-mean-square speed of the gas molecules is equal to $\sqrt{3P/\rho}$.

6. At what temperature is the root-mean-square speed of oxygen molecules (O_2) equal to the escape velocity from the earth? Take the escape velocity to be 11.2 km/s at the earth's surface.

7. A tank contains 40.0 mol of N_2 gas at 14.0 atm of pressure. If the volume of the tank is 75.0 liters, what is the rms speed of the molecules?

8. Assume the pressure in a room remains constant at 1.01×10^5 Pa and the air is composed only of nitrogen (N_2). The volume of the room is 60.0 m^3. When the temperature increases from 16 °C to 29 °C, what mass of air escapes from the room?

9. Neon (Ne) is a monatomic gas. At a temperature of 50.0 °C, what is the internal energy of two grams of neon?

10. A cylindrical tube of 4.00 cm diameter and 2.50 m length contains 5.0×10^{-14} kg of a solute diffusing through water. A concentration difference of 5.0×10^{-3} kg/m^3 is maintained between the ends of the tube and the diffusion constant of the solute is 7.5×10^{-10} m^2/s. How much time is required for all the solute to diffuse through the water?

HELPFUL SUGGESTIONS

1. One mole of **any** substance contains the same number of atoms or molecules, namely Avogadro's number. However, the mass of one mole of say, oxygen, is **not** the same mass as one mole of nitrogen. One mole of any substance has a mass in grams that is equal to the atomic or molecular mass of its constituent particles.

2. If the mass of a gas is constant during a process, $P_1V_1/T_1 = P_2V_2/T_2$. In this equation, P and V can be used in any units that are convenient (i.e., Pa or atm for pressure and m^3 or liters for volume). However, the temperature T must **always** be expressed in absolute (Kelvin) temperature.

3. When equations (14.1) or (14.2) are used, PV = nRT or PV = NkT, a consistent set of units must be employed. In the SI system, for example, pressure must be expressed in Pa (not atm), V must be in m^3 (not liters), and T must be in kelvins (not °C).

4. The average translational kinetic energy of a gas depends **only** on the absolute temperature of the gas, not on the mass of the species. However, the velocity of the particles **will** vary depending on the mass of the molecules in question.

EVERYDAY PHYSICS

1. Take an "empty" bottle and tightly attach a balloon to its neck. Heat the air in the bottle over boiling water or a small flame. What happens to the balloon? A similar experiment would be to blow up a balloon, measure its diameter, and then place it in the freezer of your refrigerator for about an hour. Measure the diameter of the balloon after removing it from the freezer. What are your results?

2. Keep in mind that the pressure in the tires on your car will vary with temperature. Not only the air temperature but also the heating due to friction between the tires and the road should be considered. Since the volume of the tires can't change very much, an increase in temperature means an increase in pressure. After you drive your car for a while, the heat due to friction causes the tire pressure to rise. So when you put air in your tires, and want to maintain the recommended pressure, fill your tires before driving a large distance. That is, the recommended tire pressure should be maintained for "cold" tires rather than hot.

CHAPTER QUIZ

1. As we heat a gas at constant pressure, its volume
   a. increases b. decreases c. stays the same

2. The volume of an ideal gas is directly proportional to its
   a. pressure b. Celsius temperature c. Kelvin temperature d. Fahrenheit temperature

3. An ideal gas is maintained at constant temperature. If the pressure on the gas is doubled, the volume is
   a. increased fourfold b. doubled c. reduced by half d. decreased by a quarter

4. Approximately how many molecules are there in 64 g of Oxygen gas (O_2)?
   a. N_A b. 2 N_A c. 32 N_A d. 64 N_A

5. A given sample of carbon dioxide, CO_2, contains 3.01 x 10^{23} molecules. What is the mass of this sample?
   a. 44 g b. 22 g c. 22 kg d. 0.5 g

6. Oxygen molecules are 16 times more massive than hydrogen molecules. At a given temperature, how does the average kinetic energy of the hydrogen molecules compare to that of the oxygen molecules, assuming both behave like ideal gases?
   a. The hydrogen molecules have 16 times the energy.
   b. The hydrogen molecules have four times the energy.
   c. The oxygen molecules have more energy.
   d. Both the oxygen and hydrogen molecules have the same kinetic energy.

7. Oxygen molecules are 16 times more massive that hydrogen molecules. At a given temperature, how does the average molecular speed of the hydrogen molecules compare to that of the oxygen molecules, assuming both behave like ideal gases??
   a. The hydrogen molecules move 16 times faster.
   b. The hydrogen molecules move four times faster.
   c. The oxygen molecules move faster.
   d. Both the oxygen and hydrogen molecules move at the same speed.

8. A container holds helium and methane gas under ideal gas conditions. The molecular weight of methane is four times that of helium. The ratio of the rms speed of helium atoms to the rms speed of methane molecules is
   a. 2 b. 4 c. 16 d. None of the above

9. A steel tank holds 0.30 m^3 of air at a pressure of 8.0 atm. If air behaves as an ideal gas, the volume this air would occupy at the same temperature but at a pressure of 2.0 atm is
   a. 0.30 m^3 b. 0.60 m^3 c. 1.2 m^3 d. 2.4 m^3

10. A sample of air at 27 °C occupies a volume of 0.30 m^3. If the temperature rises to 127 °C but the pressure is kept constant, what volume will the air now occupy? Assume air is an ideal gas.
    a. 1.4 m^3 b. 0.064 m^3 c. 0.4 m^3 d. 0.23 m^3

11. If the Kelvin temperature of an ideal gas is doubled, what happens to the rms speed of the molecules in the gas?
    a. It increases by a factor of $\sqrt{2}$.
    b. It increases by a factor of 2.
    c. It increases by a factor of 4.
    d. None of the above.

SOLUTIONS AND ANSWERS

Practice Problems

---

1. The molecular mass of $C_9H_8O_4$ is: $9(12.011 \text{ u}) + 8(1.00794 \text{ u}) + 4(15.9994 \text{ u}) = 180.16 \text{ u}$.

   Therefore, 1 mole has a mass of 180.16 g. In 30.0 g there are

   $$(30.0 \text{ g})/(180.16 \text{ g mol}^{-1}) = 0.167 \text{ mol}.$$

   Thus, the number of molecules is

   $$(6.022 \times 10^{23} \text{ mol}^{-1})(0.167 \text{ mol}) = \mathbf{1.01 \times 10^{23}}.$$

---

2. The molecular mass of $NaHCO_3$ is

   $$22.9898 \text{ u} + 1.00794 \text{ u} + 12.011 \text{ u} + 3(15.9994 \text{ u}) = 84.007 \text{ u}.$$

   Therefore, in 21.0 g of sodium bicarbonate there are

   $$(21.0 \text{ g})/(84.007 \text{ g mol}^{-1}) = \mathbf{0.250 \text{ mol}}.$$

---

3. Using the ideal gas law, $PV = nRT$, and noting that $P = P_{gauge} + 1.00 \text{ atm} = 9.00 \text{ atm} = 9.09 \times 10^5 \text{ Pa}$, we have

   (a) $$n = PV/RT = (9.09 \times 10^5 \text{ Pa})(20.0 \times 10^{-3} \text{ m}^3)/\{[8.31 \text{ J/(mol K)}](298 \text{ K})\}$$
   $$n = \mathbf{7.34 \text{ mol}}.$$

   (b) Since 1 mol contains N_A molecules, in 7.3 mol there are

   $$(7.3 \text{ mol})(6.022 \times 10^{23} \text{ mol}^{-1}) = \mathbf{4.4 \times 10^{24}}.$$

---

4. First find the pressure at the bottom of the lake, P_2, from

   $$P_2 = P_1 + \rho gh.$$

   We have,

   $$P_2 = 1.01 \times 10^5 \text{ Pa} + (1.00 \times 10^3 \text{ kg/m}^3)(9.80 \text{ m/s}^2)(30.0 \text{ m})$$
   $$p_2 = 3.95 \times 10^5 \text{ Pa} = 3.91 \text{ atm}.$$

   Since the amount of gas in the bubble is constant,

   $$P_1V_1/T_1 = P_2V_2/T_2.$$

   The volume of the bubble at the surface is therefore,

   $$V_1 = P_2V_2T_1/P_1T_2 = (3.91 \text{ atm})(0.010 \text{ m}^3)(299 \text{ K})/[(1.00 \text{ atm})(281 \text{ K})]$$
   $$V_1 = \mathbf{0.042 \text{ m}^3}.$$

---

5. We know from equation (14.6) that

$$(1/2)mv_{rms}^2 = (3/2)kT.$$

From equation (14.1)

$$PV = nRT$$

which gives

$$T = PV/nR.$$

Therefore,

$$v_{rms}^2 = 3k(PV/nR)/m = 3kPV/nmR.$$

Since

$$k = R/N_A$$

and

$$n = N/N_A$$

we have

$$k/n = (R/N_A)/(N/N_A) = R/N.$$

Now

$$v_{rms}^2 = (3PV/mR)(R/N) = 3PV/mN.$$

But mN = total mass and ρ = mass/V, so

$$v_{rms}^2 = 3P/\rho,$$

and

$$v_{rms} = \sqrt{3P/\rho}.$$

---

6. Using equation (14.6) and solving for T we get

$$T = mv_{rms}^2/3k.$$

Oxygen gas (O_2) has a molecular mass of 32.0 g/mol. Therefore,

$$m = (32.0 \text{ g/mol})/(6.022 \times 10^{23} \text{ mol}^{-1}) = 5.31 \times 10^{-23} \text{ g}$$
$$m = 5.31 \times 10^{-26} \text{ kg}.$$

The temperature is,

$$T = (5.31 \times 10^{-26} \text{ kg})(11.2 \times 10^3 \text{ m/s})^2/[3(1.38 \times 10^{-23} \text{ J/K})]$$
$$T = \mathbf{1.61 \times 10^5 \ K}.$$

---

7. Find the temperature,

$$T = PV/nR = (14.0 \text{ atm})(1.01 \times 10^5 \text{ Pa/1 atm})(75.0 \times 10^{-3} \text{ m}^3)/R(40.0 \text{ mol}).$$
$$T = 319 \text{ K}.$$

From equation (14.6),

$$v_{rms}^2 = 3kT/m.$$

Where the mass of a nitrogen molecule, N_2, is given by

$$m = (28.0 \times 10^{-3} \text{ kg/mol})/(6.022 \times 10^{23} \text{ mol}^{-1}) = 4.65 \times 10^{-26} \text{ kg}.$$

Thus, the rms speed of the molecules is

$$v_{rms}^2 = 3(1.38 \times 10^{-23} \text{ J/K})(319 \text{ K})/(4.65 \times 10^{-26} \text{ kg}).$$

This gives

$$v_{rms} = \mathbf{533\ m/s}.$$

---

8. Find the number of moles at each temperature;

$$n_1 = PV/RT_1 = (1.01 \times 10^5 \text{ Pa})(60.0 \text{ m}^3)/[R(289 \text{ K})] = 2520 \text{ mol}.$$

and

$$n_2 = PV/RT_2 = 2410 \text{ mol}.$$

The number of moles lost is 2520 - 2410 = 110 mol. The mass lost is therefore,

$$m(\text{lost}) = (110 \text{ mol})(28.01 \text{ g/mol}) = 3.0 \times 10^3 \text{ g} = \mathbf{3.0\ kg}.$$

---

9. The amount of neon present is

$$(2.00 \text{ g})/(20.179 \text{ g/mol}) = 0.0991 \text{ mol}.$$

Equation (14.7) gives the internal energy

$$U = (3/2)nRT = (3/2)(0.0991 \text{ mol})[8.31 \text{ J/(mol K)}](323 \text{ K}) = \mathbf{3.99 \times 10^2\ J}.$$

---

10. Using equation (14.8) and solving for t, we obtain,

$$t = mL/(DA\,\Delta C) = (5.0 \times 10^{-14} \text{ kg})(2.50 \text{ m})/[(7.5 \times 10^{-10} \text{ m}^2/\text{s})\pi(2.00 \times 10^{-2} \text{ m})^2(5.0 \times 10^{-3} \text{ kg/m}^3)]$$
$$t = \mathbf{27\ s}.$$

Quiz answers

1. a	3. c	5. b	7. b	9. c	11. a
2. c	4. b	6. d	8. a	10. c	

MCAT REVIEW PROBLEMS

Air pressure decreases with altitude. In fact, by assuming that air has the same temperature at all altitudes, it is possible to derive the *law of atmospheres*:

$$p = p_0 e^{-h/a},$$

where p_0 denotes the air pressure at sea level, h denotes the altitude above sea level, a denotes a constant, and p denotes the pressure at height h. For instance, many people have difficulty breathing on Mt. Everest, due in part to the lower air pressure.

---

1. According to the law of atmospheres, which graph best expresses the relationship between air pressure and altitude above sea level?

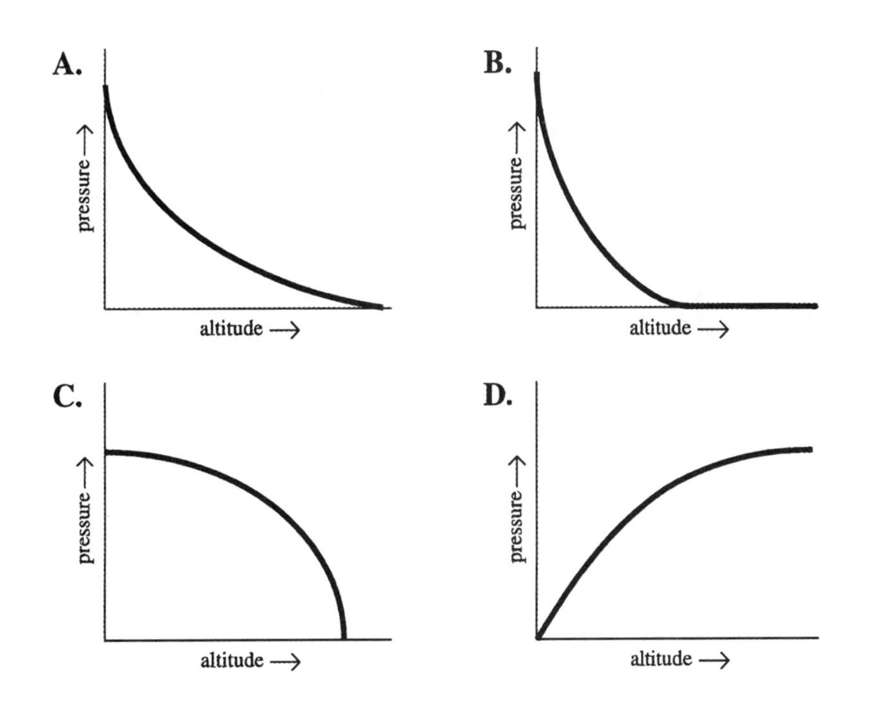

---

2. At sea level, where the air pressure is 1 atm, a glass container is filled with air at −40° C, and sealed shut. The container is taken to Mt. Everest, where the temperature is −40° C. The pressure of the air in the container is now:

   A. more than 1 atm
   B. about 1 atm
   C. less than 1 atm
   D. We cannot determine the answer without more information.

---

3. At sea level, a flexible balloon is partially filled with –40° C air, to a volume of 1 liter. The leak-proof balloon is taken to Mt. Everest, where the temperature remains –40° C. Which of the following is true about the pressure (p) and volume (V) of the air inside the balloon?

A. $p \approx 1$ atm, $V > 1$ liter
B. $p < 1$ atm, $V > 1$ liter
C. $p > 1$ atm, $V > 1$ liter
D. $p < 1$ atm, $V < 1$ liter

---

4. Which of the following is a reason why real gases deviate from the ideal gas law?

A. Mutual attractions between molecules slow them down as they approach the walls of the container.
B. Collisions between the molecules and the walls of the container are usually inelastic.
C. Molecules collide not only with the sides of the container, but also with each other.
D. In a gas, different molecules have different speeds.

ANSWERS TO MCAT REVIEW PROBLEMS

---

1. **A.** Only graph A represents an exponential decrease. According to the law of atmospheres, as the altitude increases, the pressure decreases quickly at first, but decreases more slowly at higher altitudes. So, according to this law, p never quite reaches zero, even at great heights. Only graph A satisfies these conditions.

---

2. **B.** Don't get distracted by the law of atmospheres, which applies to the air *outside* the container. On Mt. Everest, outside the container, the pressure is less than 1 atm. But according to the ideal gas law,

$$pV = nRT,$$

the pressure *inside* the container depends only on the container's volume, the number of moles of gas inside it, and the temperature. Assuming no air leaks out, n stays the same. The container doesn't change volume appreciably. And the temperature on Mt. Everest equals the temperature of the air when it entered the container at sea level. So, for the gas inside the container, n, V, and T have the same values on Mt. Everest that they had at sea level. Therefore, the pressure of that gas stays the same, too.

---

3. **B.** This differs from question 2, because the balloon's flexibility ensures that the pressure inside the balloon approximately "equalizes" with the pressure outside the balloon. Hence, when taken to Mt. Everest, the air inside the balloon decreases its pressure. According to the ideal gas law, since the temperature does not change, this decrease in pressure corresponds to an increase in volume.

Let us walk through this reasoning again, in more detail. Unlike a glass container, a flexible balloon expands or contracts until the air pressure outside the balloon roughly equals the pressure inside the balloon. To see why, suppose the pressure inside the balloon momentarily exceeds the surrounding air pressure. Then, the inner surface of the balloon gets pushed outward more forcefully than the outer surface of the balloon gets pushed inward. See figure 1.

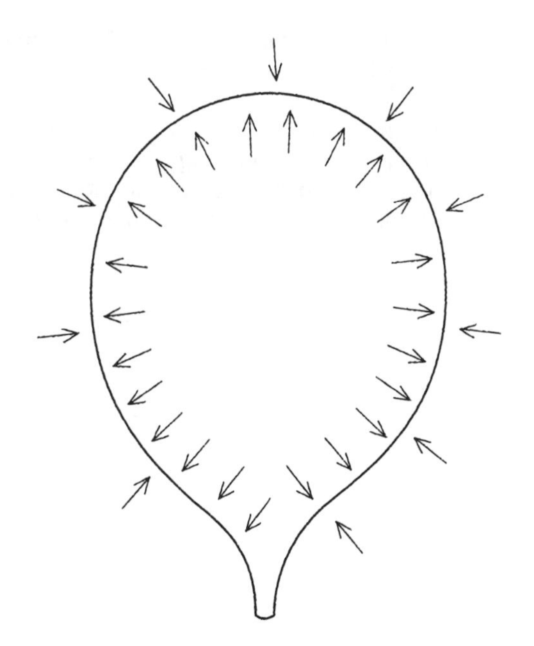

FIGURE 1

Consequently, the balloon expands.

As the balloon "climbs" from sea level to Mt. Everest, the outside air pressure decreases. Therefore, as just explained, the balloon's volume increases. According to the ideal gas law, $pV = nRT$, when the balloon's

volume increases, the pressure inside the balloon decreases. The balloon "settles" at a volume such that the pressure inside the balloon equals the outside air pressure. In summary, the balloon expands, thereby lowering the pressure of its contents.

On an airplane, sealed bags of peanuts "puff up" for this reason.

---

4. **A.** Your textbook's derivation of the ideal gas law assumes that molecules keep all of their kinetic energy as they travel back and forth across the container, and as they bump into the walls. But a molecule approaching the right wall feels attracted, by van der Waal's forces, to the nearby gas molecules. And most of those nearby molecules are to the left of the molecule under consideration.

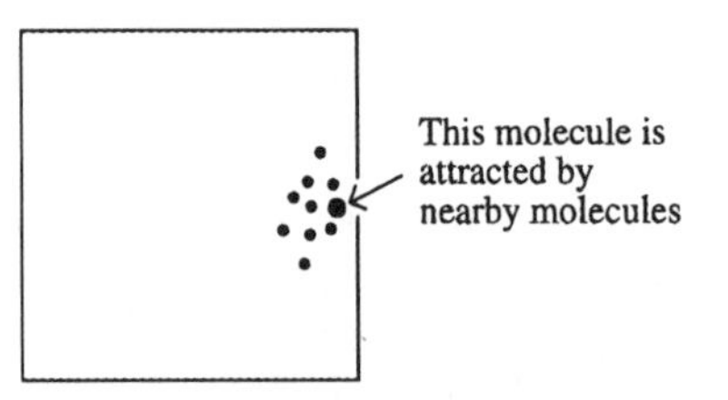

FIGURE 2

As a result, the molecule under consideration slows down slightly, and strikes the wall with less than its "full" velocity.

MCAT exams frequently test your understanding of van der Waal's forces and finite molecular volume, the two main reasons why real gases deviate from the ideal gas law. Chemistry classes cover these topics in detail.

You could also solve this problem by elimination. When molecules collide with each other or with the sides of the container, the collisions are elastic (on average), and hence, the molecules don't lose any speed (on average). Different molecules do indeed have different speeds. But since the derivation of the ideal gas law relies only on the *average* of v^2, it does not matter that some molecules travel faster than average while others travel slower than average.

CHAPTER 15
Thermodynamics

PREVIEW

In this chapter you will learn about the laws of **thermodynamics**. The zeroth law is concerned with the concept of thermal equilibrium. It requires that two systems be in thermal equilibrium if they are both in equilibrium with a third system. The first law of thermodynamics is a statement of the conservation of energy for a thermodynamic system.

You will also study the second and third laws of thermodynamics, including discussions of heat engines, Carnot's principle, and entropy. You will learn how to find the efficiency of engines, and learn about refrigerators, air conditioners, and heat pumps. Entropy, as it relates to the second law of thermodynamics, will be studied, and discussed in the context of reversible and irreversible processes.

QUICK REFERENCE

Important Terms

Thermodynamics
A branch of physics built upon the laws which govern the behavior of heat and work.

Diathermal walls
Walls which separate a system and the environment and allow heat to flow through them.

Adiabatic walls
Walls which separate the system and environment and do not allow heat to flow through them.

State of the system
The physical condition of the system which is usually described by specifying the pressure, volume and temperature.

Thermal equilibrium
The state of two or more systems in thermal contact when no heat flows between them. Systems are in thermal equilibrium when they are at the same temperature.

Isobaric process
A thermodynamic process which occurs at constant pressure.

Isochoric process
A thermodynamic process which occurs at constant volume.

Isothermal process
A thermodynamic process which occurs at constant temperature.

Adiabatic process
A thermodynamic process in which no heat flows into or out of the system.

Molar specific heat capacity
The constant, C, in the equation $Q = Cn\Delta T$ which determines the amount of heat lost or gained by n moles of a substance when its temperature changes by ΔT.

Heat engine
Any device that uses heat to perform work.

Reversible process
A process in which both the system and its environment can be returned to exactly the states they had before the process occurred.

Carnot's principle
No irreversible engine operating between two reservoirs at constant temperatures can have a greater efficiency than a reversible engine operating between the same temperatures.

Entropy
A function of state that is associated with disorder in the system and environment.

Equations

The **first law of thermodynamics** is

$$\Delta U = U_f - U_i = Q - W \qquad (15.1)$$

The **work done** during an **isobaric process** is

$$W = P\,\Delta V \qquad (15.2)$$

The **work done** during an **isothermal expansion or compression of an ideal gas** is

$$W = nRT \ln\left(\frac{V_f}{V_i}\right) \qquad (15.3)$$

The **work done** during the **adiabatic expansion** or **compression** of an monatomic **ideal gas** is

$$W = \frac{3}{2} nR(T_i - T_f) \qquad (15.4)$$

The following equation applies to an **adiabatic expansion** or **compression** of an ideal gas.

$$P_i V_i^{\gamma} = P_f V_f^{\gamma} \qquad (15.5)$$

The heat of a system Q, the molar heat capacity C, and the change in temperature ΔT are related by

$$Q = Cn\Delta T \qquad (15.6)$$

The **efficiency of a heat engine** is given by:

$$\text{Efficiency} = \frac{\text{Work Done}}{\text{Input Heat}} = \frac{W}{Q_H} \qquad (15.11)$$

We also know that:

$$Q_H = W + Q_C \qquad (15.12)$$

Which leads to:

$$\text{Efficiency} = \frac{Q_H - Q_C}{Q_H} = 1 - \frac{Q_C}{Q_H} \qquad (15.13)$$

For a **Carnot engine**,

$$\text{Efficiency} = 1 - \frac{T_C}{T_H} \qquad (15.15)$$

The entropy of a system is given by:

$$\Delta S = \left(\frac{Q}{T}\right)_R \qquad (15.18)$$

Laws

Zeroth law of thermodynamics - two systems individually in thermal equilibrium with a third system are in thermal equilibrium with each other.

First law of thermodynamics - When a system absorbs or loses an amount of heat Q, and performs work W, the internal energy of the system changes by an amount ΔU, from an initial value U_i to a final value U_f:

$$\Delta U = U_f - U_i = Q - W$$

Second Law of thermodynamics - Heat flows spontaneously from a substance at a higher temperature to a substance at a lower temperature and does not flow spontaneously in the reverse direction. In terms of entropy: the total entropy of the universe does not change when a reversible process occurs ($\Delta S_{universe} = 0$) and increases when an irreversible process occurs ($\Delta S_{universe} > 0$).

DISCUSSION OF SELECTED SECTIONS

15.3 The First Law of Thermodynamics

The first law of thermodynamics is simply a statement of the conservation of energy for a thermodynamic system which can lose or gain heat and perform negative or positive work. Any net energy gained by the system increases its store of internal energy. Also, any net energy lost by the system comes from its store of internal energy.

Keep in mind that all changes in internal energy of an ideal gas are accompanied by a change in the TEMPERATURE of the system and vice versa. In the specific case where the system consists of n moles of a monatomic ideal gas, the change in internal energy accompanying a temperature change ΔT is $\Delta U = 3/2\ nR\Delta T$.

Example 1
An ideal gas performs 550 J of work while losing 150 cal of heat energy. What are the values of W, Q and ΔU? What happens to the temperature of the system during the process?

The work done, W = + 550 J, since the system does the work. The heat transferred is

$$Q = -150 \text{ cal} = -(150 \text{ cal})(4.186 \text{ J}/1 \text{ cal}) = -630 \text{ J}.$$

The negative sign is used to indicate that the system is losing the heat.

The first law then gives the change in internal energy of the system to be

$$\Delta U = Q - W = -630 \text{ J} - 550\text{J} = \mathbf{-1180 \text{ J}}.$$

The internal energy of the system decreases during the process, so the system's **temperature decreases**.

Example 2
One mole of a monatomic ideal gas at a temperature of 57 °C absorbs 1.4 kcal of heat and does 2800 J of work on its surroundings. What is the resulting temperature of the gas?

The change in temperature of the ideal gas is

$$\Delta T = \frac{\Delta U}{\frac{3}{2} nR}$$

The first law gives the change in the internal energy to be

$$\Delta U = Q - W = (+ 1.4 \text{ kcal})(4186 \text{ J/1 kcal}) - 2800 \text{ J} = 3100 \text{ J}$$

so

$$\Delta T = \frac{3100 \text{ J}}{\frac{3}{2}(1.0 \text{ mole})(8.31 \text{ J/mole K})} = 250 \text{ K}$$

The final temperature of the gas is then

$$T_f = T_i + \Delta T = 330 \text{ K} + 250 \text{ K} = 580 \text{ K}.$$

15.4 Thermal Processes

A thermodynamic system may undergo a variety of processes. There are four basic processes, however, which are very common. These are: isobaric, isochoric, isothermal, and adiabatic processes.

Isobaric Process
This is a process in which a system maintains a constant pressure. Other quantities such as temperature, volume, and internal energy may change during the process while heat is added or taken away and work is done. The work done during an isobaric process is $W = P(V_f - V_i)$.

Example 3
When water evaporates, it does work in pushing the air away as it expands into a vapor. If 1.0 kg of water does $1.7 \text{ X } 10^5$ J of work on the atmosphere, what is the change in volume of the water as it evaporates? What is the change in internal energy of the water?

This is a constant pressure process where the pressure is that due to the atmosphere. The volume change of the system of 1.0 kg of water is

$$V_f - V_i = \frac{W}{P} = \frac{1.7 \text{ X } 10^5 \text{ J}}{1.01 \text{ X } 10^5 \text{ Pa}} = 1.7 \text{ m}^3$$

The first law requires $\Delta U = Q - W$. In order to evaporate the water an amount of energy equal to the latent heat of vaporization must be added to the water. $Q = mL_v$ so

$$\Delta U = mL_v - W = (1.0 \text{ kg})(2.26 \text{ X } 10^6 \text{ J/kg}) - 1.7 \text{ X } 10^5 \text{ J} = 2.1 \text{ X } 10^6 \text{ J}$$

Isochoric Process
No volume changes occur during this type of process. Other quantities such as pressure, temperature and internal energy may change as heat is added or taken away from the system. However, no work can be done during an isochoric process, since no mechanical motion (expansion or contraction) of the system is possible. The first law of thermodynamics then requires that $\Delta U = Q$.

Example 4
Two moles of an ideal monatomic gas at 95 °C is enclosed in a container with rigid walls. A fixed quantity of heat, Q = 5.0 kJ, is added to the gas. What is the final temperature of the gas?

The process is isochoric since the gas cannot expand against the rigid walls of the container so $\Delta U = Q$. The change in internal energy of an ideal gas is $\Delta U = 3/2\ nR(T_f - T_i)$, so

$$T_f = T_i + \frac{Q}{\frac{3}{2}nR} = 368\ K + \frac{5.0 \times 10^3\ J}{\frac{3}{2}(2.0\ mol)(8.31\ J/mol\ K)} = 5.7 \times 10^2\ K$$

Isothermal Process

The temperature remains constant in this type of process. Again, other quantities such as pressure and volume may change while heat is added to or taken away from the system and the system does work. The internal energy of the system will not change if it depends solely on the temperature of the system. This is true for an ideal gas and is covered in the next section.

Adiabatic process

No heat is allowed to flow into or out of the system in an adiabatic process. These processes occur in well insulated containers, or they happen so quickly that little heat has a chance to flow between the system and its environment. An example of the latter is when a gas, such as freon, expands rapidly from a pressurized container. The expansion in this case is at least approximately adiabatic.

The first law of thermodynamics gives $\Delta U = - W$ for an adiabatic process. Any work done on the system will increase its internal energy while work done by the system will decrease its internal energy. These internal energy changes often appear as temperature changes of the system. In the case of the freon expanding from its container, the temperature of the freon decreases as its does work to push the atmosphere out of the way.

15.5 Thermal Processes That Utilize an Ideal Gas

Isothermal Expansion or Compression

Since the internal energy of an ideal gas depends only on the temperature of the gas, it remains constant during an isothermal process. The first law of thermodynamics then gives $Q = W$.

The work done during an isothermal process involving an ideal gas system is $W = nRT \ln (V_f/V_i)$. It is the property of the natural logarithm that it is positive if the argument is greater than one, ($V_f > V_i$). This corresponds to an expansion of the gas and the work is positive indicating that work is done BY the gas. According to the first law an equal amount of heat must be ADDED to the gas to maintain the constant temperature.. The natural logarithm is negative if the argument is less than one ($V_f < V_i$). In this case the work done is negative indicating that work is being done on the gas to compress it, and an equal amount of heat must be REMOVED from the gas to maintain the constant temperature.

Example 5

A piston and cylinder arrangement is used to compress 1.0 mole of an ideal gas from an initial volume of 0.010 m^3 to a final volume of 0.0050 m^3. The temperature of the gas is maintained at a constant temperature of 150 °C. How much heat is transferred during the process? Is this heat added to or removed from the gas?

The process is isothermal since the temperature is constant throughout the process. The heat transferred is then

$$Q = W = nRT \ln (V_f/V_i)$$

$$Q = (1.0\ mol)(8.31\ J/mol\ K)(420\ K) \ln (0.0050\ m^3/0.010\ m^3) = - 2400\ J$$

The negative sign indicates that heat is lost by the gas and must be REMOVED to keep the gas temperature constant.

Adiabatic Expansion or Compression

The first law of thermodynamics requires that $\Delta U = -W$ for an adiabatic process. If the system under consideration is a monatomic ideal gas, then the work done during the adiabatic process is $W = 3/2\ nR(T_i - T_f)$. Note that if the temperature of the gas rises, then the work is negative indicating that work is done ON the gas. This corresponds to the gas being compressed by an outside agent. If the temperature of the gas falls during the process, then the work is positive and work is done BY the gas. In this case the gas has expanded.

An additional relationship applicable to the adiabatic process is $P_iV_i^\gamma = P_fV_f^\gamma$ where $\gamma = c_P/c_V$.

Example 6

A piston cylinder arrangement contains 1.5 moles of ideal air ($\gamma = 7/5$) at 27 °C. The piston is quickly pushed down compressing the gas to one-half of its original volume. What is the final temperature of the gas and how much work was done during the process?

If the piston is pushed quickly enough, the process may be considered to be adiabatic since little time is available for heat to flow. The ideal gas law can be applied to find the final temperature of the gas.

For the initial state of the air

$$P_iV_i = nRT_i$$

and for the final state

$$P_fV_f = nRT_f.$$

Dividing and solving for T_f yields

$$T_f = \left(\frac{P_f}{P_i}\right)\left(\frac{V_f}{V_i}\right)T_i$$

Equation (15.5) gives

$$\frac{P_f}{P_i} = \frac{V_i^\gamma}{V_f^\gamma}$$

so the final temperature of the air is

$$T_f = \left(\frac{V_i^{\gamma-1}}{V_f^{\gamma-1}}\right)T_i = (2)^{7/5-1}(273 + 27)\text{K} = 396\ \text{K}$$

That is, **123 °C**.

The work done during the adiabatic process is then

$$W = 3/2\ nR(T_i - T_f) = 3/2\ (1.5\ \text{mol})(8.31\ \text{J/mol K})(-96\ \text{K}) = -1800\ \text{J}.$$

1800 J of work is done **ON** the air during the process.

15.6 Specific Heat Capacities and the First Law of Thermodynamics

When a heating process occurs at constant pressure,

$$Q_{\text{constant pressure}} = nC_P\Delta T$$

where the molar specific heat capacity at constant pressure for a monatomic ideal gas, C_P is

$$C_P = 5/2\ R \qquad (15.7)$$

When the heating process occurs at constant volume,

$$Q_{\text{constant volume}} = nC_V\Delta T$$

where the molar specific heat capacity at constant volume for a monatomic ideal gas, C_V is

$$C_V = 3/2\ R \qquad (15.8)$$

The ratio of the specific heats for a monatomic ideal gas is

$$\gamma = \frac{C_P}{C_V} = \frac{\frac{5}{2}R}{\frac{3}{2}R} = \frac{5}{3} \qquad (15.9)$$

Also,

$$C_P - C_V = R \qquad (15.10)$$

Example 7
How much heat is required to change the temperature of 3.0 moles of a monatomic ideal gas by 55 K if: (a) the pressure is held constant and (b) if the volume is held constant?

(a) At constant pressure

$$Q_{\text{constant pressure}} = nC_P\Delta T = (3.0\text{ moles})(5/2)(8.31\text{ J/mole K})(55\text{ K}) = \mathbf{3400\ J.}$$

(b) At constant volume

$$Q_{\text{constant volume}} = nC_V\Delta T = (3.0\text{ moles})(3/2)(8.31\text{ J/mole K})(55\text{ K}) = \mathbf{2100\ J.}$$

15.8 Heat Engines

A heat engine is any device that uses heat to perform work and is represented schematically in the following diagram.

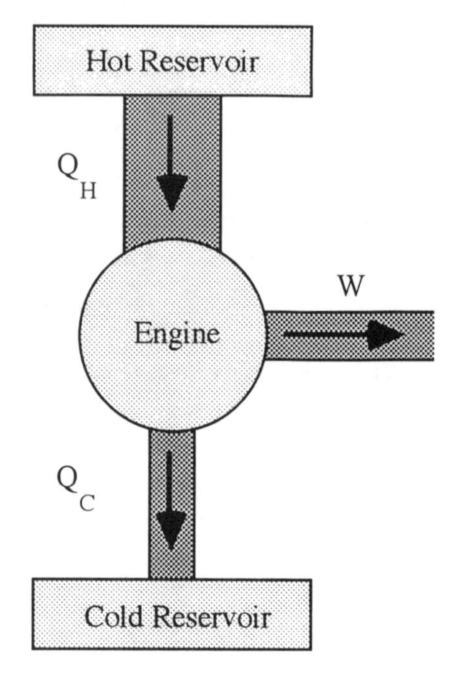

Note the main features of a heat engine:

Q_H - heat input from the hot reservoir flows into the engine.

Engine - uses available heat, Q_H, to perform work, W.

Q_C - heat rejected by the engine goes into a cold reservoir.

The efficiency of a heat engine is given as

$$\text{Efficiency} = \frac{\text{Work Done}}{\text{Input Heat}} = \frac{W}{Q_H} \qquad (15.11)$$

If there are no other losses in the engine, the principle of energy conservation requires that

$$Q_H = W + Q_C \qquad (15.12)$$

Solving for W in equation (15.12) and substituting into equation (15.11) yields,

$$\text{Efficiency} = \frac{Q_H - Q_C}{Q_H} = 1 - \frac{Q_C}{Q_H} \qquad (15.13)$$

Example 8
A heat engine does 350 J of work and rejects 750 J of heat. What is the efficiency of the engine?

In order to calculate the efficiency of the engine, we need to know what the input heat, Q_H, is. We can use equation (15.12) to find that

$$Q_H = W + Q_C = 350\text{ J} + 750\text{ J} = 1100\text{ J}.$$

The efficiency of the engine can now be obtained using either equation (15.11) or (15.13). Use equation (15.11) to get

$$\text{Efficiency} = \frac{W}{Q_H} = \frac{350\text{ J}}{1100\text{ J}} = 0.32.$$

Therefore, the efficiency of the engine is 0.32 or 32%. This means that 32% of the heat provided to the engine is converted into work which can be used for various purposes (like turning a crankshaft and providing mechanical energy to move a car).

15.9 Carnot's Principle and the Carnot Engine

If the heat engine shown above acts "reversibly", this idealized engine is called a **Carnot engine**. A reversible process is one in which both the system and its environment (the rest of the universe) can be returned to exactly the states (of P, V, and T) they had before the process occurred.

In a Carnot engine the input heat Q_H originates from a hot reservoir at a single Kelvin temperature T_H, and all rejected heat Q_C is returned to the cold reservoir at a single Kelvin temperature T_C. The ratio of the rejected heat to the input heat can be shown to be

$$\frac{Q_C}{Q_H} = \frac{T_C}{T_H} \tag{15.14}$$

Where the temperatures MUST be expressed in **Kelvins**. Using equations (15.13) and (15.14), the efficiency of a Carnot engine can therefore be written,

$$\text{Efficiency} = 1 - \frac{Q_C}{Q_H} = 1 - \frac{T_C}{T_H} \tag{15.15}$$

This relation gives the maximum possible efficiency for a heat engine operating between two Kelvin temperatures.

Example 9
In a steam turbine, steam heated to 655 °C enters the turbine and is exhausted at 115 °C. What is the maximum efficiency that this turbine can attain?

The maximum efficiency possible is that of a Carnot engine operating at the temperatures given. Converting to Kelvin temperature we have $T_H = 655 + 273 = 928$ K, and $T_C = 115 + 273 = 388$ K. Therefore,

$$\text{Efficiency} = 1 - \frac{T_C}{T_H} = 1 - \frac{388\ \text{K}}{928\ \text{K}} = 0.582.$$

Example 10
A Carnot engine has an efficiency of 0.700, and the temperature of its cold reservoir is 105 °C. (a) Determine the Kelvin temperature of its hot reservoir. (b) If 1.25 kcal of heat are rejected to the cold reservoir, what amount of heat is put into the engine?

(a) Solving equation (15.15) for T_H we obtain

$$T_H = \frac{T_C}{(1 - \text{Efficiency})} = \frac{378\ \text{K}}{(1 - 0.700)} = 1260\ \text{K}.$$

(b) Now solve equation (15.14) for Q_H to obtain

$$Q_H = Q_C\left(\frac{T_H}{T_C}\right) = (1.25\ \text{kcal})\left(\frac{1260\ \text{K}}{378\ \text{K}}\right) = 4.17\ \text{kcal}.$$

15.10 Refrigerators, Air Conditioners, and Heat Pumps

Refrigerators, air conditioners, and heat pumps are all similar devices which make heat flow from cold to hot. Of course, in order to do this, work must be done. The essential components of these devices are shown below.

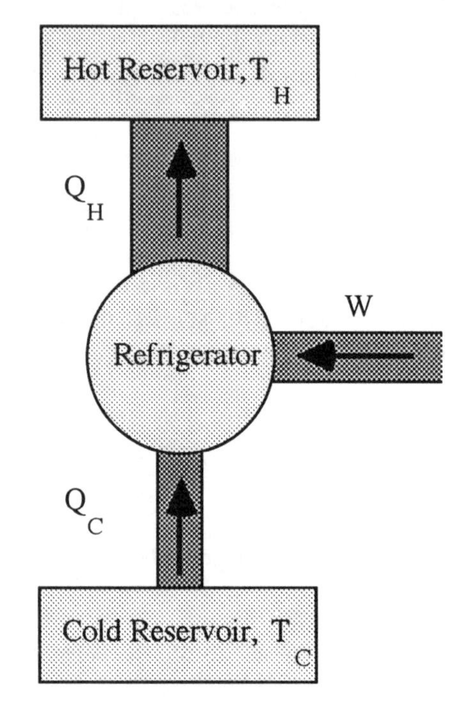

These devices use work W to "reach into" a cold reservoir, "grab onto" an amount of heat Q_C, and deposit an amount of heat Q_H into a hot reservoir. This process is called a **refrigeration process**, but applies equally as well to air conditioners and heat pumps. Notice that the directions of the arrows symbolizing heat and work in this process are opposite to those for the heat engine. If the process is a reversible one, we have a Carnot refrigerator, air conditioner, or heat pump. Under these conditions the relation

$$Q_C/Q_H = T_C/T_H$$

applies, just as it does for the Carnot engine. In a sense, these devices work to move heat "up-hill" from a lower temperature T_C to a higher temperature T_H.

Example 11
A Carnot air conditioner uses 15 400 J of electrical energy to cool a house down to 22 °C when the outdoor temperature is 36 °C. How much heat is deposited outdoors?

First find the ratio of the heats Q_C and Q_H using equation (15.14), where $T_C = 22 + 273 = 295$ K and $T_H = 309$ K.

$$\frac{Q_C}{Q_H} = \frac{T_C}{T_H} = \frac{295\text{K}}{309\text{K}} = 0.955$$

Thus, $Q_H = Q_C/0.955$. But according to equation (15.12), $Q_C = Q_H - W$, so that $Q_H = (Q_H - W)/0.955$. As a result

$$Q_H = \frac{W}{1 - 0.955} = \frac{15\,400\text{ J}}{0.045} = 3.4 \times 10^5\text{ J}$$

Example 12
Determine the coefficient of performance for the air conditioner discussed in example 11.

First we need to determine the heat removed from the cold reservoir using equation (15.12). We have

$$Q_C = Q_H - W = 3.4 \times 10^5\text{ J} - 15\,400\text{ J} = 3.2 \times 10^5\text{ J}.$$

The coefficient of performance is defined as

$$\text{Coefficient of Performance} = \frac{Q_C}{W} = \frac{3.2 \times 10^5\text{ J}}{15\,400\text{ J}} = 21.$$

15.11 Entropy and the Second Law of Thermodynamics

For irreversible processes, such as those involving friction, engines operate at less than maximum efficiency. The heat dissipated by friction causes a loss in the engine's ability to perform work. This partial loss can be expressed in terms of a concept called **entropy**.

We can define the change in entropy ΔS as

$$\Delta S = \left(\frac{Q}{T}\right)_R \qquad (15.18)$$

where the subscript R means that this relation pertains to reversible processes. The total change in entropy which occurs during some process is equal to the entropy of the final state minus the entropy of the initial state.

Example 13
If 2500 J of heat flow spontaneously from a hot reservoir at 128 °C to a cold reservoir at 52 °C, determine the total change in entropy for this irreversible process.

The spontaneous flow of heat from hot to cold is irreversible. However, the relation $\Delta S = (Q/T)_R$ can be applied to a hypothetical process whereby 2500 J of heat are taken reversibly from the hot reservoir and added to the cold reservoir. The total entropy change is the algebraic sum of the entropy changes for each reservoir.

$$\Delta S_{tot} = \Delta S_C - \Delta S_H = \frac{2500\text{ J}}{325\text{ K}} - \frac{2500\text{ J}}{401\text{ K}} = +1.5\text{ J/K}.$$

Notice that the change in entropy for the above process is greater than zero. In fact, any irreversible process always INCREASES the entropy of the universe. In other words, $\Delta S_{universe} > 0$ for an irreversible process. However, the total entropy of the universe does NOT change when a reversible process occurs. So $\Delta S_{universe} = 0$ in this case.

Example 14
Compute the change in entropy of a system consisting of 5.00 kg of ice at 0 °C which melts slowly (reversibly) to water at the same temperature. The latent heat of fusion for water is 33.5×10^4 J/kg.

Since the phase change occurs reversibly, we can use equation (15.18) to compute ΔS. We have

$$\Delta S = \frac{Q}{T} = \frac{(33.5 \times 10^4\text{ J/kg})(5.00\text{ kg})}{273\text{ K}} = +6.14 \times 10^3\text{ J/K}.$$

PRACTICE PROBLEMS

1. Three moles of an ideal gas receives 2500 J of heat energy and its temperature rises from 27 °C to 55 °C. How much work is done by the gas?

2. The P-V diagram shows three processes undergone by 1.0×10^2 mole of an ideal gas. The processes start at 1 and proceed clockwise. Identify each of the ideal gas processes and find the work done during each. (Use your best judgment to identify 1 → 2.) What is the total work done?

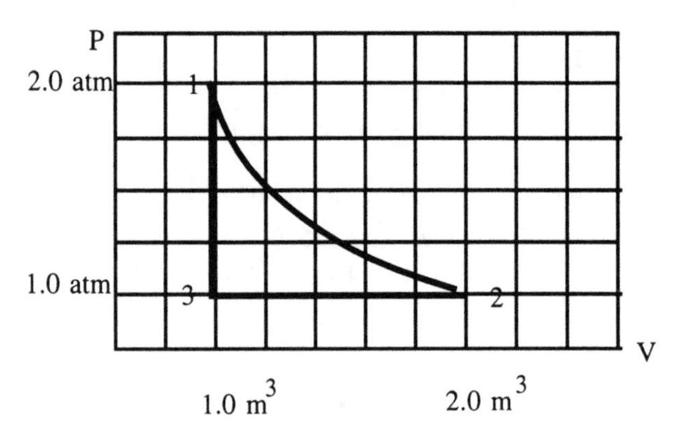

3. The internal energy of a system increases by 1350 J when the system absorbs 1150 J of heat energy at a constant pressure of 1.01×10^5 Pa. By how much does the volume of the system change?

4. Air ($\gamma = 7/5$) is compressed adiabatically to one-fifth of its original volume. By what factor did the pressure of the gas change during the compression? Assume that air behaves like an ideal gas.

5. One-half mole of an ideal gas expands isothermally at T = 250 °C to fives times its original volume. How much work was done? Was it done by or on the gas? How much heat was transferred during the process?

6. An ideal gas ($\gamma = 5/3$) expands through a nozzle of a pressurized container into the atmosphere. Assuming that the contents of the container are initially at 23°C , 10.0 atm and that the gas expands very rapidly, find the final temperature of the gas.

7. A helium filled balloon with a volume of 1.000×10^3 cm^3 and temperature of 27 °C is placed in liquid nitrogen in a open container and cools to a temperature of 77 K. Assume that helium behaves like an ideal gas at constant atmospheric pressure and find the new volume of the gas in the balloon, the change in internal energy of the gas, the work done during the process, and the heat removed from the gas.

8. A Carnot heat engine operates between 327 °C and 127 °C. It absorbs 7.50×10^4 J at the higher temperature. (a) What is the efficiency of the engine? (b) How much work per cycle is the engine capable of performing?

9. A Carnot engine operates between a hot reservoir at 455 K and a cold reservoir at 364 K. (a) If the engine absorbs 650 J of heat at the hot reservoir, how much work does it deliver? (b) If the same engine works in reverse as a refrigerator between the same two reservoirs, how much work must be supplied to remove 1200 J of heat from the cold reservoir?

10. A steam turbine takes vapor at 463 °C and exhausts it to a steam boiler at 238 °C. The turbine then receives steam from the boiler at this temperature and exhausts it to a condenser at 38.0 °C. What is the maximum efficiency of this combination?

11. In a heat pump, heat from the outdoors at -8.00 °C is transferred to a room at 22.0 °C, energy being supplied by an electric motor. How many joules of heat will be delivered to the room for each joule of electrical energy consumed, ideally?

12. A reversible engine operates in the cycle shown below, which consists of two adiabatic and two isochoric processes. The engine utilizes a monatomic ideal gas. Find the efficiency in terms of V_1 and V_2, and γ, the ratio of the heat capacities c_P and c_V.

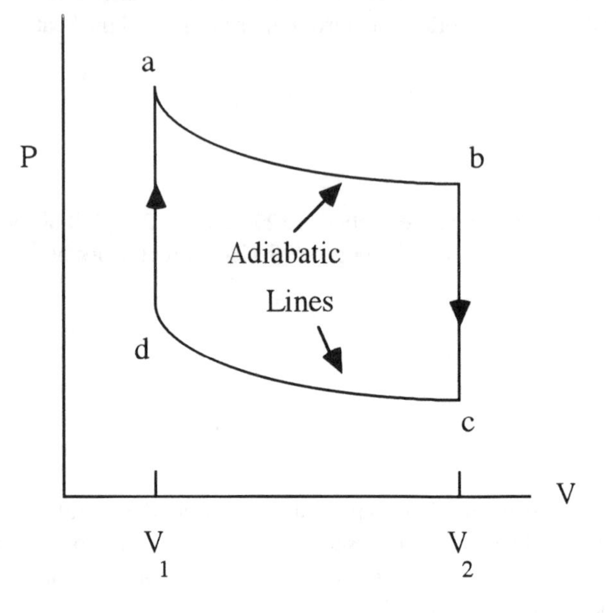

13. One end of a copper rod is in thermal contact with a reservoir at 550 K and the other end in thermal contact with a reservoir at 350 K. If 7700 J of heat are conducted from one end of the rod to the other find (a) the entropy change of each reservoir, (b) the entropy change of the rod, and (c) the entropy change of the universe.

14. If 40.0 g of ice at 0.0 °C is added to 100.0 g of water at 90.0 °C in a thermally insulated container, find the total change in entropy as the mixture comes to thermal equilibrium. Use the fact that for a change in temperature, the entropy change is $\Delta S = cm \ln(T_f/T_i)$, where c is the specific heat of the object.

HELPFUL SUGGESTIONS

1. When analyzing any problem involving multiple thermodynamic processes, look first for isothermic, isochoric, isobaric and adiabatic processes. These can be treated using the concepts in this chapter. Information about additional processes may be found if the system ultimately returns to its initial state by applying the first law and noting that $\Delta U = 0$. In this case the total change of internal energy must be zero.

2. Be sure to always use the correct sign convention when using the first law of thermodynamics, $\Delta U = Q - W$. Heat added to the system is positive while heat removed from the system is negative; and work done BY the system is positive while work done ON the system is negative.

3. For systems such as ideal gases where the internal energy depends only on the temperature, $\Delta U = 0$ for ALL isothermal processes. This means that any heat transfer produces work.

4. The temperatures used in thermodynamics are always assumed to be Kelvin temperatures.

5. Heat engines use heat from a hot reservoir to produce work. On the other hand, refrigerators (or air conditioners or heat pumps) use work to remove heat from a cold reservoir and transfer it to a hot reservoir.

6. When calculating entropy changes, ΔS is POSITIVE if heat flows INTO a region and NEGATIVE if heat flows OUT of a region. However, when using the equations for efficiency or coefficient of performance, Q and W are always taken as magnitudes only, without algebraic signs; the algebraic signs for Q and W have already been incorporated into these equations.

7. Real engines are NOT Carnot engines. However, by treating a real engine as a Carnot engine and determining the efficiency using equation (15.15), we can put an upper limit on the efficiency of any real engine.

8. Entropy can be thought of as a measure of a system's organization. Systems move from states of order to states of disorder. Thus, as melting ice goes from a very structured solid form to an unstructured liquid form, its entropy increases.

EVERYDAY PHYSICS

1. Gas expanding rapidly from the nozzle of a pressurized can, such as a hair spray can, is cooled. You may have noticed this if you happened to get part of your finger in the way of the expanding gas. The effect is demonstrated more markedly by releasing gas from a can of air-conditioner refrigerant. This latter case also involves a phase change since the refrigerant is liquid in the can and evaporates rapidly as the gas escapes. Note: Refrigerant can be VERY DANGEROUS due to the extremely low temperatures produced as it escapes from the can. If you want to see the effect, you might ask your mechanic if you can watch the next time you have your automobile air-conditioner serviced.

2. The above effect is often used by electronics technicians to diagnose bad integrated circuits (ICs) which only fail when they are hot. The hot IC is sprayed with the expanding gas from a can of refrigerant which cools it. The cooling may cause the IC to start working properly, thus identifying itself as the bad one.

3. Many examples of heat engines are present in everyday life. In an automobile engine, heating occurs as a gasoline-air mixture burns very rapidly inside the cylinders. Some of this heat energy is converted to mechanical energy (work) that ultimately turns the wheels while the unused heat is exhausted. Steam engines produce high-pressure steam in a boiler and transfer it to the cylinder at high pressure, where it expands to do work on a piston.

CHAPTER QUIZ

1. A gas in a rigid closed container receives 1500 J of heat. What is the change of internal energy of the gas?
   a. 1500 J b. 3000 J c. 750 J d. Not enough information.

2. A thermodynamic system consisting of an ideal gas is maintained at constant temperature while it does work, W, on its environment. How much heat energy is transferred during the process?
   a. W, into the system
   b. zero
   c. W, out of the system
   d. It is impossible to know without knowing the temperature.

3. A quantity of a substance undergoes a phase change at atmospheric pressure, changing in volume from 0.001 m^3 to 1.800 m^3. How much work is done during the process?
   a. 1.822×10^5 J, by the gas
   b. 2.733×10^5 J, on the gas
   c. zero
   d. It is impossible to know without knowing how much heat was transferred.

4. The work done during an isothermal expansion of an ideal gas is
   a. $P(V_f - V_i)$ b. $W = nRT \ln (V_f/V_i)$ c. $W = 3/2\, nR(T_i - T_f)$ d. zero

5. What is the change of internal energy of 1.0 mole of a monatomic ideal gas when its temperature changes from 23 °C to 123 °C?
   a. 2500 J b. 831 J c. 1250 J d. 2080 J

6. The boiling of water into steam in an open container is an example of a (an) _______ process.
   a. adiabatic b. isochoric c. isobaric d. zero work

7. A heat engine takes 8.0×10^2 J of energy at 1.0×10^3 K and exhausts 6.0×10^2 J at 4.0×10^2 K. What is the actual efficiency of the engine?
   a. 25% b. 40% c. 60% d. 75%

8. How much power is required by a refrigerator that takes 800 J of thermal energy from a cold region each second and exhausts 1200 J to a hot region?
   a. 800 W b. 1200 W c. 400 W d. 2000 W

9. When liquid water is converted to steam at 100 °C, the entropy of the water
   a. increases b. decreases c. remains the same

10. What is the change in entropy that occurs when 1.00 kg of water freezes into ice under normal conditions?
    a. +1230 J/K b. -1230 J/K c. -900 J/K d. 0 J/K

11. What is the maximum theoretical efficiency of a heat engine operating between temperatures of 27 °C and 59 °C?
    a. 53% b. 47% c. 9.6% d. 90%

12. In doing 2500 J of work, an engine rejects 750 J of heat. What is the efficiency of the engine?
    a. 30% b. 70% c. 23% d. 77%

13. A Carnot engine does 2500 J of work and rejects 750 J of heat into the cold reservoir at 57 °C. What is the Kelvin temperature of the hot reservoir?
    a. 160 K b. 330 K c. 430 K d. 1400 K

SOLUTIONS AND ANSWERS

Practice Problems

---

1. The first law gives $\Delta U = Q - W$. For an ideal gas, $\Delta U = 3/2\ nR(T_f - T_i)$ so

$$W = Q - 3/2\ nR(T_f - T_i) = 2500\ J - 3/2\ (3.00\ mol)(8.31\ J/mol\ K)(28\ K) = \mathbf{1500\ J}.$$

---

2. The process 1 → 3 is clearly **isochoric**. The work done is $W = \mathbf{0\ J}$ since the volume doesn't change. The process 2 → 3 is **isobaric**. The work done is

$$W = P(V_f - V_i) = (1.0 \times 10^5\ Pa)(1.0\ m^3 - 2.0\ m^3) = \mathbf{-\ 1.0 \times 10^5\ J}.$$

The process 1 → 2 looks isothermic. In order to decide if it is, calculate the temperatures at points 1 and 2. The ideal gas law gives $PV = nRT$ for any point. At point 1, $T_1 = 240$ K. At point 2, $T_2 = 240$ K. The process is probably **isothermal**. The work done during the isothermal process is

$$W = nRT \ln (V_f/V_i) = (1.0 \times 10^2\ mol)(8.31\ J/mol\ K)(240\ K) \ln (2.0) = \mathbf{+\ 1.4 \times 10^5\ J}.$$

The total work done is then $0\ J - 1.0 \times 10^5\ J + 1.4 \times 10^5\ J = \mathbf{+\ 0.4 \times 10^5\ J}$. The positive sign indicates that the work is done by the gas.

---

3. The work done during an isobaric process is $W = P(V_f - V_i)$ so $V_f - V_i = W/P$. The first law requires that $W = Q - \Delta U$ so

$$V_f - V_i = (Q - \Delta U)/P = (1150\ J - 1350\ J)/(1.01 \times 10^5\ Pa) = \mathbf{-\ 2.0 \times 10^{-3}\ m^3}.$$

The negative sign indicates that the system was compressed.

---

4. Equation (15.5) applies to an adiabatic process. Solving for the ratio of the pressures results in

$$P_f/P_i = (V_i/V_f)^{\gamma} = (5)^{7/5} = \mathbf{9.5}$$

---

5. The work done during an isothermal process is given by equation (15.3).

$$W = nRT \ln (V_f/V_i) = (1/2\ mol)(8.31\ J/mol\ K)(523\ K) \ln (5) = \mathbf{+\ 3.5 \times 10^3\ J}.$$

The positive sign indicates that the work is done **BY the gas**.
The internal energy of an ideal gas depends ONLY on the temperature so $\Delta U = 0$ for an isothermal process. The first law then requires $Q = W = \mathbf{+\ 3.5 \times 10^3\ J}$. The positive sign shows that heat is added to the gas.

---

6. Since the gas expands quickly, the process of expansion may be considered adiabatic. The ideal gas law applied to the initial and final states of the gas gives $T_f/T_i = (P_f/P_i)(V_f/V_i)$. In addition, equation (15.5) gives for the adiabatic process, $V_f/V_i = (P_i/P_f)^{1/\gamma}$. Now

$$T_f = (P_f/P_i)(P_i/P_f)^{1/\gamma}\ T_i = (1/10)(10)^{3/5}\ (296\ K) = \mathbf{120\ K = -\ 160\ °C}.$$

---

7. The pressure of the helium is essentially atmospheric throughout the process. The ideal gas law gives for the isobaric process

$$V_f = (T_f/T_i)V_i = (77\text{ K}/3.00 \times 10^2\text{ K})(1.00 \times 10^3\text{ cm}^3) = \mathbf{260\ cm^3\ or\ 2.60 \times 10^{-4}\ m^3}.$$

The internal energy change of the helium is $\Delta U = 3/2\, nR(T_f - T_i)$ and the number of moles of helium $n = PV/RT = 0.0405$ mol, so

$$\Delta U = \mathbf{-113\ J.}$$

The work done is

$$W = P(V_f - V_i) = (1.01 \times 10^5\text{ Pa})(-7.40 \times 10^{-4}\text{ m}^3) = \mathbf{-74.7\ J}.$$

The heat removed from the helium is found from the first law to be

$$Q = \Delta U + W = (-113\text{ J}) + (-74.7\text{ J}) = \mathbf{-188\ J}.$$

---

8. (a) Equation (15.15) gives;

$$\text{Efficiency (e)} = 1 - T_C/T_H = 1 - (400\text{ K})/(600\text{ K}) = \mathbf{0.33}.$$

(b) Equation (15.11) says $e = W/Q_H$ so

$$W = e \times Q_H = (0.33)(7.50 \times 10^4\text{ J}) = \mathbf{2.5 \times 10^4\ J}.$$

---

9. Calculate the efficiency, $e = 1 - T_C/T_H = 1 - (364\text{ K})/(455\text{ K}) = 0.20$.

(a) $W = e \times Q_H = (0.20)(650\text{ J}) = \mathbf{130\ J}$.

(b) Equation (15.13) gives $e = 1 - Q_C/Q_H$ so

$$Q_H = Q_C/(1 - e) = (1200\text{ J})/(0.80) = 1500\text{ J}.$$

Therefore, the work is

$$W = e \times Q_H = (0.20)(1500\text{ J}) = \mathbf{3.0 \times 10^2\ J}.$$

---

10. The efficiency of the engine is simply

$$e = 1 - T_C/T_H = 1 - (311\text{ K})/(736\text{ K}) = \mathbf{0.577}.$$

---

11. We know

$$Q_H = W/(1 - T_C/T_H) = (1.00\text{ J})/(1 - 265\text{ K}/295\text{ K}) = \mathbf{9.83\ J}.$$

---

12. We know $e = 1 - Q_C/Q_H$. The heat enters the system during step d-a and leaves at step b-c. No heat is exchanged during segments a-b or c-d since these processes are adiabatic. Using the first law of thermodynamics we know

$$Q_H = \Delta U_{da} = (3/2)nR(T_a - T_d) \text{ and } Q_C = \Delta U_{bc} = (3/2)nR(T_b - T_c).$$

Note: We used the fact that W_{da} and W_{bc} are both zero since there is no volume change over segments d-a and b-c. The efficiency is $e = 1 - (T_b - T_c)/(T_a - T_d)$. Now using the fact that steps a-b and c-d are adiabatic, we have

$$T_aV_a^{\gamma-1} = T_bV_b^{\gamma-1} \text{ and } T_cV_c^{\gamma-1} = T_dV_d^{\gamma-1}.$$

Using $V_a = V_d = V_1$ and $V_b = V_c = V_2$, we can substitute for T_b and T_c and the volumes to obtain

$$T_b - T_c = T_a\left(\frac{V_a}{V_b}\right)^{\gamma-1} - T_d\left(\frac{V_d}{V_c}\right)^{\gamma-1} = (T_a - T_d)\left(\frac{V_1}{V_2}\right)^{\gamma-1}$$

The efficiency is therefore,

$$e = 1 - \frac{(T_b - T_c)}{(T_a - T_d)} = \mathbf{1 - \left(\frac{V_1}{V_2}\right)^{\gamma - 1}}$$

---

13. The entropy change is given by equation (15.18), $\Delta S = Q/T$.

(a) For the hot reservoir, $\Delta S_H = Q/T_H = (-7700 \text{ J})/(550 \text{ K}) = \mathbf{-14\ J/K}$.
For the cold reservoir, $\Delta S_C = Q/T_C = (+7700 \text{ J})/(350 \text{ K}) = \mathbf{+22\ J/K}$.

(b) Since the rod hasn't changed in any way, $\Delta S_{rod} = \mathbf{0}$.
(c) The entropy change of the universe is therefore,

$$\Delta S_{universe} = \Delta S_H + \Delta S_C + \Delta S_{rod} = \mathbf{+8\ J/K}.$$

---

14. First find the equilibrium temperature of the system. We have $\Delta Q_{ice} = \Delta Q_{water}$. We have $(mL_f)_{ice} + (cm\Delta T)_{ice} = (cm\Delta T)_{water}$ so that,

$$(0.040 \text{ kg})(33.5 \times 10^4 \text{ J/kg}) + [4186 \text{ J/(kg C°)}](0.040 \text{ kg})T_f = [4186 \text{ J/(kg C°)}](0.100 \text{ kg})(90.0\ °\text{C} - T_f).$$

From which we obtain $T_f = 41.4$ °C. Now we can find the entropy change for each heat change.

$\Delta S_{ice} = \Delta Q_{ice}/T = (mL_f)_{ice}/T = (0.0400 \text{ kg})(33.5 \times 10^4 \text{ J/kg})/(273 \text{ K}) = +49.1$ J/K.

$\Delta S_{ice\ water} = m_{ice}c_{water} \ln(T_f/T_i) = (0.0400 \text{ kg})[4186 \text{ J/(kg C°)}] \ln(314.4 \text{ K}/273 \text{ K}) = +23.6$ J/K.

$\Delta S_{water} = m_{water}c_{water} \ln(T_f/T_i) = (0.100 \text{ kg})[4186 \text{ J/(kg C°)}] \ln(314.4 \text{ K}/363 \text{ K}) = -60.2$ J/K.

Therefore, $\Delta S_{universe} = \Delta S_{ice} + \Delta S_{ice\ water} + \Delta S_{water} = +49.0 \text{ J/K} + 23.6 \text{ J/K} - 60.2 \text{ J/K} = \mathbf{+12.4\ J/K}$.

Quiz answers

1. a	4. b	7. a	10. b	13. d
2. a	5. c	8. c	11. c	
3. a	6. c	9. a	12. d	

MCAT REVIEW PROBLEMS

Consider an insulated cylinder containing nitrogen gas and sealed on top by a heavy metal piston, as drawn in figur 1. The piston is free to slide up and down the cylinder, with negligible friction.

Initially, the piston is 0.20 m above the bottom of the cylinder, and the gas is at room temperature, 300 K But then, a student ignites a small flame under the cylinder. As a result, the metal piston slowly rises until reachinĝ a height 0.30 m above the bottom of the cylinder, at which point the student extinguishes the flame.

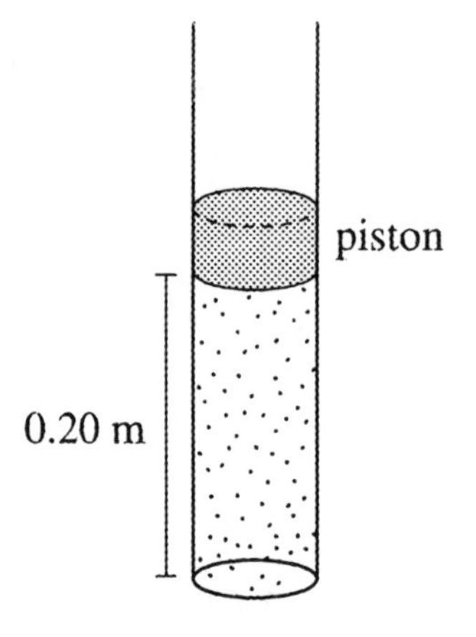

FIGURE 1

1. The air pressure outside the cylinder is 1 atm. Immediately before the flame gets ignited, the pressure of the nitrogen inside the cylinder is:

   A. greater than 1 atm
   B. equal to 1 atm
   C. less than 1 atm
   D. We cannot determine the answer without more information.

2. When the piston reaches a height of 0.30 m, the temperature of the nitrogen inside the cylinder is most nearly:

   A. 200 K
   B. 300 K
   C. 450 K
   D. 900 K

3. Let W denote the work done by the nitrogen on the piston as it rises from 0.20 m to 0.30 m. During that time interval, the net heat absorbed by the nitrogen is:

   A. Equal to W, because of conservation of energy.
   B. Equal to W, because with pistons, work equals heat by definition.
   C. Greater than W, because the entropy of the nitrogen increases.
   D. Greater than W, because the internal energy of the nitrogen increases.

4. Which graph best expresses the relationship between the pressure and volume of the nitrogen, as the piston rises from 0.20 m to 0.30 m?

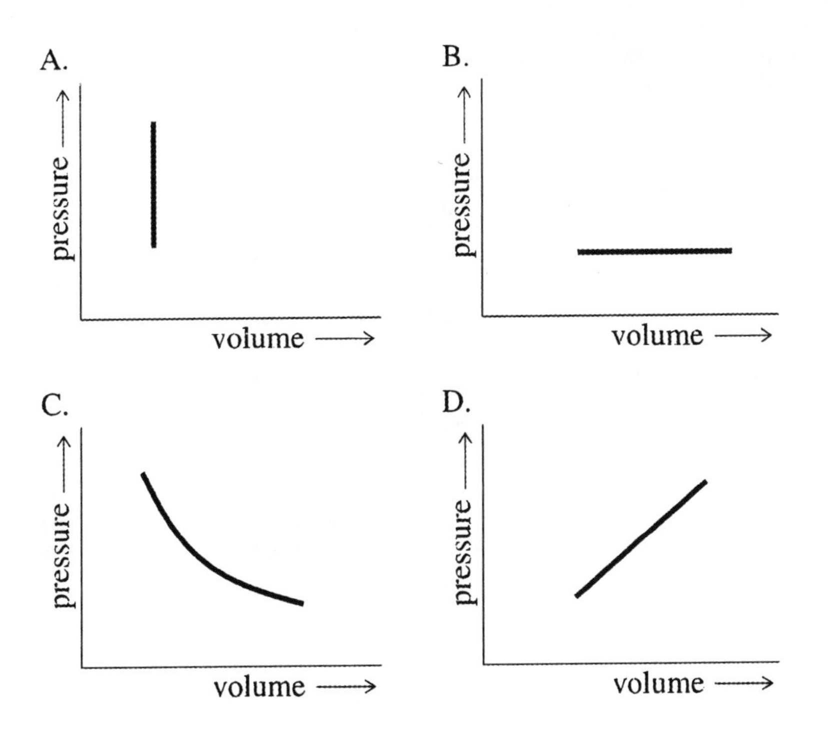

ANSWERS TO MCAT REVIEW PROBLEMS

---

1. **A.** If the piston were massless, then the nitrogen's pressure would equal 1 atm. But in addition to counterbalancing the outside air pressure, the nitrogen must also hold up the heavy piston. So, the pressure of the nitrogen must exceed the outside air pressure.

To flesh out this intuitive argument, draw a free-body diagram of the piston. The forces on the piston must balance.

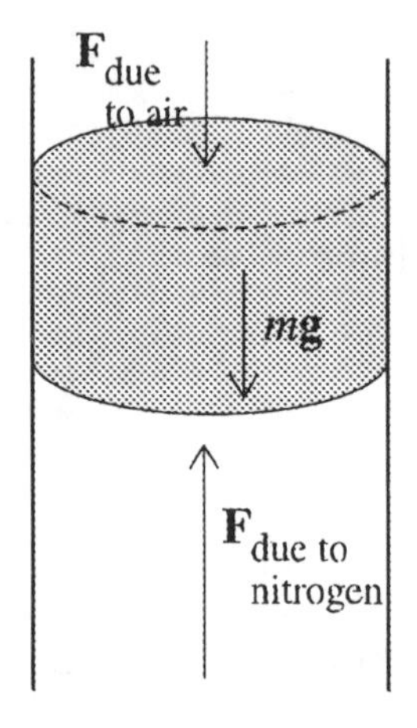

FIGURE 2

Hence,

$$F_{\text{due to nitrogen}} = mg + F_{\text{due to air pressure}}.$$

So, the upward force due to nitrogen must *exceed* the downward force due to the outside air. Therefore, since pressure is force per area, the nitrogen's pressure inside the cylinder must exceed the outside air pressure.

---

2. **C.** To solve for T using the ideal gas law,

$$PV = nRT,$$

we must think about how the pressure and volume change as the nitrogen expands. (The number of moles of nitrogen, n, stays constant unless the cylinder leaks.)

Since the piston rises from 0.20 m to 0.30 m above the bottom of the cylinder, the nitrogen's volume increases by a factor of $\frac{0.30\text{ m}}{0.20\text{ m}} = 1.5$. In other words, the volume goes up 50%. *But the pressure stays constant.* To see why, look again at the free-body diagram in figure 2. Those forces—and hence, the corresponding pressures—do *not* depend on the piston's height. Therefore, since the outside air pressure stays constant, and the pressure caused by the piston's weight stays constant, so does the nitrogen's pressure.

In summary, as the nitrogen expands, its volume increases by 50%, while its pressure stays the same. Therefore, according to the ideal gas law, the temperature also increases by 50%. Otherwise, the left- and right-hand sides of $PV = nRT$ would not stay equal.

---

3. **D.** In thermodynamics, conservation of energy shows up as the *first law of thermodynamics*:

$$\Delta U = Q - W,$$

where ΔU denotes the *change* in internal energy of the system, Q denotes the net heat absorbed by the system, and W denotes the work done by the system. For instance, if the nitrogen absorbs 10 joules of heat but performs only 8 joules of work, then the nitrogen "keeps" those extra 2 joules as internal energy.

In question 2, we found that the nitrogen's temperature rises during the expansion. Since temperature is proportional to internal energy, the nitrogen gained internal energy; ΔU is positive. Therefore, Q must be *greater* than W. Intuitively, the nitrogen absorbs enough heat to do the work, plus some additional heat that causes its temperature to rise.

The nitrogen's entropy also increases during the expansion. But that's not *why* $Q > W$. The entropy of a gas can increase even when $Q = W$, as exemplified by isothermal expansions.

---

4. **B.** In question 2, we found that the volume increases while the pressure stays the same. Only graph B fits this description. Remember, the pressure stays constant because the nitrogen inside the cylinder always "fights" the same force, namely the force due to the outside air pressure plus the piston's weight.

CHAPTER 16
Waves and Sound

PREVIEW

This is the first of two chapters covering waves. In this chapter you will become acquainted with the concept of waves, in particular a special wave referred to as a periodic wave. You will learn to describe a periodic wave in space by its wavelength and amplitude and in time by its amplitude and period. Combining these descriptions allows us to find the speed of a wave in terms of its frequency and wavelength. The speed of waves on a stretched string will be investigated. A mathematical description of harmonic waves is given. This is very important since any complicated wave can be described as a sum of harmonic waves, as you will learn in a later chapter.

In this chapter you will also study the nature of sound. The wave properties of sound are examined, including the speed of sound in gases, liquids, and solids. You will be introduced to the concepts of sound intensity, loudness, and the decibel scale. Applications of sound are also discussed, such as sonar, ultrasound, and the Doppler effect.

QUICK REFERENCE

Important Terms

Wave
A traveling disturbance which carries energy. Mechanical waves require a medium.

Transverse wave
The wave disturbance is in a direction perpendicular to the direction of wave travel.

Longitudinal wave
The wave disturbance is parallel to the line of wave travel.

Periodic wave
The wave disturbance is repeated continuously by the source of the disturbance.

Amplitude (A)
The maximum excursion of a particle of the medium from the particle's undisturbed position.

Wavelength (λ)
The distance along the length of a wave between two equivalent points.

Period (T)
The time required for a wave to travel one wavelength.

Frequency (f)
The number of wavelengths per second which pass a given point. Note: $f = 1/T$.

Speed of a wave
The product of the wavelength and the frequency of a wave.

Condensation
A region of slightly increased air pressure produced by a vibrating source.

Rarefaction
A region of slightly reduced air pressure immediately following a condensation.

Frequency of a sound wave
The number of cycles per second (each cycle being composed of one condensation and one rarefaction) that pass a given location.

Pure tone
A sound containing a single frequency.

Pitch
A subjective quality related to the frequency of sound detected by the human ear. High pitch is high frequency and low pitch is low frequency.

Pressure amplitude
The maximum amount of change in pressure produced by a sound wave, measured relative to the undisturbed pressure.

Loudness
A subjective quality related to the pressure amplitude of a sound wave.

Sound intensity
The sound power per unit area that passes perpendicularly through a surface. Units are W/m^2.

Threshold of hearing
The smallest sound intensity that the human ear can detect.

Intensity level
A quantity which relates the intensity of a sound to some reference intensity level. The intensity level is measured in decibels.

Doppler effect
A change in the frequency of sound detected by an observer because the sound source and the observer have different velocities with respect to the medium.

Equations

The **speed of a periodic wave** which has a wavelength, λ, and a frequency, f, is

$$v = \lambda f \qquad (16.1)$$

The **speed of a wave traveling on a string** under a tension, F, and having a linear density m/L is

$$v = \sqrt{\frac{F}{m/L}} \qquad (16.2)$$

The displacement of a particle of the medium from its undisturbed position when a harmonic wave is moving along the + x axis is

$$y = A \sin (2\pi ft - 2\pi x/\lambda) \qquad (16.3)$$

and for motion of the wave toward - x

$$y = A \sin (2\pi ft + 2\pi x/\lambda) \qquad (16.4)$$

The **speed of sound in a gas** is:

$$v = \sqrt{\frac{\gamma kT}{m}} \qquad (16.5)$$

The **speed of sound in a liquid** is:

$$v = \sqrt{\frac{B_{ad}}{\rho}} \qquad (16.6)$$

The **speed of sound in a long, thin, solid** bar is:

$$v = \sqrt{\frac{Y}{\rho}} \qquad (16.7)$$

The **sound intensity** is defined as:

$$I = \frac{P}{A} \qquad (16.8)$$

If a sound is emitted uniformly in all directions, the **intensity at a distance r from the source** is:

$$I = \frac{P}{4\pi r^2} \tag{16.9}$$

The **intensity level** in **decibels** is defined as:

$$\beta \text{ (in dB)} = 10 \log\left(\frac{I}{I_0}\right) \tag{16.10}$$

The **Doppler shifted frequency** (general case) is:

$$f' = f\left(\frac{1 \pm \frac{v_o}{v}}{1 \mp \frac{v_s}{v}}\right) \tag{16.15}$$

DISCUSSION OF SELECTED SECTIONS

16.2 Periodic Waves

Three quantities are needed to describe periodic waves: period, wavelength, and amplitude. The wavelength and amplitude can be found from a space representation of the wave which is essentially a "snapshot" taken of the wave at an instant in time.

The period and amplitude can be found from a time representation of the wave. This is constructed by making a graph of the motion of a point in the medium. The graph is the displacement of the point from its undisturbed position versus time. This is equivalent to making a vertical position versus time graph of a small cork bobbing up and down in water as a wave passes by.

Example 1
A snapshot is taken of a periodic wave traveling on a string as shown below. What is the amplitude and wavelength of the wave?

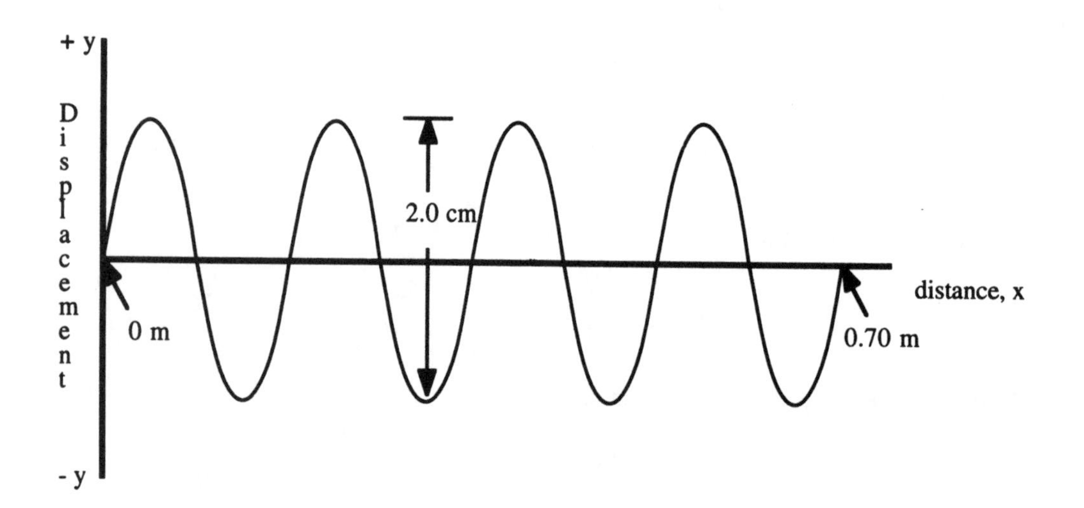

The height of the wave is seen from the "snapshot" to be 2.0 cm. The amplitude is only one-half of this, so

$$A = 1.0 \text{ cm}.$$

The wave is seen to have four wavelengths spread over a distance of 0.70 m so the wavelength

$$\lambda = 0.70 \text{ m}/4 = 0.18 \text{ m}.$$

Example 2

A point in the medium carrying the wave in example 1 is observed and a displacement versus time graph is made and shown below. Find the period, frequency, and amplitude of the wave from this graph. What is the speed of the wave?

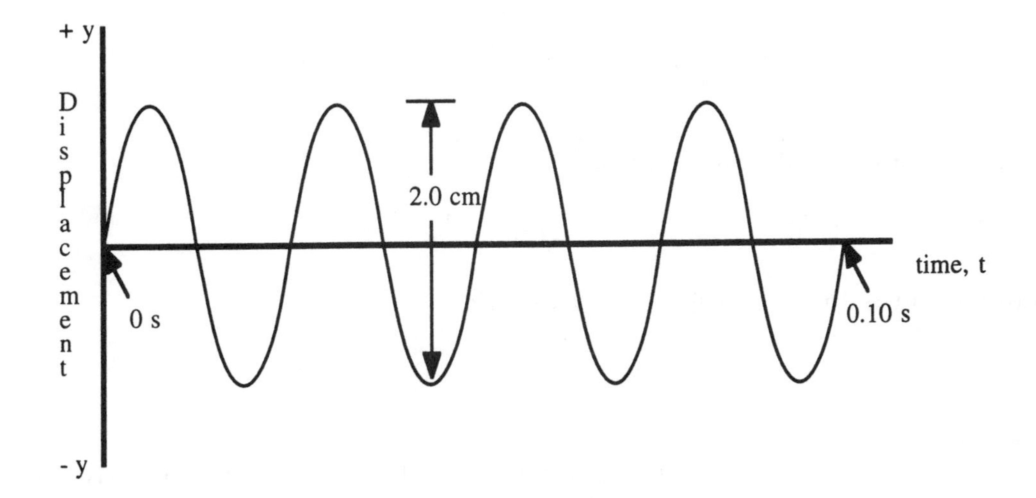

Again, the amplitude is one-half of the total height of the sine wave,

$$A = 1.0 \text{ cm}.$$

The period of the wave is found by observing that four complete oscillations occur in 0.10 s, so

$$T = (0.10 \text{ s})/4 = 0.025 \text{ s}.$$

The frequency of the wave is then

$$f = 1/T = 1/0.025 \text{ s} = 4.0 \text{ X } 10^1 \text{ Hz}.$$

The speed of the wave is

$$v = \lambda f = (0.18 \text{ m})(4.0 \text{ X } 10^1 \text{ Hz}) = 7.2 \text{ m/s}.$$

Example 3

Radio waves travel with a speed of $3.00 \text{ X } 10^8$ m/s. What is the wavelength of AM radio waves whose frequency is 640 kHz?

Equation (16.1) is true for waves of all kinds, including radio waves. The wavelength of the AM waves is

$$\lambda = v/f = (3.00 \text{ X } 10^8 \text{ m/s})/(640 \text{ X } 10^3 \text{ Hz}) = 470 \text{ m}.$$

16.3 The Speed of a Wave on a String

The speed of transverse waves on a string of uniform linear density is

$$v = \sqrt{\frac{F}{m/L}}$$

Example 4
A string of linear density, 2.0 gm/m, is stretched to a tension of 4.9 N. What is the speed of the waves on the string? If the tension is increased so that the speed of the waves doubles, what is the new tension?

$$v = \sqrt{\frac{F}{m/L}} = \sqrt{\frac{4.9\ \text{N}}{(2.0 \text{ X } 10^{-3}\ \text{kg/m})}} = 49\ \text{m/s}$$

Solving the above equation for the tension results in

$$F = (m/L)v^2$$

If the speed of the waves doubles to 2v, then the new tension is

$$F' = 4\ (m/L)v^2 = 4F = 2.0 \text{ X } 10^1\ \text{N}$$

Example 5
A uniform wire carries waves whose frequency and wavelength are 450 Hz and 1.2m, respectively. If the string is known to be under a tension of 250 N, what is the linear density of the wire?

The linear density can be found from equation (16.2). $m/L = F/v^2$.
The speed of the waves on the wire can be found from $v = \lambda f = (1.2\ \text{m})(450\ \text{Hz}) = 540\ \text{m/s}$, so

$$m/L = (250\ \text{N})/(540\ \text{m/s})^2 = 8.57 \text{ X } 10^{-4}\ \text{kg/m}.$$

16.4 The Mathematical Description of a Wave

If a periodic wave results from simple harmonic motion at the source, then, ideally, EACH point in the medium oscillates with simple harmonic motion. The displacement at a particular point, x =0 (time view) is
$y = A \sin (2\pi ft)$. As we saw before, the wave frozen in time looks like a sine wave, too. The displacement of each point at time zero (space view) can be written $y = A \sin (\pm 2\pi x/\lambda)$. It is customary to combine the space and time views into a single equation

$$y = A \sin (2\pi ft - 2\pi x/\lambda)$$

for a wave traveling towards +x and

$$y = A \sin (2\pi ft + 2\pi x/\lambda)$$

for a wave traveling towards -x.

Example 6
Show that equation (16.3) represents a wave traveling toward +x and that equation (16.4) represents a wave traveling toward -x.

Imagine watching a single point on a wave as it travels. This point appears to move with the speed of the wave and always has the same displacement. That is, the value of the sine function is constant. The only way for a sine function to remain constant is for its argument (the angle) to remain constant.

Equation (16.3) then gives $2\pi ft - 2\pi x/\lambda = C$, a constant. At some time, t_1, the point has a position, x_1, so $2\pi ft_1 - 2\pi x_1/\lambda = C$. At a later time, t_2, the point has a new position, x_2, so $2\pi ft_2 - 2\pi x_2/\lambda = C$. Subtracting and rearranging yields the velocity of the point on the wave

$$\frac{x_2 - x_1}{t_2 - t_1} = + \lambda f = + v$$

which is positive, indicating that the wave is traveling toward +x.

A similar treatment using equation (16.4) gives

$$\frac{x_2 - x_1}{t_2 - t_1} = - \lambda f = - v$$

which is negative. This wave is traveling toward the -x direction.

Example 7
The wave traveling on the string in example 4 has a wavelength of 0.50 m and an amplitude of 2.5 mm. Write an equation describing the wave as it travels toward the +x axis.

Equation (16.3) applies to a wave traveling toward +x. We need to find the frequency.

$$f = v/\lambda = (49 \text{ m/s})/(0.50 \text{ m}) = 98 \text{ Hz}.$$

Now

$$y = A \sin (2\pi ft - 2\pi x/\lambda) = (2.5 \text{ mm}) \sin [2\pi(98 \text{ s}^{-1})t - 2\pi x(2.0 \text{ m}^{-1})]$$

where x is in meters and t is in seconds.

Example 8
What are the wavelength, frequency and speed of the wave described by

$$y = (5.0 \text{ mm}) \sin 2\pi[(450 \text{ s}^{-1})t - (25 \text{ m}^{-1})x]?$$

Comparing our wave with equation (16.3) shows that $2\pi f = 450 \text{ s}^{-1}$ and $2\pi/\lambda = 25 \text{ m}^{-1}$, so

$$f = 72 \text{ Hz and } \lambda = 0.25 \text{ m}$$

The speed of the wave is then

$$v = \lambda f = 18 \text{ m/s}.$$

16.6 The Speed of Sound

Sound travels through different materials at different speeds. For example, at room temperature sound travels at about 343 m/s in air. However, sound travels at about 1482 m/s through water and 5960 m/s through bulk steel at the same temperature. In general, sound travels slowest in gases, faster in liquids, and fastest in solids.

For a **gas**, the speed of sound depends on the temperature of the gas, and the type of molecules making up the gas. If we consider the situation for an ideal gas, we find that the speed of sound in that gas can be written

$$v = \sqrt{\frac{\gamma k T}{m}} \tag{16.5}$$

Where $\gamma = c_P/c_V$ is the ratio of specific heat capacity at constant pressure to the specific heat capacity at constant volume, m is the mass of a molecule of the gas, k is Boltzmann's constant, and T is the Kelvin temperature.

Example 9

Determine the speed of sound in air at 20.0 °C assuming the mean molecular mass of air to be 28.9×10^{-3} kg/mol and the ratio of specific heats to be $\gamma = 1.4$.

First we need to find the mass of an air molecule based on the mean molecular mass and Avogadro's number.

$$m = \frac{28.9 \times 10^{-3}\ \text{kg/mol}}{6.022 \times 10^{23}\ \text{mol}^{-1}} = 4.80 \times 10^{-26}\ \text{kg}$$

Now use equation (16.5) to determine the speed of sound in air

$$v = \sqrt{\frac{\gamma k T}{m}} = \sqrt{\frac{(1.4)(1.38 \times 10^{-23}\ \text{J/K})(293\ \text{K})}{4.80 \times 10^{-26}\ \text{kg}}} = 343\ \text{m/s}$$

In a **liquid**, the speed of sound depends on the density of the material ρ and the adiabatic bulk modulus B_{ad}.

$$v = \sqrt{\frac{B_{ad}}{\rho}} \tag{16.6}$$

In a **solid**, in the shape of a long slender bar, the speed of sound depends on the density and Young's modulus for the material;

$$v = \sqrt{\frac{Y}{\rho}} \tag{16.7}$$

Example 10

A steel rod extends 2500 m under the ocean water. A sound is produced at one end of the rod and travels through both the rod and the water. What is the delay time between the arrival of the sound waves in the rod and the water? Take the bulk modulus for sea water to be $B_{ad} = 2.3 \times 10^9$ Pa, the density of sea water to be $\rho_w = 1025\ \text{kg/m}^3$, Young's modulus for steel to be $Y = 2.0 \times 10^{11}$ Pa, and the density of steel to be $\rho_s = 7860\ \text{kg/m}^3$.

$$v_w = \sqrt{\frac{B_{ad}}{\rho_w}} = \sqrt{\frac{2.3 \times 10^9\ \text{Pa}}{1025\ \text{kg/m}^3}} = 1.5 \times 10^3\ \text{m/s}; \quad v_s = \sqrt{\frac{Y}{\rho_s}} = \sqrt{\frac{2.0 \times 10^{11}\ \text{Pa}}{7860\ \text{kg/m}^3}} = 5.0 \times 10^3\ \text{m/s}$$

The time it takes for each sound to travel 2500 m can now be calculated:

$$t_w = \frac{s}{v_w} = \frac{2500\ \text{m}}{1.5 \times 10^3\ \text{m/s}} = 1.7\ \text{s}; \qquad t_s = \frac{s}{v_s} = \frac{2500\ \text{m}}{5.0 \times 10^3\ \text{m/s}} = 0.50\ \text{s}$$

Therefore, the delay time between the arrival of the two waves is

$$\Delta t = t_w - t_s = 1.7\ \text{s} - 0.50\ \text{s} = 1.2\ \text{s}.$$

16.7 Sound Intensity

The **sound intensity I** is defined as the power P that passes perpendicularly through a surface divided by the area A of the surface:

$$I = \frac{P}{A} \tag{16.8}$$

The unit of sound intensity is W/m^2. If the source emits sound uniformly in all directions, the radiated power P passes through the spherical surface area $A = 4\pi r^2$, where r is the distance to the sound source. The intensity is:

$$I = \frac{P}{4\pi r^2} \tag{16.9}$$

Example 11
At a distance of 3.8 m from a siren, the sound intensity is $3.6 \times 10^{-2}\ W/m^2$. Assuming that the siren radiates sound uniformly in all directions, find the total power radiated.

Using equation (16.9) and solving for the power P gives

$$P = 4\pi r^2 I = 4\pi\,(3.8\ m)^2\,(3.6 \times 10^{-2}\ W/m^2) = 6.5\ W.$$

16.8 Decibels

The **decibel (dB)** is a measurement unit encountered frequently in connection with audio equipment. The main application of the decibel concept is for comparing two sound intensities. The **intensity level** β is defined as:

$$\beta\ (\text{in dB}) = 10 \log\left(\frac{I}{I_0}\right) \tag{16.10}$$

where I_0 is the intensity of the reference level to which I is being compared, and is often taken as the threshold of hearing, $I_0 = 1.00 \times 10^{-12}\ W/m^2$. Note that β is a dimensionless quantity and does NOT have units of W/m^2.

Example 12
The intensity level of the sound produced at a live rock concert is 120 dB above the threshold of hearing. What is the sound intensity I?

Use equation (16.10) with $I_0 = 1.00 \times 10^{-12}\ W/m^2$ and solve for the intensity I. We obtain

$$I = I_0 \times 10^{\beta/10} = (1.00 \times 10^{-12}\ W/m^2) \times 10^{12} = 1.00\ W/m^2.$$

This represents a sound intensity that is one trillion times the threshold of hearing! However, the **loudness** will only be about 4100 times louder because each increase of 10 dB corresponds to a doubling of the loudness level. That is, a 10 dB sound will sound twice as loud as one which is 0 dB. A 20 dB sound is twice as loud as a 10 dB sound (four times louder than a 0 dB sound), and so on. Therefore, a 120 dB sound will appear about 2^{12} or approximately 4100 times as loud as a 0 dB sound.

16.10 The Doppler Effect

When a source of sound is in motion, or when a listener is moving, the frequency of the sound that is observed will change. This effect, known as the **Doppler effect**, is the change in pitch or frequency of the sound detected by an observer because the sound source and the observer have different velocities with respect to the medium of sound propagation. For example, when you hear the siren of an approaching fire truck you notice a definite change in pitch of the siren as it passes. As the truck approaches, the pitch is relatively high (that is, the frequency is high). However, as the truck passes you and moves away, the pitch drops (frequency lowers).

Let f represent the frequency of the sound emitted by a source. Sound travels at a speed v in the particular medium of propagation. The source may be moving with a speed v_s, and the observer may be moving with a speed v_o. The frequency observed, f', will be different from the emitted frequency according to the relation

$$f' = f\left(\frac{1 \pm \frac{v_o}{v}}{1 \mp \frac{v_s}{v}}\right) \qquad (16.15)$$

In the numerator of the equation, the plus sign applies when the observer moves toward the source, and the minus sign applies when the observer moves away from the source. In the denominator, the minus sign is used when the source moves toward the observer, and the plus sign is used when the source moves away from the observer. The symbols v, v_o, and v_s denote numbers without algebraic sign, because the direction of travel has been taken into account by the plus and minus signs that appear directly in the equation.

Example 13

Two trucks travel at the same speed. They are far apart on adjacent lanes and thus, approach each other essentially head-on. One driver hears the horn of the other truck at a frequency that is 1.20 times the frequency he hears when the trucks are stationary. The speed of sound is 343 m/s. At what speed is each truck moving?

In this particular problem, the observer and source move with the same speed, so $v_o = v_s$. If the observed frequency is 1.20 times the emitted frequency, than f'/f = 1.20. Equation (16.15) then yields

$$\frac{f'}{f} = \frac{1 + \frac{v_o}{v}}{1 - \frac{v_s}{v}} = \frac{v + v_o}{v - v_s} = 1.20$$

Note that the plus sign was used in the numerator (since the observer is approaching the source) and a minus sign was used in the denominator (since the source is approaching the observer). Solving for v_o leads to

$$v + v_o = 1.20\,(v - v_o)$$

which gives

$$2.20\,v_o = 0.20\,v.$$

$$v_o = \frac{0.20\,v}{2.20} = (0.091)\,(343 \text{ m/s}) = 31 \text{ m/s}.$$

Thus, in order for the frequency to increase by a factor of 1.20, the observer and source are **each** moving at 31 m/s along the road.

PRACTICE PROBLEMS

1. A light wave has a wavelength of 4.55×10^{-7} m and travels with a speed of 3.00×10^{8} m/s. What is the frequency of the light wave?

2. A wave on a string has a frequency of 440 Hz and a wavelength of 1.3 m. How fast does the wave travel?

3. A transverse wave on a string travels with a speed of 550 m/s and has a wavelength of 0.62 m and an amplitude of 0.50 cm. What is the maximum speed and the maximum acceleration of a point on the string?

4. A wire with a linear density of 2.5×10^{-3} kg/m is under 35 N of tension. How fast would a wave travel on the wire?

5. The waves traveling on a stretched wire travel with a speed of 180 m/s when the tension in the wire is 110 N. What is the linear density of the wire?

6. A wave has an amplitude of 0.70 mm, a wavelength of 5.5 m and travels with a speed of 150 m/s. Write an equation describing the wave.

7. A wave is described by $y = (1.0\ \text{cm}) \sin [(550\ \text{s}^{-1})t + (35\ \text{m}^{-1})x]$. What are the amplitude, wavelength, frequency and velocity of the wave?

8. A wave is described by $y = (1.5\ \text{cm}) \sin [(150\ \text{m}^{-1})x - (760\ \text{s}^{-1})t]$. How does this wave differ from one described by $y' = (1.5\ \text{cm}) \sin[(760\ \text{s}^{-1})t - (150\ \text{m}^{-1})x]$?

9. The speed of sound in a diatomic gas (= 1.40) is measured at 27 °C to be 280 m/s. What is the molecular mass of the gas?

10. Brass has a Young's modulus of 9.0×10^{10} Pa and a density of 8470 kg/m^3. How much time would it take for the sound to travel the length of a 5.5-m brass rod?

11. At the bottom of a lake, a sound pulse is generated which is received by an observer 15.0 m above the surface of the lake in 0.100 s. Both the lake and air are at a temperature of 20.0 °C. How deep is the lake?

12. Two sounds have intensities of 2.0×10^{-5} W/m^2 and 4.0×10^{-8} W/m^2. What is the difference in the decibel levels of the two sounds?

13. At a certain distance from a sound source, the intensity level is 75 dB. (a) What is the intensity of the sound in W/m^2? (b) How much energy falls on a 5.0 m^2 area in 25 s?

14. A listener triples his distance from a source that emits sound uniformly in all directions. By how many decibels does the sound intensity level change?

15. At 35 m from a uniformly radiating source of sound, the intensity level is 65 dB. At what distance from this source will its sound be just barely perceptible if absorption of sound by air is neglected?

16. A train whistle emits sound at a frequency of 555 Hz. A person standing next to the train track hears the whistle at a frequency of 488 Hz. What is the speed of the train? Is the train approaching or receding? Take the speed of sound to be 343 m/s.

17. A whistle of frequency 444 Hz rotates in a circle of radius 1.00 m with an angular speed of 25.0 rad/s. What is (a) the lowest frequency, and (b) the highest frequency heard by a listener a long distance away at rest with respect to the center of the circle?

18. An ambulance travels down the highway at a speed of 120 km/h. Its siren emits sound at a frequency of 660 Hz. A passenger car traveling at 88 km/h is on the other side of the highway (i.e., it's traveling in the opposite direction as the ambulance). What is the frequency of the siren heard by a passenger in the car (a) as the car and ambulance are approaching each other, and (b) as the car and ambulance are receding from one another?

HELPFUL SUGGESTIONS

1. Do not confuse the motion of a sinusoidal wave disturbance through a medium with the motion of the particles of the medium itself. As time passes the wave disturbance moves through the medium, whereas the particles of the medium only oscillate and undergo no NET displacement.

2. Remember that $\lambda f = v$ is valid for ANY sinusoidal wave regardless of the medium in which the wave travels. That is, it is equally valid for waves on a spring or string, water waves, sound waves, light waves, etc.

3. The key to "reading" a wave description to find the direction of wave travel is in the signs in the argument of the sine function. If the signs are opposite, then the wave travels toward +x and if they are the same, then the wave travels toward -x. Both $y = A \sin(2\pi ft - 2\pi x/\lambda)$ and $y = A \sin(2\pi x/\lambda - 2\pi ft)$ describe waves traveling toward +x while both $y = A \sin(2\pi ft + 2\pi x/\lambda)$ and $y = A \sin(-2\pi ft - 2\pi x/\lambda)$ describe waves traveling toward -x.

4. Remember, when dealing with intensity levels measured in decibels, that the dB is a logarithmic function of the ratio of intensities. Thus, a 0 dB intensity level does NOT mean that the sound intensity is zero; it simply means that the intensity $I = I_0$, the threshold of hearing.

5. Note that a 3-dB increase in loudness corresponds to a doubling of the sound intensity. However, experiments show that if the intensity level increases by 10 dB, the new sound appears approximately twice as loud as the original sound.

6. When applying the equations for the Doppler effect, be careful to use the proper signs for approach and recession. Equation (16.15) is the general expression and can be used for ANY situation.

$$f' = f\left(\frac{1 \pm \frac{v_o}{v}}{1 \mp \frac{v_s}{v}}\right) \qquad (16.15)$$

A rule to remember when applying this equation is as follows: the **top** sign in BOTH the numerator and denominator refer to velocity of **approach** for observer and source. The **bottom** sign in BOTH the numerator and denominator refer to velocity of **recession** for source and observer.

EVERYDAY PHYSICS

1. Waves abound in our everyday world. Most of us are so familiar with them that we never stop to really look at their characteristics. Investigate the waves on the surface of the water in a pool or bathtub created by moving a small object (such as your finger) periodically up and down. Can you estimate the frequency, wavelength and speed of the waves?

2. Can you retrieve a floating object (such as a cork) in a body of water by throwing rocks to create waves? Try it. Does the size of the rocks have any effect? Why or why not?

3. Why do the ocean waves at the seashore sometimes knock you down? Does this contradict Helpful Suggestion 1?

4. You can investigate the relationship between the speed of waves on a rope and the tension in the rope by tying one end of the rope to a fixed support. pulling the other end to produce a tension and introducing a "kink" in the rope with your thumb and forefinger. When the "kink" is released, a disturbance will travel down to the fixed end and reflect back to you. You can judge the speed of the disturbance by noting the time it takes for the disturbance to return to you. Apply more tension and try it again.

5. To illustrate how frequency and pitch are related try one of the following experiments:

   - Hold a card against the spokes of a bicycle wheel and spin the wheel slowly; then increase the speed of the wheel. What do you notice?
   - Play a 33 or 45 rpm record at the wrong speed on a phonograph. What happens to the sound? Explain.

6. To illustrate the Doppler effect place a small round whistle into a funnel. Tie a string to the funnel and have someone rotate the funnel-whistle in a horizontal circle. Notice how the pitch changes as the whistle approaches and recedes from you.

7. The tremor of the ground from a distant explosion or from thunder can be felt before the sound can be heard. Why is this the case?

8. Police radar used in detecting speeding motorists is based on the Doppler effect. How might this work?

9. Some animals rely on an acute sense of hearing for survival, and the visible part of the ear on such animals is often relatively large. Explain how this anatomical feature helps to increase the sensitivity of the animal's hearing for low-intensity sounds.

CHAPTER QUIZ

1. A periodic wave is produced on a stretched string. Note the following quantities associated with the wave and the string:

   I. frequency
   II. wavelength
   III. tension in the string
   IV. linear density of the string

   Which of the quantities influences the speed of the waves?
   a. I only
   b. I and II only
   c. III and IV only
   d. I, II, III, and IV

2. A wave travels through a medium with a speed of 340 m/s. If the frequency of the wave is doubled, what happens to the wave speed? What happens to the wavelength?
   a. remains constant, doubles
   b. remains constant, reduced by a factor of two
   c. doubles, reduced by a factor of two
   d. reduced by a factor of two, remains, constant

3. The tension in a wire is tripled. What happens to the speed of waves on the wire?
   a. It increases by $\sqrt{3}$.
   b. It triples
   c. It is reduced by $\sqrt{3}$
   d. It is reduced by 1/3.

4. What is the speed of a 150 Hz wave whose wavelength is measured to be 0.30 m?
   a. 45 m/s
   b. 5.0×10^{2} m/s
   c. 0.0020 m/s
   d. It cannot be found.

5. A 440 Hz sound wave travels with a speed of 340 m/s. What is the wavelength of the wave?
   a. 1.5×10^{5} m
   b. 0.77 m
   c. 1.3 m
   d. 1.1 m

6. A wave is described by $y = (1.0\text{ cm}) \sin[2\pi ft - 15\text{ m}^{-1})x]$ and travels with a speed of 15 m/s. What is the frequency of the wave?
   a. 6.3 Hz
   b. 0.50 Hz
   c. 3.1 Hz
   d. 36 Hz

7. A wave travels with a speed of 450 m/s on a wire stretched to a tension of 150 N. What is the linear density of the wire?
   a. 0.33 kg/m
   b. 3.3×10^{-3} kg/m
   c. 7.4×10^{-4} kg/m
   d. 0.57 kg/m

8. Which of the following represents a wave traveling toward +x? Note the units have been suppressed for clarity.
   a. $y = (2.0) \sin(6.0t + 1.5x)$
   b. $y = (-8.5) \sin(0.5x - 60t)$
   c. $y = (6.1) \sin(-15.0t - 25x)$
   d. None of the above.

9. The frequency of the particles oscillating in a medium is ______ the frequency of waves in the medium.
   a. greater than
   b. less than
   c. unrelated to
   d. the same as

10. A wave has an amplitude of 25 cm, a frequency of 150 Hz and a wavelength of 0.50 m and moves toward +x. which of the following describes the wave? Note the units of x and t have been suppressed for clarity.
    a. $y = (25\text{ cm}) \sin(150t - 0.50x)$
    b. $y = (0.50\text{ m}) \sin(150t - 25x)$
    c. $y = (25\text{ cm}) \sin(940t - 13x)$
    d. $y = (0.50\text{ m}) \sin(940t - 0.25x)$

11. The atomic mass of argon is twice that of neon. Both argon and neon are monatomic gases. At a given temperature, the ratio of the speed of sound in argon to that in neon is
    a. 2
    b. $\sqrt{2}$
    c. 1/2
    d. $1/\sqrt{2}$

12. What happens to the velocity of sound in an ideal gas when the absolute temperature of the gas is doubled?
    a. It doubles
    b. It quadruples
    c. It increases to 1.4 times its original value
    d. None of the above

13. Consider a source of sound of constant power which radiates equally in all directions. If you double your distance from this sound source, what happens to the intensity of the sound. It reduces to
    a. one-half its original value
    b. one-fourth its original value
    c. one-eighth its original value.
    d. None of the above.

14. If, as in question 13, you were to double your distance from a constant sound source which radiates equally in all directions, what would happen to the intensity level of the sound? It drops by
    a. 2 dB b. 4 dB c. 6 dB d. 8 dB

15. The intensity level of normal conversation (one person speaking) is 65 dB. What is the intensity level if two people are speaking?
    a. 65 dB b. 68 dB c. 75 dB d. 130 dB

16. An intensity level of 0 dB corresponds to a sound intensity of
    a. 0 W/m^2 b. 1 W/m^2 c. $1.0 \times 10^{-12} W/m^2$ d. $1.0 \times 10^{-6} W/m^2$

17. When a source of sound moves away from you at a constant speed, the frequency of the sound
    a. continually increases as it gets farther away.
    b. continually decreases as it gets farther away.
    c. shifts downward depending on how far away it is.
    d. is higher and constant.
    e. is lower and constant.

18. A constant sound source (stationary frequency 444 Hz) approaches a stationary observer at one-half the speed of sound. The observer hears a frequency of
    a. 888 Hz b. 666 Hz c. 444 Hz d. 222 Hz

19. An observer approaches a stationary sound source (emitted frequency 444 Hz) at one-half the speed of sound. The observer hears a frequency of
    a. 888 Hz b. 666 Hz c. 444 Hz d. 222 Hz

20. The frequency of sound observed by a person moving toward a stationary source is 15% higher than the emitted frequency. If the speed of sound is 340 m/s, what is the velocity of the moving person?
    a. 51 m/s b. 15 m/s c. 44 m/s d. 295 m/s

21. What is the intensity of a 15 W source of sound at a distance of 2.0 m from the source? Assume the sound source radiates uniformly in all directions.
    a. 7.5 W/m^2 b. 3.8 W/m^2 c. 0.30 W/m^2 d. 0.13 W/m^2

SOLUTIONS AND ANSWERS

Practice Problems

---

1. $f = v/\lambda = (3.00 \text{ X } 10^8 \text{ m/s})/(4.55 \text{ X } 10^{-7} \text{ m}) = \mathbf{6.59 \text{ X } 10^{14} \text{ Hz}}$.

---

2. $v = \lambda f = (1.3 \text{ m})(440 \text{ Hz}) = \mathbf{570 \text{ m/s}}$.

---

3. The points on the string oscillate with the same frequency as the wave frequency. The maximum speed of points doing simple harmonic motion is

$$v_{max} = 2\pi fA = 2\pi(v/\lambda)A = 2\pi(550 \text{ m/s})/(0.62 \text{ m})](0.50 \text{ X } 10^{-2} \text{ m}) = \mathbf{28 \text{ m/s}}.$$

The maximum acceleration is

$$a_{max} = (2\pi f)^2 A = \mathbf{1.6 \text{ X } 10^5 \text{ m/s}^2}.$$

---

4. Equation (16.2) gives
$$v = \sqrt{F/(m/L)} = \sqrt{35\text{N}/(2.5 \times 10^{-3} \text{ kg/m})}$$
$$v = \mathbf{120 \text{ m/s}}.$$

---

5. Equation (16.2) gives
$$m/L = F/v^2 = (110 \text{ N})/(180 \text{ m/s})^2$$
$$m/L = \mathbf{3.4 \text{ X } 10^{-3} \text{ kg/m}}.$$

---

6. Equation (16.3) describes a wave moving toward +x. In our case,

$$f = v/\lambda = (150 \text{ m/s})/(5.5 \text{ m}) = 27 \text{ Hz},$$

so

$$\mathbf{y = (0.70 \text{ mm}) \sin [(170 \text{ s}^{-1})t - (1.1 \text{ m}^{-1})x]}.$$

---

7. Compare the given equation with equation (16.4) for a wave traveling toward -x.

$$A = \mathbf{1.0 \text{ cm}}, f = (550 \text{ s}^{-1})/2\pi = \mathbf{88 \text{ Hz}}, \lambda = 2\pi/(35 \text{ m}^{-1}) = \mathbf{0.18 \text{ m}}$$

and the speed of the wave is

$$v = \lambda f = 16 \text{ m/s}.$$

The velocity is then

$$\mathbf{v = - 16 \text{ m/s}}.$$

---

8. The first of the equations can be written

$$y = -(1.5 \text{ cm}) \sin [(760 \text{ s}^{-1})t - (150 \text{ m}^{-1})x]$$

due to the property of the sin function that $\sin(-\theta) = -\sin\theta$. The two waves are identical and both travel toward +x. The first is the negative of the second. That is, it is "flipped" over with respect to the second wave.

---

9. Using equation (16.5),

$$v = \sqrt{(\gamma kT/m)}$$

so that

$$m = \gamma kT/v^2 = (1.4)(1.38 \times 10^{-23}\ J/K)(300\ K)/(280\ m/s)^2$$

which yields, $m = 7.4 \times 10^{-26}$ kg. The molecular mass is

$$(7.4 \times 10^{-23}\ g)(6.02 \times 10^{23}) = \mathbf{45\ g}.$$

---

10. Using equation (16.7),

$$v = \sqrt{(Y/\rho)} = \sqrt{[(9.0 \times 10^{10}\ Pa)/(8470\ kg/m^3)]} = 3.3 \times 10^3\ m/s.$$

The time it takes to travel a distance of 5.5 m is therefore,

$$t = s/v = (5.5\ m)/(3.3 \times 10^3\ m/s) = \mathbf{1.7 \times 10^{-3}\ s}.$$

---

11. The distance traveled in the water is $s_w = v_w t$. The distance in air is $s_a = v_a (0.100\ s - t) = 15.0$ m. Using $v_a = 343$ m/s we find t = 0.056 s. Therefore,

$$s_w = v_w t = (1482\ m/s)(0.056\ s) = \mathbf{83\ m}.$$

---

12. The difference in intensity levels is

$$\beta = 10 \log (I_2/I_1) = 10 \log [(2.0 \times 10^{-5}\ W/m^2)/(4.0 \times 10^{-8}\ W/m^2)]$$
$$\beta = \mathbf{27\ dB}.$$

---

13. (a)

$$\beta = 10 \log (I/I_0)$$

so that

$$I = I_0 \times 10^{\beta/10} = (1.0 \times 10^{-12}\ W/m^2) \times 10^{7.5} = \mathbf{3.2 \times 10^{-5}\ W/m^2}.$$

(b)

$$P = IA = E/t$$

gives

$$E = IAt = (3.2 \times 10^{-5}\ W/m^2)(5.0\ m^2)(25\ s) = \mathbf{4.0 \times 10^{-3}\ J}.$$

---

14. Using equation (16.9) we have

$$I_2/I_1 = (r_1/r_2)^2 = (r_1/3r_1)^2 = 1/9.$$

Therefore, the sound intensity changes by

$$\beta = 10 \log (I_2/I_1) = 10 \log (1/9) = \mathbf{-\ 9.5\ dB}.$$

---

15. Since

$$I_2/I_1 = (r_1/r_2)^2$$

then

$$\beta = 10 \log (I_2/I_1) = 10 \log (r_1/r_2)^2 = 20 \log (r_1/r_2) = 65 \text{ dB}.$$

Solving for r_1 yields

$$r_1 = r_2 \times 10^{\beta/20} = (35 \text{ m}) \times 10^{65/20} = (35 \text{ m}) \times 10^{3.25} = \mathbf{6.2 \times 10^4\ m}.$$

---

16. Since the observed frequency decreases, this means the train is **receding** from the observer. Using

$$f' = f\ [1/(1 + v_s/v)]$$

we get

$$v_s = v\ (f/f' - 1) = (343 \text{ m/s})\{[(555 \text{ Hz})/(488 \text{ Hz})] - 1\} = \mathbf{47.1\ m/s}.$$

---

17. To find the speed of the sound source use

$$v_s = r\omega = (1.00 \text{ m})(25.0 \text{ rad/s}) = 25.0 \text{ m/s}.$$

So

$$v_s/v = 0.0729.$$

(a) For recession,

$$f' = f\ [1/(1 + v_s/v)] = (444 \text{ Hz})[1/(1 + 0.0729)] = \mathbf{414\ Hz}.$$

(b) For approach,

$$f' = f\ [1/(1 - v_s/v)] = (444 \text{ Hz})[1/(1 - 0.0729)] = \mathbf{479\ Hz}.$$

---

18. (a) Both the source and the observer are approaching, so using equation (16.15) with a + sign in the numerator and a - sign in the denominator we have,

$$f' = f\ (1 + v_o/v)/(1 - v_s/v) = (660 \text{ Hz})(1 + 0.0711)/(1 - 0.0971)$$
$$f' = \mathbf{780\ Hz}.$$

(b) Now the source and observer are receding, so use a - sign in the numerator and a + sign in the denominator. We have $f' = f\ (1 - v_o/v)(1 + v_s/v) = (660 \text{ Hz})(1 - 0.0711)(1 + 0.0971) = \mathbf{560\ Hz}$.

Quiz Answers

1. c	5. b	9. d	13. b	17. e	21. c
2. b	6. d	10. c	14. c	18. a	
3. a	7. c	11. d	15. b	19. b	
4. a	8. b	12. c	16. c	20. a	

MCAT REVIEW PROBLEMS

Light waves and sound waves are similar in several respects. For instance, the *intensity* of light and sound both decrease with the square of the distance from the source to the observer. For light, intensity corresponds to brightness.

Another similarity is the Doppler shift. Just as the pitch of sound goes up or down depending on the motion of the source and the observer, the frequency of light also depends on the speed with which the light source moves towards or away from the observer. For instance, when the source moves towards the observer (or vice versa), the light gets "blue shifted" to a higher frequency. When the source moves away from the observer (or vice versa), the light gets red shifted to a lower frequency. But for light, unlike sound, a Doppler shift also occurs when the source moves at right angles to an imaginary line connecting the source to the observer. This "transverse" Doppler shift always lowers the frequency; the frequency perceived by the observer is lower than the frequency emitted by the source.

These phenomena are illustrated by binary stars, which are actually two separate stars orbiting each other. In figure 1, a star travels counterclockwise in uniform circular motion, and is observed by an astronaut in a spaceship at point W. (This diagram does not include the second star in the binary system.) The total power emitted by the star stays constant, and the average wavelength of its light is 500 nanometers.

Point W is a distance $2R$ from the center of the circle traced out by the star, where R denotes the circle's radius.

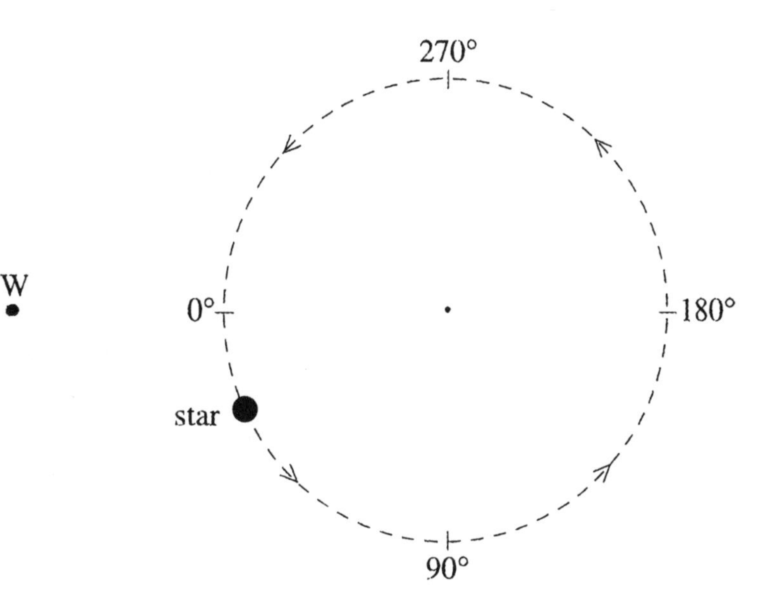

FIGURE 1

1. As the star moves from 0° to 180°, its brightness as seen by the astronaut at point W decreases by a factor of:

   A. 2 **B.** 3 **C.** 4 **D.** 9

2. When the star is at 180°, the astronaut perceives its starlight as having an average wavelength of:

   A. 0 **B.** less than 500 nm **C.** 500 nm **D.** more than 500 nm

3. At which point in its orbit is the star's light most blue shifted, according to the astronaut?

A. 240° **B.** 270° **C.** 300° **D.** 0°

---

4. The astronaut moves his spaceship to the *center* of the circle traced out by the star, and observes the star's brightness and average wavelength. According to his instruments:

A. the brightness and average wavelength both stay constant.
B. the brightness stays constant, but the average wavelength varies with time.
C. the brightness varies with time, but the average wavelength stays constant.
D. the brightness and average wavelength both vary with time.

ANSWERS TO MCAT REVIEW PROBLEMS

---

1. **D.** Let d denote the distance from the astronaut to the star. At 0°, $d = R$. At 180°, $d = 3R$. So, d triples. Since intensity (brightness) decreases with d^2, the brightness decreases by a factor of 9.

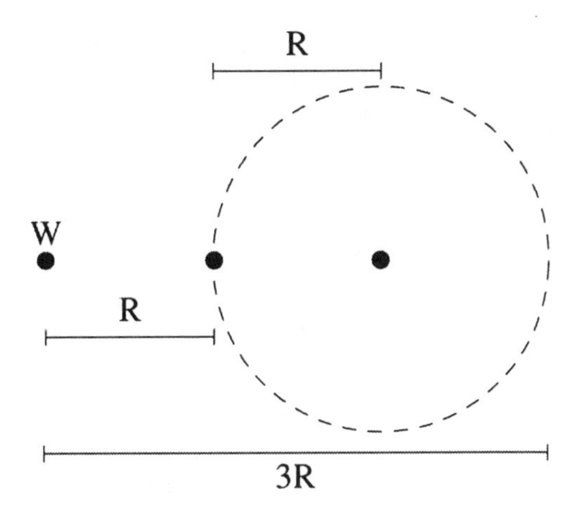

---

2. **D.** At 180°, the star moves neither towards nor away from the astronaut. Instead, it moves perpendicular to an imaginary line connecting the astronaut to the star.

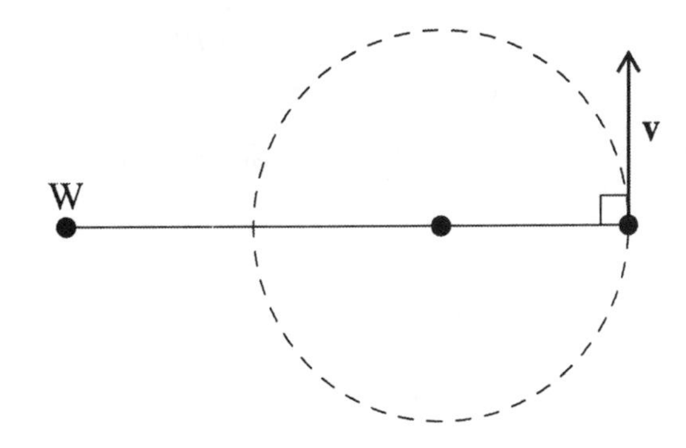

Therefore, according to the passage, a "transverse" Doppler shift kicks in. The transverse Doppler shift lowers the frequency perceived by the observer. But what about the wavelength?

To figure this out, recall the *general* wave formula,

$$v = \lambda f,$$

where v denotes the speed of light, λ denotes wavelength, and f denotes frequency. The speed of light does not depend on the motion of the source. So, when the transverse Doppler shift makes f decrease, v stays the same. Therefore, the wavelength λ must increase, to keep the above equation balanced.

---

3. **C.** According to the passage, the blue shift stems from the source's motion towards the observer. As with sound, the size of the Doppler shift depends on the *speed* with which the source approaches the observer (or vice versa). Here, the star travels with constant speed. But at 300°, *all* of the star's motion is directed towards the observer. So, at that moment, all of its speed "feeds into" the blue shift.

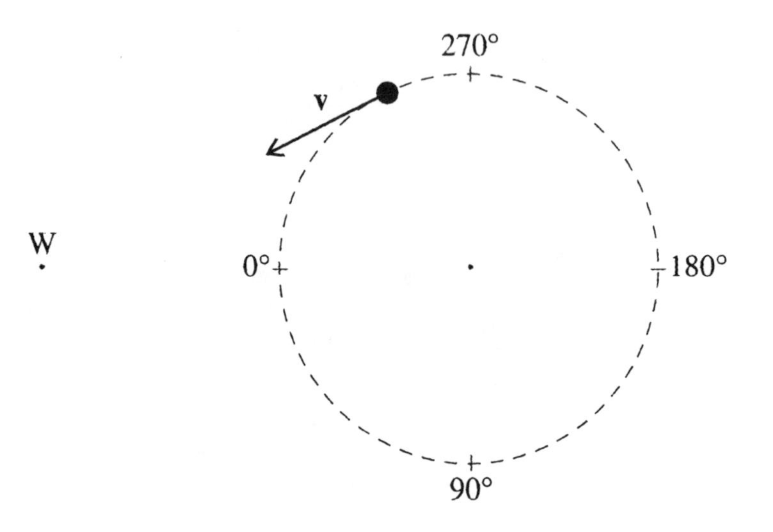

By contrast, at 240° or 270°, only a component of the star's velocity points toward the astronaut. The blue shift depends on that component. Therefore, the blue shift is smaller at 240° or 270° than it is at 300°. (And at 0°, the star isn't moving towards the astronaut at all.)

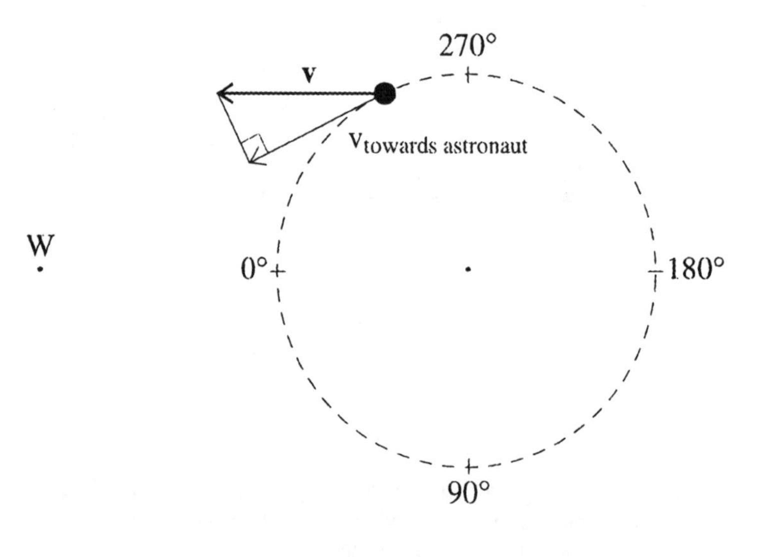

4. **A.** Because the star now remains at constant distance R from the astronaut, its brightness (as seen by the astronaut) neither increases nor decreases.

What about Doppler shift? Since the star moves perpendicular to an imaginary line connecting it to the astronaut, a transverse Doppler shift kicks in. The astronaut sees an average wavelength longer than 500 nm, as explained in question 2. But the size of this Doppler effect stays constant in time, because the star travels at constant speed, and all of its motion always points perpendicular to the imaginary line from the star to the astronaut.

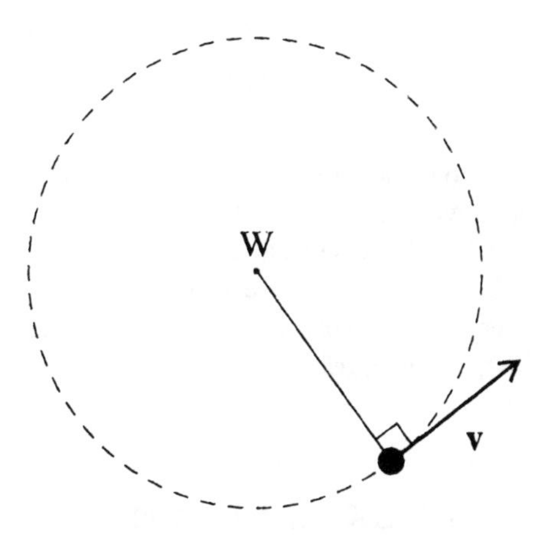

Chapter 17
The Principle of Linear Superposition and Interference Phenomena

PREVIEW

When waves meet they combine in the region where they overlap to form a new wave. The new wave disturbance is the sum of the disturbances of the original waves. This is called the Principle of Linear Superposition. The result of the addition or superposition of the wave disturbances is called interference.

In this chapter you will learn how to superimpose waves and will study the resulting interference of two waves generated by point sources or very narrow slits (interference patterns), many waves which have diffracted through a single slit (diffraction patterns), two identical waves traveling in opposite directions (standing waves), and two waves of different frequencies (beats).

QUICK REFERENCE

Important Terms

Superposition
The adding together of two or more wave disturbances to form a net disturbance.

Interference
The result of the superposition of two or more waves.

Constructive interference
The result of the superposition of two waves which meet exactly in phase. That is, crest to crest and trough to trough.

Destructive interference
The result of the superposition of two waves which meet exactly out of phase. That is, crest to trough and trough to crest.

Diffraction
The bending of a wave around an obstacle or edges of an opening.

Beats
A periodic variation in the amplitude of a wave resulting from the linear superposition of waves that have slightly different frequencies.

Standing wave
The stationary pattern of disturbance produced when identical waves traveling in opposite directions superimpose.

Node
A point on a standing wave where the wave disturbance is zero.

Antinode
A point on a standing wave where the wave disturbance is maximum.

Natural frequencies
The frequencies of waves necessary to produce resonance.

Harmonics
A set of natural frequencies of a system which are related by being integer multiples of the lowest frequency, that is, f_1, $2f_1$, $3f_1$, etc, where f_1 is the lowest frequency called the fundamental or first harmonic. The frequency $2f_1$ is called the second harmonic, the frequency $3f_1$ is called the third harmonic, etc.

Equations

The position of the **first minimum** of a **single slit diffraction pattern** can be found from

$$\sin\theta = \frac{\lambda}{D} \tag{17.1}$$

The position of the **first minimum** of the **diffraction pattern of a circular opening** can be found from

$$\sin\theta = 1.22\frac{\lambda}{D} \tag{17.2}$$

The set of **natural frequencies** of a **string fixed at both ends** is given by

$$f_n = n\left(\frac{v}{2L}\right) \qquad n = 1, 2, 3, \ldots \tag{17.3}$$

The set of **natural frequencies** of a **tube open at both ends** is given by

$$f_n = n\left(\frac{v}{2L}\right) \qquad n = 1, 2, 3, \ldots \tag{17.4}$$

The set of **natural frequencies** of a **tube open at one end** is given by

$$f_n = n\left(\frac{v}{4L}\right) \qquad n = 1, 3, 5, \ldots \tag{17.5}$$

Principles

The Principle of Linear Superposition - when two or more waves are present simultaneously at the same place, the resultant wave is the sum of the individual waves.

DISCUSSION OF SELECTED SECTIONS

17.2 Constructive and Destructive Interference of Sound Waves

Any time that two or more waves overlap in space an interference pattern results. The pattern may be very complicated if many waves of different amplitudes and frequencies overlap, or relatively simple if only two waves with the same amplitude and frequency are involved. Constructive interference occurs when the waves are exactly in phase and destructive interference occurs when they are exactly out of phase.

Two waves originally in phase can get out of phase if they are required to travel different distances before combining. If one wave travels λ, 2λ, 3λ, etc. farther than the other, then the two waves will be in phase and constructively interfere. If one travels $\lambda/2$, $3\lambda/2$, $5\lambda/2$, etc. farther than the other, then they will be out of phase and will destructively interfere.

Example 1
Two in-phase speakers are located 2.0 m apart and produce identical 550 Hz tones. A person stands 2.5 m from the left speaker and hears no sound. What are the possible distances of the person from the right speaker?

If the person hears no sound at all, completely destructive interference of the two sound waves arriving at his ear must be occurring. One of the sound waves must have traveled some odd multiple of $\lambda/2$, say $n\lambda/2$, farther than the other. The difference in the distance traveled is $s_R - s_L = n\lambda/2$, so

$$s_R = s_L + n\lambda/2.$$

The wavelength of the sound wave is $\lambda = v/f$. Then

$$s_R = s_L + nv/2f \; ; \; n = 1, 3, 5, \ldots$$

$$s_R = 2.5 \text{ m} + n[343/2(550) \text{ m}]$$

$$s_R = 2.5 \text{ m} + n(0.31 \text{ m}).$$

$$s_R = 2.8 \text{ m}, 3.4 \text{ m}, 4.1 \text{ m, etc.}$$

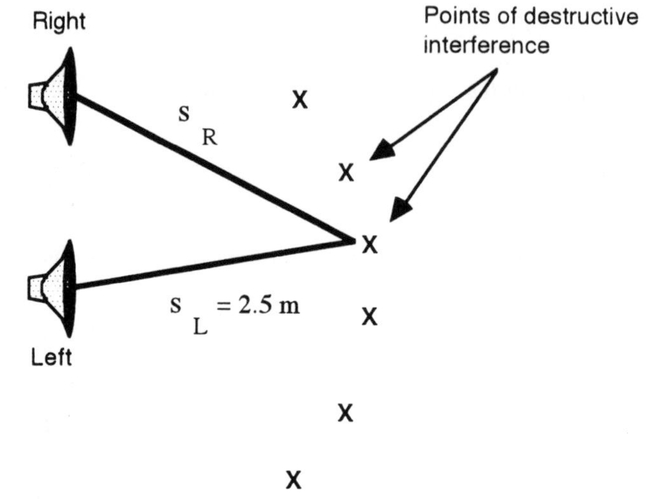

The person could be 2.8 m, 3.4 m, 4.1 m, . . . from the right speaker. The points of destructive interference are shown as **x**'s in the figure.

Example 2
Two in-phase radio towers 1.2 mi apart are broadcasting identical radio waves of frequency 550 kHz. Locate the points of constructive and destructive interference along the line that joins the two towers. Take the speed of radio waves to be 186 300 mi/s.

The wavelength is

$$\lambda = v/f = (186\,300 \text{ mi/s})/(550\,000 \text{ Hz})$$
$$\lambda = 0.34 \text{ mi}.$$

The maximum path difference of the two radio waves is 1.2 mi. along the line joining the towers. This corresponds to

$1.2 \text{ mi}/\lambda = 3.5$ wavelengths of path difference.

There will then be 7 points of constructive interference and 6 points of destructive interference on the line between the towers, as shown in the drawing.

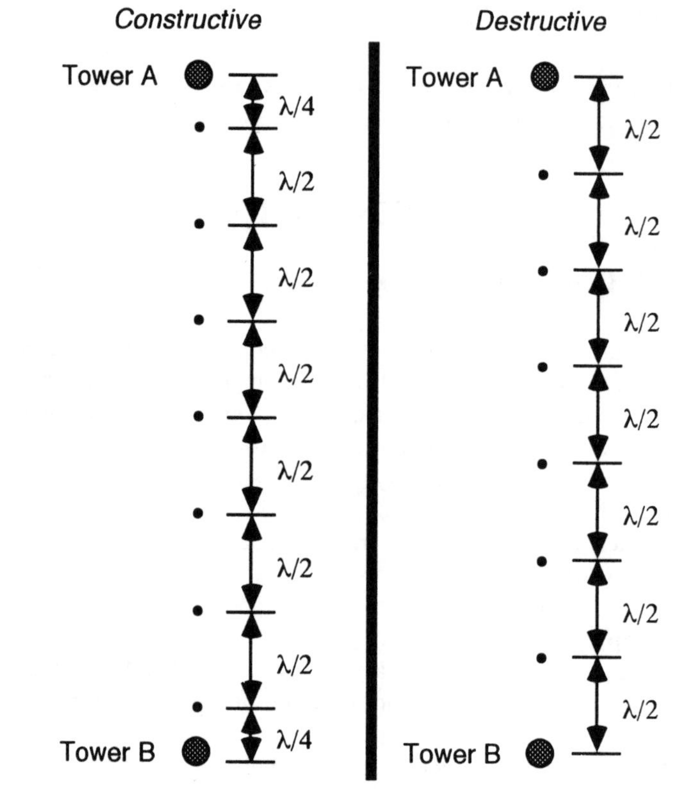

17.3 Diffraction

Whenever a wave passes close to an obstacle or edge it "bends" around the obstacle or edge. This is diffraction. It is the phenomenon which enables you to hear people on the other side of a house from you.

All waves exhibit diffraction, but sometimes the effect is so small that it escapes notice. A general rule of thumb is that diffraction of waves will be obvious when the wavelength of the wave is about the same size as the dimensions of the obstacles it encounters. This is why we hear the effect of diffracting low frequency (long wavelength) sound waves but do not see the effects of the diffraction of light (wavelength about 5×10^{-7} m) in everyday life.

Example 3
Green light whose wavelength is 5.5×10^{-7} m and sound of frequency 760 Hz pass through identical doors that are 0.92 m wide. What is the angle θ that locates the first minimum in the diffraction pattern for each wave? Take the speed of sound to be 343 m/s.

The angle θ that locates the first minimum in the diffraction pattern is given by (17.1). For the green light,

$$\theta = \sin^{-1} (\lambda/D)$$
$$\theta = \sin^{-1} [(5.5 \times 10^{-7} \text{ m})/(0.92 \text{ m})]$$
$$\theta = 3.4 \times 10^{-5}{}^\circ.$$

Repeating the procedure for the sound which has a wavelength of $\lambda = v/f = (343 \text{ m/s})/(760 \text{ Hz}) = 0.45$ m we obtain

$$\theta = \sin^{-1} (0.45 \text{ m}/0.92 \text{ m}) = 29^\circ.$$

The angular spread is much greater for sound than for light.

17.4 Beats

Beats are heard when two sound waves of slightly different frequencies overlap. An example of this is when two slightly out of tune guitar strings are plucked simultaneously. The resulting sound intensity rises and falls at a rate which is the difference between the two sound wave frequencies. All waves exhibit the beat phenomenon. It is just much easier to detect beats in sound than in other waves such as light.

Example 4
An A-note (440.0 Hz) tuning fork is struck along with the corresponding A key on a piano. The resulting sound level rises and falls 3 times each second. The tension on the piano wire is increased by a small amount and the beat frequency increases. What was the out of tune frequency of the piano's A-note?

Since the piano and tuning fork produced 3 beats per second, the piano produced either 437 Hz or 443 Hz. Which is the correct value can be determined by considering what happens to the frequency of a wire as the tension is increased. Recall that $f = v/\lambda$ for any kind of wave and

$$v = \sqrt{\frac{F}{m/L}}$$

for the waves traveling on a wire.

As the tension, F, increases the speed of the waves on the wire increases and the frequency will increase. If the piano's note had been low, then a small increase in tension would have made it closer to 440.0 Hz and the beat frequency would have decreased. This was not observed to be the case. Hence, the piano's note was high to begin with and must have been **443 Hz**.

17.5 Transverse Standing Waves

Standing wave patterns are produced when identical waves traveling in opposite directions combine. Most often one of the waves is generated by a source and the other is derived from reflection at a boundary such as a wall. In this circumstance, the relationship between the distance from the generator to the boundary and the wavelength determines whether the waves maintain the proper phase relationship to reinforce or cancel each other.

Example 5
Waves are introduced on one end of a string and meet waves which have previously been reflected from a clamped end of the string 2.0 m from the generator. Assume that the generator itself behaves approximately as a clamped end and find the possible wavelengths which can produce standing waves on the string.

Both ends of the string are fixed so any standing wave which exists on it must have stationary points (nodes) at each end. The longest wavelength which corresponds to this condition is shown in the diagram and is

$$\lambda_1 = 2L = 4.0 \text{ m}$$

The next longest wavelength is

$$\lambda_2 = L = 2.0 \text{ m}$$

and the next is

$$\lambda_3 = 2/3\ L = 1.3 \text{ m, etc.}$$

In general,

$$\lambda_n = 2L/n \quad n = 1, 2, 3, \text{etc.}$$

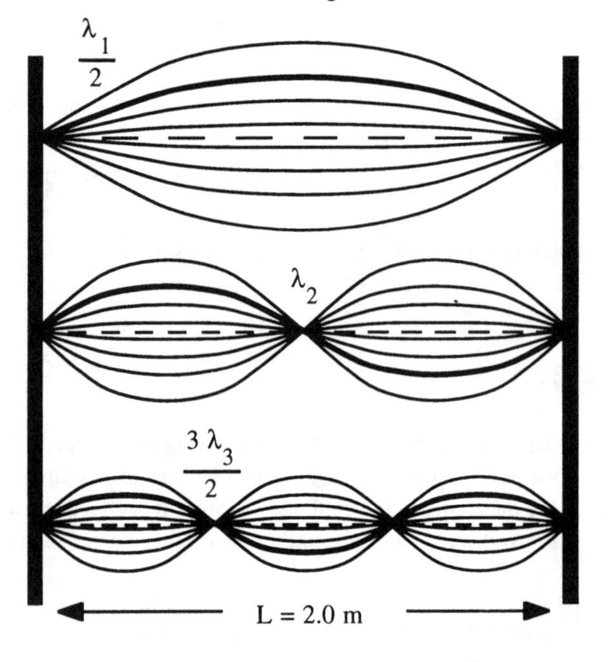

Example 6
A 30.0 m power line is observed to vibrate in its second harmonic due to excitation by the wind. The period T of the oscillation of the line is determined to be 1.5 s. What is the speed of the waves on the line?

The speed of the waves is

$$v = \lambda f = \lambda/T.$$

The wavelength of the second harmonic is just equal to the length of the line so, λ = 30.0 m. Now

$$v = (30.0 \text{ m})/(1.5 \text{ s}) = 2.0 \times 10^1 \text{ m/s}.$$

17.6 Longitudinal Standing Waves

Longitudinal waves such as sound waves can also produce standing waves by reflection from a boundary. The boundary can be any change in the environment of the sound wave such as that at the end of a tube. If the tube end is closed, then the standing wave must have a displacement node there since the air is unable to move. If the end of the tube is open then the standing wave will have a displacement antinode at the end.

Example 7
A 1.5 m tube which is closed at one end supports a standing sound wave. What are the three longest wavelengths of sound for which standing waves can exist in the tube? What are the frequencies?

Since one end of the tube must contain a node and the other end an antinode the longest wavelength which will "fit" in the tube is such that $L = \lambda_1/4$. This wavelength is

$$\lambda_1 = 4L = 6.0 \text{ m}.$$

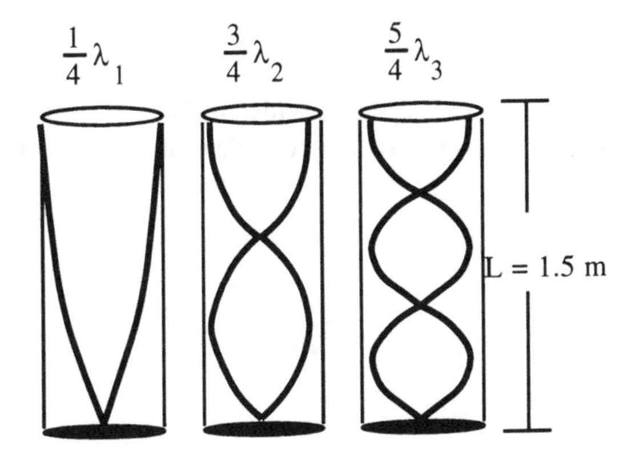

The wavelength corresponds to a frequency,

$$f_1 = v/\lambda_1 = (343 \text{ m/s})/(6.0 \text{ m}) = 57 \text{ Hz}.$$

The second longest wavelength which will fit in the tube is such that $L = 3/4\,\lambda_2$ so

$$\lambda_2 = 4/3\ (1.5 \text{ m}) = 2.0 \text{ m}.$$

This corresponds to a frequency,

$$f_2 = (343 \text{ m/s})/(2.0 \text{ m}) = 170 \text{ Hz}.$$

The third longest wavelength is such that $L = 5/4\,\lambda_3$, so

$$\lambda_3 = 4/5\ L = 1.2 \text{ m}.$$

The corresponding frequency is

$$f_3 = (343 \text{ m/s})/1.2 \text{ m}) = 290 \text{ Hz}.$$

Example 8
Repeat example 7 assuming that both ends of the tube are open.
The procedure is the same as in example 7. The main difference is that the standing wave must have an antinode at EACH end of the tube. The diagram shows the situation for the three longest wavelengths.

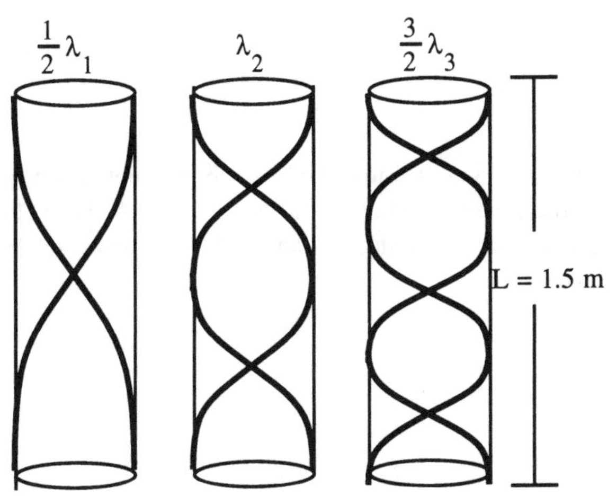

$$\lambda_1 = 2L = 3.0 \text{ m}$$
$$f_1 = 110 \text{ Hz}$$

$$\lambda_2 = L = 1.5 \text{ m}$$
$$f_2 = 230 \text{ Hz}$$

$$\lambda_3 = 2/3\ L = 1.0 \text{ m}$$
$$f_3 = 340 \text{ m}$$

PRACTICE PROBLEMS

In all of the following the speed of sound is assumed to be 343 m/s unless otherwise stated.

1. Two speakers generate identical 650 Hz sound waves which arrive at a person's ear after traveling 2.0 m and 2.5 m, respectively. Describe the sound heard by the person.

2. Two in-phase speakers produce identical 440 Hz sound waves. A person stands 2.0 m directly in front of one speaker and hears the loudest possible sound. What is the minimum nonzero separation of the speakers?

3. A young boy stands in a 3.5 m wide alley between two very tall buildings and shouts to a friend standing 15 m in front of one of the buildings and 5.0 m to the side of the alley. What sound frequencies of the first boy's shout will the second boy hear?

4. What must be the diameter of a speaker to produce a 15 kHz tone with a diffraction angle of 35°?

5. The same note is sounded on two slightly out of tune guitars, and a beat frequency of 7.00 Hz is heard. One of the guitars is then sounded along with a 440.0 Hz tuning fork and a beat frequency of 4.00 Hz is heard. It is judged that the tuning fork has the higher frequency. What are the possible frequencies of the note on each guitar?

6. A 1.5 m long string, fixed at both ends, has a fundamental frequency of 120 Hz. What is the speed of waves traveling on the string?

7. A 0.90 m string is stretched between rigid supports. What are the wavelengths of the four longest wavelength that can lead to standing waves on the string?

8. A 1.50 m long string, whose mass is 2.50 g, is stretched between fixed supports to a tension of 9.00 N. What are the natural frequencies of the string?

9. A loudspeaker is facing a perfectly reflecting wall located 3.50 m away. How many frequencies of sound in the audible range (20.0 Hz to 20.0 kHz) produced by the speaker would result in standing waves between the wall and the speaker?

10. A "pop" bottle is partially filled with pop so that a 10.0 cm air column is left at the top. A person blows across the top of the bottle producing sound waves of many frequencies in the audible range (20 Hz to 20 kHz). What is the lowest frequency of sound which will produce standing waves in the bottle, thus sounding louder. Note: The neck of a pop bottle is small and the shape of the bottle may have some effect. Ignore these in your calculation.

HELPFUL SUGGESTIONS

1. Waves which originally start in phase can get out of phase by traveling different distances. A way to visualize this is to think of a formation of marching soldiers. Each soldier begins marching by stepping forward with the same foot as his fellows. As long as each soldier in the formation takes the same length of step (wavelength) in the same time (period) and travels in a straight line, the formation will remain in step. As the formation makes a turn, however, the soldiers on the outside of the turn must travel farther than the soldiers on the inside of the turn, and the formation will fall out of step. Of course, the soldiers avoid this by decreasing their step length if they are on the inside of the turn and increasing their step length if they are on the outside of the turn. Waves cannot make a similar adjustment of the wavelength and get out of phase (step).

2. Interference occurs ONLY where waves overlap. Two waves may move toward the same point and completely cancel each other if they arrive at the point out of phase. After the waves pass through the point, they no longer affect each other and continue on as if nothing had happened.

3. The number of loops in a standing wave pattern corresponds to the harmonic number of the natural frequency for a string fixed at both ends or a tube open at both ends.

4. The tube open at only one end can support standing waves which correspond to ODD harmonics only.

5. The term beat frequency refers to the frequency at which the amplitude of a wave formed by the superposition of waves varies in time. It does not refer to the frequency of a new wave.

EVERYDAY PHYSICS

1. Interference of sound abounds in everyday life. We usually don't notice these effects since most sources of sound produce a large number of frequencies each with its own interference pattern. The combined effects of this large number of patterns is to make each much less obvious.
   You can experiment with interference of sound if you have a stereo system and a method of producing a single frequency of sound. The source can be as sophisticated (and expensive) as a signal generator or as simple as a monaural tape recording of pure tones. The latter can often be made with the aid of a sympathetic physics professor or a physics major. For most purposes, tones of 100, 500 and 1000 Hz are sufficient. Try these in a large room:

   a. Play the 100 Hz tone from the tape with your speakers about 3 ft apart and facing the same direction. Walk in front of the speakers and see if you can detect places of constructive and destructive interference. You might even make a map of these. Repeat for the 500 Hz and 1000 Hz frequencies and compare the maps. How does the pattern depend on frequency?
   b. Choose the frequency which shows the easiest to hear pattern and experiment with changing the distance between the speakers. What effect does this have?
   c. Stereo speakers are normally connected so that they produce sound waves which are in phase. You can see the difference between the in phase patterns and out of phase patterns by switching the wires on the back of ONE of the speakers. What effect does this have on the interference pattern? Please turn OFF the stereo BEFORE you switch the wires.
   Before reconnecting the wires in their proper position you might listen to some familiar music. Does it sound strange? Why?

2. You can use the above equipment to investigate diffraction of sound through an opening. Place one of your speakers inside the house so that it beams sound through an open doorway. It is best if you place the speaker quite a distance behind the doorway rather than in it. It is also preferable that the doorway be the only opening in the wall and that the wall be as rigid as possible, perhaps brick. Play one of your tones while you walk parallel to the wall on the outside. Note the region in which you can hear the sound. Repeat for the other tones and compare the results. Which frequency appears to diffract best? Does this result agree with equation (17.1)?

3. Beats are easy to hear with a six string guitar. If the sixth string is pushed against the fifth fret, it should produce the same frequency as the fifth string played open on a well tuned guitar. Strike the fretted sixth string and the open fifth string together. If the guitar is in good tune you should hear little, if any, rise and fall in the loudness of the resulting sound (beats). Decrease the tension of the sixth string a small amount and sound again. Do you now hear the beats? Play with the tension of the sixth string and see how the beat frequency changes. Can you make sense of what you observe based on what you have learned in this chapter?

4. The speed of objects is often measured using the Doppler effect of waves along with the beat phenomenon. The principles are the same whether it is an ultrasonic fetal heart monitor counting the heartbeats of a fetus by determining how many times per minute the fetus' heart wall changes direction, or a ship's sonar system measuring the speed of a submarine, or even the police clocking your car's speed as you travel down the highway.
In any case, waves (sound, radio, etc) of a known frequency, f, are generated and transmitted toward the object in question. The waves reflected from the moving object are "Doppler shifted" to a new frequency, f', and return to a receiver where they are converted into electrical waves of the same frequency. Electronic circuitry then adds this wave to an electrical wave coming directly from the generator to produce beats of frequency, f_B . The beat frequency is used to determine f'. For sound waves, the Doppler formula (16.15) can then be used to find the speed of the object.

5. Standing waves can be produced on a rope by simply tying one end to a door handle, tree, etc, pulling to provide some tension and shaking the free end of the rope up and down. Your hand will soon produce a frequency which will result in a standing wave on the rope. After the standing wave is started, it is easy to maintain since the rope guides your hand to shake at the correct frequency. See how many different harmonics you can produce on the rope.

6. Standing sound waves can be produced very easily by humming into a paper towel or wrapping paper roll. When your humming produces one of the harmonics of the tube, then a standing wave results which causes an enhanced sound to leak from the tube. Did you try this as a child? Try it again.

CHAPTER QUIZ

1. Two in-phase speakers produce identical 150 Hz sound waves. How much farther would one of the waves have to travel for the two to arrive at the same point and interfere destructively?
   a. 2.3 m b. 3.4 m c. 4.6 m d. 1.1 m

2. A standing sound wave in a tube open at only one end has a displacement ________at the open end and a displacement_________at the closed end.
   a. node, antinode b. node, node c. antinode, node d. antinode, antinode

3. The lowest frequency of standing waves which can exist in a tube open at both ends is 250 Hz. What is the length of the tube?
   a. 0.69 m b. 0.34 m c. 1.4 m d. 2.8 m

4. The second lowest frequency of standing waves which can exist in a tube open at only one end is 440 Hz. What is the length of the tube?
   a. 0.78 m b. 0.39 m c. 1.6 m d. 0.58 m

5. A certain string can support standing waves of frequency 248 Hz and 1240 Hz. Which of the following could not be the fundamental frequency of the string?
   a. 248 Hz b. 124 Hz c. 165 Hz d. 62 Hz

6. Two otherwise identical waves cross paths so that one wave has a crest and the other has a trough where they meet. Which of the following is true?
   a. The waves do not interact with each other at all.
   b. The waves destroy each other.
   c. The waves add together to produce a large wave.
   d. The waves cancel each other only where they overlap.

7. Which of the following would produce the largest diffraction angle.
   a. A 0.20 m wavelength wave through a 2.0 m wide opening.
   b. A 0.10 m wavelength wave through a 2.0 m wide opening.
   c. A 0.20 m wavelength wave through a 1.0 m wide opening.
   d. A 0.10 m wavelength wave through a 1.0 m wide opening.

8. A guitar string is sounded together with a 440.0 Hz tuning fork and the loudness of the resulting sound is heard to rise and fall 5.0 times each second. Which of the following is true about our knowledge of the frequency of sound produced by the guitar string?
   a. It is 440.0 Hz
   b. It is 445.0 Hz
   c. It is 435.0 Hz
   d. It could be either 445.0 or 435.0 Hz.

9. Standing waves are produced by
   a. The superposition of identical waves traveling in opposite directions.
   b. The superposition of identical waves traveling in the same direction.
   c. The superposition of otherwise identical waves of slightly different frequencies.
   d. The superposition of waves which travel with different speeds.

10. Sound is produced by a small speaker with a diameter of 2.54 cm. What frequency of sound produced by the speaker would have a diffraction angle of 60.0°?
    a. 15.6 kHz b. 19.0 kHz c. 9.5 kHz d. 5.4 kHz

SOLUTIONS AND ANSWERS

Practice Problems

---

1. The wavelength of the sound is $\lambda = v/f = (343 \text{ m/s})/(650 \text{ Hz}) = 0.53$ m and the difference in the path lengths of the two waves is 2.5 m - 2.0 m = 0.5 m. Comparison of the wavelength and path difference shows that one wave traveled approximately one wavelength farther than the other. The waves will arrive at the person's ear in phase and will produce a **loud sound** due to constructive interference.

---

2. The minimum non-zero separation, d, of the speakers which will produce constructive interference corresponds to a one wavelength difference in the path lengths traveled by the two waves.

$$\sqrt{d^2 + (2.0 \text{ m})^2} - 2.0 \text{ m} = (343 \text{ m/s})/(440 \text{ Hz})$$

Solving for d gives

$$d = \mathbf{1.9 \text{ m}}.$$

---

3. In order for the sound to reach the second boy, it must spread through at least an angle

$$\theta = \tan^{-1} (5.0 \text{ m})/(15.0 \text{ m}) = 18.4°$$

Equation (17.1) and $\lambda f = v$ give

$$f = v/(D \sin \theta) = (343 \text{ m/s})/[(3.5 \text{ m})(\sin \theta)] = \mathbf{310 \text{ Hz}}.$$

This is the **maximum frequency** which the second boy can hear.

---

4. Equation (17.2) and $\lambda = v/f$ give

$$D = 1.22(v/f)/(\sin \theta) = 1.22[(343 \text{ m/s})/(15 \times 10^3 \text{ Hz})]/(\sin 35°) = \mathbf{0.049 \text{ m}}.$$

---

5. The frequency produced by the guitar which is sounded with the tuning fork must be 440.0 Hz + 4 Hz = 444 Hz or 440.0 Hz - 4 Hz = 436 Hz. The tuning fork is higher so this guitar must produce **436 Hz**. The other guitar produces either 436 Hz + 7 Hz = **443 Hz** or 436 Hz - 7 Hz = **429 Hz**. It is impossible to know which frequency the guitar actually produces.

---

6. The fundamental standing wave on the string corresponds to $1/2\, \lambda = L$. Then $\lambda = 2(1.5 \text{ m}) = 3.0$ m. The traveling waves on the string have this wavelength, therefore, their speed is

$$v = \lambda f = (3.0 \text{ m})(120 \text{ Hz}) = \mathbf{360 \text{ m/s}}.$$

---

7. The longest standing wave (fundamental) which can exist on the string must satisfy the condition

$$1/2\, \lambda_1 = L \text{ or } \lambda_1 = \mathbf{1.8 \text{ m}}.$$

The second longest standing wave (second harmonic) which can exist on the string must satisfy the condition

$$2/2\ \lambda_2 = L \text{ or } \lambda_2 = \mathbf{0.90\ m}.$$

The third longest standing wave (third harmonic) which can exist on the string must satisfy the condition

$$3/2\ \lambda_3 = L \text{ or } \lambda_3 = \mathbf{0.60\ m}.$$

The fourth longest standing wave (fourth harmonic) which can exist on the string must satisfy the condition

$$4/2\ \lambda_4 = L \text{ or } \lambda_4 = \mathbf{0.45\ m}.$$

---

8. The natural frequencies of the string are given by equation (17.3). We need the speed of the waves.

$$v = \sqrt{\frac{F}{(m/L)}} = \sqrt{\frac{9.00\ \text{N}}{(2.50 \times 10^{-3}\ \text{kg})/(1.50\ \text{m})}} = 73.5\ \text{m/s}$$

Now

$$f_n = \frac{nv}{2L} = n\left(\frac{73.5\ \text{m/s}}{2(1.5\ \text{m})}\right) = n(24.5\ \text{Hz})$$

The natural frequencies are **{24.5 Hz, 49.0 Hz, 73.5 Hz, etc.}**

---

9. The situation described here is similar to a tube open at one end (speaker) and closed at the other (wall). Equation (17.5) gives the frequencies of the standing waves to be

$$f_n = \frac{nv}{4L} = \frac{n(343\ \text{m/s})}{4(3.50\ \text{m})} = n(24.5\ \text{Hz})$$

Standing waves can result from all frequencies with odd n from 1 to n = $(20.0 \times 10^3\ \text{Hz})/(24.5\ \text{Hz})$ = **816.**

---

10. The partially filled pop bottle is a tube open at one end, so equation (17.5) applies. The lowest frequency which will produce standing waves is corresponds to n = 1.

$$f_1 = \frac{v}{4L} = \frac{343\ \text{m/s}}{4(10.0 \times 10^{-2}\ \text{m})} = \mathbf{858\ Hz}$$

Quiz answers

1. d	3. a	5. c	7. c	9. a
2. c	4. b	6. d	8. d	10. b

MCAT REVIEW PROBLEMS

Consider this passage from a hypothetical 10th grade textbook.

Airplane passengers hate the loud noise produced by airplane engines. Like all sound waves, these noises travel at about 340 meters per second. When they reach your ear, the sound waves make your ear drum vibrate.

Some airlines have started experimenting with anti-sound headphones. These headphones contain speakers that give off "anti-sound" waves. The anti-sound waves are designed to cancel out the sound waves coming from the engines. As a result, the passenger hears softer engine noise. Early testing shows that these anti-sound headphones work fairly well.

---

1. Anti-sound headphones reduce the noise level heard by the passenger primarily by taking advantage of:

   A. Absorption **B.** Standing waves **C.** Interference **D.** Diffraction

---

2. In air at 20° C, lower-frequency sound waves travel 340 meters per second. In air at 20° C, higher-frequency sound waves travel:

   A. More than 340 meters per second **B.** About 340 meters per second
   C. Fewer than 340 meters per second **D.** We cannot determine the answer.

---

3. From the passage, we can infer that an anti-sound headphone cannot work unless it incorporates some

   A. Absorbing surfaces to "soak up" excess sound from the engines
   B. Small fans to circulate air near the passenger's ears
   C. Reflecting surfaces to redirect sound from the engines
   D. Microphones to detect sound from the engines.

---

4. When sound waves of wavelength 0.34 m enter your ears, what is the approximate period of oscillation of your ear drums, in seconds?

   A. 1000 **B.** 100 **C.** .01 **D.** .001

ANSWERS TO MCAT REVIEW PROBLEMS

---

1. **C.** The "anti-sound" waves must be sound waves that destructively interfere with the sound waves from the engine. If "anti-sound" were some *other* kind of wave, then it couldn't cancel sound waves. Destructive interference occurs only when the overlapping waves are of the same type.

   If the headphones worked simply by absorbing engine sound, then they wouldn't need speakers. And those speakers do not produce standing sound waves inside the passenger's ear, because the passenger would hear them—loudly! Diffraction occurs when waves bend or spread out upon encountering an obstacle of comparable size to the wavelength. Nothing in the passage suggests that anti-sound headphones exploit this effect.

---

2. **B.** The speed of sound depends on the temperature of the air, but not on the wavelength or frequency of the sound waves. In general, the speed of a wave depends only on the properties of the medium through which it travels, *not* on the frequency, wavelength, amplitude, or shape of the wave itself. For instance, red light and blue light travel through a vacuum at the same speed. In the ocean, big and small water waves move equally fast. To see this for yourself, stretch out a slinky and shake one end. The speed of the waves you create won't depend on how rapidly you shake.

---

3. **D.** The speakers in the headphones produce sound waves that destructively interfere with the sound waves from the engine. So, the sound waves from the speaker must be "out of phase" with the engine's sound waves. To generate such waves, the speakers need to "know" the wavelength, amplitude, and phase of the sound waves from the engine. For this reason, the headphones must contain microphones and electronic circuitry that detect and analyze the sound waves coming from the engine.

---

4. **D.** To address this problem, you need to know two things. First, the frequency of a wave relates to its speed and wavelength according to

   $$v = \lambda f.$$

   Second, frequency (cycles per time) and period (time per cycle) are inversely related:

   $$T = \frac{1}{f},$$

   where T stands for period.

   Solve the first equation for f, to get

   $$f = \frac{v}{\lambda} = \frac{340 \text{ m / s}}{0.34 \text{ m}} = 1000 \text{ Hz}.$$

   So, the sound waves make your ear drum vibrate back and forth at a frequency of 1000 cycles per second. Therefore, each cycle takes only one thousandth of a second:

   $$T = \frac{1}{f} = \frac{1}{1000 \text{ Hz}} = .001 \text{ s}.$$

CHAPTER 18
Electric Forces and Electric Fields

PREVIEW

In this chapter you will be introduced to the concepts of electric charge, electric forces and electric fields. You will use Coulomb's law to find the forces and fields associated with point charges and charged plates. You will also learn how to "map" electric field lines and how electric fields are affected by the presence of conductors.

QUICK REFERENCE

Important Terms

Electric charge
An intrinsic property of protons (positive charge) and electrons (negative charge). It is found that like charges (same polarity) repel one another while unlike charges (opposite polarity) attract one another. The SI unit of charge is the coulomb (C).

Electrical conductors
Materials which allow electric charge to flow freely through them.

Electrical insulators
Materials which do not allow electric charge to flow freely through them.

Coulomb's law
The magnitude of the electrical force exerted by one point charge on another is directly proportional to the magnitudes of the charges and inversely proportional to the square of the distance between the charges. The direction of the force is along the line joining the charges and is repulsive for like charges and attractive for unlike charges.

Electric field
At a point in space, the electrostatic force experienced by a small test charge placed at that point divided by the charge itself. Electric field is a vector quantity. SI unit of electric field is the newtons per coulomb (N/C).

Electric field lines
A visual representation of the electric field. It can be viewed as a "map" providing information about the direction and strength of the electric field in various regions.

Gaussian surface
A hypothetical *closed* surface in the vicinity of an electric field or a charge distribution. The Gaussian surface may enclose a net charge.

Electric flux
The product of the magnitude of the normal component of the electric field at any point on a surface element and the area of the surface element. It is assumed that the magnitude of the normal component of the electric field is constant over the surface element. The SI unit of electric flux is the newton square meter per coulomb ($N \cdot m^2/C$).

Gauss' Law
The mathematical expression that relates the electric flux through a closed (Gaussian) surface to the net charge enclosed by the surface. Gauss' Law can be used to calculate the magnitude of the electric field surrounding charge distributions that possess spatial symmetry.

Equations

Coulomb's law for the magnitude of the force that one *point charge* exerts on a second *point charge* is written as

$$F = k\frac{q_1 q_2}{r^2} \tag{18.1}$$

where q_1 and q_2 are the magnitudes of the two point charges, and r is the distance between them.

The **electric field** can be written

$$\mathbf{E} = \frac{\mathbf{F}}{q_0} \tag{18.2}$$

The magnitude of the **electric field** due to a **point charge** is

$$E = \frac{kq}{r^2} \tag{18.3}$$

The magnitude of the **electric field** between the plates of a **parallel plate capacitor** is given by

$$E = \frac{q}{\varepsilon_0 A} = \frac{\sigma}{\varepsilon_0} \tag{18.4}$$

The **electric flux** through a closed surface can be written as

$$\Phi_E = \Sigma(E \cos \phi)\Delta A \tag{18.6}$$

where E is the magnitude of the electric field at a point on the surface, and ϕ is the angle between the electric field and the normal at that point.

Gauss' Law for a closed surface is written as

$$\Sigma(E \cos \phi)\Delta A = \frac{Q}{\varepsilon_0} \tag{18.7}$$

where Q is the net charge enclosed by the surface and ε_0 is the permittivity of free space.

DISCUSSION OF SELECTED SECTIONS

18.1 The Origin of Electricity
18.2 Charged Objects and the Electric Force

An intrinsic property of protons and electrons, which make up all matter, is **electric charge**. Only two types of electric charge have been discovered, positive and negative; a proton has a positive charge, and an electron has a negative charge.

The SI unit for measuring the magnitude of electric charge is the **coulomb (C)**. Each proton carries a charge +e, and each electron carries a charge -e, where

$$e = 1.60 \times 10^{-19} \text{ C}$$

This charge e is the smallest amount of free charge that is possible. Charges larger than this are integer multiples of e, that is, q = Ne where N is the number of electrons (or protons) present.

Example 1

How many electrons are contained in 3.20×10^{-6} coulombs of charge?

Since the total charge q = Ne we can write

$$N = \frac{q}{e} = \frac{3.20 \times 10^{-6} \text{ C}}{1.60 \times 10^{-19} \text{ C}} = 2.00 \times 10^{13} \text{ electrons}$$

Because any electric charge, q, occurs as integer multiples of the elementary charge e, electric charge is said to be **quantized**.

Since the charges on the electron and proton have identical magnitudes but opposite signs, the algebraic sum of the two charges is zero. Since most materials have equal numbers of protons and electrons, most matter is electrically neutral. However, it is possible to transfer electric charge from one object to another. Usually electrons are transferred, and the body that gains electrons acquires an excess of negative charge. The body that loses these electrons then acquires an excess of positive charge. In this sense, charge can be "separated". However, during any process, the net electric charge of an entire isolated system remains constant (is conserved). This is referred to as the **law of conservation of electric charge**.

Charge separation is important in the operation of electrical equipment. For example, batteries, alternators in cars, and electric power generators all depend on the separation of electric charges for their operation.

It is easy to demonstrate that two electrically charged objects exert a force on one another. If the two objects contain charge of the same sign, the force will be one of repulsion, i.e., the objects push one another apart. If the objects contain charge of the opposite sign, the force will be one of attraction and the objects will pull each other closer together. It is a fundamental characteristic of electric charges that **like charges repel and unlike charges attract each other**. The magnitude of the force exerted by one point charge on another was experimentally determined by the French physicist Charles-Augustin Coulomb. His result, which has become known as Coulomb's law, is stated in the next section.

18.5 Coulomb's Law

Consider two "point charges", q_1 and q_2, separated by a distance r, as shown in the diagram at the right. If the charges have unlike signs, as in part (a) of the diagram, each charge is attracted to the other by a force that is directed along the line between them; **+F** is the electric force exerted on charge 1 by charge 2 and **-F** is the electric force exerted on charge 2 by charge 1. If the charges have the same sign, as in (b), each charge is repelled by the other. The forces **+F** and **-F** form an action-reaction pair as predicted by Newton's third law.

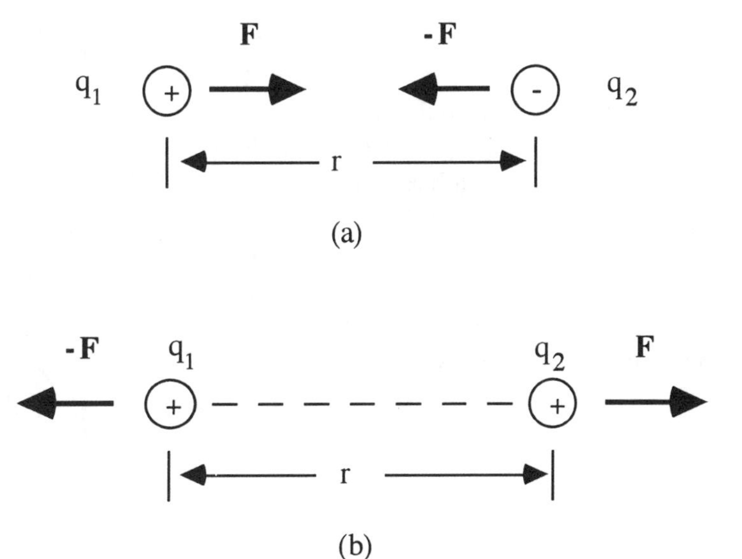

The magnitude of the force (either attractive or repulsive) is given by Coulomb's law, i.e.,

$$F = k\frac{q_1 q_2}{r^2} \qquad (18.1)$$

where k is a constant of proportionality whose value in SI units is $k = 8.99 \times 10^9$ N m^2/C^2.

Example 2
What is the force between charges $q_1 = +\,1.0\ \mu C$ and $q_2 = -\,2.0\ \mu C$ which are separated by a distance of 2.0 m? Equation (18.1) indicates the magnitude of the force is given by

$$F = k\frac{q_1 q_2}{r^2} = \frac{(8.99 \times 10^9\ \text{N m}^2/\text{C}^2)(1.0 \times 10^{-6}\ \text{C})(2.0 \times 10^{-6}\ \text{C})}{(2.0\ \text{m})^2} = 4.5 \times 10^{-3}\ \text{N}.$$

Since the two charges have opposite signs, the force is one of attraction., q_1 will exert a force of 4.5×10^{-3} N which attracts q_2. According to Newton's third law, q_2 will exert a force of 4.5×10^{-3} N on q_1, which again is a force of attraction.

Example 3 *Force on a Point Charge due to Two or More Other Point Charges*

Consider four point charges arranged in the corners of a square, as shown in the diagram. Using,

$q_1 = +\,3.00\ \mu C$ $\qquad q_2 = -\,1.20\ \mu C$

$q_3 = -\,2.00\ \mu C$ $\qquad q_4 = -\,1.20\ \mu C$

$a = 25.0$ cm

Find the net force acting on charge q_3 due to charges q_1, q_2, and q_4. Assume that all the point charges are fixed so that no motion occurs.

To find the net force acting on q_3, we need to find the vector sum of the forces acting on it. Note that q_3 will be repelled by q_2 and q_4, while q_3 will be attracted to q_1. If the force of charge q_1 acting on q_3 is $\mathbf{F}_{13}$, and the forces on q_3 due to charges 2 and 4 are $\mathbf{F}_{23}$ and $\mathbf{F}_{43}$, respectively, we can use the following diagram.

We begin by calculating the magnitudes of the three forces acting on q_3. We will then use the rules of vector addition to find the desired resultant.

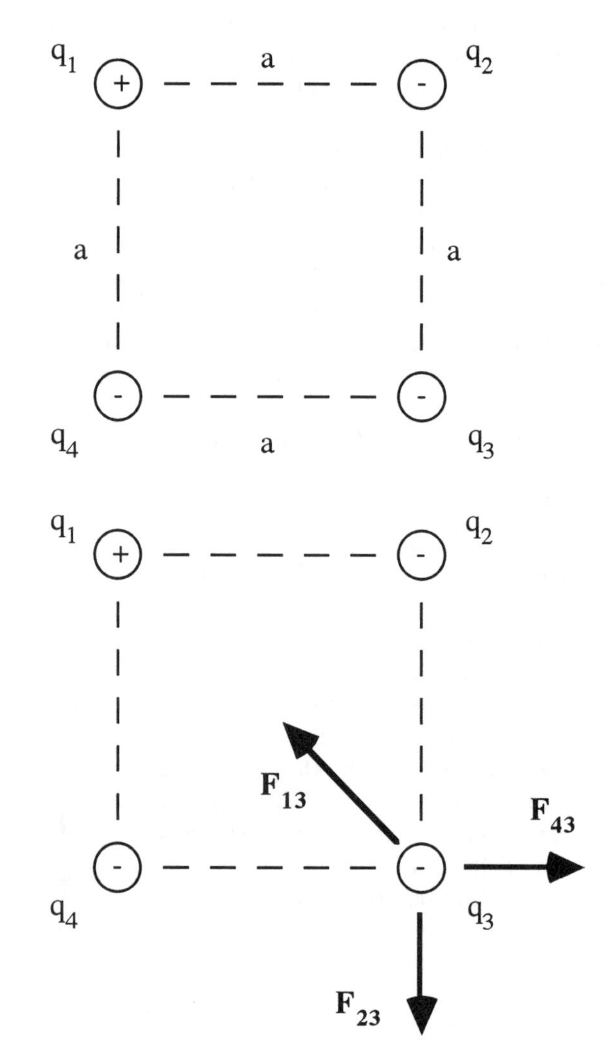

The force exerted on the charge q_3 by the charge q_1 is, from Coulomb's law,

$$F_{13} = k\frac{q_1 q_3}{r_{13}^2} = \frac{(8.99 \times 10^9 \text{ N m}^2/\text{C}^2)(3.00 \times 10^{-6} \text{ C})(2.00 \times 10^{-6} \text{ C})}{2\,(0.250 \text{ m})^2} = 0.432 \text{ N}$$

Similarly, for F_{23} and F_{43} we obtain,

$$F_{23} = k\frac{q_2 q_3}{r_{23}^2} = 0.345 \text{ N} \qquad F_{43} = k\frac{q_4 q_3}{r_{43}^2} = 0.345 \text{ N}$$

We can now add forces F_{13}, F_{23}, and F_{43} using the following vector diagram.

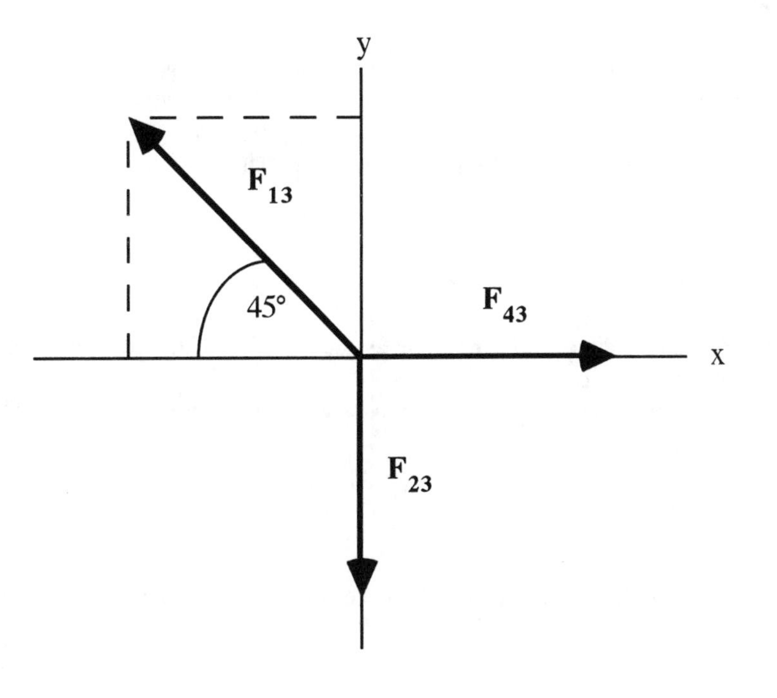

Force	x component	y component
F_{13}	$-(0.432 \text{ N}) \cos 45°$	$+(0.432 \text{ N}) \sin 45°$
	$= -0.305$ N	$= +0.305$ N
F_{23}	$=$ 0	$= -0.345$ N
F_{43}	$= +0.345$ N	$=$ 0
F	$= +0.040$ N	$= -0.040$ N

The magnitude F of the resultant is given by

$$F = \sqrt{F_x^2 + F_y^2} = \sqrt{(+0.040 \text{ N})^2 + (-0.040 \text{ N})^2} = 0.057 \text{ N}$$

The angle θ that the resultant force makes with the x axis is

$$\theta = \tan^{-1}\left(\frac{F_y}{F_x}\right) = \tan^{-1}\left(\frac{-0.040\ \text{N}}{+0.040\ \text{N}}\right) = -45°$$

where the negative angle means that the direction is clockwise with respect to the +x axis. In other words, the resultant force on q_3 is directed *below* the positive x axis.

So the general procedure for finding the force on a point charge due to two or more other point charges is to find the vector sum of the individual forces. The magnitudes of the forces are calculated from Coulomb's law. The directions of the forces are determined from the sign of each individual charge. That is, the force is either one of attraction or repulsion, depending on whether each pair of charges has unlike or like signs, respectively.

18.6 The Electric Field

A charge can experience an electrostatic force due to the presence of other charges in the environment. If a positive charge q_0 is placed at a point in space, and if other charges are present, the charge q_0 feels a force **F**, which is the vector sum of the forces exerted on it by all the other charges present. We can define the electric field **E**, that exists at this point, as the electrostatic force **F** divided by the test charge q_0. That is,

$$\mathbf{E} = \frac{\mathbf{F}}{q_0} \qquad (18.2)$$

The electric field is a vector, and its direction is the same as the direction of the force **F** on a positive test charge. The SI unit of electric field is newtons per coulomb (N/C).

Consider the electrostatic force exerted by a point charge q on a test charge q_0. The force is given by equation (18.1),

$$F = k\frac{q\,q_0}{r^2}$$

Using equation (18.2), we can see that the magnitude of the electric field due to a point charge can be written as

$$E = \frac{kq}{r^2} \qquad (18.3)$$

Example 4
Two point charges, $q_1 = -8.0\ \mu\text{C}$ and $q_2 = +4.0\ \mu\text{C}$, are separated by a distance of 0.50 m. Find the electric field (both magnitude and direction) at a point P on the line between the charges 0.20 m from q_2.

We can see from the following diagram that the net electric field at point P is due to the combined effects of both charges. Note the directions of the fields E_1 and E_2; these based on the fact that the direction of the field is the same as the direction of the force on a POSITIVE test charge.

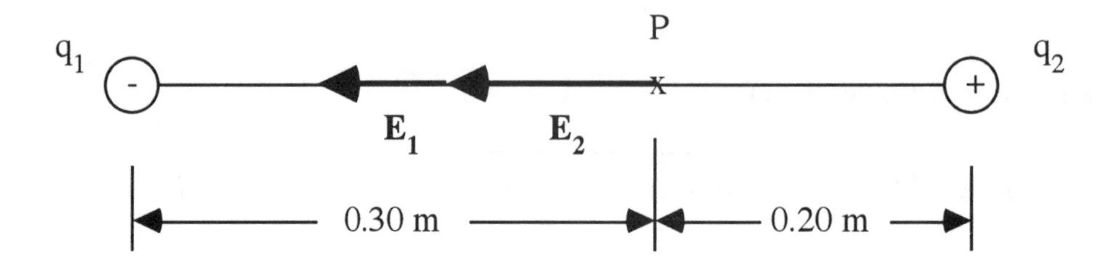

Using equation (18.3) we have,

$$E_1 = \frac{k\,q_1}{r_1^2} = \frac{(8.99 \times 10^9\ \mathrm{N\,m^2/C^2})(8.0 \times 10^{-6}\ \mathrm{C})}{(0.30\ \mathrm{m})^2} = 8.0 \times 10^5\ \mathrm{N/C}$$

$$E_2 = \frac{k\,q_2}{r_2^2} = \frac{(8.99 \times 10^9\ \mathrm{N\,m^2/C^2})(4.0 \times 10^{-6}\ \mathrm{C})}{(0.20\ \mathrm{m})^2} = 9.0 \times 10^5\ \mathrm{N/C}$$

Since E_1 and E_2 are in the same direction, the total electric field is obtained by adding them together. Thus

$$E = E_1 + E_2 = 8.0 \times 10^5\ \mathrm{N/C} + 9.0 \times 10^5\ \mathrm{N/C} = 1.7 \times 10^6\ \mathrm{N/C}.$$

Which is directed to the left, or towards q_1.

Now consider a **parallel plate capacitor** which consists of two parallel metal plates, each with area A. A charge +q is spread uniformly over one plate, while a charge -q is spread uniformly over the other plate. In the region between the plates and away from the edges, the electric field points from the positive plate toward the negative plate and has a magnitude of

$$E = \frac{q}{\varepsilon_0 A} = \frac{\sigma}{\varepsilon_0} \qquad (18.4)$$

where ε_0 is the **permittivity of free space** and has a value of $\varepsilon_0 = 8.85 \times 10^{-12}\ \mathrm{C^2/(N\,m^2)}$, and $\sigma = q/A$ is the surface charge density of the plates.

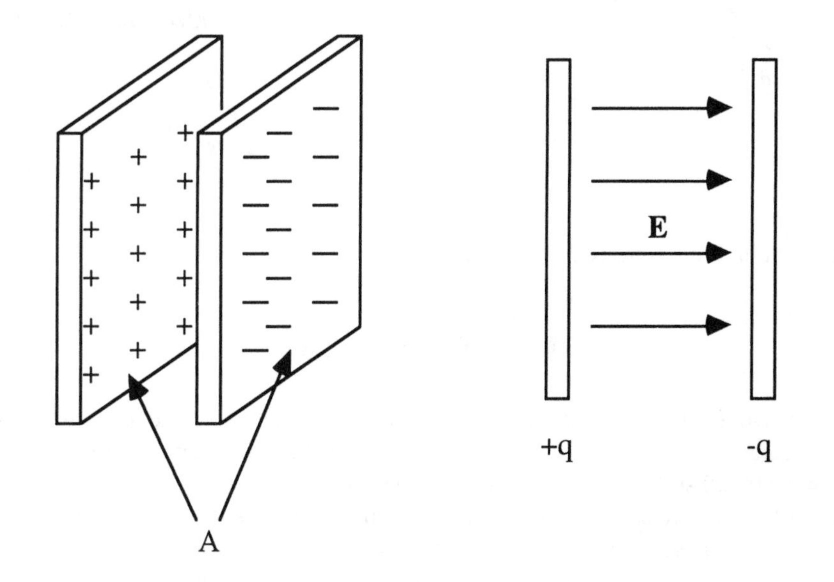

Example 5

The magnitude of the electric field between the plates of a parallel plate capacitor is 2.4×10^5 N/C. Each plate carries a charge whose magnitude is 0.15 μC. What is the area of the plates?

Using equation (18.4) and solving for A we obtain

$$A = \frac{q}{\varepsilon_0 E} = \frac{1.5 \times 10^{-7}\ \text{C}}{[8.85 \times 10^{-12}\ \text{C}^2/(\text{N m}^2)](2.4 \times 10^5\ \text{N/C})} = 0.071\ \text{m}^2$$

18.9 Gauss' Law

Gauss' law relates the electric flux through a closed surface to the net charge enclosed by the surface. In order to fully understand Gauss' law, we must first discuss the concept of electric flux.

Consider a surface in the vicinity of an electric field as shown in part (a) of the figure below. Imagine dividing the surface into a number of smaller surface elements, each of area ΔA. The size of each element is chosen so that the magnitude and direction of the electric field is constant over each surface element. We now imagine constructing the line that is normal to each surface element. In general, the electric field at any point on the surface element will make an angle ϕ with the normal, as shown in part (b) of the figure below.

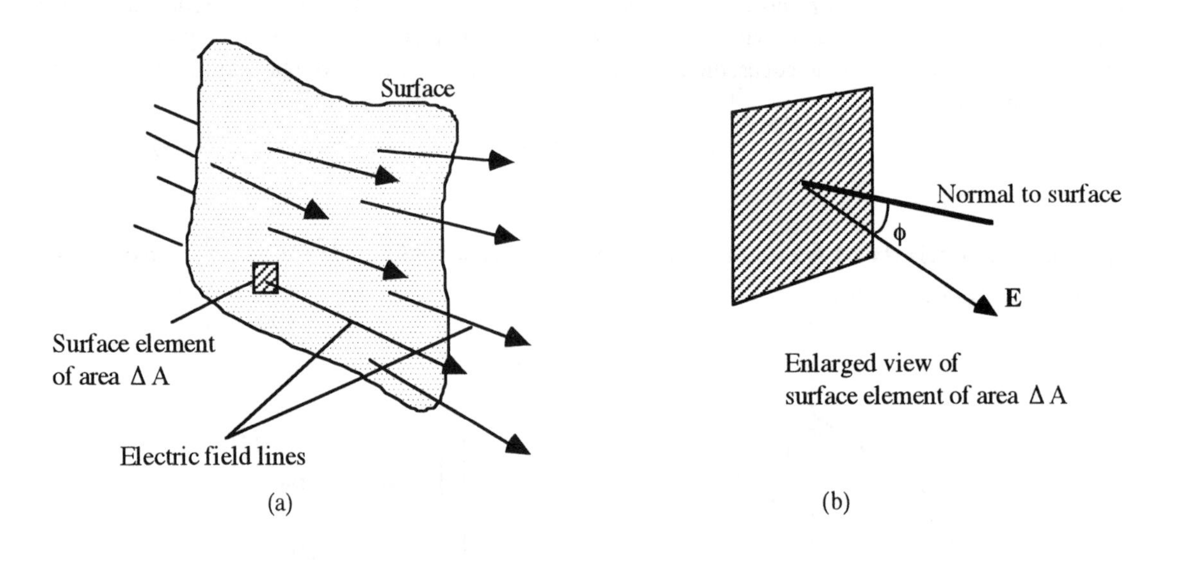

The electric flux through the entire surface is defined as

$$\Phi_E = \sum(E \cos \phi)\Delta A \qquad (18.6)$$

where E is the magnitude of the electric field at a point on the surface, and ϕ is the angle between the electric field and the normal at that point. The summation is carried out over the entire surface; in other words, the quantity $(E \cos \phi)\Delta A$ is summed for all surface elements that comprise the surface.

In general, surfaces can be *open* or *closed.* An open surface always has an edge, a rim, or some boundary. Examples include a sheet of paper, a lamp shade, and half of an egg shell. Closed surfaces, on the other hand, do not posses an edge or boundary. Examples of closed surfaces include, a sphere and an unbroken egg shell. Gauss's law pertains only to *closed* surfaces, appropriately named *Gaussian surfaces.* Clearly, the surface elements on a closed surface have both an inward and an outward normal. In applying Equation (18.6) to closed surfaces, the angle ϕ is, by convention, the angle that the electric field makes with the outward normal. The electric flux through a closed surface will be positive if the electric field lines leave the surface and negative if the electric field lines enter the surface.

Gauss' law states that the electric flux Φ_E through a Gaussian surface is equal to the net charge Q enclosed by the surface divided by ε_0, the permittivity of free space:

$$\sum(E \cos \phi)\Delta A = \frac{Q}{\varepsilon_0}. \qquad (18.7)$$

From Equation (18.7), we see that the flux through a closed surface is positive if it encloses a net positive charge, and negative if the surface encloses a net negative charge.

Gauss's law is an important calculational tool in many circumstances:

1. It can be used to determine the electric flux through a closed surface without performing the summation in Equation (18.6) if the net enclosed charge Q is known.

2. It can be used to determine the net charge within a closed surface, if the flux through the surface is known (or can be determined).

3. It can be used to determine the magnitude of the electric field for charge distribution that possess a high degree of spatial symmetry.

The next two examples illustrate the first two circumstances described above.

Example 6

A point charge of + 5.0 μC is placed at the center of a cube of side length 0.25 m. Determine the electric flux through the cube.

From Gauss' law

$$\Phi_E = \frac{Q}{\varepsilon_0} = \frac{+5.0 \times 10^{-6}\ \text{C}}{8.85 \times 10^{-12}\ \text{C}^2/(\text{N} \cdot \text{m}^2)} = 5.6 \times 10^5\ \text{N} \cdot \text{m}^2/\text{C}\ .$$

While this problem can be solved using Equation (18.6), the method is tedious and requires the use of integral calculus. Gauss' law provides a quick and convenient method that requires only simple arithmetic. Notice that the answer does not depend on the fact that the charge is at the center of the cube. In fact, the electric flux through the cube will be the same regardless of the position of the charge as long as it remains inside the cube. Certainly, if the charge is moved to a different position in the cube, the magnitude of the electric field at points on the surface of the cube will change; however, Gauss' law predicts that the electric field will change in such a way that the flux through the cube remains the same.

Example 7

One day, the electric field in the atmosphere near ground level is 110 N/C. Assume that the magnitude of this field is the same everywhere around the earth and that the direction of the field is radially inward. The radius of the earth is 6.38×10^6 m. Calculate the net charge (magnitude and sign) on the earth.

For the purposes of this problem, imagine that the earth is *just* surrounded by a spherical Gaussian surface (i.e., the radius of the Gaussian surface is essentially the same as that of the earth). The electric flux through the Gaussian surface is given by

$$\Phi_E = \Sigma(E \cos \phi)\Delta A$$

where ϕ is the angle between the electric field and the *outward* normal to the surface element ΔA. Combining this with Gauss' law, we have

$$\frac{Q}{\varepsilon_0} = \Sigma(E \cos \phi)\Delta A.$$

Since the electric field is uniform in magnitude and directed radially inward, the angle between the electric field and the outward normal is 180° for each surface element ΔA. Furthermore, since the electric field is uniform in magnitude, E can be factored out of the summation. The sum $\sum \Delta A$ must equal the surface area of the spherical Gaussian surface. Solving for Q gives

$$Q = \varepsilon_0 (E \cos \phi) \sum \Delta A = \varepsilon_0 (E \cos \phi)(4\pi R_E^2)$$

$$Q = [8.85 \times 10^{-12}\ C^2/(N \cdot m^2)](110\ N/C)(\cos 180°)[4\pi(6.38 \times 10^6\ m)^2] = -5.0 \times 10^5\ C$$

As described above, Gauss' law can also be used to calculate the magnitude of the electric field in the vicinity of a charge distribution provided that the charge distribution (and hence the electric field) possesses sufficient symmetry. The symmetry of the charge distribution is the key factor. Imagine that we surround a point charge with an arbitrarily shaped Gaussian surface. We can deduce from Example 6 that the flux through this surface is the same regardless of its shape; however, we cannot, in general, use Gauss' law to solve for the electric field on the Gaussian surface. For an arbitrarily shaped Gaussian surface, the magnitude of the electric field will, in general, vary from point to point on the surface. Hence, we cannot solve for E in Equation (18.7). In order to solve for E, we must be able to "factor" the E out of the summation. This can only be done if the magnitude of E is *constant* over the Gaussian surface.

We must then use the symmetry of the charge distribution as a guide in constructing a Gaussian surface over which E is constant. For a charge distribution that possesses spherical symmetry, we can deduce that the magnitude of the electric field should be the same at all points equidistant from the distribution. Therefore, if we surround such a distribution with a spherical Gaussian surface that is concentric with the distribution, we can assume that E is constant on the surface. For a line of charge (text Example 17), we can deduce that the magnitude of the electric field should be the same at all points equidistant from the line of charge. In this case, such points would lie on the surface of a cylinder that is concentric with the line of charge; hence we would construct a cylindrical Gaussian surface.

Example 8
A solid conducting sphere of radius R carries a total charge -q uniformly distributed over the surface of the sphere. Determine the magnitude and direction of the electric field for any point at a distance r from the center of the sphere if (a) $r > R$ (i.e., at any point outside the sphere), and (b) $r < R$ (i.e., at any point inside the sphere).

Since we want to find E, we must perform the calculation with a Gaussian surface over which E is constant. We can assume that since the charge distribution possesses spherical symmetry, the magnitude of the electric field (if it exists), should be the same at all points equidistant from the center of the sphere. Thus, in both parts (a) and (b) we will use spherical Gaussian surfaces.

(a) To find the magnitude of the electric field outside the charged sphere, we construct a spherical Gaussian surface of radius r ($r > R$) concentric with the charged sphere. The flux through this Gaussian surface is

$$\Phi_E = \Sigma(E \cos \phi)\Delta A.$$

Since the charge is spread uniformly over the surface of the sphere, the charge distribution possesses spherical symmetry. Therefore, we expect the electric field to be directed radially. Since the charge is negative, the field is radially inward; $\phi = 180°$ for each surface element, and $(E \cos 180°) = -E$. Furthermore, since the charge distribution is uniform, we expect the electric field to be uniform in magnitude over the Gaussian surface. Hence, E can be factored out of the summation.

$$\Phi_E = -\Sigma E\Delta A = -E\Sigma \Delta A,$$

where $\Sigma\Delta A$ is the sum of the area elements that make up the Gaussian surface. This sum must equal the surface area of the spherical Gaussian surface or

$$\Phi_E = -E\Sigma\Delta A = -E(4\pi r^2)$$

The explicit statement of Gauss' law becomes

$$-E(4\pi r^2) = \frac{Q}{\varepsilon_0},$$

where Q is the net charge enclosed by the Gaussian surface. From the statement of the problem, the net charge Q enclosed by the Gaussian surface is -q. Thus we have

$$-E(4\pi r^2) = \frac{-q}{\varepsilon_0}$$

Solving for E gives

$$E = \frac{q}{4\pi\varepsilon_0 r^2}. \qquad (r > R)$$

Since the charge is negative the direction of the field must be radially *inward.* Note that the direction of the electric field is deduced from the fact that the charge is negative. Gauss' law is used only to find the magnitude of the field.

(b) To find the magnitude of the electric field inside the charged sphere, we construct a spherical Gaussian surface of radius r ($r < R$) concentric with the charged sphere. We can argue, as we did in part (a), that the field (if it exists) should possess spherical symmetry. Using the arguments similar to those of part (a), we find that the explicit statement of Gauss' law for this situation is

$$E(4\pi r^2) = \frac{Q}{\varepsilon_0}.$$

Since we have a conducting sphere, however, all of the charge resides on the outer surface of the sphere. The Gaussian surface encloses zero net charge so $Q = 0$. Thus we obtain

$$E = 0. \qquad (r < R)$$

PRACTICE PROBLEMS

1. How many electrons must be removed from an electrically neutral silver dollar to give it a charge of +3.8 μC?

2. Three charges are located along a straight line, as shown below. What is the net electric force on charge q_2?

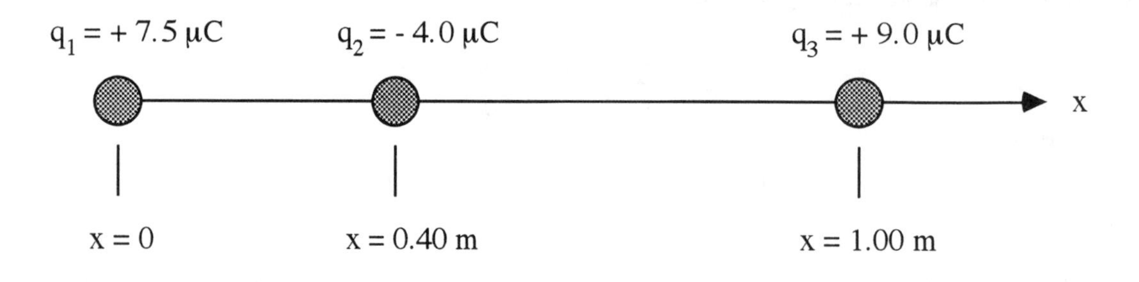

3. Three charges are located at the corners of an equilateral triangle, as shown below. What is the net electric force (magnitude and direction) acting on the top charge, q_1? Take d = 0.50 m.

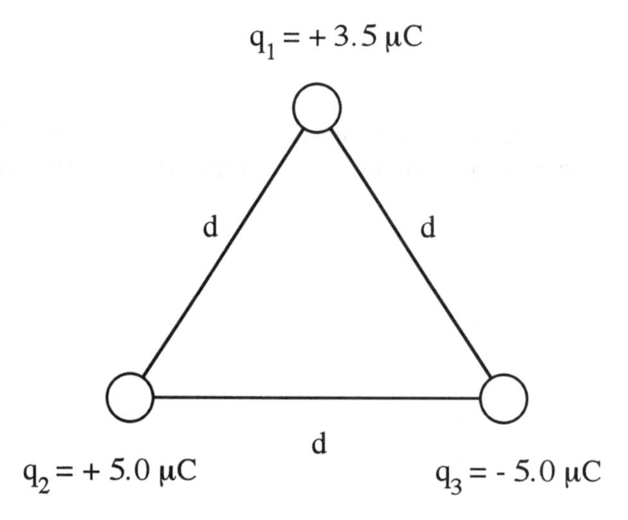

4. Two charges, $q_1 = +5.0\ \mu C$ and $q_2 = -3.0\ \mu C$, are separated by a distance of 1.0 m. Find the spot along the line between the charges where the net electric field is zero.

5. Four charges are arranged at the corners of a square, as shown below. Take $a = 0.25$ m. If all the charges have the same magnitude of 6.0 μC, what is the magnitude and direction of the electric field at the center of the square, at P?

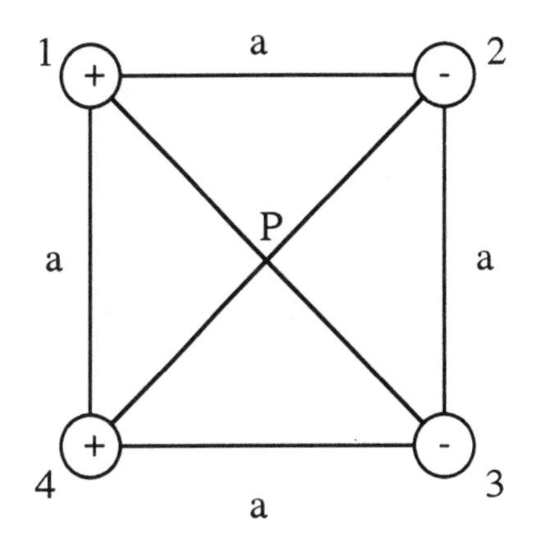

6. A negative charge - q is fixed to one corner of a rectangle, as in the drawing. What positive charge must be fixed to corner A and what positive charge must be fixed to corner B, so the total electric field at the remaining corner is zero? Express your answer in terms of q.

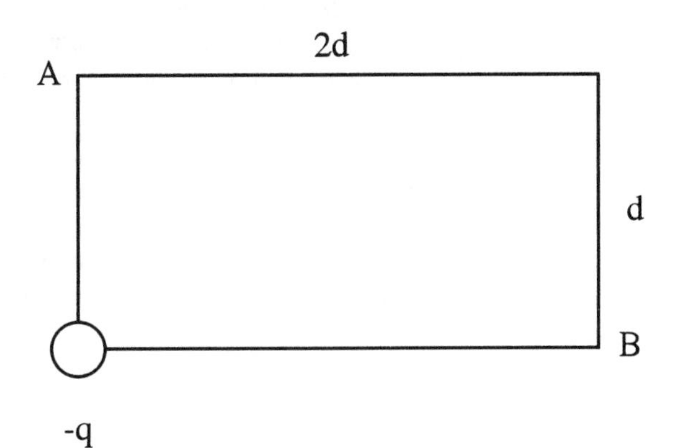

HELPFUL SUGGESTIONS

1. Coulomb's law is used to find the **magnitude** of the electric force between point charges. The sign of the charges is used to determine the **direction** of the force. That is, if charges have like signs, the force is one of repulsion; if charges have unlike signs, the force is one of attraction. The sign of the charge should be ignored when substituting q into Coulomb's law, equation (18.1), take the direction into account separately. Remember, however, that the electric force is a **vector** and therefore must be treated like any other vector in terms of addition and subtraction.

2. To determine the force on a point charge, q, due to a number of other point charges q_1, q_2, . . . , use Coulomb's law to find the force $\mathbf{F_1}$ due to q_1, the force $\mathbf{F_2}$ due to q_2, etc., for all the charges. The net force on q is then the **vector** sum of all the forces acting on q, $\Sigma \mathbf{F_i}$.

3. The electric field is also a **vector** quantity with both magnitude and direction. Its direction is the same direction as the force on a **positive** charge placed in the field. Thus, if a charge is negative, the electric field at a point due to that charge would be directed toward the charge.

4. To calculate the electric field due to a number of charges q_1, q_2, . . . , calculate the fields due to each charge $\mathbf{E_1}$, $\mathbf{E_2}$. . . . etc., and then add all the **E** s as vectors.

5. When using Gauss' law to find the magnitude of the electric field in the vicinity of a charge distribution, use the symmetry of the problem to construct a Gaussian surface over which E is constant.

EVERYDAY PHYSICS

1. You have probably experienced static electricity by walking across a carpet and getting shocked by touching another person or some metal surface. Comb your hair in the dark sometime (the effect works best on a dry day) and you might see flashes of light accompanied by a slight crackling sound. The same effect, on a much larger scale, is what occurs during a lightning storm.

2. Comb your hair (or rub the comb briskly on a woolen garment) and place the comb near a small stream of running water. Is the water charged?

3. Place some tiny pieces of paper on a table. Hold a plastic ruler or comb directly above and close to the bits of paper. What happens? Now rub the ruler or comb with a clean, dry paper towel or piece of wool and again place it directly above and close to the bits of paper. What happens now?

4. You can build your own electroscope from a small jar sealed with a cork. Pass a metal wire (without insulation) through the center of the cork and bend the lower end of the wire to support two thin strips of aluminum foil. Charge the electroscope by touching it with some object you have charged by friction. The aluminum leaves should separate. Hold an object, such as a pencil, near the top of the electroscope and touch the wire (the wire should stick out of the cork). Do the leaves fall? Try other objects, including some metal ones, to test for conductivity.

5. Why are the tires of trucks which carry gasoline or other flammable fluids manufactured to conduct electricity?

CHAPTER QUIZ

1. Two point charges are separated by a distance d and feel a force F. What would the force be if the two charges are moved a distance d/3 apart?
   a. The force would be 3F.
   b. The force would be F/3.
   c. The force would be 9F.
   d. The force would be F/9.

2. How many electrons are required to produce a charge of - 6.4 μC?
   a. 4.0×10^{13}
   b. 6.3×10^{18}
   c. 2.0×10^{13}
   d. 4.0×10^{19}

3. Two charged objects attract each other with a certain force F. The objects may be treated as point charges. If the charges on both objects are doubled, with no change in separation, the force between them becomes
   a. F
   b. 2F
   c. 4F
   d. 8F

4. Consider two point charges of +q and +4q which are separated by a distance of 3 m. At what point, on the line between the charges, is the electric field equal to zero?
   a. 1 m from the +4q charge
   b. 1 m from the +q charge
   c. 3/4 m from the +q charge
   d. There is no such point possible.

5. A force of 12 N acts on a charge of 2.0×10^{-6} C when it is placed in a uniform electric field. What is the magnitude of this electric field?
   a. 24×10^{-6} N/C
   b. 1.7×10^{-7} N/C
   c. 24×10^{6} N/C
   d. 6.0×10^{6} N/C

6. Electric field lines
   a. circle clockwise around positive point charges.
   b. circle counterclockwise around positive point charges.
   c. radiate outward from negative point charges.
   d. radiate inward toward negative point charges.

7. A **charged** insulator and an **uncharged** conductor
   a. always repel each other electrically.
   b. always attract each other electrically.
   c. exert no electrostatic force on each other.
   d. may attract or repel, depending on whether the charge is positive or negative.

8. Three charges are arranged as shown below. The magnitude of the force of the +2 μC on the +4 μC charge is F_{24} and the magnitude of the force of the +3 μC charge on the +4 μC charge is F_{34}. The ratio F_{34}/F_{24} is
   a. 3/2
   b. $1/\sqrt{2}$
   c. 3
   d. 2

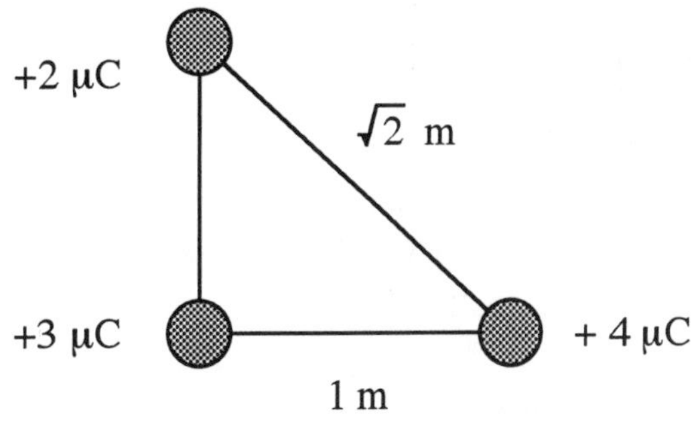

+2 μC

$\sqrt{2}$ m

+3 μC

+ 4 μC

1 m

SOLUTIONS AND ANSWERS

Practice Problems

---

1. We know that $q = Ne$ so,

$$N = q/e = (3.8 \times 10^{-6}\ C)/(1.6 \times 10^{-19}\ C)$$

$$N = \mathbf{2.4 \times 10^{13}}.$$

---

2. The forces on q_2 are shown in the diagram below.

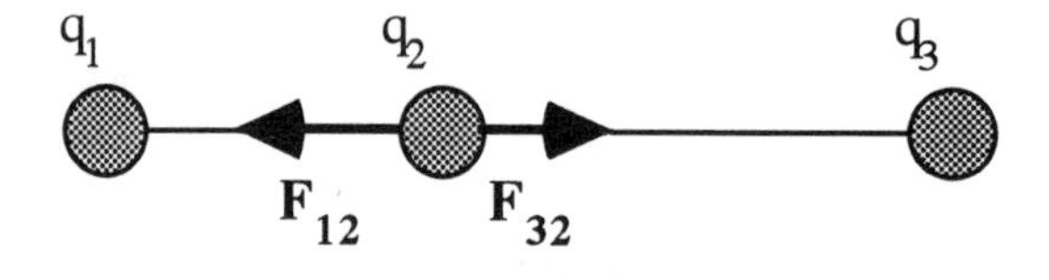

Using Coulomb's law:

$$F_{12} = kq_1q_2/r_{12}^2 = (8.99 \times 10^9\ N\ m^2/C^2)(7.5 \times 10^{-6}\ C)(4.0 \times 10^{-6}\ C)/(0.40\ m)^2 = 1.7\ N.$$
$$F_{32} = kq_3q_2/r_{32}^2 = (8.99 \times 10^9\ N\ m^2/C^2)(9.0 \times 10^{-6}\ C)(4.0 \times 10^{-6}\ C)/(0.60\ m)^2 = 0.90\ N.$$

The net electric force on q_2 is therefore, $F = F_{12} - F_{32} =$ **0.8 N to the left**.

---

3. The forces acting on q_1 are shown in the following diagram,

We have:

$$F_{21} = kq_2q_1/r_{21}^2$$
$$F_{21} = (9.0 \times 10^9\ N\ m^2/C^2)(5.0\ C)(3.5\ C)/(0.50\ m)^2$$
$$F_{21} = 0.63\ N$$

Similarly,

$$F_{31} = kq_3q_1/r_{31}^2 = 0.63\ N.$$

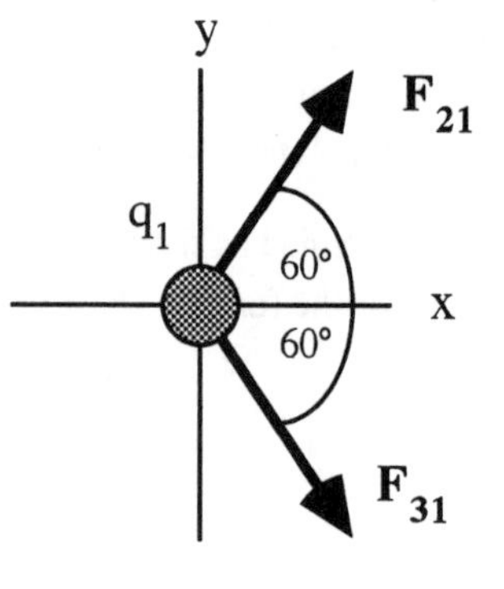

So,

$$F_x = (F_{21})_x + (F_{31})_x$$
$$F_x = 2(0.63\ N)\cos 60° = 0.63\ N.$$

and

$$F_y = (F_{21})_y + (F_{31})_y = 0.$$

Therefore, the net force acting on q_1 is **0.63 N in the +x direction**.

---

4. Since q_1 and q_2 are of opposite sign, the electric field can not be zero in the region between the charges. Instead, $E = 0$ along the line joining the charges but away from the smaller charge, as in the figure below.

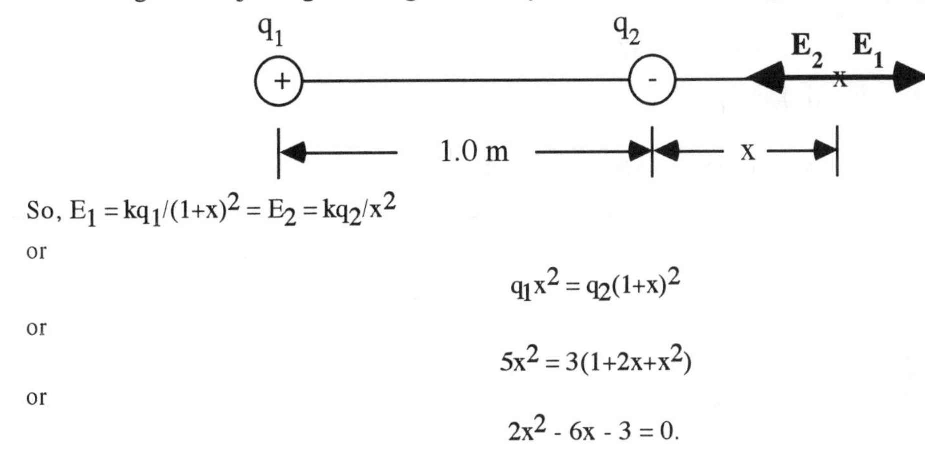

So, $E_1 = kq_1/(1+x)^2 = E_2 = kq_2/x^2$

or

$$q_1x^2 = q_2(1+x)^2$$

or

$$5x^2 = 3(1+2x+x^2)$$

or

$$2x^2 - 6x - 3 = 0.$$

Using the quadratic formula we find the solution to be $x = \mathbf{3.4\ m}$.

---

5. The total field at the center of the square is the vector sum of the fields due to each point charge.

Since each charge is equidistant from the center of the square, and since each charge has the same magnitude, the magnitudes of E_1, E_2, E_3, and E_4 are all the same. We have

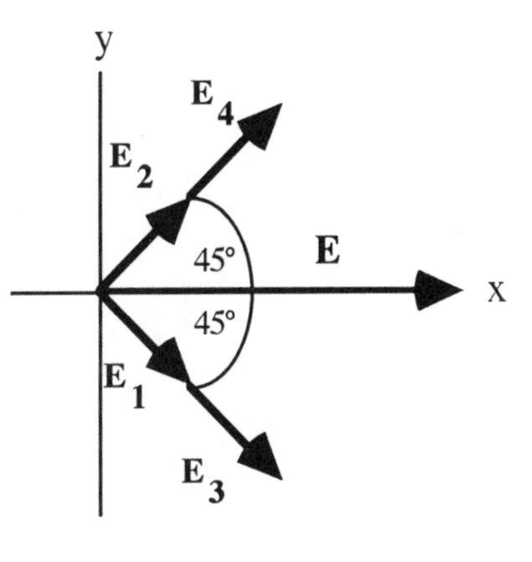

$E = kq/r^2 = (8.99 \times 10^9\ N\ m^2/C^2)(6.0 \times 10^{-6}\ C)/(0.18\ m)^2$
$E = 1.7 \times 10^6\ N/C.$

$\mathbf{E_{tot}} = \mathbf{E_1} + \mathbf{E_2} + \mathbf{E_3} + \mathbf{E_4}$. From the diagram we see that $(\mathbf{E_{tot}})_y = 0$ and

$(\mathbf{E_{tot}})_x = 4E \cos 45° = \mathbf{4.8 \times 10^6\ N/C}$.

---

6. The field at the empty corner of the rectangle is the vector sum of the fields due to q_A, q_B, and q_-. We can see from the diagram that $\theta = \tan^{-1}(d/2d) = 26.57°$. Since the diagonal is of length $d\sqrt{5}$,

$$E_- = kq/(5d^2),$$

Also,

$$E_A = kq_A/(4d^2),$$

directed in the +x direction, and

$$E_B = kq_B/d^2,$$

directed in the +y direction. From the diagram we see that

$$(E_-)_x = E_A, \text{ and } (E_-)_y = E_B.$$

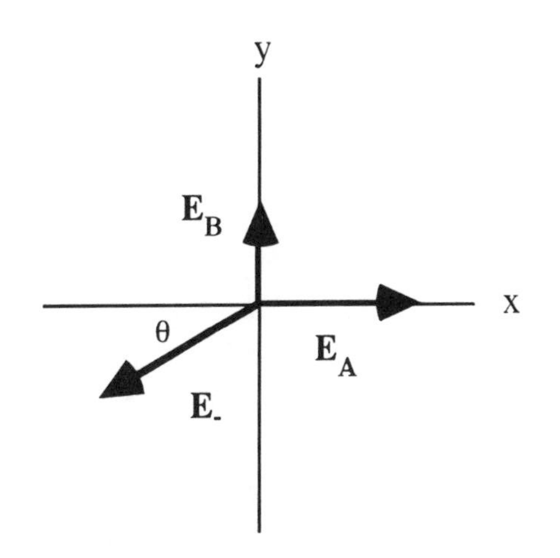

Thus,

$(E_-)_x = E_- \cos 26.57° = (kq \cos 26.57°)/(5d^2)$
$(E_-)_x = kq_A/(4d^2)$

So,

$$q_A = (4/5)q \cos 26.57° = \mathbf{0.716\ q}.$$

Now

$$(E_-)_y = E_- \sin 26.57° = (kq \sin 26.57°)/(5d^2)$$

$$(E_-)_y = kq_B/d^2$$

So,

$$q_B = (1/5)q \sin 26.57° = \mathbf{0.0895\ q}.$$

Chapter Quiz

1. c	3. c	5. d	7. b
2. a	4. b	6. d	8. c

MCAT REVIEW PROBLEMS

A simple electric dipole consists of a positive and negative charge of equal magnitude held very close to one another. The component of the electric field pointing away from a dipole has magnitude

$$E = \frac{2kp\cos\theta}{d^3},$$

where d is the distance from the center of the dipole to the point in question, $k = 9.0\times 10^9\ \mathrm{N\cdot m^2/C^2}$ is a universal constant, and p is the magnitude of the dipole moment vector, which specifies the strength and direction of the dipole. Here, θ denotes the angle between the dipole moment vector and **d**, the displacement vector (from the dipole to the point in question).

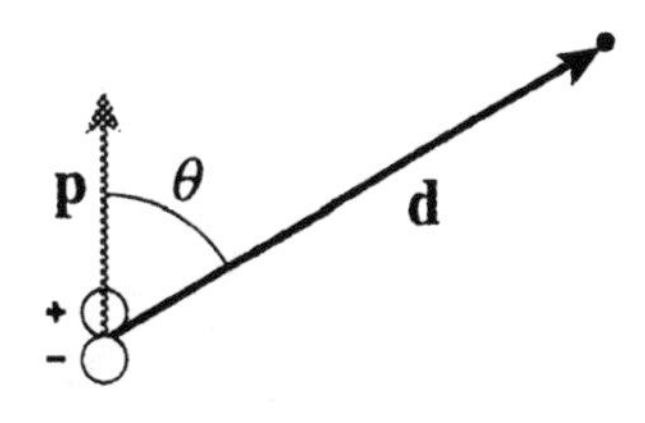

A student performs an experiment to determine if a mystery object is an electric dipole. (The mystery object is only a few millimeters long.) Using a sophisticated instrument, the student measures the component of the electric field pointing away from the object, at various distances from the center of the object. By taking each measurement along an imaginary line emanating outward from the center of the mystery object, he ensures that "θ" stays the same throughout the experiment. Table 1 shows the electric field he found at various distances.

TABLE 1

Trial	distance (m)	field (N/C)
1	0.010	3.6×10^{-10}
2	0.020	9.0×10^{-11}
3	0.030	4.0×10^{-11}

1. Assuming the data is free from major errors, what can the student conclude about the mystery object?

   I. It might be a dipole.
   II. It might be (approximately) a point charge.

   A. I only **B.** II only **C.** Both I and II **D.** Neither I nor II

2. From the given information, can we calculate the electrostatic force that would act on a point charge $q = 2.5\times 10^{-7}$ C, held at the location where the student measured the electric field in trial 2?

   A. Yes, because we can use $F = \frac{kq^2}{r^2}$, with $r = 0.02$ m.

   B. Yes, because we know the electric field at the relevant point.

   C. No, because the formula $F = \frac{kq_1q_2}{r^2}$ doesn't apply, and we don't know the dipole moment.

   D. No, because even though $F = \frac{kq_1q_2}{r^2}$ applies, we don't know both q_1 and q_2.

---

3. Which of these graphs best expresses how the electric field measured by the student varies with distance from the mystery object?

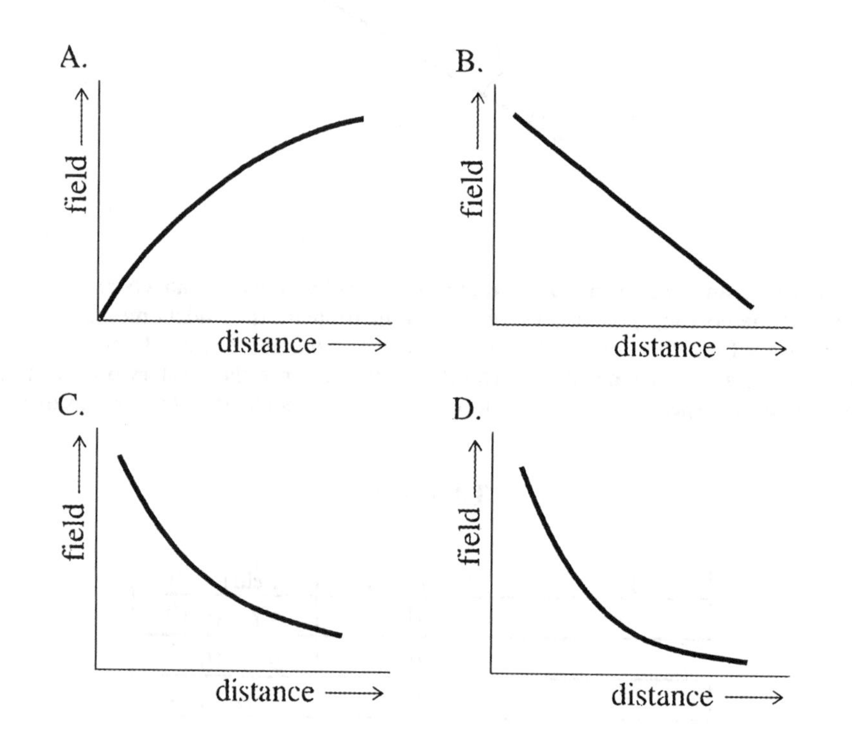

---

4. Consider the electric field produced by a dipole. If the dipole moment and the distance from the dipole are both doubled, while θ is kept the same, the electric field component pointing away from the dipole decreases by a factor of:

   A. 2 **B.** 4 **C.** 8 **D.** 16

ANSWERS TO MCAT REVIEW PROBLEMS

---

1. **B.** The data indicate an *inverse square* relationship between distance and field strength. For instance, when the distance doubles (from .01 m to .02 m), the field strength decreases by a factor of $2^2 = 4$, from 3.6×10^{-10} to 9.0×10^{-11} N/C. When the distance triples (from .01 m to .03 m), the field decreases by a factor of $3^2 = 9$, from 3.6×10^{-10} to 4.0×10^{-11} N/C.
   This inverse square relationship characterizes point charges; please review Coulomb's law for details. By contrast, if the mystery object were a dipole, the field would decrease with the distance *cubed.* For instance, doubling the distance would decrease the field by a factor of $2^3 = 8$. The data don't fit this "inverse cube" relationship.

---

2. **B.** The electrostatic force generated by an electric field E on a point charge q is *always*

$$F = qE,$$

   no matter what object generates the electric field. Think of the mystery object as creating an electric field, which then pushes on q. If we know the field, we need not know what kind of object created the field.

---

3. **D.** As discussed in question 1, the data indicate an inverse square relationship; doubling the distance decreases the field by a factor of 4. Only graph D captures this insight. Graph C represents a regular inverse proportionality, in which doubling the distance cuts the field in half.

---

4. **B.** According to the formula $E = \dfrac{2kp\cos\theta}{d^3}$, doubling the distance from the dipole *decreases* the field by a factor of $2^3 = 8$. (Mathematically, that's because doubling d increases the denominator by 8.) By similar reasoning, doubling the dipole moment, p, *increases* the field by a factor of 2. So, when we "turn on" both of these effects at once, the field decreases by a factor of 4.
   If this quick-and-dirty reasoning doesn't make sense, you can reason in steps as follows. Suppose the electric field has strength E_0. First, we double p. This increases the field strength to $2E_0$. Next, we double d. This decreases the field by a factor of 8, to $\dfrac{2E_0}{8} = \dfrac{E_0}{4}$. The order of the steps makes no difference; you get the same answer either way. Therefore, this step-by-step reasoning works even if, in real life, the dipole moment and the distance get doubled simultaneously.

CHAPTER 19
Electric Potential Energy and the Electric Potential

PREVIEW

In the previous chapter you were introduced to the electric force and electric field. Both of these are vector quantities. Since the electric force is conservative, it is possible to define a scalar electric potential energy of a charged object in an electric field in a way similar to that of defining a gravitational potential energy.

Two other useful concepts, electric potential and electric potential difference, closely related to electric potential energy, will be developed in this chapter. A discussion of how charged objects lose or gain electric energy as they move through electric potential differences is presented.

The capacitance of a capacitor will also be discussed.

QUICK REFERENCE

Important Terms

Electric potential energy (EPE)
The energy a charged object has due to its position in an electric field.

Electric potential
The electric potential energy per unit charge that is available at a given point in an electric field.

Electrical potential difference
The difference in the electric potential between two points in an electric field.

Equipotential surface
A surface within a region of electric field on which all points are at the same electric potential. No work is done in moving a charged object at constant speed along an equipotential surface.

Capacitor
A device that can store charge, and hence electric energy.

Capacitance
A measure of the ability of a capacitor to store charge. It is the charge stored per unit of applied voltage.

Dielectric
An insulating material. Dielectrics are often used between the plates of a capacitor to raise the capacitance.

Equations

The **work done** by the electric force when a charged particle moves from a point A to a point B in the electric field is

$$W_{AB} = EPE_A - EPE_B \qquad (19.1)$$

The **work done by the electric force per unit charge** when an object of charge q_0 moves from point A to point B in an electric field is

$$\frac{W_{AB}}{q_0} = \frac{EPE_A - EPE_B}{q_0} \qquad (19.2)$$

The **electric potential** at a point in an electric field is found by placing a small charge, q_0, at the point and dividing its electric potential energy by the charge.

$$V = \frac{EPE}{q_0} \tag{19.3}$$

The **electric potential difference** between two points A and B in an electric field is

$$V_B - V_A = \frac{EPE_B - EPE_A}{q_0} = \frac{-W_{AB}}{q_0} \tag{19.4}$$

or in "delta" notation

$$\Delta V = \frac{\Delta(EPE)}{q_0} = \frac{-W_{AB}}{q_0} \tag{19.4}$$

The **potential difference** between two points A and B in the electric field due to a point charge is

$$V_B - V_A = \frac{kq}{r_B} - \frac{kq}{r_A} \tag{19.5}$$

The **electric potential** at a point a distance, r, from a **point charge** is

$$V = \frac{kq}{r} \tag{19.6}$$

The **electric field** is related to the **gradient of the electric potential** by

$$E = -\frac{\Delta V}{\Delta s} \tag{19.7}$$

The magnitude of the **charge** on each plate of a **capacitor** is

$$q = CV \tag{19.8}$$

The **capacitance of a parallel plate capacitor** is

$$C = \frac{\kappa \varepsilon_o A}{d} \tag{19.10}$$

The **energy stored in a capacitor** is

$$\text{Energy} = \frac{1}{2} CV^2 \tag{19.11}$$

The **energy density** between the plates **of a capacitor** is

$$\text{Energy density} = \frac{1}{2} \kappa \varepsilon_o E^2 \tag{19.12}$$

DISCUSSION OF SELECTED SECTIONS

19.1 Potential Energy

A very close analogy can be made between the theories of gravity and electricity since the force laws have the same mathematical form. The main difference to keep in mind is that while there exists only one kind of mass and the gravitational force is always attractive, there exist two kinds electric charge and the electric force can be either attractive or repulsive. In either theory the concept of potential energy is of central importance. Equation (19.1) can be viewed as a defining relationship for the change in electric potential energy of a charge as it moves from point A to point B.

Example 1

A proton, located at point A in an electric field, has an electric potential energy of $EPE_A = 3.20 \times 10^{-19}$ J. The proton experiences an average electric force of 0.80×10^{-9} N, directed to the right. The proton then moves to point B, which is a distance of 1.00×10^{-10} m to the right of point A. What is the electric potential energy of the proton at point B ?

The work done by the electric force is, from Equation 6.1, the product of the magnitude of the force and the magnitude of the displacement:

$$W_{AB} = Fs = (0.80 \times 10^{-9}\ N)(1.00 \times 10^{-10}\ m) = 0.80 \times 10^{-19}\ J$$

Equation (19.1) gives the EPE at the final position to be

$$EPE_B = -W_{AB} + EPE_A$$
$$EPE_B = -0.80 \times 10^{-19}\ J + 3.20 \times 10^{-19}\ J = 2.40 \times 10^{-19}\ J$$

19.2 The Electric Potential Difference

The electric potential difference between two points in an electric field is related by equation (19.4) to the difference in the potential energy that a charge would have if moved between the points. The existence of a difference in potential, of course, implies that there is such a thing as electric potential at a point in the electric field. The electric potential is given by (19.3).

It is important to understand that to speak of a change in EPE one must actually have a charge and move it. We can, however, speak of the electric potential difference between two points in an electric field without actually moving a charge between them. In other words, electric potential energy is something that "belongs" to a charge, while electric potential is something which "belongs" to the electric field. Why this is true will become more apparent when we discuss the electric potential due to a point charge. The common name for electric potential difference is "voltage".

Example 2
What is the potential difference between the initial and final positions of the proton in example 1 above? Which point has the higher potential?

Equation (19.4) gives the difference in electric potential between points A and B to be

$$\Delta V = -W_{AB}/q_0 = -(0.80 \times 10^{-19}\ J)/(1.60 \times 10^{-19}\ C) = -0.50\ V$$

Now $\Delta V = V_B - V_A = -0.50$ V. The potential at point B is lower by 0.50 V than the potential at point A.

A few rules regarding the relationship between the electric field, potential, potential energy and the motion of charges placed in the field are in order at this point.

If a POSITIVE charge is placed in an electric field and released, it will
1. move to decrease its potential energy.
2. move toward a region of lower potential.
3. move ALONG the electric field lines.

If a NEGATIVE charge is placed in an electric field and released, it will
1. move to decrease its potential energy.
2. move toward a region of higher potential.
3. move OPPOSITE the electric field lines.

Example 3

A charged object, $q = +2.0\ \mu C$ and $m = 1.5 \times 10^{-15}$ kg, is placed at rest at point B and released. Which point will it move toward? How much kinetic energy will it have if it arrives at the point? How fast will it be traveling then?

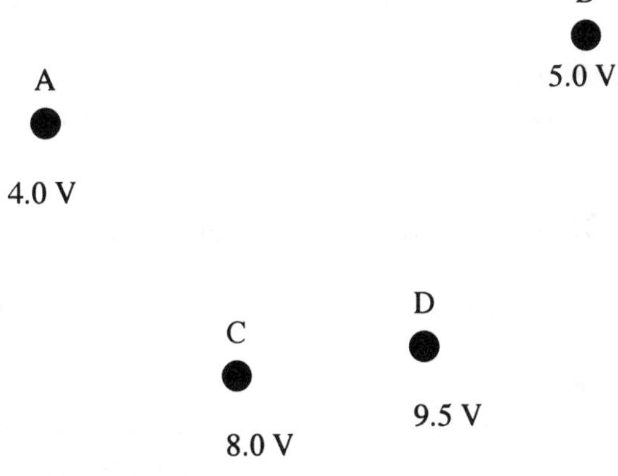

The charge is positive and must move toward a region of lower potential. Point A is the only region of lower potential. The charge will move towards point A.

Energy is conserved as the charge moves from point B to point A.

$$KE_A + EPE_A = KE_B + EPE_B$$

The kinetic energy at point B is zero and equation (19.3) applies at each point so

$$KE_A = EPE_B - EPE_A = qV_B - qV_A$$

$$KE_A = (2.0 \times 10^{-6}\ C)(5.0\ V - 4.0\ V)$$

$$KE_A = 2.0 \times 10^{-6}\ J$$

The speed of the charged object is

$$v = \sqrt{\frac{2KE}{m}} = \sqrt{\frac{2(2.0 \times 10^{-6}\ J)}{1.5 \times 10^{-15}\ kg}} = 5.2 \times 10^{4}\ m/s$$

19.3 The Electric Potential Difference Created by Point Charges

The potential difference between two points in the region around a point charge is given by (19.5) and the electric potential a distance, r, from the charge is given by (19.6). Note that neither of these equations depends on there being a test charge, q_0, in the field of the point charge, q, since q_0 divided out.

The potential at a point due to any number of point charges can be found by simply finding the potential at the point due to each alone and adding the potentials.

The electric potential energy of a group of charges is found by assembling the group, one charge at a time, and determining the electric potential energy at each step. The electric potential energy of the group is the sum of the electric potential energies determined for each step.

19.4 Equipotential Surfaces and Their Relation to the Electric Field

The electric field and electrical potential are alternative ways of describing the effect a charge or charge distribution has on the space around it. If one is known at each point in space, then it is possible to find the other at each point and vice versa. In general this requires using the calculus, but for the special case where the electric field is constant or an average field is desired, equation (19.7) can be used. Keep in mind that the electric field always points in a direction opposite to the change of the potential and perpendicular to equipotential surfaces.

Example 4
The metal contacts of an electric wall socket are about 1.0 cm apart and are maintained at a potential difference of 120 V. What is the average electric field strength between the contacts? What is the direction of the electric field if the left contact is the higher potential. The lower potential? Treat the potential difference between the contacts as being constant in time.

The magnitude of the average electric field between the contacts is given by equation (19.7) without the minus sign.

$$E = \Delta V/\Delta s = 120\ \text{V}/(1.0 \times 10^{-2}\ \text{m}) = 1.2 \times 10^{4}\ \text{V/m}.$$

The electric field is directed toward decreasing potential, so it would point right since the left contact is the higher potential and left if the left contact is the lower potential.

19.5 Capacitors and Dielectrics

A capacitor can store charge $q = CV$ when potential difference (voltage), V, is applied across its plates.

Example 5
How much charge can be placed on the plates of a $1.0 \times 10^{3}\ \mu\text{F}$ capacitor by a 12 V battery? How large a voltage would be required to place the same amount of charge on a $1.0\ \mu\text{F}$ capacitor?

Equation (19.8) gives the charge on the $1.0 \times 10^{3}\ \mu\text{F}$ capacitor to be

$$q = CV = (1.0 \times 10^{-3}\ \text{F})(12\ \text{V}) = 1.2 \times 10^{-2}\ \text{C}.$$

The voltage required to place the same charge on a $1.0\ \mu\text{F}$ capacitor is

$$V = q/C = (1.2 \times 10^{-2}\ \text{C})/(1.0 \times 10^{-6}\ \text{F}) = 1.2 \times 10^{4}\ \text{V}.$$

The capacitance of a capacitor depends on its GEOMETRY and the material between its plates. The insertion of a material with a higher dielectric constant between the plates of a capacitor results in a higher capacitance. If the same charge is retained on the plates, then the potential difference across the capacitor plates will drop.

Example 6
An empty parallel plate capacitor ($C_0 = 25\ \mu\text{F}$) is charged with a 12 V battery. The battery is disconnected and the region between the plates of the capacitor is filled with pure water. What is the capacitance, charge, and voltage for the water-filled capacitor?

The geometry of the capacitor has not changed, so the new capacitance is

$$C = \kappa\varepsilon_0 A/d = \kappa C_0 = (80.4)(25\ \mu\text{F}) = 2.0 \times 10^{3}\ \mu\text{F}.$$

The charge on the capacitor cannot change since the battery has been disconnected, so

$$q = q_0 = C_0V_0 = (25\ \mu F)(12\ V) = 3.0 \times 10^2\ \mu C.$$

The new voltage across the capacitor plates is

$$V = q/C = (3.0 \times 10^2\ \mu C)/(2.0 \times 10^3\ \mu F) = 0.15\ V.$$

Energy is stored in a charged capacitor as electric potential energy due to the separation of charge on the plates of the capacitor. The amount of energy stored can be expressed several ways. Energy = $\frac{1}{2}qV$ or $\frac{q^2}{2C}$ or $\frac{1}{2}CV^2$. One form may be derived from any other by using $q = CV$.

Example 7
How much energy is stored by a 25 μF capacitor charged to a potential difference of 12 V?

The energy stored is

$$\text{Energy} = 1/2\ CV^2 = 1/2\ (25 \times 10^{-6}\ F)(12\ V) = 1.5 \times 10^{-4}\ J.$$

Example 8
A capacitor is to be built which will store enough energy to light a 75 W light bulb for 15 min. If the available potential difference is 150 V, what capacitance is needed?

The needed capacitance is

$$C = 2(\text{energy})/V^2 = 2(75\ W)(15\ \text{min})(60\ s/\ 1\ \text{min})/(150\ V)^2 = 6.0\ F.$$

This is a very large capacitance. Such capacitors are quite large in size and are not generally available.

PRACTICE PROBLEMS

1. An electron can move from a point of 35.0 V potential to a point of 125 V potential. How much kinetic energy can the electron gain? If it started from rest, how fast would it be traveling when it arrives at the 125 V point?

2. Location A is 3.00 m from a charge of $- 3.00 \times 10^{-8}$ C, while location B is 2.00 m from the charge. Find the potential difference $V_B - V_A$ between the two points, and state which point is at the higher potential.

3. What is the electrical potential difference between a point 1.00×10^{-10} m from an electron and another point 5.00×10^{-10} m from the electron?

4. A 0.47 μF parallel plate capacitor has 7.1×10^{-5} C of charge on its plates which are separated by 0.052 mm. What is the magnitude of the electric field between the plates?

5. An empty 15 μF parallel plate capacitor is charged by a 12 V battery. The battery remains connected, and the region between the plates of the capacitor is filled with pure water. What is the new capacitance of the capacitor and the new charge on the plates of the capacitor?

6. A positive and a negative charge have the same magnitude of charge and are located 2.0 mm apart. Where in the region around the charges is the electrical potential zero?

7. Sketch the electric field lines and equipotential surfaces in the region around two positive charges of equal magnitude separated by a small distance. Label the regions of relatively high and low potential.

8. The electron in the Bohr model of the hydrogen atom is assumed to move in a circular orbit while the proton in the nucleus is stationary. The total mechanical energy E of the electron is the sum of the kinetic energy and electric potential energy EPE. What is the ratio EPE/E?

9. An air filled parallel plate capacitor has a plate separation of 0.15 mm. What plate area is required for the capacitor to have a capacitance of 33 μF?

10. Two parallel plate capacitors, one air filled and the other filled with paper, are otherwise identical and hold the same charge, q. Which capacitor has stored more energy? How much more?

HELPFUL SUGGESTIONS

1. Electric potential, potential difference and potential energy are scalar quantities and can often be used more easily than the vector quantities of force and electric field. When you are confronted with a problem in electricity, check to see if the scalar quantities can be conveniently used. The idea of electrical potential energy is especially important when used with the principle of conservation of energy.

2. The concept of electric potential is related to the concept of electric fields. If the electric potential at each point in space is known, then the electric field is in principle known and vice versa. An electric field is manifested by the force exerted on a charge placed in the field. The electric potential shows itself as the work done on a charge moving through a potential DIFFERENCE.

3. Electric fields ALWAYS point from regions of HIGH potential toward regions of LOW potential. A region of high potential is a region of high potential energy for a positive charge placed in the field and a region of low potential energy for a negative charge placed in the field. Remember that in the analogy with gravity, positive charges "fall down" while negative charges "fall up". An external force is required to make a charge move in the opposite direction.

4. The zero of electric potential is arbitrary and can be chosen for your convenience. It is customary, however, to choose the zero of potential for a point charge to be at infinity. This means that the electric potential due to a negative charge has negative values while the electric potential due to a positive charge has positive values.

5. Equipotential surfaces can be likened to contour lines on a map. Contour lines are lines of equal height, thus lines of equal gravitational potential energy for a mass. If an object is placed on a mountain and allowed to fall, then it will fall perpendicular to the contour lines. This is quite similar to a charge "falling" perpendicular to an equipotential surface along an electric field line.

6. The effect of inserting a dielectric material between the plates of a capacitor is always to increase the capacitance of the capacitor by a factor equal to the dielectric constant of the material. The capacitor can then hold more charge per volt of potential difference between its plates. For a given charge on the plates the effect of the dielectric is to reduce the potential difference needed since the induced electric field in the dielectric causes a reduction in the total electric field between the plates.

EVERYDAY PHYSICS

1. We all use the concepts of electric potential and potential difference (voltage) each day whether it is the potential difference supplied by a 9 V battery in a calculator, a 12 V battery in a car or the 120 V across the contacts in a wall outlet. The importance of each of these is that the "voltage" refers to the amount of energy per coulomb of charge that is available to be given to a charge. If this energy is supplied to charges such as electrons, then they move and can do useful things for us like running hair dryers.

2. What are the "real" potentials of the posts on your car battery? Is the negative post 0 V while the positive post is + 12 V or is the negative post - 12 V and the positive post 0 V or is the negative post 973 V and the positive 985 V? How much energy can be given to an electron by the battery in each case?

3. Your body and clothes can store an amount of charge which depends on the state of your body and the type of clothes you are wearing and the potential difference between you and a carpet that you have walked across. The potential difference built up as you walk across the carpet can be estimated by noting the length of electric arc produced when you bring your hand close to a water pipe or other grounded object. Experiment with this on dry and humid days. Be sure to wear the same clothes. Are the results as you would expect?

CHAPTER QUIZ

1. An electric force does 6.4 μJ of work in moving a 2.0 μC charge from point A to point B. What is the potential difference, $V_B - V_A$, between the points?

   a. + 3.2 V   b. - 3.2 V   c. + 0.31 V   d. - 0.31 V

2. Point A is at a potential of + 5.0 V while point B is at a potential of - 2.5 V. How much kinetic energy could a +1.0 μC charge gain if allowed to move from point A to point B?

   a. 2.5 μJ   b. - 2.5 μJ   c. 7.5 μJ   d. zero

3. Three points in a electric field A, B, C lie in a line and have potentials of 8.0 V, 4.0 V and - 2.0 V, respectively. A negative charge is placed at point B and released at rest. Which way will it move if it moves at all?

   a. toward A   b. toward B   c. It will not move.

4. Two point charges - 2.0 μC and + 3.0 μC are separated by a distance of 1.5 mm. Where along the line through the charges is the electric potential zero?

   a. Between the charges 0.90 mm from the negative charge.
   b. Between the charges 0.60 mm from the negative charge.
   c. 2.0 mm from the positive charge and 0.5 mm from the negative charge.
   d. 2.1 mm from the negative charge and 0.60 mm from the positive charge.

5. The negative post of a 12 V battery is held at a potential of 75 V above ground while the negative post of a second 12 V battery is held at ground potential, that is 0 V . Compare the work that the batteries could do on an electron.

   a. The first battery can do more work.
   b. Neither battery can do work.
   c. The second battery can do more work.
   d. The batteries can do the same work.

6. How far from a + 8.5 μC point charge is its + 2.5 V equipotential surface?

   a. 3.2 m   b. 3.3×10^{-5} m   c. 3.1×10^{4} m   d. 2.5 m

7. Electric field lines are ________their equipotential surfaces.

   a. parallel to   b. anti-parallel to   c. tangent to   d. perpendicular to

8. Which of the following is NOT true. If placed in an electric field and released a______charge will move to a region of_______.

   a. positive, lower potential
   b. positive, lower potential energy
   c. negative, higher potential
   d. negative, higher potential energy

9. A capacitor is charged by a 9.0 V battery and receives a charge of 3.0 μC. What is the capacitance of the capacitor?

   a. 0.33 μF   b. 27 μF   c. 3.0 μF   d. It cannot be determined.

10. Two parallel plate capacitors are identical except that one is air-filled and the other is filled with a dielectric material with $\kappa = 10.5$. How much larger is the capacitance of the dielectric-filled capacitor than the air-filled capacitor?

    a. 10.5   b. 0.0095   c. 1.0095   d. 11.5

SOLUTIONS AND ANSWERS

Practice Problems

---

1. The kinetic energy gained by the electron is equal to the EPE lost by the electron.

$$K = eV_2 - eV_1 = (1.60 \text{ X } 10^{-19} \text{ C})(125 \text{ V} - 30.0 \text{ V}) = \mathbf{1.52 \text{ X } 10^{-17} \text{ J}}.$$

The speed acquired by the electron is

$$v = \sqrt{\frac{2K}{m}} = \sqrt{\frac{2(1.52 \text{ X } 10^{-17} \text{ J})}{9.11 \text{ X } 10^{-31} \text{ kg}}} = \mathbf{5.78 \text{ X } 10^{6} \text{ m/s}}.$$

---

2. The potential at point A is $V_A = kq/r_A$ and the potential at point B is $V_B = kq/r_B$.
Now

$$V_B - V_A = kq[1/r_B - 1/r_A]$$
$$V_B - V_A = (8.99 \text{ X } 10^9 \text{ N m}^2/\text{C}^2)(-3.00 \text{ X } 10^{-8} \text{ C})[1/(2.00 \text{ m}) - 1/(3.00 \text{ m})] = \mathbf{-45 \text{ V}}.$$

Point A is at the higher potential since the difference, $V_B - V_A$, is negative.

---

3. Equation (19.5) gives for a point charge

$$V_B - V_A = kq[1/r_B - 1/r_A]$$
$$V_B - V_A = (8.99 \text{ X } 10^9 \text{ N m}^2/\text{C}^2)(-1.60 \text{ X } 10^{-19} \text{ C})[1/(5.00 \text{ X } 10^{-10} \text{ m}) - 1/(1.00 \text{ X } 10^{-11} \text{ m})] = \mathbf{11.5 \text{ V}}$$

---

4. The electric field between the plates is constant and its magnitude is $E = V/d$. The potential across the plates is $V = q/C$ so

$$E = q/(Cd)$$
$$E = (7.1 \text{ X } 10^{-5} \text{C})/[(0.47 \text{ X } 10^{-6} \text{ F})(0.052 \text{ X } 10^{-3} \text{ m})] = \mathbf{2.9 \text{ X } 10^{6} \text{ V/m}}.$$

---

5. The new capacitance is given by equation (19.10) $C = \kappa \varepsilon_0 A/d$, but the capacitance of the empty capacitor is

$$C_0 = \varepsilon_0 A/d \text{ so } C = \kappa C_0 = (80.4)(15\,\mu\text{F}) = \mathbf{1210\,\mu F}.$$

The new charge on the plates is

$$q = CV = (1210 \text{ X } 10^{-6} \text{ F})(12 \text{ V}) = \mathbf{1.45 \text{ X } 10^{-2} \text{ C}}.$$

---

6. Let r_1 be the distance from the negative charge to a point where the electrical potential is zero and r_2 be a similar distance from the positive charge. Then

$$kq/r_1 = kq/r_2, \text{ or } r_1 = r_2.$$

All points equidistant from the charges have zero potential.

---

7. Place a small positive test charge in the space around the two charges, release it and trace out its path as shown in the first drawing. The dark arrows represent the repulsive forces exerted on the test charge by the positive charges and the light arrows represent the path of the test charge, hence an electric field line. Repeat this for different electric field lines to obtain the pattern shown in the second drawing.

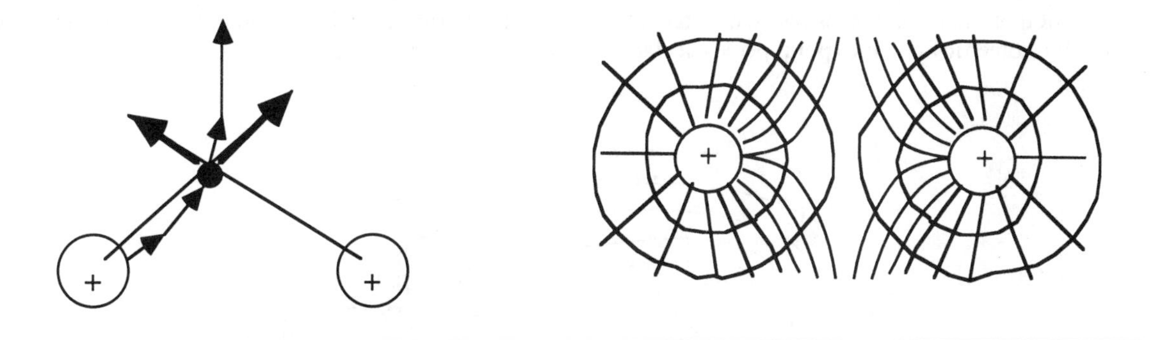

---

8. The EPE of the electron in its circular orbit is

$$EPE = k(e)(-e)/r = -ke^2/r \text{ where } e = 1.6 \times 10^{-19}\ C.$$

The total energy of the electron is

$$E = KE + EPE = 1/2\ mv^2 - ke^2/r.$$

Since the electron is in a circular orbit, the electrical force must be the centripetal force, so

$$mv^2/r = ke^2/r^2.$$

Hence, $KE = 1/2\ mv^2 = ke^2/(2r)$. Now

$$EPE/E = (-ke^2/r)/[-ke^2/(2r)] = \mathbf{2}.$$

---

9. Equation (19.10) applies if we take $\kappa = 1$ for air.

$$A = Cd/\varepsilon_0 = (33 \times 10^{-6}\ F)(0.15 \times 10^{-3}\ m^2)/(8.85 \times 10^{-12}\ C^2/N\ m^2) = \mathbf{560\ m^2}.$$

---

10. The energy stored in a capacitor is given by equation (19.11). Using equation (19.8) we obtain

$$\text{energy} = 1/2\ CV^2 = q^2/2C.$$

From this we see that a smaller capacitor will hold more energy for the same charge than a larger capacitor. The capacitance of the air-filled capacitor is $C_a = \varepsilon_0 A/d$ and the capacitance of the paper filled capacitor is $C_p = \kappa\varepsilon_0 A/d$. Hence, $C_p = \kappa C_a = (3.3)C_a$. The air capacitor has the smallest capacitance and will store more energy than the paper capacitor. The ratio of the energy stored in the air capacitor to the energy stored in the paper capacitor is $C_p/C_a = 3.3$. **The air capacitor stores 3.3 times more energy.**

Quiz answers

1. b	3. a	5. d	7. d	9. a
2. c	4. b	6. c	8. d	10. a

MCAT REVIEW PROBLEMS

An asymmetric blob creates an electric field corresponding to the equipotential lines in figure 1. (These lines are a cross section of the three-dimensional equipotential surfaces.) Each line differs in potential from its nearest neighbor by 5 volts. Electric field lines point perpendicular to equipotential lines.

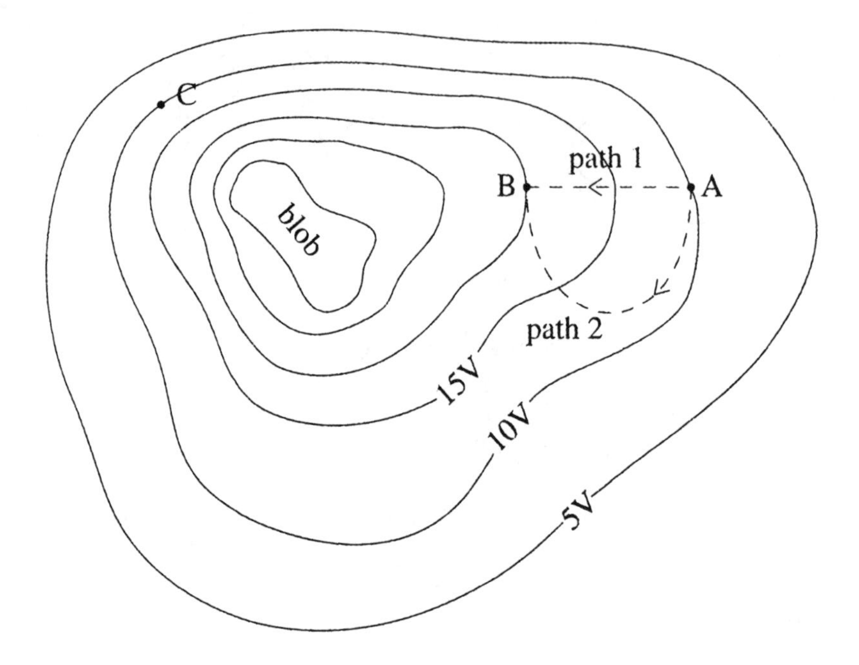

FIGURE 1

A positive point charge is going to be pushed from point A to point B. Two students argue about which requires less work: Pushing the charge along path 1, or pushing it along path 2.

Student 1

Pushing the charge along path 1 requires less work, because work is proportional to distance. Since path 1 is shorter, the work is less.

Student 2

But work also depends on force. Along path 1, you're pushing the charge directly against the electric field. This requires a large force. By contrast, along path 2, you're not fighting the field head on. Since you can push the particle with a much smaller force, the total work is less along path 2, even though more distance is covered.

---

1. Which student, if either, is correct about the work required to push the point charge along path 1 vs. path 2?

A. Student 1
B. Student 2
C. Neither student 1 nor student 2
D. We cannot determine who is correct without more information.

2. What additional information is needed to calculate the work done while pushing the point charge at constant speed from A to B along path 1?

   I. The charge of the point charge
   II. The charge of the asymmetric blob
   III. The electric field along that path

   A. I only **B.** I and II **C.** I and III **D.** I, II, and III

---

3. The point charge has charge 0.01 coulombs. How much work is needed to push it from A to C along the equipotential line containing those two points?

   A. –0.2 joules **B.** 0 joules
   C. 0.2 joules **D.** We need more information.

---

4. At which of the three labeled points is the electric field strongest?

   A. Point A **B.** Point B **C.** Point C **D.** We cannot determine the answer.

---

5. The net charge of the asymmetric blob is:

   A. Positive **B.** Negative **C.** Zero **D.** We cannot determine the answer.

ANSWERS TO MCAT REVIEW PROBLEMS

1. **C.** The same work gets done *no matter what path* is used. As student 1 points out, path 1 is shorter, which tends to lower the work. But as student 2 emphasizes, along path 2, you push the particle with less force, because you are not fighting the electric field head-on. Hence, along path 1, you exert a greater force, but over a smaller distance. These two effects "cancel"; work does not depend on path.

   To see why this conclusion applies to *any* path, recall that potential (V) is electric potential energy (EPE) *per charge*:

$$V = \frac{\text{EPE}}{q},$$

   and hence, $\Delta\text{EPE} = q\Delta V$. So, when you push the particle from A to B, its potential energy changes by an amount that depends *only* on the particle's charge and on the potential difference between A and B. The path does not matter. Since this change in potential energy *is* the work done on the point charge,

$$W = \Delta\text{EPE},$$

   the work doesn't depend on path.

   You have seen this kind of path-independence before, with gravity. The work needed to lift a book a height h off the floor is mgh, no matter whether you lift it straight up, slide it up a ramp, or use another path.

---

2. **A.** As just explained, the work depends only on the particle's charge and on the potential difference between A and B. Since $V_A = 10$ V and $V_B = 20$ V,

$$W = \Delta\text{EPE} = q\Delta V = q(20\text{ V} - 10\text{ V}),$$

   where q denotes the charge of the particle getting pushed, not the charge of the blob generating the electric field. Given q, we can immediately calculate W.

---

3. **B.** Since A and C sit on the same equipotential line ($V_A = V_C = 10$ V), the particle has the same electric potential energy at those two points:

$$\text{EPE}_A = \text{EPE}_C = qV = (.01\text{ C})(20\text{ V}) = 0.2\text{ joules}.$$

   So, between A and C, the particle's potential energy doesn't change: $\Delta\text{EPE} = 0$. Consequently, pushing the particle from A to C requires no work: $W = \Delta\text{EPE} = 0$. This situation resembles rolling a ball along the floor, as opposed to rolling it up a ramp.

---

4. **C.** The electric field is strongest wherever the equipotential lines are most densely spaced. To see why, imagine pushing a 1-coulomb charge from the $V = 10$ V to the $V = 15$ V equipotential line. "Fighting" the electric field requires $W = \Delta\text{EPE} = q\Delta V = 5$ joules of work. If the equipotential lines are closely spaced, then you "expend" those 5 joules over a small distance. In other words, you do lots of work, even though you cover a small distance. Therefore, the electric field against which you're fighting must be strong.

   By contrast, if the equipotential lines are far apart, then you perform those 5 joules of work over a long distance. Therefore, you must be exerting a smaller force, which implies that you are fighting a weaker electric field. This goes to show that closely spaced equipotential lines correspond to stronger fields, while sparse equipotential lines indicate weaker fields.

Notice that the size of the potential makes no difference. What matters is the rate at which the potential *changes* over distance.

---

5. **A.** Closer to the blob, V gets larger. Therefore, since EPE = qV, the potential energy of a positive charge gets higher as it approaches the blob. But charges "want" to *lower* their electric potential energy. In other words, a positive charge released from rest would "fall" *away* from the blob, because doing so would lower its EPE. Since like charges repel, this proves that the blob carries positive charge.

CHAPTER 20
Electric Circuits

PREVIEW

In this chapter you will be introduced to the basic concepts needed to analyze electric circuits. Current is defined and Ohm's law is used in simple circuits. The concepts of resistance and resistivity are examined in detail along with electric power, and dc and ac circuits.

This chapter also deals with circuits containing more than one device. The devices used in the circuits are resistors, voltage sources and capacitors. Methods of analyzing circuits containing devices connected in series and parallel by constructing equivalent circuits are investigated. In addition a general method of circuit analysis utilizing Kirchhoff's laws is presented. Kirchhoff's laws can be applied to any circuit, but are particularly useful when the devices cannot be regarded as being connected in either series or parallel.

The construction of ammeters, voltmeters, and ohmmeters is discussed along with how they are connected to a circuit to measure current, voltage and resistance.

QUICK REFERENCE

Important Terms

Electromotive force (emf)
The maximum electric potential difference (in volts) that exists between the terminals of an electric generator, such as a battery.

Electric current
The rate of flow of electric charge, $I = \Delta q/(\Delta t)$. The SI unit for current is the ampere (A).

Direct current (dc)
The electric current that results when charges flow in only one direction.

Alternating current (ac)
The electric current that results when the direction of charge flow changes from moment to moment.

Conventional current
The hypothetical flow of positive charges that would accomplish the same effect in the circuit as the movement of negative electrons that actually does occur in the wires.

Electrical resistance
The ratio of the voltage applied across a piece of material to the current that flows in that material. The SI unit of resistance is the ohm(Ω).

Ohm's law
The relation between current, voltage and a constant resistance for a material, $V = IR$.

Resistivity
The proportionality constant ρ for a material which relates the resistance R, the length of the material L, and the material's cross-sectional area A. That is, $R = \rho L/A$.

Electric power
The power P delivered to a circuit in which a current I results from a voltage V is $P = IV$.

Series wiring
Circuit devices are wired together so that each device MUST carry the same current.

Parallel wiring
Circuit devices are wired together so that each MUST have the same voltage drop.

Equivalent resistance
A single resistance which would demand the same current from the same source as does a network of resistors.

Internal resistance
The resistance to current flow through a voltage source such as a battery which arises from the resistance of the materials from which the source is made.

Terminal voltage
The actual voltage output of a voltage source such as battery. It is the emf of the source minus the voltage drop across the internal resistance of the source.

Equivalent capacitor
A single capacitor which would store the same charge or energy as does a network of capacitors.

Galvanometer
An electrical meter which contains a moving coil and is used as the basic component of analog ammeters and voltmeters.

Ammeter
A galvanometer in parallel with a shunt resistance. The ammeter measures current.

Voltmeter
A galvanometer in series with a large resistance. The voltmeter measures potential difference (voltage).

Time constant (τ)
The product of resistance and capacitance, RC. In a discharging series RC circuit the time constant represents the time needed for the charge on the capacitor to fall to 1/e (37 %) of its initial value at time zero. In a charging RC circuit the time constant represents the time required for the charge on the capacitor to rise to 63 % of its fully-charged value.

Equations

Electric current,

$$I = \frac{\Delta q}{\Delta t} \tag{20.1}$$

Ohm's law,

$$V = IR \tag{20.2}$$

Resistivity,

$$R = \rho\frac{L}{A} \tag{20.3}$$

Temperature coefficient of resistivity,

$$\rho = \rho_0[1 + \alpha(T - T_0)] \tag{20.4}$$

$$R = R_0[1 + \alpha(T - T_0)] \tag{20.5}$$

Electric power,

$$P = IV \tag{20.6}$$

For **alternating currents,**

$$V = V_0 \sin 2\pi ft \tag{20.7}$$

$$I = I_0 \sin 2\pi ft \tag{20.8}$$

$$P = I_0 V_0 \sin^2 2\pi ft \tag{20.9}$$

$$\overline{P} = \frac{1}{2} I_0 V_0 \tag{20.10}$$

$$\overline{P} = \left(\frac{I_0}{\sqrt{2}}\right)\left(\frac{V_0}{\sqrt{2}}\right) = I_{rms} V_{rms} \tag{20.11}$$

$$I_{rms} = \frac{I_0}{\sqrt{2}} \qquad (20.12)$$

$$V_{rms} = \frac{V_0}{\sqrt{2}} \qquad (20.13)$$

$$V_{rms} = I_{rms}R \qquad (20.14)$$

$$\bar{P} = I_{rms}V_{rms} \qquad (20.15a)$$

$$\bar{P} = I^2_{rms}R \qquad (20.15b)$$

$$\bar{P} = \frac{V^2_{rms}}{R} \qquad (20.15c)$$

Series Circuits

The **equivalent resistance** of a network of **resistors** connected in series is

$$R_S = R_1 + R_2 + R_3 + \ldots \qquad (20.16)$$

The **equivalent capacitance** of a network of **capacitors** connected in series is

$$\frac{1}{C_S} = \frac{1}{C_1} + \frac{1}{C_2} + \frac{1}{C_3} + \ldots \qquad (20.19)$$

Parallel Circuits

The **equivalent resistance** of a network of **resistors** connected in parallel is

$$\frac{1}{R_P} = \frac{1}{R_1} + \frac{1}{R_2} + \frac{1}{R_3} + \ldots \qquad (20.17)$$

The **equivalent capacitance** of a network of **capacitors** connected in parallel is

$$C_P = C_1 + C_2 + C_3 + \ldots \qquad (20.18)$$

RC Circuits

The **charge on the plates** of a **charging capacitor** is

$$q = q_0\left[1 - e^{-t(RC)}\right] \qquad (20.20)$$

The time constant of an RC circuit is

$$\tau = RC \qquad (20.21)$$

The **charge on the plates** of a **discharging capacitor** is

$$q = q_0 \, e^{-t/(RC)} \qquad (20.22)$$

Kirchhoff's Rules

Loop rule - the sum of potential drops equals the sum of potential rises around any closed loop.

Junction rule - the sum of the magnitudes of the currents entering a junction equals the sum of the magnitudes of currents leaving the junction.

DISCUSSION OF SELECTED SECTIONS

20.2 Ohm's Law

The current that a battery can "push" through a wire is analogous to the water flow through a pipe. Long and narrow pipes offer greater resistance to the moving water and lead to smaller flow rates. In the electrical case, the **resistance** (R) is defined as the ratio of the voltage V applied across a piece of the material to the current I through the material, or $R = V/I$. If the ratio V/I for a given material is constant over a wide range of voltages and currents, the resistance is constant. This relation is referred to as **Ohm's law**, after the German physicist Georg Simon Ohm, who discovered it.

Example 1
A resistor is connected to a 6.0-V battery. If the current flowing from the battery is 1.2 A, what is the resistance of the resistor?

The resistance of the resistor is given by equation (20.2),

$$R = \frac{V}{I} = \frac{6.0\ \text{V}}{1.2\ \text{A}} = 5.0\ \Omega$$

20.3 Resistance and Resistivity

In the case of the water pipe, the length and cross-sectional area of the pipe determine the resistance the pipe offers to the flow of water; the longer the pipe, and the smaller the cross-sectional area, the greater the resistance. In the electrical case, an analogous effect occurs. If L is the length of the material, and A its cross-sectional area, we have

$$R = \rho\frac{L}{A} \qquad (20.3)$$

where ρ is a proportionality constant known as the **resistivity** of the material. The resistivity is an inherent property of the material, while the resistance may vary depending on the geometry of the material.

The resistivity of the material depends on temperature. In metals, the resistivity normally increases with increasing temperature according to the relation

$$\rho = \rho_0[1 + \alpha(T - T_0)] \qquad (20.4)$$

In equation (20.4) ρ and ρ_0 are the resistivities at temperatures T and T_0, respectively. The term α is known as the **temperature coefficient of resistivity**.

Example 2

A 4.0 m long piece of copper wire having a diameter of 2.0 mm is heated from 20.0 °C to 352 °C. If the resistivity of the copper at 20.0 °C is $\rho_0 = 1.72 \times 10^{-8}\ \Omega$ m, and the temperature coefficient of resistivity is $3.93 \times 10^{-3} (°C)^{-1}$, what is the resistance of the wire at 352 °C?

Using equation (20.4) we can find the resistivity of the copper at 352 °C. We have,

$$\rho = \rho_0[1 + \alpha(T - T_0)]$$

$$\rho = (1.72 \times 10^{-8}\ \Omega\ \text{m})[1+(3.93 \times 10^{-3}/\ °C)(352\ °C - 20.0\ °C)] = 3.96 \times 10^{-8}\ \Omega\ \text{m}$$

The resistance of the wire can now be obtained using equation (20.3). Note that the cross-sectional area of the copper wire is $A = \pi r^2 = \pi(1.0 \times 10^{-3}\ \text{m})^2 = 3.1 \times 10^{-6}\ \text{m}^2$. Therefore, the resistance is

$$R = \rho\frac{L}{A} = \frac{(3.96 \times 10^{-8}\ \Omega\ \text{m})(4.0\ \text{m})}{3.1 \times 10^{-6}\ \text{m}^2} = 5.1 \times 10^{-2}\ \Omega$$

20.4 Electric Power

When there is a current I in a circuit as a result of a voltage V, the electric power delivered to the circuit is

$$P = IV \qquad (20.6a)$$

Using V = IR and substituting in equation (20.6), alternate expressions for P can be derived. These are

$$P = IV = I(IR) = I^2R \qquad (20.6b)$$

$$P = IV = \left(\frac{V}{R}\right)V = \frac{V^2}{R} \qquad (20.6c)$$

Example 3

A cigarette lighter in a car is just a resistor that, when activated, is connected across the 12-V battery. Suppose the lighter dissipates 33 W of power. Find (a) the resistance of the lighter and (b) the current that the lighter draws from the battery.

(a) Use equation (20.6c) to find the resistance,

$$R = V^2/P = (12\ \text{V})^2/(33\ \text{W}) = 4.4\ \Omega.$$

(b) The current, from equation (20.6a), is

$$I = P/V = (33\ \text{W})/(12\ \text{V}) = 2.8\ \text{A}.$$

20.5 Alternating Current

While many electric circuits involve direct currents (dc), considerably more operate with alternating current (ac), in which the direction of charge flow reverses itself periodically. The most common kind of ac generator produces a voltage that fluctuates sinusoidally between positive and negative values as time passes. The voltage that exists at time t can be expressed as

$$V = V_0 \sin 2\pi ft \qquad (20.7)$$

where V_0 is the maximum or peak value of the voltage, and f is the frequency at which the voltage oscillates. In the United States, the voltage present at most home wall outlets has a peak value of approximately 170 volts and oscillates with a frequency of 60 Hz. The current in an ac circuit also varies sinusoidally according to

$$I = I_0 \sin 2\pi ft \qquad (20.8)$$

where I_0 is the peak current in the circuit. The power delivered to the circuit by an ac generator is P = IV, so

$$P = I_0 V_0 \sin^2 2\pi ft \qquad (20.9)$$

Of more significance in an ac circuit is the average power of the circuit, which is just one-half the peak power,

$$\overline{P} = \frac{1}{2} I_0 V_0 \qquad (20.10)$$

The relationship between the peak power, I_0V_0, and the average power is shown in the following figure.

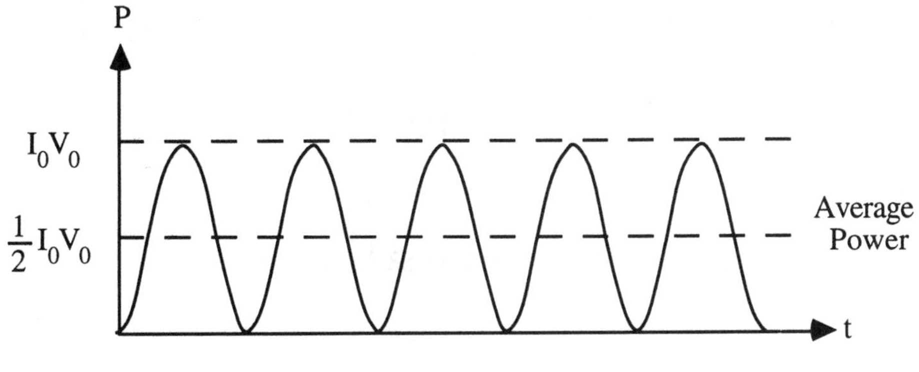

On the basis of the above expressions, we can rearrange terms to reveal that

$$\overline{P} = \left(\frac{I_0}{\sqrt{2}}\right)\left(\frac{V_0}{\sqrt{2}}\right) = I_{rms} V_{rms} \qquad (20.11)$$

where I_{rms} and V_{rms} are called the **root mean square** (rms) current and voltage, respectively. We see that

$$I_{rms} = \frac{I_0}{\sqrt{2}} \qquad (20.12) \qquad\qquad V_{rms} = \frac{V_0}{\sqrt{2}} \qquad (20.13)$$

So in the United States the maximum ac voltage is $V_0 = 170$ volts, and the corresponding rms voltage is $V_{rms} = 170\text{ volts}/\sqrt{2} = 120$ volts. Electrical appliances usually specify 120 V, so the rms voltage is indicated. In terms of the rms voltage, current, and power, the above equations may be written

$$V_{rms} = I_{rms} R \tag{20.14}$$

$$\overline{P} = I_{rms} V_{rms} \tag{25.15a}$$

$$\overline{P} = I_{rms}^2 R \tag{25.15b}$$

$$\overline{P} = \frac{V_{rms}^2}{R} \tag{25.15c}$$

Example 4
An ac generator applies a peak voltage of 75 V to an electrical device whose resistance is 22 Ω. Determine the (a) rms voltage, (b) rms current, (c) average power (d) and the peak power of this circuit.

(a) The peak value of the voltage is 75 V, so the corresponding rms value is

$$V_{rms} = \frac{V_0}{\sqrt{2}} = \frac{75\text{ V}}{\sqrt{2}} = 53\text{ V}$$

(b) The rms current is obtained from Ohm's law

$$I_{rms} = \frac{V_{rma}}{R} = \frac{53\text{ V}}{22\ \Omega} = 2.4\text{ A}$$

(c) The average power is therefore

$$\overline{P} = I_{rms} V_{rms} = (2.4\text{ A})(53\text{ V}) = 130\text{ W}$$

(d) The peak power is

$$P = I_0 V_0 = (I_{rms}\sqrt{2})(V_{rms}\sqrt{2}) = 2I_{rms}V_{rms} = 2(130\text{ W}) = 260\text{ W}$$

20.6 Series Wiring

The hallmark of devices connected in series is that the current through each device MUST be the same under all circumstances. Two or more devices in a circuit may happen to have the same current through them for certain values of resistance and applied voltage, but not be in series. In looking at a circuit and trying to determine whether the devices are in series or not, one may look for devices which are connected together at one end with no other devices being connected to that end. The entire current through one device has no other choice than to go through the second device.

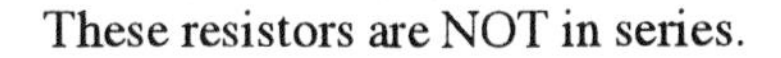

These resistors are in series. These resistors are NOT in series.

The equivalent resistance of a network of series resistors may be found by simply adding the individual resistances according to equation (20.16). Note that the equivalent resistance is ALWAYS larger than the largest series resistor.

Example 5
What is the equivalent resistance of a 15.0 Ω resistor and a 26.0 Ω resistor connected in series? How much current would a 12.0 V battery deliver to the resistor network, and what is the voltage drop across each resistor?
The equivalent resistance of the two resistors is

$$R_S = 15.0\ \Omega + 26.0\ \Omega = 41.0\ \Omega.$$

The battery "sees" only the equivalent resistance. Ohm's law then requires that the battery send a current

$$I = (12.0\ V)/(41.0\ \Omega) = 0.293\ A.$$

The current through each resistor is the same and equal to the current supplied by the battery. Ohm's law gives for the 15 Ω resistor

$$V_{15} = (0.293\ A)(15.0\ w) = 4.4\ V$$

and for the 26 Ω resistor

$$V_{26} = (0.293\ A)(26.0\ \Omega) = 7.6\ V$$

Note that the voltage drops add to 12.0 V, the emf of the battery.

Example 6
The resistor bin in an electronics lab has a quantity of the following values of resistors 47 Ω, 68 Ω and 225 Ω. A resistor is needed which has a value of 1500 Ω. Which of the resistors (all of the same value connected in series) could you use to do the job?

The equivalent resistance of N resistors each of resistance R connected in series is R_S = NR. We want R_S = 1500 Ω to within significant figures. We need

$$N = (1500\ \Omega)/(47\ \Omega) = 32$$

of the 47 Ω resistors and

$$N = (1500\ \Omega)/(68\ \Omega) = 22$$

of the 68 Ω resistors.

We cannot use the 225 Ω resistors at all since we need $N = (1500\ \Omega)/(225\ \Omega) = 6.7$ of them. Using 6 would give a resistance of $6(225\ \Omega) = 1350\ \Omega$ which is too little and using 7 would give $7(225\ \Omega) = 1575\ \Omega$ which is too much.

20.7 Parallel Wiring

The hallmark of parallel wiring is that the voltage across each of the devices connected in parallel MUST be the same under all circumstances. We recognize that devices may happen to have the same voltage across them under special circumstances, but not be wired in parallel.

In order to be in parallel BOTH ends of the devices must be connected directly together with no other devices inserted between them.

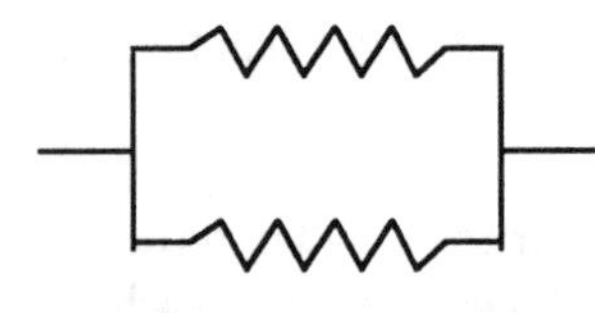

These resistors are in parallel.

NONE of these resistors are in parallel.

The equivalent resistance of a network of resistors connected in parallel can be found from equation (20.17). Note that the equivalent resistance is ALWAYS less than the smallest parallel resistor.

Example 7
A 47.0 Ω and a 33.0 Ω resistor are connected in parallel. What is the equivalent resistance of the resistors? How much current would a 12.0 V battery supply to the network and how much current would flow through each resistor?

Equation (20.17) gives

$$\frac{1}{R_p} = \frac{1}{47.0\ \Omega} + \frac{1}{33.0\ \Omega}$$

which gives

$$R_p = 19.4\ \Omega.$$

Note that this is less than either of the resistor values.

The battery "sees" only the equivalent resistance and will supply a current consistent with Ohm's law of

$$I = (12.0\ \text{V})/(19.4\ \Omega) = 0.619\ \text{A}.$$

The voltage across each of the resistors is the same and equal to the emf supplied by the battery. Ohm's law applied to the 47.0 Ω resistor gives

$$I_{47} = (12.0\ \text{V})/(47.0\ \Omega) = 0.255\ \text{A}$$

and to the 33.0 Ω resistor gives

$$I_{33} = (12.0\ \text{V})/(33.0\ \Omega) = 0.364\ \text{A}.$$

The currents through each of the resistors total to equal the current supplied by the battery.

Example 8
You have on hand several resistors of each of the following values: 24 Ω, 36 Ω and 6.0 Ω. How could you use these (all of one value) in a parallel circuit to make an equivalent resistance of 8.0 Ω?

First we see that the 6.0 Ω resistors are useless in making an 8.0 Ω resistance since any parallel combination of resistors with a 6.0 Ω resistor will result in an equivalent resistance of LESS than 6.0 Ω.

If we wish to use a network of all one value, we would need N resistors each of value R to give an equivalent $R_P = R/N$. The number of resistors needed is then $N = R/R_P$. We would need N = (24 Ω)/(8.0 Ω) = 3 of the 8.0 Ω resistors or N = (36 Ω)/(8.0 Ω) = 4.5 of the 36 Ω resistors. Four of these gives an equivalent resistance of 9.0 Ω which is too much while five gives 7.2 Ω which is too little. We would have to use three 24 Ω resistors.

20.8 Circuits Wired Partially in Series and Partially in Parallel

Some circuits can be analyzed by reducing the circuit to an equivalent one by combining series resistors and parallel resistors using the previous methods. This is feasible ONLY if you can identify the devices as being wired in series and/or wired in parallel. Otherwise you will need to use the Kirchhoff rules given later in this chapter.

Example 9
Identify the series and parallel combinations of resistors in the circuit diagram. Reduce the circuit to an equivalent resistance and find the current supplied by the battery.

The two 4.0 Ω resistors are in series with each other and both together are in parallel with the 8.0 Ω resistor. The remaining resistors are in series with each other and with the network containing the two 4.0 Ω and the 8.0 Ω resistors.

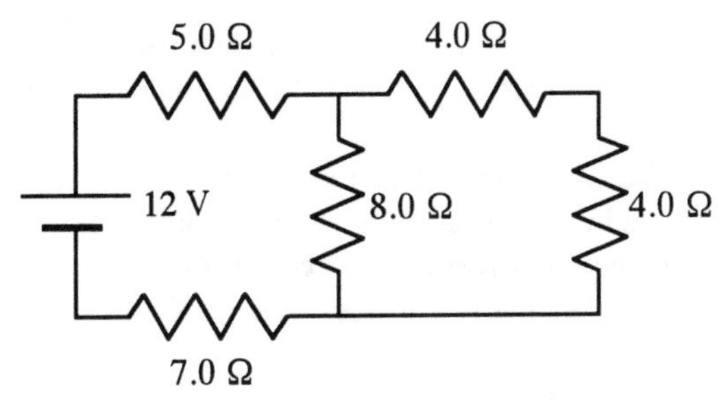

First reduce the circuit by combining the two 4.0 Ω resistors into a 4.0 Ω + 4.0 Ω = 8.0 Ω equivalent.

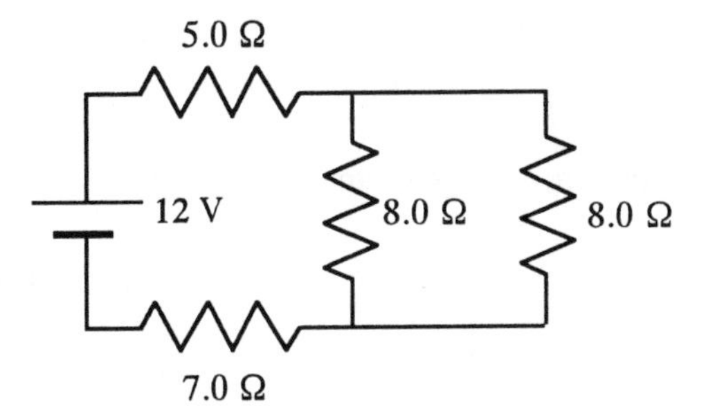

Next reduce the resulting parallel combination of the two 8.0 Ω resistances to an equivalent of 4.0 Ω.

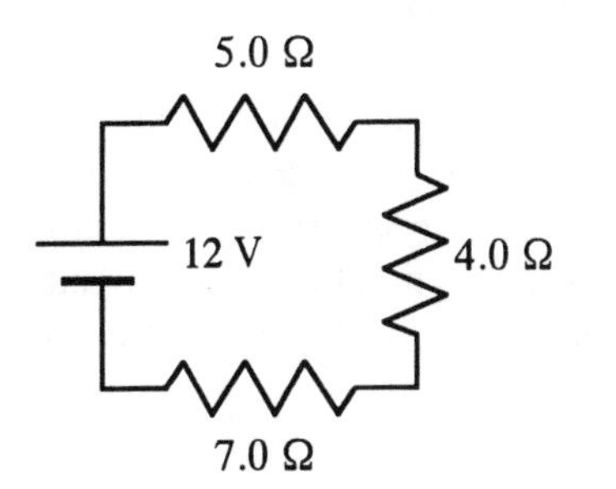

Now the three series resistances can be added to give the final equivalent 5.0 Ω + 4.0 Ω + 7.0 Ω = 16.0 Ω

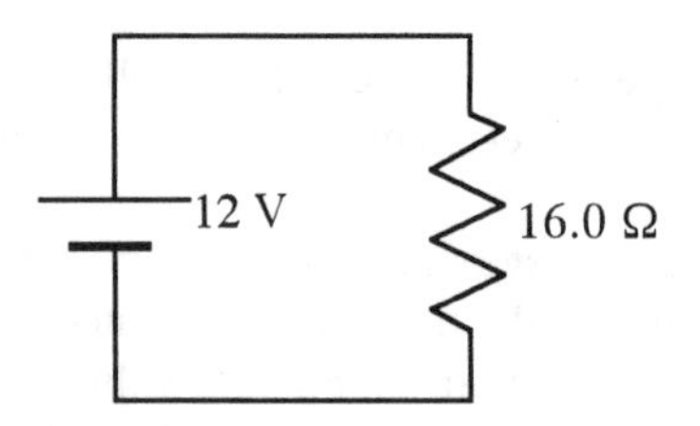

The battery "sees" only the final equivalent resistance and will provide a current consistent with Ohm's law.

$$I = (12\ \text{V})/(16.0\ \Omega) = 0.75\ \text{A}$$

20.9 Internal Resistance

All real voltage sources such as generators and batteries have their own "internal resistance" since they contain wires, devices, plates, etc which have resistance. Whenever a voltage source is required to deliver current, a voltage drop (due to Ohm's law) will appear across this internal resistance and reduce the terminal voltage or output of the source. A real voltage source can be thought of as an ideal voltage source in series with the internal resistance.

Example 10
A laboratory signal generator has an internal resistance of 600 Ω and delivers 0.050 A of current to a load circuit whose equivalent resistance is also 600 Ω. What is the emf of the generator? What is the voltage drop across the generator?
The emf of the generator is the terminal voltage plus the drop across its internal resistance, r. The terminal voltage is just the voltage drop across the load circuit since the generator and load circuit are connected in parallel.

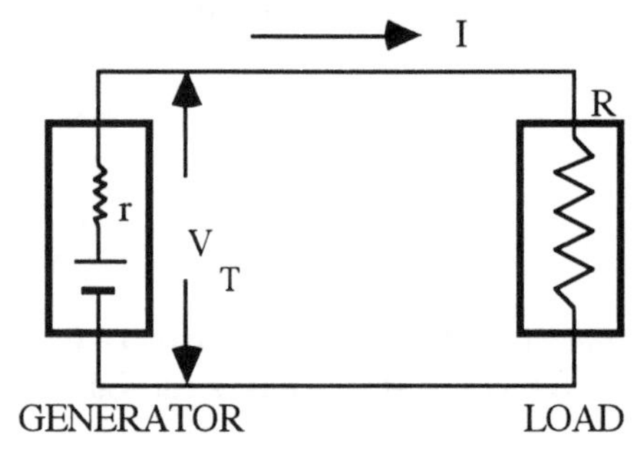

$$\text{emf} = V_T + Ir = I(R + r)$$

$$\text{emf} = (0.050\ \text{A})(600\ \Omega + 600\ \Omega) = 60\ \text{V}$$

The voltage drop across the generator is

$$Ir = (0.050\ \text{A})(600\ \Omega) = 30\ \text{V}$$

20.10 Kirchhoff's Rules

The Kirchhoff rules are basically statements of two conservation laws. The loop rule is simply a statement of the conservation of energy for charges going around a closed loop in a circuit, while the junction rule is a statement of the conservation of charge entering and leaving a junction.

Kirchhoff's rules are applicable to ALL circuits and must be used when the circuit in question cannot be reduced to an equivalent circuit by identifying parallel and series combinations of devices.

The method of applying Kirchhoff's rules to a circuit is basically the same for all circuits and involves the following steps.

1. Identify the independent loops and junctions in the circuit.
2. Choose current directions for the currents flowing in the circuit. Your choice is arbitrary. If you choose wrong, Kirchhoff's laws will tell you.
3. Pick a loop and proceed clockwise around the loop adding up potential drops and adding up potential rises. Set the drops equal to the rises. Repeat for each independent loop in the circuit. You should now have one equation for each loop.
4. Choose a junction and apply the junction rule by setting the total current into the junction equal to the total current out of the junction. Repeat for each independent junction to obtain an equation for each junction.
5. Solve the loop equations and junction equations simultaneously for the unknown quantities. You can find one unknown quantity for each of the equations that you have written.

Example 11
Find each of the three currents indicated in the circuit diagram.

This circuit is a good candidate for Kirchhoff's rules since none of the devices can be identified as being wired in series or parallel.

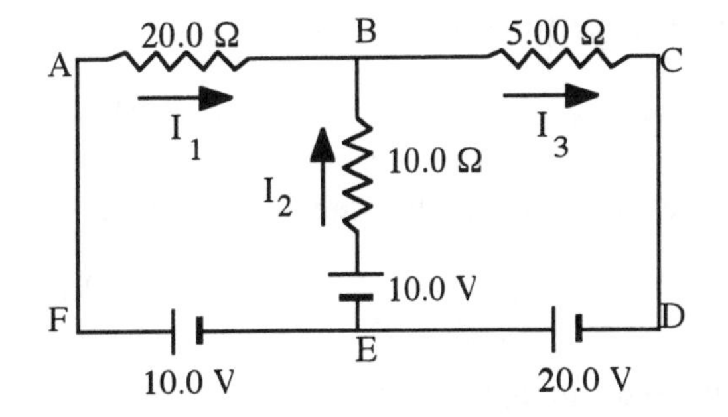

Step 1: The circuit has three possible loops ABEF, BCDE, and ACDF (only two will be independent) and two possible junctions B and E (only one will be independent).

Step 2: The current directions have already been chosen. We are not guaranteed that they are correct, however.

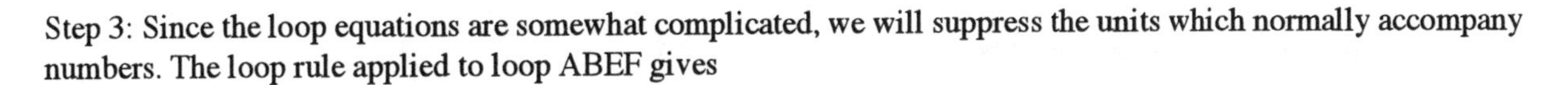

Step 3: Since the loop equations are somewhat complicated, we will suppress the units which normally accompany numbers. The loop rule applied to loop ABEF gives

$$(20.0)I_1 + 10.0 = (10.0)I_2 + 10.0$$

or

$$(20.0)I_1 - (10.0)I_2 = 0 \qquad (1)$$

The loop rule applied to the loop BCDE gives

$$(5.00)I_3 + (10.0)I_2 = 20.0 + 10.0$$

or

$$(10.0)I_2 + (5.00)I_3 = 30.0 \qquad (2)$$

The loop rule applied to loop ACDF gives

$$(20.0)I_1 + (5.00)I_3 = 20.0 + 10.0$$

or

$$(20.0)I_1 + (5.00)I_3 = 30.0$$

NOTE: The last equation can be derived by adding equations (1) and (2). Only two of the equations are independent and can be used. Which two? It's your choice.

Step 4: The junction rule applied at B gives

$$I_1 + I_2 = I_3$$

or

$$I_1 + I_2 - I_3 = 0 \qquad (3)$$

The junction rule applied at E gives

$$I_3 = I_1 + I_2$$

or

$$I_1 + I_2 - I_3 = 0$$

NOTE: The second junction equation is identical to the first, therefore only one junction equation can be used.

Step 5: This step is pure algebra. Use your favorite technique to solve the three equations simultaneously for the currents.
Add (1) and (2) to obtain

$$(20.0)I_1 + (5.00)I_3 = 30.0 \qquad (4)$$

Multiply (3) by 10 and add the result to (1) to get

$$(30.0)I_1 - (10.0)I_3 = 0 \qquad (5)$$

Multiply (4) by 2 and add the result to (5) and find

$$(70.0)I_1 = 60.0 \text{ or } I_1 = +\,0.857 \text{ A}$$

Use this result in (5) to find

$$I_3 = +\,2.57 \text{ A}.$$

Equation (3) then gives the last current to be

$$I_2 = +\,1.71 \text{ A}.$$

All of the currents are positive, indicating that we guessed the correct directions.

20.11 The Measurement of Current and Voltage

The moving coil galvanometer is the heart of analog voltmeters and ammeters. The galvanometer is capable of measuring small currents flowing through its coil.

The key to constructing an ammeter is to allow only a small known fraction of the current to be measured to flow through the galvanometer. This is accomplished by "shunting" the bulk of the current around the galvanometer via a shunt resistor placed in parallel with the galvanometer. The ammeter constructed this way must be INSERTED INTO the circuit so that the current can flow through it. Since the ammeter contributes a series resistance to the circuit, its very presence changes the current in the circuit. It is desired that the equivalent resistance of the ammeter be as small as possible so that it affects the current to be measured in only a small way.

Example 12
A perfect 9.0 V battery is connected to a 1800 Ω load resistance so that 5.0 mA flows in the circuit. An ammeter with a resistance of 100 Ω is inserted in the circuit and used to measure the current. What error does the ammeter introduce in the measurement?

When the ammeter is inserted, the battery "sees" a new equivalent resistance of

$$1800\ \Omega + 100\ \Omega = 1900\ \Omega.$$

Ohm's law then demands that the battery produce a current of

$$(9.0\ V)/(1900\ \Omega) = 4.7\ mA$$

which is the current that the ammeter will register. The error introduced by the ammeter is then

$$(100\%)(5.0\ mA - 4.7\ mA)/(5.0\ mA) = 6\ \%.$$

The voltmeter is constructed by placing a large resistor value in series with the galvanometer. When the voltmeter is connected ACROSS the device whose voltage is to be measured, a small current will flow through the galvanometer and cause a deflection of the pointer. This current, hence the deflection of the pointer, is proportional to the voltage across the voltmeter. Again, the introduction of the meter changes the very thing it is supposed to measure. To ensure accuracy, the resistance of the voltmeter is usually very high so that little current is "drawn" by the voltmeter.

Example 13
A 1.0 MΩ voltmeter is used to measure the voltage drop across one of two 25 kΩ resistors connected in series to a perfect 12 V battery. How much error in the measured voltage is introduced by the voltmeter?

Before the voltmeter is connected the voltage drop across each resistor is just 6.0 V, since the resistors are identical and connected in series. After the voltmeter is connected in parallel with one resistor, the battery "sees" a new equivalent resistance of

$$R_S = 25\ k\Omega + R_P$$

where R_P is given by

$$1/R_P = 1/(25 \times 10^3\ \Omega) + 1/(1.0 \times 10^6\ \Omega)$$

so $R_P = 24\ k\Omega$ and $R_S = 49\ k\Omega$.

The battery now delivers a new current to the circuit of

$$(12\ V)/(49 \times 10^3\ \Omega) = 0.24\ mA.$$

The new voltage drop across the resistor-voltmeter combination is then

$$IR_P = (0.24 \times 10^{-3}\ A)(24 \times 10^3\ \Omega) = 5.8\ V.$$

The error is

$$(100\ \%)(6.0\ V - 5.8\ V)/6.0\ V) = 3\ \%.$$

20.12 Capacitors in Series and Parallel

The magnitude of the charges on the plates of capacitors connected in series MUST be the same due to the conservation of charge. The voltages across the capacitors may be different if the capacitors have different capacitances. An equivalent capacitance can be defined for series capacitors and is given by (20.19).

The voltage across capacitors connected in parallel MUST be the same. This allows different charges to be on the capacitors if their capacitances are not equal. An equivalent capacitance for parallel capacitors is given by (20.18).

Circuit analysis for capacitors is very similar to the analysis for resistors. Often you can reduce a circuit containing parallel and series capacitors to a circuit containing one equivalent capacitor. If this cannot be done, you may use the Kirchhoff loop rule along with the principle of conservation of charge.

Example 14
Find the equivalent capacitance, the charge separated by the battery and the charge on each capacitor in the circuit shown.

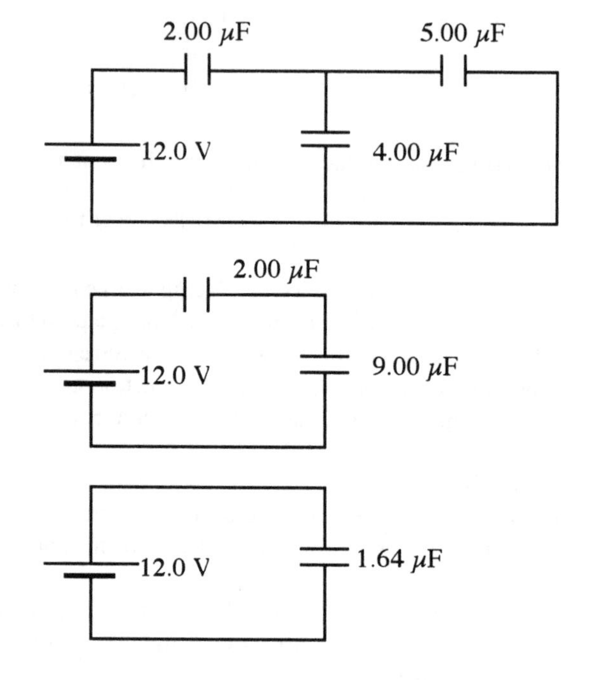

The 5.00 μF and the 4.00 μF capacitors are in parallel and have an equivalent capacitance of

$$C_P = 5.00\ \mu F + 4.00\ \mu F = 9.00\ \mu F.$$

This combination is in series with the 2.00 μF capacitor so the final equivalent capacitance of the circuit is found from

$$1/C_S = 1/(2.00\ \mu F) + 1/(9.00\ \mu F)$$

and is $C_S = 1.64\ \mu F$.

The battery "sees" only the equivalent capacitance and separates an amount of charge

$$q = (1.64 \times 10^{-6}\ F)(12.0\ V)$$
$$q = 1.97 \times 10^{-5}\ C.$$

Now the 2.00 μF capacitor and the 9.00 μF equivalent capacitance are in series and must have the same charge on their plates. That is, $1.97 \times 10^{-5}\ C = 19.7\ \mu C$

The 5.00 μF and 4.00 μF share the 19.7 μC in such a way that the voltage across each of them is the same.

$$V_4 = V_5$$

or

$$q_4/q_5 = C_4/C_5 = 0.800.$$

We also know that

$$q_4 + q_5 = 19.7\ \mu C$$

so

$$q_5 = 10.9\ \mu C \text{ and } q_4 = 8.8\ \mu C.$$

20.13 RC Circuits

Example 15
What fraction of charge is lost by a discharging capacitor during the first five time constants?

Let the charge on the capacitor at time = 0 s be $q_0 = CV_0$. Equation (20.22) gives for $t = 5\tau$

$$q = q_0 e^{-t/(RC)} = q_0 e^{-5} = q_0 (6.7 \times 10^{-3})$$

The fraction of the initial charge remaining after 5τ is q/q_0 so that the fraction lost is

$$1 - q/q_0 = 1 - (6.7 \times 10^{-3}) = 0.993.$$

PRACTICE PROBLEMS

1. A 75-W light bulb draws a current of 0.80 A. If the light bulb stays on for 3.0 hours (a) how much charge flows through the light bulb? (b) How many electrons have passed through the bulb?

2. A certain material used for a wire is 25 m long and has a diameter of 3.0 mm. The wire carries a 2.5 A current when a 12-V potential difference is applied between its ends. What is the resistivity of the wire?

3. At what temperature would the resistance of a silver conductor be double its resistance at 20.0 °C? Take the temperature coefficient of resistivity for silver to be 3.80×10^{-3} $(C°)^{-1}$.

4. A wire whose resistance is 12.0 Ω is melted down. From the melted material a new wire is made that is 1/4 the diameter of the original wire. What is the resistance of the new wire?

5. An electric alarm clock uses a 5.0-W motor and runs all day, every day. If electricity costs $0.10 per kWh, determine the yearly cost of running the clock.

6. An ac generator produces a voltage given by V = 283 sin 188t, where V is in volts and t in seconds. (a) What is the peak voltage produced by the generator? (b) What is its frequency? (c) If the voltage supplied by the generator produces an rms current of 2.0 A in a device in the circuit, what is the resistance of this device?

7. A device is rated at 750 watts when used with a 120-V ac line. (a) What peak current flows through it? (b) What peak current would flow through it if connected to a 240-V ac line? (c) In which case is more power utilized? (Assume the resistance of the device is constant).

8. Find the equivalent resistance of the circuit shown.

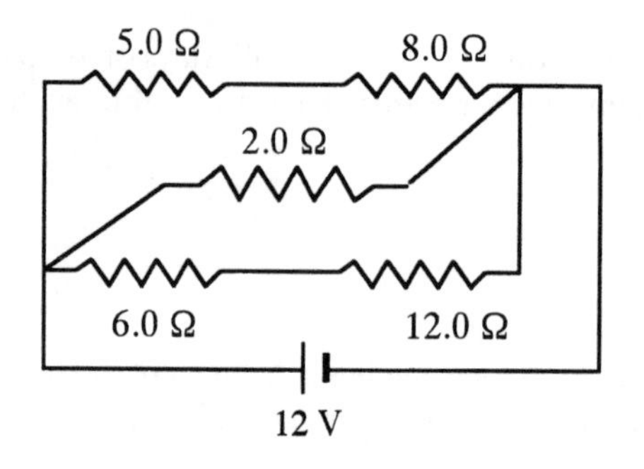

9. A battery with an emf of 12.0 V is used to turn the starter motor on a car. When the starter motor is engaged, the battery applies only 8.2 V to the motor while it delivers a current of 120 A. What is the internal resistance of the battery and the resistance of the starter motor?

10. A galvanometer whose full scale current is 150.0 μA and coil resistance is 75.0 Ω is used with a 1.50 Ω shunt resistor to make an ammeter. What is the largest current which can be measured by the ammeter?

11. An ammeter whose resistance is 9.0 Ω is inserted into a circuit containing a 95 Ω resistor and a 9.0 V battery. What percentage error does the ammeter introduce into the current it measures?

12. A galvanometer whose full scale current is 150 μA and coil resistance is 75 Ω is used with a 25 KΩ series resistor to make a voltmeter. What is the largest voltage which can be measured by the voltmeter?

13. Apply Kirchhoff's rules to find the currents in the circuit shown.

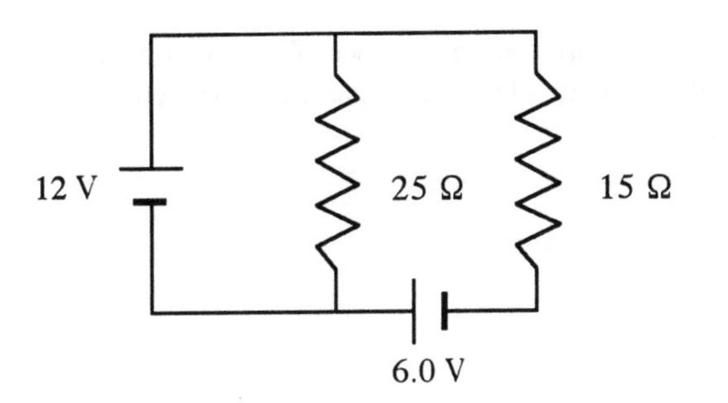

14. Find the current through each of the resistors in the diagram shown. Also, find the current delivered by the battery.

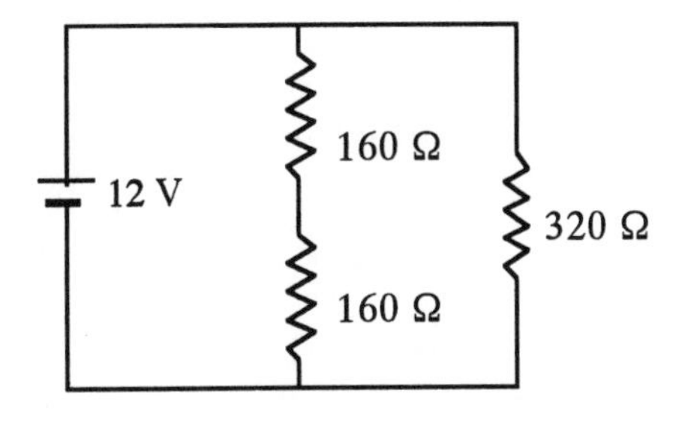

15. Find the equivalent capacitance of the circuit shown in the diagram. How much charge is separated by the battery?

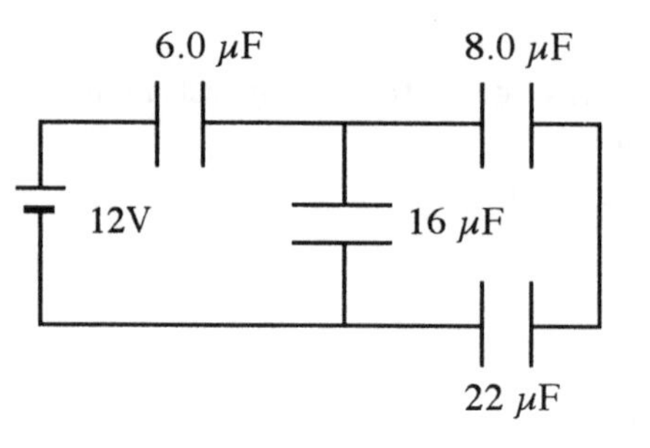

16. A 1500 Ω resistor a 25 μF capacitor are connected together to form a circuit. The capacitor has been previously charged to a voltage of 35 V. How long will it take for the voltage on the capacitor to drop to 15 V?

17. A 1500 μF capacitor is charged through a 25 kΩ resistor by a 12 V battery. How long does it take for the battery to charge the capacitor to a voltage of 8.0 V?

HELPFUL SUGGESTIONS

1. The direction of the conventional current through a circuit is defined as the direction of flow of **positive** charge. This current flows from higher to lower potential.

2. The resistivity ρ of a material is an intrinsic property of that material, and does NOT depend on the geometry (i.e., the length and cross-sectional area) of the material. However, the resistance R does depend on L and A.

3. Although we call electric utility companies "power companies", we buy energy from them, not power. Power is the rate of using energy; the energy used divided by the time it was used. The bills we receive are for kilowatt-hours, which is, in fact, a unit of energy. 1 kW-h = 3.6×10^6 J.

4. In discussing ac circuits, it is common to use rms values rather than peak values of voltage and current. This is the case because it is generally the **average power** utilization that is of interest. Remember that the rms value of voltage or current is the peak value divided by $\sqrt{2}$.

5. The resistance of a material changes with temperature according to equation (20.5). In general, the resistance of a material increases with temperature (i.e., the temperature coefficient of resistivity α is positive). However, in certain materials, called **semiconductors**, the resistance decreases with increasing temperature (hence α is negative).

6. Keep in mind that the currents through series devices MUST be the same under all circumstances, while the voltages across parallel devices MUST be the same.

7. When you are confronted with a circuit problem first check to see which devices are connected in series and which are connected in parallel. It may be advantageous to reduce these series and parallel combinations to their equivalents and apply Ohm's law.

8. If a circuit cannot be regarded as series and parallel combinations of devices, then Kirchhoff's rules are in order. They work for any circuit. Follow the steps outlined in the discussion to generate a set of equations. Make sure that the ones you eventually use are independent.

9. When using the Kirchhoff rules you may choose the current directions for your convenience. If you have chosen wrong, the Kirchhoff equations will tell you by making that current negative.

10. You MUST use a consistent direction of travel around loops when writing the equations for the loop rule. A good choice is to always traverse a loop clockwise. If you cross a resistor in the assumed current direction or cross a battery from + to -, then the potential difference will be a drop. If you cross a resistor against the assumed current direction or a battery from - to + then the potential difference will be a rise.

11. A convenient way to evaluate expressions like $1/R_P = 1/(5.0\ \Omega) + 1/\ 2.0\ \Omega) + 1/(8.0\ \Omega)$ is NOT to find a common denominator and combine the fractions, but to use the reciprocal function (1/x) of the calculator.

12. The capacitance of a capacitor depends ONLY on it geometry and the material between its plates. It does NOT depend on how much voltage is across it or the amount of charge on its plates.

13. The time constant of a discharging RC circuit is the amount of time required for the charge (and voltage) on the capacitor to fall from its original value to 1/e = (0.368) of that value. This is true regardless of the value of the original charge or voltage.

EVERYDAY PHYSICS

1. Calculate how much energy (in kilowatt hours) you use in an average day around your home. You will need to find out the power consumption (the "wattage") of each device used; it's usually found somewhere on the device. For example, a 50-W light bulb consumes 50 J of energy each second. If only the current rating of the device is given, you can find the power from $P = IV$, where V is the rms voltage (normally 120 V). To determine the total energy consumption, you'll also need to estimate how long each appliance is used during a typical 24 hour period. Find out the cost per kW-h charged by your electric company. Does your estimate match your utility bill? You can check your estimates by reading the electric meter.

2. Current through your body causes electric shocks. It is the **current** that produces the electric shock, not the voltage as such. The voltage, of course, drives the current through your body, but no current will result unless your body is part of a complete circuit. This means that different parts of your body must be at different potentials, since the current flows from higher to lower potential. So, for example, if you are standing on a dry insulated surface (like a wooden floor or a carpet) and you touch a frayed electrical cord plugged into the wall outlet, you won't get a shock. This is because your entire body is at a potential difference of 120 volts, so no current will flow. However, if you touched a grounded electrical appliance while also touching the frayed electrical cord, you would receive a substantial shock. The grounded electrical appliance is at zero voltage, so a 120 volt potential difference would exist between the two parts of your body, a current would flow, and **ZAP**.

3. The two types of Christmas tree lights provide a good lesson in the advantages and disadvantages of series and parallel wiring. A 120 V series string of lights of 100 bulbs requires that the voltage drop across each bulb be 1.2 V. Low voltage bulbs can be used in this kind of string. The problem is that the total current in the circuit must pass through each bulb. If one bulb burns out, no current at all flows through the circuit and all of the bulbs are out. The above problem can be avoided by wiring the bulbs in parallel. This method requires that each bulb drop 120 V, hence high voltage bulbs must be used. However, if one bulb burns out, the rest will stay on since the circuit is still complete. Most series connected bulb sets sold now use bulbs which contain a shunt which keeps the current flowing even if the bulb is burned out as long as it is still in the circuit. Low voltage bulbs can still be used in this string.

4. A "dead" battery can be checked by using a voltmeter and appear to be all right. A voltmeter is a very high resistance device and "draws' little current from the battery. The potential drop (Ir) across the internal resistance of the battery is very small due to the small current even though the internal resistance of the battery is high enough to qualify it to be "dead". A battery tester, such as those found in department stores, always connect a load resistor to the battery so that reasonably large currents flow from the battery. Then if the internal resistance of the battery is in the "dead" range, the terminal voltage of the battery will be low and will produce a low reading on the voltmeter.

CHAPTER QUIZ

1. A wire carries a current of 5.0 A. How many electrons cross a given area of the wire each second?
   a. 5 b. 5.0×10^6 c. 3.1×10^{19} d. 8.0×10^{19}

2. A battery charger delivers a current of 2.0 A for 6.0 hours to a 12-V battery. What is the total charge that passes through the battery?
   a. 12 C b. 4.3×10^4 C c. 8.6×10^4 C d. 5.2×10^5 C

3. A 24-Ω resistor is connected to a 12-V battery. How much current flows through the resistor?
   a. 0.50 A b. 2.0 A c. 6.0 A d. 288 A

4. A 24-Ω resistor is connected to a 12-V battery. How much power is used by the resistor?
   a. 0.5 W b. 2.0 W c. 6.0 W d. 288 W

5. Consider two silver wires. One has twice the length and twice the cross-sectional area of the other. How do the resistances of the two wires compare?
   a. Both wires have the same resistance.
   b. The longer wire has twice the resistance of the shorter wire.
   c. The longer wire has four times the resistance of the shorter wire.
   d. The shorter wire has the greater resistance.

6. If the resistance in a constant voltage circuit is cut in half, the power dissipated by that circuit
   a. is cut in half. c. is quadrupled.
   b. is doubled. d. remains unchanged.

7. The resistance of a copper wire (resistivity $\rho = 1.7 \times 10^{-8}$ Ω m) is 25 Ω. If the wire's cross-sectional area is 3.4×10^{-6} m^2, what is the length of the wire?
   a. 2.0×10^{-4} m b. 8.0 m c. 2.0×10^2 m d. 5.0×10^3 m

8. The peak voltage delivered by an electric generator is 277 V. The rms voltage delivered is
   a. 196 V b. 277 V c. 392 V d. 554 V

9. The peak voltage delivered to a circuit is 220 V and the peak current which results is 3.0 A. The average power delivered to the circuit is
   a. 660 W b. 73 W c. 330 W d. 220 W

10. A generator delivers an ac voltage given by V = 120 sin 553t. The peak voltage of the generator is
    a. 120 V b. 553 V c. 88 V d. 4.4 V

11. A generator delivers an ac voltage given by V = 120 sin 553t. The frequency of the generator is
    a. 120 Hz b. 553 Hz c. 88 Hz d. 4.4 Hz

12. A generator delivers an ac voltage given by V = 120 sin 553t, producing an rms current of 2.0 A. The average power delivered to the circuit is
    a. 240 W b. 1100 W c. 60 W d. 170 W

13. Three resistors, 5.0 Ω, 255 Ω and 55 Ω, are connected in series with each other and in series with a 12 battery. How much current flows from the battery?
    a. 2.7 A b. 38 mA c. 47 mA d. 0.21 A

14. Two resistors, 4.0 Ω and 8.0 Ω are connected in series to a 9.0 V battery. How much power is dissipated in the 8.0 Ω resistor?
    a. 4.5 W b. 2.3 W c. 6.8 W d. 110 W

15. What is the equivalent resistance of a parallel combination of 12 Ω, 6.0 Ω and a 25 Ω resistor?
 a. 0.26 Ω b. 11 Ω c. 43 Ω d. 3.4 Ω

16. A large number of 150 Ω, 220 Ω and 47 Ω resistors are on hand in an electronics lab. Which of the resistor values could not possibly be used in a parallel combination to produce a 53 Ω resistor?
 a. 150 Ω b. 220 Ω c. 47 Ω d. They could ALL be used.

17. A 120 V string of Christmas lights contains fifty identical bulbs in series. The entire string consumes 95 W of power. What is the resistance of each bulb?
 a. 150 Ω b. 3.0 Ω c. 0.20 Ω d. 25 Ω

18. A circuit loop, part of a larger circuit, contains a 5.0 Ω resistor, a 25 Ω resistor and a battery. The current through the 5.0 Ω resistor is 0.50 A in a clockwise sense and the current through the 25 Ω resistor is 0.75 A in a counterclockwise sense. What is the voltage of the battery?
 a. 13 V b. 9.4 V c. 21 V d. 16 V

19. Two capacitors, 47 μF and 33 μF, are connected in parallel to a 5.0 V voltage source. How much energy is stored by the fully charged 47 μF capacitor?
 a. 1.7×10^{-3} J b. 1.0×10^{-4} J c. 5.9×10^{-4} J d. 4.1×10^{-4} J

20. A/an ________ must be inserted INTO the circuit while a/an________ can be attached to the circuit.
 a. voltmeter, ammeter b. voltmeter, ohmmeter c. ammeter, voltmeter d. voltmeter, galvanometer

21. The voltage across a capacitor in a discharging RC circuit at some time is 3.5 V. What is the voltage across the capacitor one time constant later?
 a. 1.3 V b. 2.3 mV c. 0.47 V d. 3.1 V

22. What is the current through the 36 Ω resistor shown in the drawing?
 a. 0.25 A c. 0.10 A
 b. 0.33 A d. 0.17 A

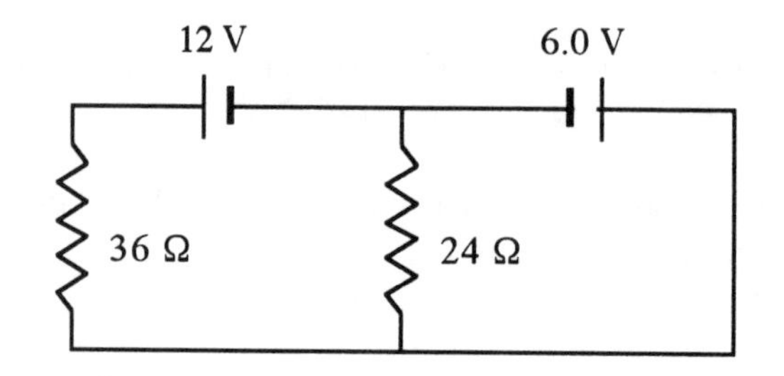

SOLUTIONS AND ANSWERS

Practice Problems

---

1. (a) Using equation (20.1),

$$\Delta q = I\Delta t = (0.80 \text{ C/s})(3 \text{ h})(3600 \text{ s/1 h}) = \mathbf{8.6 \times 10^3\ C}.$$

(b) We know,

$$N = (\Delta q)/e = (8.6 \times 10^3 \text{ C})/(1.6 \times 10^{-19} \text{ C}) = \mathbf{5.4 \times 10^{22}\ electrons}.$$

---

2. Equation (20.2) gives,

$$R = V/I = (12 \text{ V})/(2.5 \text{ A}) = 4.8\ \Omega$$

The area is

$$A = \pi r^2 = \pi(1.5 \times 10^{-3} \text{ m})^2 = 7.1 \times 10^{-6} \text{ m}^2.$$

Using equation (20.3), $R = \rho L/A$ gives

$$\rho = RA/L = (4.8\ \Omega)(7.1 \times 10^{-6} \text{ m}^2)/(25 \text{ m}) = \mathbf{1.4 \times 10^{-6}\ \Omega\ m}.$$

---

3. Using equation (20.5) we have,

$$R = R_0[1 + \alpha(T - T_0)].$$

In this case we have $R = 2R_0$, so that

$$R/R_0 = 2 = 1 + \alpha(T - T_0)$$

or

$$T = T_0 + 1/\alpha = 20.0\ °\text{C} + 1/(3.80 \times 10^{-3}/\text{C}°) = \mathbf{283\ °C}.$$

---

4. The volume of the wire is a constant so $A_1L_1 = A_2L_2$ gives

$$L_2/L_1 = A_1/A_2.$$

Now use equation (20.3) to get,

$$R_2/R_1 = (\rho L_2/A_2)/(\rho L_1/A_1) = (L_2/L_1)(A_1/A_2) = A_1^2/A_2^2.$$

Since $A = \pi d^2/4$, we have

$$R_2 = R_1(A_1^2/A_2^2) = R_1(d_1/d_2)^4 = (12.0\ \Omega)(4)^4 = \mathbf{3070\ \Omega}.$$

---

5. Since power is $P = E/t$, then

$$E = Pt = (5.0 \text{ J/s})(365 \text{ days})(24 \text{ h/1 day})(3600 \text{ s/1 h}) = 1.6 \times 10^8 \text{ J}.$$

So the cost is,

$$\text{Cost} = (\$0.10/1 \text{ kWh})(1 \text{ kWh}/3.6 \times 10^6 \text{ J})(1.6 \times 10^8 \text{ J}) = \mathbf{\$4.4}.$$

---

6. (a) The equation is in the form $V = V_0 \sin 2\pi ft$, where V_0 is the peak voltage and f is the frequency. Therefore, since $V = 283 \sin 188t$, we see that the peak voltage is **283 V**.

(b) The frequency is obtained from $2\pi f = 188$ or

$$f = 188/2\pi = \mathbf{29.9\ Hz}.$$

(c) We know that

$$R = V_{rms}/I_{rms} = (283\ V/\sqrt{2})/(2.0\ A) = \mathbf{1.0 \times 10^2\ \Omega}.$$

---

7. (a) The rms voltage is 120 V, and the average power is 750 W. The rms current is obtained from $P = I_{rms}V_{rms}$, so

$$I_{rms} = P/V_{rms} = (750\ W)/(120\ V) = 6.3\ A.$$

The peak current is

$$I_0 = I_{rms}\sqrt{2} = \mathbf{8.9\ A}.$$

(b) If connected to a 240-V line, the power drawn is no longer 750 W. We need the resistance of the device,

$$R = V_{rms}/I_{rms} = (120\ V)/(6.3\ A) = 19\ \Omega.$$

The new current is therefore,

$$I'_{rms} = V'_{rms}/R = (240\ V)/(19\ \Omega) = 13\ A.$$

The peak current is therefore,

$$I'_0 = I'_{rms}\sqrt{2} = \mathbf{18\ A}.$$

(c) The power drawn from the 240-V line is, $P' = I'_{rms}V'_{rms} = (13\ A)(240\ V) = 3100\ W.$

So the power drawn from the 240-V line is **4 times as great** as that drawn from the 120-V line.

---

8. The 5.0 Ω and 8.0 Ω are in series with an equivalent resistance of 13.0 Ω. The 6.0 Ω and 12.0 Ω are also in series and have an equivalent resistance of 18.0 Ω. The 2.0 Ω resistor is in parallel with the 13.0 Ω and 18.0Ω equivalents so the equivalent resistance of the entire circuit is

$$1/R = 1/(13.0\ \Omega) + 1/(18.0\ \Omega) + 1/(2.0\ \Omega) \text{ or } R = \mathbf{1.6\ \Omega}.$$

---

9. The terminal voltage of a battery is $V_T = \text{emf} - Ir$, so

$$r = (\text{emf} - V_T)/I = (12.0\ V - 8.2\ V)/(120\ A) = \mathbf{0.032\ \Omega}.$$

The resistance of the starter motor is

$$R = V_T/I = (8.2\ V)/(120\ A) = \mathbf{0.068\ \Omega}.$$

---

10. The shunt resistor and the galvanometer are in parallel so each will have the same voltage drop. The drop across the galvanometer is

$$V = I_gR_g = (150.0 \times 10^{-6}\ A)(75.0\ \Omega) = 1.13 \times 10^{-2}\ V.$$

The equivalent resistance of the ammeter is

$$1/R = 1/(75.0\ \Omega) + 1/(1.50\ \Omega)$$

which gives

$$R = 1.47\ \Omega.$$

The current through the ammeter at full scale is

$$I = V/R = (1.13 \times 10^{-2}\ V)/(1.47\ \Omega) = \mathbf{7.69 \times 10^{-3}\ A}.$$

---

11. The current in the circuit before the ammeter is inserted is the "true" value and is

$$I = (9.0\text{ V})/(95\ \Omega) = 0.095\text{ A}.$$

After the ammeter is inserted, the current in the circuit is reduced to

$$I' = (9.0\text{ V})/(104\ \Omega) = 0.087\text{ A}.$$

The percentage error introduced by the ammeter is

$$\%\text{ error} = (0.095\text{ A} - 0.087\text{ A})/(0.095\text{ A}) \times 100\ \% = \mathbf{8\ \%}.$$

---

12. A full scale voltage across the voltmeter produces a current of $I_g = 150\ \mu A$ through the voltmeter. The equivalent resistance of the voltmeter is $R = 75\ \Omega + 25\text{ k}\Omega$, so

$$V = (150 \times 10^{-6}\text{ A})(75\ \Omega + 25 \times 10^{3}\ \Omega) = \mathbf{3.8\ V}.$$

---

13. Label the currents and choose the current directions. It is not necessary to "guess" the true current directions. Let I_1 be up through the 12 V battery, I_2 be down through the 15 Ω resistor, and I_3 be down through the 25 Ω resistor. Apply the loop rule (drops = rises)to the left loop starting at the upper left corner and proceeding clockwise.

$$I_3(25\ \Omega) = 12\text{ V},$$

then $\mathbf{I_3 = 0.48\ A}$. Repeat for the outside loop

$$I_2(15\ \Omega) = 6.0\text{ V} + 12\text{V}, \text{ then } \mathbf{I_2 = 1.2\ A}.$$

Apply the junction rule (currents in = currents out) at the top junction.

$$I_1 = I_2 + I_3$$

which gives $\mathbf{I_1 = 1.7\ A}$.

Note that all of the currents are positive, indicating that we "guessed" their directions correctly.

---

14. The two 160 Ω resistors are in series and in parallel with the battery. Their equivalent resistance is

$$160\ \Omega + 160\ \Omega = 320\ \Omega$$

and the voltage across them is 12 V, so the current through them is

$$I_{160} = (12\text{ V})/(320\ \Omega) = \mathbf{3.8 \times 10^{-2}\ A}.$$

The 320 Ω resistor is in parallel with the battery, hence "draws" a current

$$I_{320} = (12\text{ V})/(320\ \Omega) = \mathbf{3.8 \times 10^{-2}\ A}.$$

The total current delivered by the battery is $I = I_{160} + I_{320} = \mathbf{7.6 \times 10^{-2}\ A}$.

---

15. The 8.0 μF and 22 μF capacitors are in series and have an equivalent capacitance given by

$$1/C_1 = 1/(8.0\ \mu F) + 1/(22\ \mu F)$$

or

$$C_1 = 5.9\ \mu F.$$

The equivalent capacitance, C_1, is in parallel with the 16 μF capacitor with an new equivalent capacitance of

$$C_2 = 5.9\ \mu F + 16\ \mu F = 22\ \mu F.$$

The equivalent capacitance is in series with the 6.0 μF capacitor, so the equivalent capacitance of the circuit is found from

$$1/C_3 = 1/(22\ \mu F) + 1/(6.0\ \mu F)$$

Then

$$\mathbf{C_3 = 4.7\ \mu F.}$$

The charge separated by the battery is

$$q = C_3V = (4.7 \times 10^{-6}\ F)(12\ V) = \mathbf{5.6 \times 10^{-5}\ C}.$$

---

16. The charge on the capacitor plates at any time, t, is given by equation (20.22)

$$q = CV_0\, e^{-t/(RC)}.$$

The charge on the plates of the capacitor is $q = CV$ so

$$V/V_0 = e^{-t/(RC)}.$$

Taking the natural logarithm of both sides and solving for t yields

$$t = -(RC)\ln(V/V_0) = -(1500\ \Omega)(25 \times 10^{-6}\ F)\ln(15\ V/35\ V) = \mathbf{3.2 \times 10^{-2}\ s}.$$

---

17. Equation (20.20) applies to the charging of a capacitor in a series RC circuit: $q = CV_0\,(1 - e^{-t/(RC)})$.

The charge on the plates of the capacitor is $q = CV$, so

$$V = V_0\,(1 - e^{-t/(RC)}).$$

Rearranging, taking the natural log of both sides, and solving for t yields

$$t = -(RC)\ln(1 - V/V_0) = -(25 \times 10^3\ \Omega)(1500 \times 10^{-6}\ F)\ln(1 - 8.0\ V/12.0\ V)$$
$$t = \mathbf{41\ s}.$$

Quiz answers

1. c
2. b
3. a
4. c
5. a
6. b
7. d
8. a
9. c
10. a
11. c
12. d
13. b
14. a
15. d
16. c
17. b
18. d
19. c
20. c
21. a
22. d

MCAT REVIEW PROBLEMS

A 12-volt battery is connected to two light bulbs, as drawn in figure 1. Light bulb 1 has resistance 3 ohms, while light bulb 2 has resistance 6 ohms. The battery has essentially no internal resistance, and all the wires are essentially resistanceless, too.

When a light bulb is unscrewed, no current flows through that branch of the circuit. For instance, if light bulb 2 is unscrewed, current flows only around the lower loop of the circuit, which consists of the battery and light bulb 1. The more current flows through a light bulb, the brighter it shines

When two resistors are wired in series, their equivalent resistance is $R_{eq} = R_1 + R_2$. By contrast, when two resistors are wired in parallel, their net resistance is given by

$$\frac{1}{R_{eq}} = \frac{1}{R_1} + \frac{1}{R_2}.$$

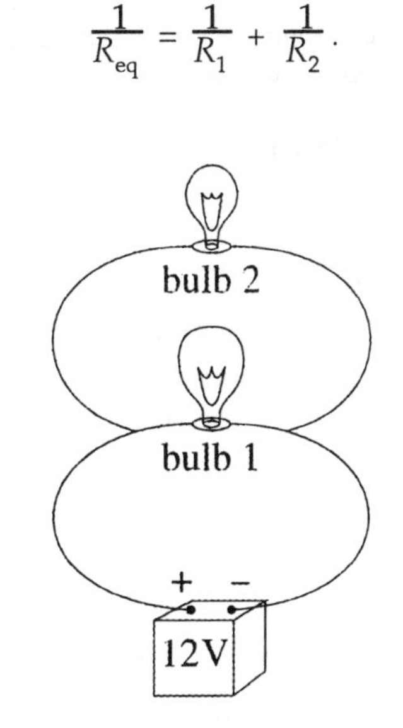

FIGURE 1

---

1. When bulb 1 is screwed in, but bulb 2 is unscrewed, the power generated in bulb 1 is:

   A. 4 watts **B.** 12 watts **C.** 36 watts **D.** 48 watts

---

2. Which of the following is FALSE?

   A. Some of the energy produced by the light bulb takes the form of heat.
   B. The battery is the source of all the electrons flowing around the circuit.
   C. The current entering the light bulb equals the current leaving the light bulb.
   D. The potential in the wire to the left of the light bulb differs from the potential in the wire to the right of that bulb.

3. Bulb 2 is now screwed in. As a result, bulb 1:

A. turns off
B. becomes dimmer
C. stays about the same brightness
D. becomes brighter

---

4. With both light bulbs screwed in, the current through the battery is:

A. 1.2 amperes
B. 2 amperes
C. 4 amperes
D. 6 amperes

---

5. With only light bulb 1 screwed in, a NeverQuit 12-volt battery goes dead in 24 days. With both light bulbs screwed in, a NeverQuit 12-volt battery goes dead in:

A. 12 days
B. 14 days
C. 16 days
D. 18 days

ANSWERS TO MCAT REVIEW PROBLEMS

---

1. **D.** Since the power generated in the bulb equals the voltage across the bulb times the current through the bulb ($P = IV$), we need to know the current and voltage. (Equivalently, you can use $P = I^2R$ or $P = V^2/R$, in which case you need to know the current *or* the voltage.)

Because the lower circuit loop contains only the battery and bulb 1, the bulb "feels" the battery's full voltage, 12 volts. So, by Ohm's law ($V = IR$), a current

$$I = \frac{V}{R} = \frac{12\text{ V}}{3\ \Omega} = 4\text{ A}$$

flows through that bulb. Therefore, the bulb generates power

$$P = IV = (4\text{ A})(12\text{ V}) = 48\text{ W}.$$

---

2. **B.** Most of the charges flowing around the circuit are valence electrons stripped off the metal atoms in the wires and light bulbs. A battery doesn't "supply" all of the charges. It merely pushes around charges already present in the circuit.

Statements C and D are both true. All charges flowing into the light bulb also flow back out; no current gets "used up." But inside the bulb, those charges lose energy. This lost electrical energy converts into light and heat. So, the current has lower "potential" after flowing through the bulb.

---

3. **C.** If the battery had internal resistance, bulb 1 would dim. But here, screwing in bulb 2 doesn't prevent bulb 1 from "feeling" the full 12 volts produced by the battery. Both bulbs get a full "dose" of that voltage. So, bulb 2 turns on without lessening the brightness of bulb 1.

If you picked B, you probably thought that the battery can "supply" only a certain amount of current. According to this reasoning, bulb 2 draws current at the expense of bulb 1. But the battery doesn't "supply" all of the electrons flowing around the circuit. It merely pushes around electrons already present in the metal wires and bulb filaments, as mentioned in question 2. Screwing in bulb 2 allows the valence electrons in that branch of the circuit to "join the current."

You can also solve this problem by calculating the current through each bulb. But on an MCAT, you'll often need to reason more quickly and intuitively, as demonstrated here.

---

4. **D.** The total circuit current, i.e., the current through the battery, is simply

$$I = \frac{V}{R_{eq}},$$

where R_{eq} denotes the equivalent resistance of the whole circuit. These two light bulbs are wired in parallel; any given electron flowing around the circuit passes through one bulb *or* the other, but not both. So, we must use the in-parallel formula

$$\begin{aligned}\frac{1}{R_{eq}} &= \frac{1}{R_1} + \frac{1}{R_2}\\ &= \frac{1}{3\ \Omega} + \frac{1}{6\ \Omega}\\ &= \frac{2}{6\ \Omega} + \frac{1}{6\ \Omega}\\ &= \frac{3}{6\ \Omega}\\ &= \frac{1}{2\ \Omega},\end{aligned}$$

and hence, $R_{eq} = 2\ \Omega$. So, $I = \frac{V}{R_{eq}} = \frac{12\text{ V}}{2\ \Omega} = 6$ amperes.

Intuitively, the equivalent resistance of two in-parallel resistors is *less* than either individual resistance, because the current can "split" itself between two paths. It's like opening an extra lane on a highway; cars can flow by more easily, because they can spread themselves out among more paths.

---

5. **C**. The battery loses energy at rate P, the battery's power. So, the rate of energy loss is $P = IV$, where I denotes the current through the battery.

With only bulb 1 screwed in, $I = 4$ amperes, as found in question 1. With both bulbs screwed in, $I = 6$ amperes, as found in question 4. Therefore, screwing in the second bulb causes the current, and hence the power, to increase by a factor of $\frac{6}{4}$. Since the battery's energy gets used up $\frac{6}{4}$ times as fast, it takes $\frac{4}{6}$ as many days to go dead:

$$\frac{4}{6}(24\text{ days}) = 16\text{ days}.$$

Here is another way to reach the same answer. Start by calculating the total chemical energy stored in a NeverQuit battery. With only bulb 1 screwed in, the energy gets used at a rate of 48 watts. (See question 1.) At this rate, the energy runs out in 24 days. So, a NeverQuit battery stores total energy E_{stored} = (48 watts)(24 days). We could convert this to joules, but need not do so to solve the problem.

With both bulbs screwed in, that same total energy get used at a rate (12 V)(6 amperes) = 72 watts. Let t denote the time the battery takes to go dead, when used at that higher power. Then, E_{stored} = (72 watts)t. So,

$$E_{stored} = (48\text{ W})(24\text{ days}) = (72\text{ W})t.$$

Solve for t to get

$$t = \frac{48\text{ W}}{72\text{ W}}(24\text{ days}) = \frac{4}{6}(24\text{ days}) = 16\text{ days}.$$

CHAPTER 21
Magnetic Forces and Magnetic Fields

PREVIEW

In this chapter you will study the properties of magnetic fields. You will learn how a magnetic field exerts a force on moving charges and currents, how electric currents produce magnetic fields, and about the properties of magnetic materials. You will see how magnetic fields play important roles in such devices as mass spectrometers, stereo speakers and tape recorders, galvanometers, and electric motors.

QUICK REFERENCE

Important Terms

Magnetic field
A vector field which exists in the region around a magnet. The direction of the field at any point is the direction indicated by the north pole of a small compass needle placed at that point. The SI unit is the tesla (T).

Magnetic force
The force exerted on a **moving** charge due to a magnetic field. The magnetic force is always perpendicular to both the direction of the magnetic field and the velocity vector.

Mass spectrometer
An instrument which uses magnetic fields to determine the relative masses of atoms and molecules.

Hall effect
An effect where moving charges in a current-carrying metal or semiconductor are deflected by an external magnetic field. The deflected charge carriers create an emf, the polarity of which can be used to determine whether the charge carriers are positive or negative.

Ferromagnetic materials
Materials, such as iron, which can be permanently magnetized. These materials are made up of tiny regions that behave like small magnets. In a permanent magnet, these regions are aligned, and a high degree of magnetism results.

Equations

The **magnitude** B of the magnetic field at any point in space is defined as

$$B = \frac{F}{q_0 (v \sin\theta)} \tag{21.1}$$

The **radius** of the circular **path of a charged particle** moving in a **magnetic field** is given by

$$r = \frac{mv}{qB} \tag{21.2}$$

The **magnetic force** on a **current-carrying wire** of length L is

$$F = ILB \sin\theta \tag{21.3}$$

The **torque τ on a current-carrying loop of wire in a magnetic field** can be written as

$$\tau = NIAB \sin \phi \tag{21.4}$$

The **magnetic field** produced by a **long straight wire** has a magnitude

$$B = \frac{\mu_o I}{2\pi r} \tag{21.5}$$

The **magnetic field** at the **center of a current-carrying loop** of wire has a magnitude

$$B = N\frac{\mu_o I}{2R} \tag{21.6}$$

The **magnetic field in the interior of a long solenoid** is

$$B = \mu_0 nI \tag{21.7}$$

DISCUSSION OF SELECTED SECTIONS

21.2 The Force That a Magnetic Field Exerts on a Moving Charge

We are all familiar with simple magnets, and we know that magnets attract certain metallic objects (like iron). In fact, magnets can exert forces on each other. A magnet has two poles, a **north pole** and a **south pole**. When two magnets are placed near one another it is found that like poles repel each other, and unlike poles attract each other. This behavior is similar to that of like and unlike electric charges.

We know that an electric field exists in the space around electric charges. Similarly, a **magnetic field** exists in the region around a magnet. The magnetic field is a vector that has both magnitude and direction. The direction of the magnetic field at any point in space is the direction indicated by the north pole of a small compass needle placed at that point.

When a charge is placed in a magnetic field, does it experience a magnetic force? Yes, if two conditions are met:

1. The charge must be moving, for no magnetic force acts on a stationary charge.
2. The velocity of the moving charge must have a component that is perpendicular to the direction of the field.

Consider a test charge $+q_0$ moving with a velocity **v** through a magnetic field **B**. If the charge is moving at an angle θ with respect to the magnetic field, the field will exert a force **F** on the test charge. The magnitude B of the magnetic field at any given point in space is then given by

$$B = \frac{F}{q_0 (v \sin \theta)} \tag{21.1}$$

where F and v are the magnitudes of the force and velocity, respectively. The SI unit of magnetic field is:

$$\frac{\text{newton second}}{\text{coulomb meter}} = 1 \text{ tesla (T)}$$

The direction of the magnetic force **F** is perpendicular to both **v** and **B**; in other words, **F** is perpendicular to the plane defined by **v** and **B**. The directions of the force, magnetic field, and velocity are related by a convenient rule called **Right-Hand Rule No. 1** (RHR-1). This can be stated as

> **Right-Hand Rule No. 1.** Extend the right hand so the fingers point along the direction of the magnetic field **B** and the thumb points along the velocity **v** of the charge. The palm of the hand then faces in the direction of the magnetic force **F** that acts on a positive charge.

If the moving charge is negative instead of positive, the direction of the magnetic force is opposite to that predicted by RHR-1. So, to find the force on a negative charge, use RHR-1 as if the charge were positive, and find the direction of the force. Then, reverse this direction to find the direction of the force that acts on the negative charge.

Example 1

An electron traveling at a velocity of 5.0×10^6 m/s west enters a region of uniform magnetic field of magnitude 2.0×10^{-3} T directed upward. What are the magnitude and direction of the force that acts on this electron?

We can use equation (21.1) to find the magnitude of the force. We have

$$F = qvB \sin \theta = (1.6 \times 10^{-19}\ \text{C})(5.0 \times 10^6\ \text{m/s})(2.0 \times 10^{-3}\ \text{T}) \sin 90° = 1.6 \times 10^{-15}\ \text{N}.$$

RHR-1 tells us that the force on a positive charge would be north (place your fingers upward with your thumb to the west, your palm then faces north). Therefore, the force on the electron (which is negative) would be **south**.

21.3 The Motion of a Charged Particle in a Magnetic Field

Consider the case in which a particle has a velocity that is perpendicular to a uniform magnetic field. As shown in the figure below, the magnetic force serves to move the particle in a circular path.

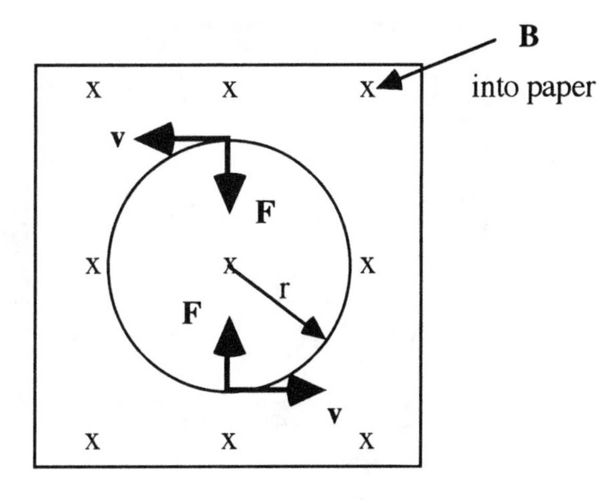

Note that with the magnetic field pointing into the paper (x's represent **B** into the page) and with the velocity **v** at each position shown, RHR-1 predicts the force **F** at each point to be directed in toward the center of a circle of radius r. That is, the magnetic force always remains perpendicular to the velocity and is directed toward the center of the circular path shown. The radius of the circular path is obtained by equating the magnetic force to the centripetal force.

We can therefore write,

$$F_c = \frac{mv^2}{r} = qvB \Rightarrow r = \frac{mv}{qB} \tag{21.2}$$

Example 2

In a magnetic field of B = 0.75 T, for what path radius will an electron circulate at a speed of 3.0 X 10^7 m/s? What will its kinetic energy be?

Using equation (21.2), we can find the radius of the orbit for v = 3.0 x 10^7 m/s.

$$r = \frac{mv}{qB} = \frac{(9.11\text{x}10^{-31}\text{kg})(3.0\text{x}10^{7}\text{m/s})}{(1.6\text{x}10^{-19}\text{C})(0.75\text{T})} = 2.3\text{x}10^{-4}\text{ m}$$

The kinetic energy of the electron is

$$KE = \frac{1}{2}mv^2 = \frac{1}{2}(9.11 \text{ x } 10^{-31}\text{ kg})(3.0 \text{ x } 10^{7}\text{ m/s})^2 = 4.1 \text{ x } 10^{-16}\text{ J}$$

21.4 The Mass Spectrometer

A mass spectrometer is a device for determining the relative masses and abundances of atoms and molecules. In the type illustrated below, the atoms or molecules are vaporized and ionized by the ion source, S. The charged ion is given a charge +e by the removal of one electron from the atom/molecule. The positive ions are then accelerated through a potential difference V and are injected into a region of constant magnetic field **B**, where they are deflected in semicircular paths of radius r.

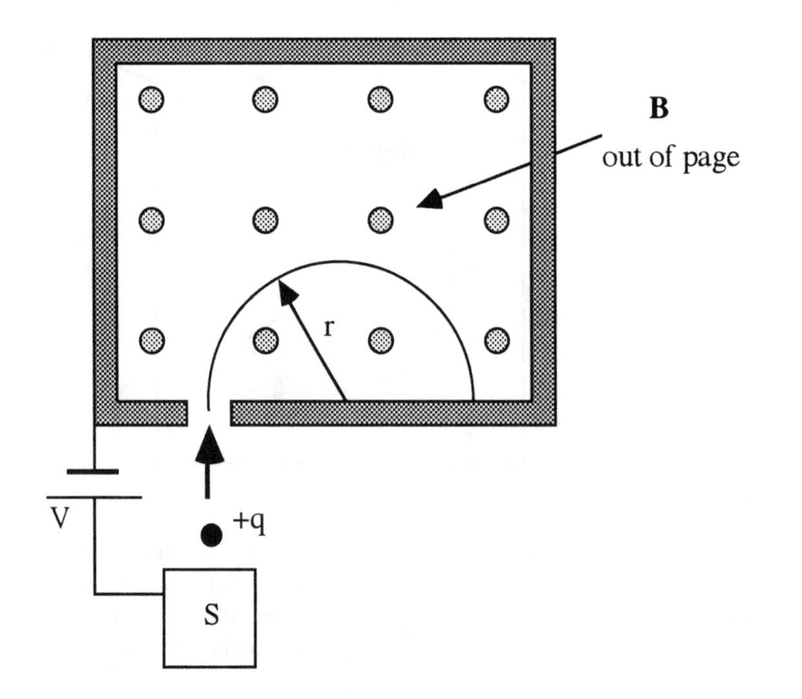

The ion has a kinetic energy $KE = (1/2)mv^2 = eV$, and since $r = mv/qB$ we can combine equations to obtain

$$m = \left(\frac{er^2}{2V}\right) B^2$$

Example 3
A mass spectrometer uses a potential difference of 2.00 kV to accelerate a singly ionized ion to the proper speed. A 0.400-T magnetic field then bends the ion into a circular path of radius 0.226 m. What is the mass of the ion?

Using the above equation we obtain

$$m = \frac{er^2 B^2}{2V} = \frac{(1.60 \times 10^{-19}\ C)(0.226\ m)^2(0.400\ T)^2}{2(2.00 \times 10^3\ V)} = 3.27 \times 10^{-25}\ kg$$

21.5 The Force on a Current in a Magnetic Field

We have seen that a moving charge in a magnetic field experiences a magnetic force. Since a current is nothing more than a stream of moving charges, it is natural to expect that a current-carrying wire will experience a magnetic force when placed in a magnetic field. For a wire of length L carrying a current I placed in a magnetic field B, the force exerted is

$$F = ILB \sin \theta \qquad (21.3)$$

where θ is the angle between the direction of the current and the direction of the magnetic field.

Example 4
An electric power line carries a current of 1400 A in a location where the earth's magnetic field is 0.50×10^{-4} T. The line makes an angle of 75° with respect to the field. Determine the magnitude of the magnetic force on a 120-m length of line.

Using equation (21.3) we have

$$F = ILB \sin \theta = (1400\ A)(120\ m)(0.50 \times 10^{-4}\ T) \sin 75° = 8.1\ N.$$

21.7 Magnetic Fields Produced by Currents

In the previous section we saw that a current-carrying wire experiences a magnetic force when placed in a magnetic field. We now consider a phenomena in which **a current-carrying wire produces a magnetic field**.

Consider a long, straight wire carrying a current. If we place compass needles at different points near the wire, we find that the magnetic field lines form a circular pattern about the wire. If the direction of the current is reversed, we find that the direction of the magnetic field has also reversed. The direction of the magnetic field can be obtained by using **Right-Hand Rule No. 2** (RHR-2). It can be stated as follows:

Right-Hand Rule No. 2. Curl the fingers of the right hand into the shape of a half-circle. Point the thumb in the direction of the conventional current I, and the tips of the fingers will point in the direction of the magnetic field **B**.

Right-Hand Rule No. 2 is illustrated in the following diagram.

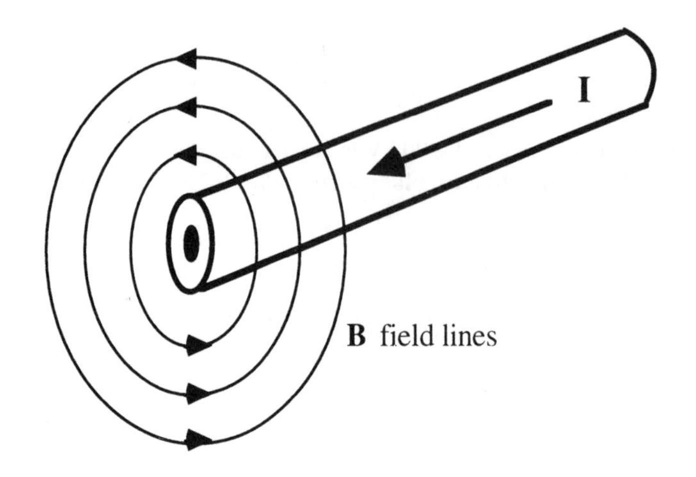

The magnitude of the magnetic field at a distance r from a long, straight wire is given by the expression;

$$B = \frac{\mu_0 I}{2\pi r} \qquad (21.5)$$

where μ_0 is a constant known as the **permeability of free space**, and its value is $\mu_0 = 4\pi \times 10^{-7}$ T m/A.

Example 5
A 6.00-μC charge is moving with a velocity of 7.50×10^6 m/s parallel to a long, straight wire. The wire carries a current of 67.0 A in the same direction as the moving charge, and is 5.00×10^{-2} m away from the charge. Find the magnitude and direction of the force on the charge.

First calculate the magnetic field produced by the wire and then the force that this field exerts on the moving charge.

$$B = \frac{\mu_0 I}{2\pi r} = \frac{(4\pi \times 10^{-7}\ \text{T m/A})(67.0\ \text{A})}{2\pi(5.00 \times 10^{-2}\ \text{m})} = 2.68 \times 10^{-4}\ \text{T}$$

The force on the charge is given by equation (21.1) with $\theta = 90°$. Therefore,

$$F = qvB \sin\theta = (6.00 \times 10^{-6}\ \text{C})(7.50 \times 10^{6}\ \text{m/s})(2.68 \times 10^{-4}\ \text{T}) \sin 90° = 1.21 \times 10^{-2}\ \text{N}$$

The direction of the force on the moving charge is obtained by applying **each** of the right-hand rules. As seen in the following diagram, if the current in the wire is directed OUT of the page, the magnetic field produced by this wire circles counterclockwise about the wire, according to RHR-2. The charge is moving in the same direction as the current, i.e., OUT of the page. At the site of the moving charge, the magnetic field is directed UP. By applying RHR1, our fingers point upward (in the direction of the B-field), our thumb points out of the page (in the direction of the velocity) and the palm of our hand (which represents the direction of the magnetic force) points toward the wire. The force on the moving charge is therefore directed **toward the wire**.

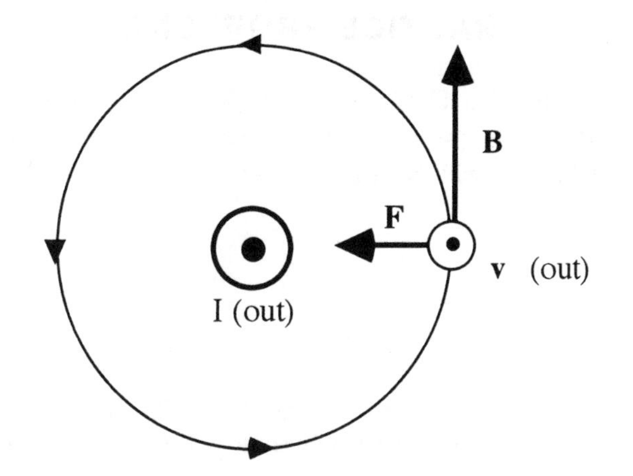

If we take a wire and bend it in the shape of a circular loop, the magnetic field at the center of the loop is given by

$$B = N\frac{\mu_o I}{2R} \qquad (21.6)$$

Where the loop contains N turns of wire.

Another device of interest is the **solenoid**. It is made up of a long coil of wire in the shape of a helix. If the wire is wound so the turns are packed close to each other and the solenoid is long compared to its diameter, the magnetic field inside the solenoid is nearly constant in magnitude and directed parallel to the axis. The magnitude of the field in the interior of a long solenoid is

$$B = \mu_0 n I \qquad (21.7)$$

where n is the number of turns per unit length of the solenoid and I is the current. If, for example, the solenoid contains 500 turns and has a length of 0.1 m, the number of turns per unit length is n = 5000 turns/m.

Example 6
A solenoid is 0.50 m long, has three layers of windings of 750 turns each, and carries a current of 4.0 A. What is the magnetic field at the center of the solenoid?

Since there are three layers of windings, there are (3 x 750) turns in 0.50 m. So, n = (3 x 750)/0.50 = 4500 turns/m. The magnetic field at the center of the solenoid is obtained using equation (21.7),

$$B = \mu_0 \, n \, I = (4\pi \times 10^{-7} \text{ T m/A})(4500 \text{ turns/m})(4.0 \text{ A}) = 2.3 \times 10^{-2} \text{ T}.$$

PRACTICE PROBLEMS

1. The electrons in the beam of a television tube have an energy of 15 keV. The tube is oriented so that the electrons move horizontally from east to west. The earth's magnetic field points down and has a magnitude $B = 4.5 \times 10^{-5}$ T. (a) In what direction will the beam deflect? (b) What is the acceleration of a given electron?

2. Protons in a magnetic field of 0.80 T follow a circular trajectory of 75-cm radius. (a) What is the speed of the protons? (b) If electrons traveled at the same speed in this field, what would the radius of their trajectory be?

3. Two isotopes of carbon, carbon-12 and carbon-13, have masses of 19.92×10^{-27} kg and 21.59×10^{-27} kg, respectively. These two isotopes are singly ionized and each is given a speed of 6.667×10^{5} m/s. The ions then enter the bending region of a mass spectrometer where the magnetic field is 0.8500 T. Determine the spatial separation between the two isotopes after they have traveled through a half-circle.

4. A weighing scale supports a battery, to which a rigid wire loop is attached, as shown in the figure. The lower part of the loop is in a magnetic field of 0.20 T. If the combined mass of the battery and wire loop is 0.25 kg, what current in the wire is necessary for the scale to indicate zero weight? Which pole of the battery is positive?

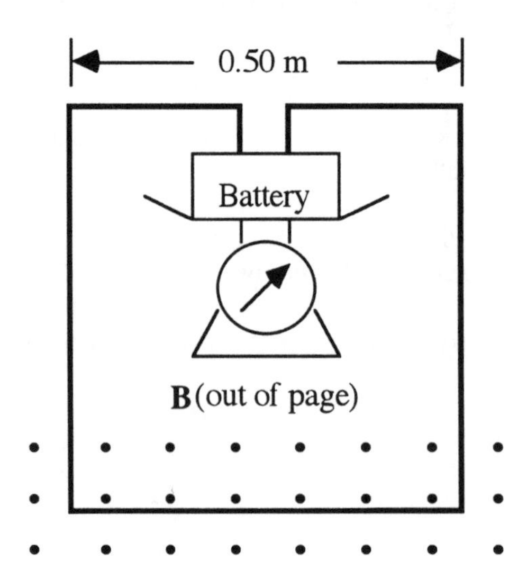

5. Two wires parallel to one another are separated by 1.0 m, as shown in the figure. Each carries a current of 3.0 A, but in opposite directions. Find the magnitude and direction of the magnetic field at a point P midway between the wires.

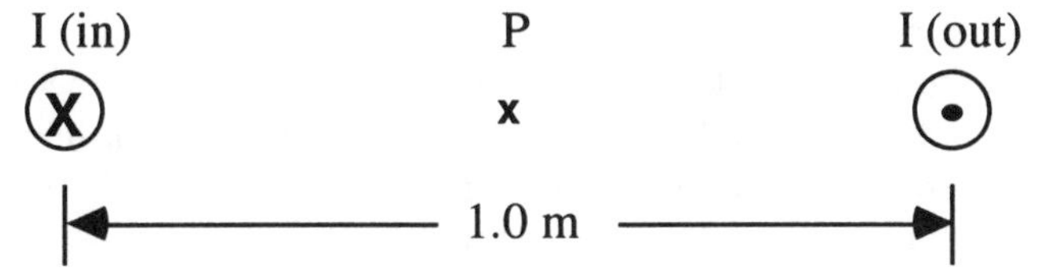

6. Two parallel straight wires are separated by a distance 0f 0.75 m and carry currents $I_1 = 25$ A and $I_2 = 35$ A. Find the force (magnitude and direction) that each wire exerts on a 2.5 m length of the other wire if the currents are (a) in the same direction and (b) in the opposite directions.

7. You are given 251 m of copper wire that can carry a current of 0.500 A without excessive heating. This wire is used to make a solenoid of 25.0 cm length and 3.00 cm diameter. Determine the magnetic field that can be achieved at the center of the solenoid.

8. Four long, parallel conductors all carry currents of 5.0 A. An end view of the conductors is shown in the figure below. The current directions for wires A and D are into the page (indicated by crosses) while the current directions for B and C are out of the page (indicated by dots). Calculate the magnitude and direction of the magnetic field at point P, located at the center of the square of edge length 0.40 m.

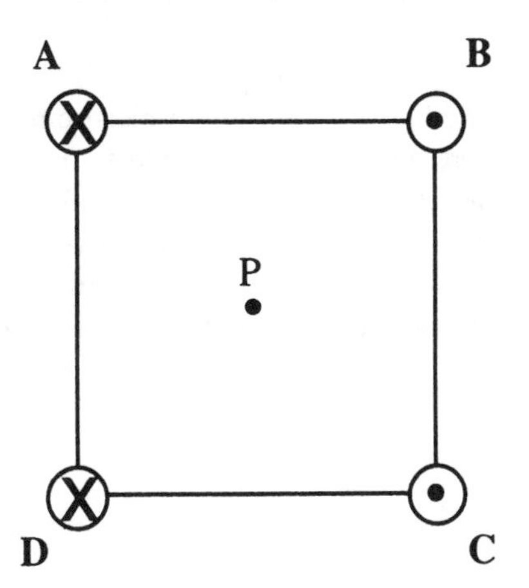

HELPFUL SUGGESTIONS

1. Magnetic forces act only on moving charges. A magnetic field will not exert a force on a stationary charge.

2. The magnetic force on a charged particle is always perpendicular to the velocity of the particle AND the magnetic field. Use Right-Hand Rule No. 1 to find the relationship between the directions of **B**, **v**, and **F**. Note that if B and v are parallel, no force is felt by the moving charge since $F = qvB \sin\theta = 0$ because θ (the angle between the velocity and the magnetic field) is zero and $\sin 0° = 0$.

3. Remember that RHR-1 for finding the direction of the force applies to positive charges. If the charge is negative, the direction of the force is reversed from that obtained from RHR-1.

4. The magnetic field is a VECTOR quantity. Therefore, vector addition techniques must be used to obtain the resultant field due to more than one magnetic field.

5. To find the direction of the magnetic field associated with a current-carrying wire, use Right-Hand Rule No. 2. Curl the fingers of the right hand into the shape of a half-circle. Point your thumb in the direction of the conventional current I, and the tips of your fingers will point in the direction of the magnetic field **B**.

EVERYDAY PHYSICS

1. Test for the presence of magnetic fields using a compass. Place one at different points around a bar or horseshoe magnet and "map" the magnetic field lines. Hook a wire up to a battery, place the compass near the wire, and notice the magnetic field that results in this case. Try using your "magnetic field probe" to detect the presence of fields near your TV or radio, household appliances, under the hood of your car, etc.

2. A magnet will distort the image on the screen of an oscilloscope or television. What does this tell you about how the image on the screen was produced?

3. The earth's magnetic field exerts a force on charged particles that impinge on the outer atmosphere. High energy electrons and protons from the sun, stars, and other regions of the universe passing near the earth are deflected by the earth's magnetic field, and many are trapped by it. The trapped particles move back and forth between the earth's magnetic poles, following spiral paths around the magnetic field lines. The trapped radiation is mainly confined to two doughnut-shaped regions surrounding the earth. These are known as the Van Allen radiation belts, named after James Van Allen, who determined their existence in 1958 from data gathered by the Explorer I satellite.

4. Magnetic fields can be used as "magnetic bottles" to trap plasmas (ionized gases). For example, physicists are attempting to carry out nuclear fusion reactions in the laboratory, under controlled conditions. The hydrogen plasma required for the reaction must be heated to temperatures of millions of degrees. There is the problem as to what material can be used to contain the plasma, since all known materials melt and vaporize below 4000°C. However, a strong magnetic field, which can exist at any temperature, can exert forces on charged particles in motion. Powerful electromagnets can be used to create a sort of "magnetic straitjacket", containing the plasma and keeping it away from the walls of the container.

CHAPTER QUIZ

1. An electron moving with velocity **v** to the right enters a region of uniform magnetic field that points into the paper. After the electron enters this region, it will be
   a. deflected out of the plane of the paper.
   b. deflected into the plane of the paper.
   c. deflected upward.
   d. deflected downward.

2. A charged particle moves in a circular path in a uniform magnetic field. If the speed of the particle is doubled, the radius of the circular path would be
   a. doubled b. quadrupled c. cut in half d. unchanged

3. A charged particle moves in a circular path in a uniform magnetic field. If the charge on the particle were doubled, the radius of the circular path would be
   a. doubled b. quadrupled c. cut in half d. unchanged

4. A horizontal wire carries a current directly away from you. From your viewpoint, the magnetic field produced by this current
   a. points directly away from you.
   b. points directly toward you.
   c. circles the wire in a clockwise direction.
   d. circles the wire in a counterclockwise direction.

5. Two long parallel wires are placed side-by-side on a horizontal, frictionless table. If the wires carry currents in opposite directions
   a. the wires would move toward one another.
   b. the wires would move away from one another.
   c. the wires would be lifted off the table.
   d. nothing at all would happen.

6. A 2 m long straight wire carrying a current of 5 A is placed perpendicular to a magnetic field of B = 0.1 T. The magnitude of the force on the wire is
   a. 0 b. 0.1 N c. 1 N d. 10 N

7. The magnetic field 2 m away from a long straight wire carrying a current of 2 A is
   a. 1 T b. 4 T c. 4×10^{-7} T d. 2×10^{-7} T

8. A solenoid of 10 000 turns/m carries a current of 1000 A. The magnetic field at the center of the solenoid is
   a. 1×10^{7} T b. $4\pi \times 10^{7}$ T c. $4\pi \times 10^{-7}$ T d. 4π T

9. A positively charged particle is moving parallel to, and in the same direction as the current in a long straight wire. In what direction will the charge be deflected?
   a. toward the wire b. away from the wire c. parallel to the wire d. it won't be deflected.

10. The fact that the magnetic force on a moving charge is perpendicular to the velocity of the charge implies that
   a. the moving charge cannot accelerate at all.
   b. no work is done by the magnetic field on the moving charge.
   c. the right-hand rule cannot be used in this situation.
   d. the moving charge gains or loses kinetic energy.

SOLUTIONS AND ANSWERS

Practice Problems

---

1. (a) The direction of the beam is obtained using RHR-1. With B down (fingers point down), and the velocity to the west (thumb to the west), the palm of the hand (F) is facing south. For an electron, force is directed **north**.
(b) The kinetic energy is

$$KE = (1/2)mv^2 = (15\,000 \text{ eV})(1.6 \times 10^{-19} \text{ J}/1 \text{ eV}) = 2.4 \times 10^{-15} \text{ J}.$$

Solving for v yields,

$$v^2 = 2KE/m = 2(2.4 \times 10^{-15} \text{ J})/(9.11 \times 10^{-31} \text{ kg}) \Rightarrow v = 7.3 \times 10^7 \text{ m/s}.$$

The force is $F = qvB \sin\theta$ ($\theta = 90°$). The acceleration can therefore be obtained using

$$a = F/m = qvB \sin 90°/m = (1.6 \times 10^{-19} \text{ C})(7.3 \times 10^7 \text{ m/s})(4.5 \times 10^{-5} \text{ T})/(9.11 \times 10^{-31} \text{ kg}) = \mathbf{5.8 \times 10^{14} \text{ m/s}^2}.$$

---

2. (a) Solving equation (21.2) for the velocity yields

$$v = qrB/m = (1.6 \times 10^{-19} \text{ C})(0.75 \text{ m})(0.80 \text{ T})/(1.67 \times 10^{-27} \text{ kg}) = \mathbf{5.7 \times 10^7 \text{ m/s}}.$$

(b) For an electron,

$$r = mv/qB = (9.11 \times 10^{-31} \text{ kg})(5.7 \times 10^7 \text{ m/s})/[(1.6 \times 10^{-19} \text{ C})(0.80 \text{ T})] = \mathbf{4.1 \times 10^{-4} \text{ m}}.$$

---

3. The spatial separation of the two isotopes is the difference between the diameter of their trajectories. Therefore,

$$r_{12} = m_{12}v/qB = (19.92 \times 10^{-27} \text{ kg})(6.667 \times 10^5 \text{ m/s})/[(1.602 \times 10^{-19} \text{ C})(0.8500 \text{ T})] = 9.753 \times 10^{-2} \text{ m}.$$

$$r_{13} = m_{13}v/qB = (21.59 \times 10^{-27} \text{ kg})(6.667 \times 10^5 \text{ m/s})/[(1.602 \times 10^{-19} \text{ C})(0.8500 \text{ T})] = 10.57 \times 10^{-2} \text{ m}.$$

The spatial separation is therefore, $s = 2(r_{13} - r_{12}) = \mathbf{1.6 \times 10^{-2} \text{ m}}$.

---

4. To support the weight of the battery, the magnetic force on the wire must equal the weight of the battery. That is $F = ILB \sin\theta = mg$. Using $\theta = 90°$ we can solve for the current to obtain,

$$I = mg/LB = (0.25 \text{ kg})(9.80 \text{ m/s}^2)/[(0.50 \text{ m})(0.20 \text{ T}) = \mathbf{25 \text{ A}}.$$

In order to balance the weight, the magnetic force must be directed UP. Since the field is OUT of the page, and the force is up, RHR-1 tells us I is to the left. Therefore, the **right-hand terminal is positive**.

---

5. At point P, the field due to each wire is pointing DOWN, as determined from RHR-2. Since each wire carries the same current and is at the same distance from P, the total field is

$$B_{TOT} = 2B = 2(\mu_0 I/2\pi r) = \mu_0 I/\pi r = (4\pi \times 10^{-7} \text{ T m/A})(3.0 \text{ A})/\pi(0.50 \text{ m}) = \mathbf{2.4 \times 10^{-6} \text{ T, DOWN}}.$$

---

6. The magnitude of the force between the wires (see example 9 in the text) is

$$F = \mu_0 I_1 I_2 L/2\pi r = (4\pi \times 10^{-7}\ \text{T m/A})(25\ \text{A})(35\ \text{A})(2.5\ \text{m})/2\pi(0.75\ \text{m}) = \mathbf{5.8 \times 10^{-4}\ N}.$$

(a) If the currents are in the same direction, by RHR-2 and RHR-1, the force is one of **attraction**.
(b) If the currents flow in opposite directions through each wire, the force is one of **repulsion**.

---

7. To find the number of turns of wire, divide the total length of the wire (251 m) by the length per turn, which is just $l = 2\pi r = 2\pi(1.50 \times 10^{-2}\ \text{m}) = 0.0942$ m per turn. Thus, N = (251 m)/(0.0942 m/turn) = 2660 turns. Since the solenoid is 25.0 cm in length, n = 2660 turns/0.250 m = 10 600 turns/m. The magnetic field at the center is

$$B = \mu_0 nI = (4\pi \times 10^{-7}\ \text{T m/A})(10\,600\ \text{m}^{-1})(0.500\ \text{A}) = \mathbf{6.66 \times 10^{-3}\ T}.$$

---

8. The diagram showing the directions of the fields due to each wire is shown below.

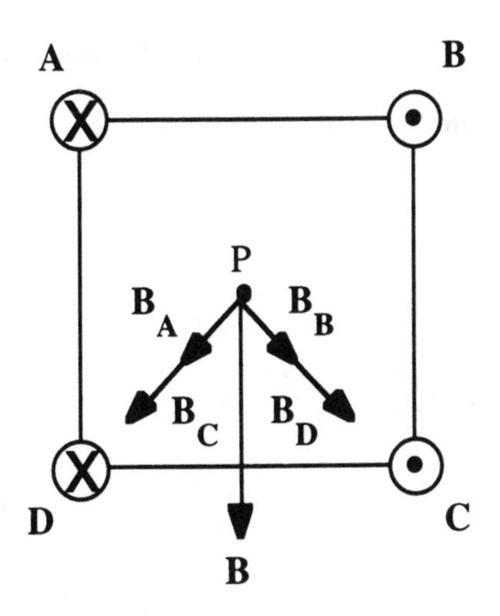

Since each wire carries the same current, and each is equidistant from the center, $r^2 = (0.20\ \text{m})^2 + (0.20\ \text{m})^2$, so r = 0.28 m. The field due to each wire is

$$B = \mu_0 I/2\pi r = (4\pi \times 10^{-7}\ \text{T m/A})(5.0\ \text{A})/2\pi(0.28\ \text{m}) = 3.6 \times 10^{-6}\ \text{T}.$$

Note that the x-components of the fields vanish, while the y-components add together. The total field is thus,

$$B_{tot} = 4B \sin 45° = 4(3.6 \times 10^{-6}\ \text{T}) \sin 45° = \mathbf{1.0 \times 10^{-5}\ T,\ DOWN}.$$

Quiz answers

1. d
2. a
3. c
4. c
5. b
6. c
7. d
8. d
9. a
10. b

MCAT REVIEW PROBLEMS

A cyclotron is a device used by physicists to study the properties of subatomic particles. Small charged particles are deposited at high speed into a circular pipe. Electric and magnetic fields then accelerate the particle to an even higher speed. Finally, when the particle reaches the desired speed, a magnetic field keeps it moving in a circle at constant speed.

In figure 1, a magnetic field pointing "into the page" keeps a charged particle traveling in a counterclockwise circle at constant speed inside the cyclotron. The magnetic force on the particle points towards the center of the circle, and has strength

$$F_{mag} = qvB,$$

where q is the particle's charge, v is its speed, and B is the magnetic field strength. As a result of this force, the particle circles the cyclotron at frequency

$$f = \frac{qB}{2\pi m},$$

where m denotes the particle's mass. The frequency (in hertz) is the number of revolutions completed by the particle per second.

In the following questions, neglect gravity.

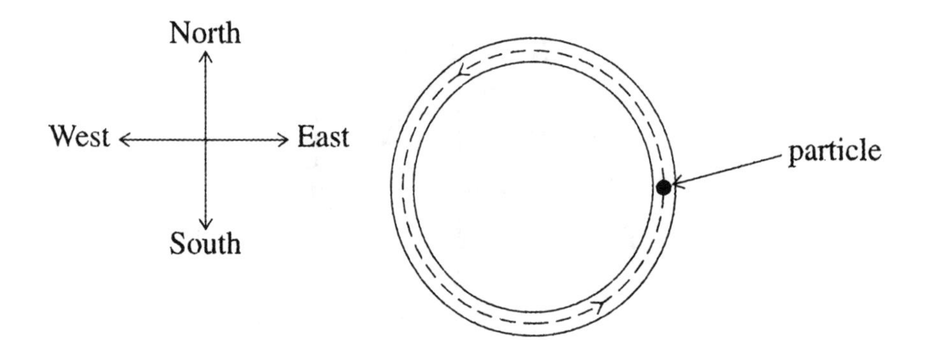

FIGURE 1

---

1. In figure 1, what is the direction (if any) of the particle's acceleration?

   A. North **B.** East **C.** West **D.** It has no acceleration

---

2. If the magnetic field in figure 1 were turned off, in which direction would the particle travel (until crashing)?

   A. North **B.** East **C.** West **D.** Nowhere; it would stop moving

---

3. An alpha particle has charge $2e$ and mass 4 amu. A proton has charge e and mass 1 amu. Let f_{alpha} and f_{proton} denote the frequencies with which those particles circle a cyclotron. If both particles experience the same magnetic field in the same cyclotron, what is $\frac{f_{\text{alpha}}}{f_{\text{proton}}}$?

   A. 2 **B.** 1 **C.** 1/2 **D.** 1/4.

---

4. In order for a cyclotron to work properly, the magnetic field must make the particle move in a circle. Which of the following particles would *not* work in a cyclotron?

   A. lithium atom (Li)
   B. positive lithium ion (Li^+)
   C. negative lithium ion (Li^-)
   D. all of the above particles would "work."

---

5. Which of the following *cannot* create a magnetic field?

   A. Electrical current flowing through a well-insulated straight metal wire.
   B. A beam of electrons moving across a cathode ray tube.
   C. Electric current flowing around a superconducting ring.
   D. Static electricity (i.e., extra electrons) built up on a stationary door knob.

ANSWERS TO MCAT REVIEW PROBLEMS

---

1. **C.** Whenever an object travels in a circle at constant speed, its acceleration points towards the center of the circle. Here, the passage reminds you of this fact, by mentioning that the magnetic force points towards the center. Since the magnetic force is (approximately) the net force, Newton's 2nd law, $\mathbf{F}_{net} = m\mathbf{a}$, implies that the acceleration also points towards the center.

---

2. **A.** When the magnetic field switches off, the particle no longer feels any forces (neglecting gravity and the "induced" electric force). In the absence of forces, an object keeps traveling at constant velocity in whatever direction it was moving. The particle in figure 1 was moving north.

   A common mistake is to say the particle would get "thrown outward" (eastward). When a car turns in a tight circle, the passenger does indeed feel thrown outward against the car door. But as your textbook explains, the passenger doesn't experience an outward force. He "tries" to travel in a straight line, while the car veers off from that straight line. So, the passenger doesn't get thrown into the car door. Instead, the passenger travels straight, and the car door veers into him! Space doesn't permit me to explain this more fully. Please see MCAT review chapter 8 and your professor for additional clarification.

---

3. **C.** From the formula $f = \frac{qB}{2\pi m}$, the frequency is proportional to q/m, the ratio of charge to mass (for fixed B). For an alpha particle, $\frac{q_{alpha}}{m_{alpha}} = \frac{2}{4}e$ per amu. For a proton, $\frac{q_{proton}}{m_{proton}} = \frac{1}{1}e$ per amu. So, compared to a proton, an alpha particle has *half* as large a charge-to-mass ratio. It therefore circles the cyclotron with half the frequency of a proton, when both particles experience the same field.

---

4. **A.** Only *charged* particles work in a cyclotron. Remember, the particle moves in a circle *because* of the "centripetal" force provided by the magnetic field, $F_{mag} = qvB$. But for a chargeless particle ($q = 0$), this force vanishes.

---

5. **D.** Moving charges create a magnetic field, but stationary charges do not (neglecting quantum effects). Stationary charges produce electric fields only.

   Importantly, any moving charges create a magnetic field, no matter whether those charges flow in a straight line or around a ring, and no matter whether they flow through a solid or through a vacuum tube (as a "beam"). Furthermore, wire insulation or other materials can't "block" the magnetic fields generated by those moving charges.

CHAPTER 22
Electromagnetic Induction

PREVIEW

Whenever the magnetic flux changes through a loop, an emf is generated in the loop. This observation is given substance in the form of Faraday's Law of Electromagnetic Induction which is presented in this chapter. Electromagnetic induction is used to explain the operation of electric generators and useful circuit devices called inductors along with the operation of the common transformer.

QUICK REFERENCE

Important Terms

Induced current
The current produced in a conducting coil when it is exposed to a changing magnetic field or its area or orientation changes relative to a magnetic field.

Induced emf
The emf in a coil or loop when it is exposed to a changing magnetic field or its area or its orientation changes in a magnetic field. The induced emf MAY produce an induced current if the coil or loop is a conductor.

Motional emf
The emf produced across a conductor due to its motion through a magnetic field.

Magnetic flux
The product of the area of a loop, the magnetic field through the loop and the cosine of the angle that the magnetic field makes with a line drawn normal to the plane of the loop. It is the change of magnetic flux through a loop which produces the induced emf.

Induced magnetic field
The magnetic field which is produced by an induced current.

Back emf
The induced emf produced in the rotating coils of a motor whose polarity is opposite the polarity of the emf applied to the motor.

Mutual induction
The process by which a changing current in one circuit produces a changing magnetic field which induces a current in another circuit.

Self-induction
The process by which a changing current in a circuit induces an emf in the same circuit.

Inductors
Coils which are used as circuit elements to exploit their property of self-induction.

Transformer
A device which uses mutual inductance between primary and secondary coils to increase or decrease an ac voltage.

Induced emfs

The emf induced in a **rod moving perpendicularly with respect to a magnetic field** is

$$\text{emf} = vBL \qquad (22.1)$$

The emf induced in a **coil of N loops** (Faraday's Law)

$$\text{emf} = -N\frac{\Delta\Phi}{\Delta t} \tag{22.3}$$

The emf produced by a **generator**

$$\text{emf} = NAB\omega \sin \omega t \tag{22.4}$$

The emf due to **mutual induction**

$$(\text{emf})_2 = -M\frac{\Delta I_1}{\Delta t} \tag{22.7}$$

The emf due to **self-induction**

$$\text{emf} = -L\frac{\Delta I}{\Delta t} \tag{22.9}$$

Magnetic Flux

The magnetic flux through a **plane area or coil**

$$\Phi = BA \cos \phi \tag{22.2}$$

Inductance

Mutual inductance of **two coils**

$$M = \frac{N_2\Phi_2}{I_1} \tag{22.6}$$

Self-induction of a coil

$$L = \frac{N\Phi}{I} \tag{22.8}$$

Motors

The **current** drawn by a motor (emf is the "back emf" of the motor)

$$I = \frac{V - \text{emf}}{R} \tag{22.5}$$

Energy in Inductors

The **energy** stored in an inductor due to **self inductance** is

$$\text{Energy} = \frac{1}{2}LI^2 \tag{22.10}$$

The **energy density** in a **magnetic field**

$$\text{Energy density} = \frac{1}{2\mu_0}B^2 \tag{22.11}$$

Transformers

The **transformer equation**

$$\frac{V_s}{V_p} = \frac{N_s}{N_p} \tag{22.12}$$

DISCUSSION OF SELECTED SECTIONS

22.1 Induced EMF and Induced Current

It can be demonstrated that relative motion between a magnet and a coil which forms part of a circuit will lead to an induced current flowing in the circuit containing the coil. The induced current is seen to change direction when the direction of the relative motion is reversed. (Perhaps your instructor will demonstrate this.) Further investigation indicates that two additional actions can lead to an induced current in the coil: A change in the area of the coil and a change in the orientation of the coil with respect to the magnetic field. The emf which is produced in the coil which drives the induced current is called the "induced emf". The induced emf exists whether or not the coil is part of a closed circuit. However, a closed circuit is necessary for the induced current to flow.

22.2 Motional EMF

Motional emf is an emf induced in a conductor moving through a magnetic field. The emf arises from a charge separation in the conductor due to the magnetic forces exerted on charges in the conductor. It depends on the component of magnetic field perpendicular to the motion of the conductor, the speed and length of the conductor. This emf is present whether or not the conductor is part of a closed circuit.

Example 1 .
The horizontal component of the earth's magnetic field at a certain location is 2.6×10^{-5} T and points due north. What motional emf appears in the 0.80 m vertical radio antenna of a car traveling east with a speed of 95 km/h? What motion emf appears in the antenna if the car turns north?

Equation (22.1) applies to the emf induced in a conductor moving perpendicular to a magnetic field. If the car is going east, then its vertical antenna is moving perpendicular to the horizontal component of the earth's magnetic field. The induced emf is

$$\text{emf} = vBL = (95 \times 10^3 \text{ m}/3600 \text{ s})(2.6 \times 10^{-5} \text{ T})(0.80 \text{ m}) = 5.5 \times 10^{-4} \text{ V}.$$

If the car is moving north, then the vertical antenna is moving parallel to the magnetic field and no emf will be generated.

Example 2
A rectangular loop of wire is moving at 22 m/s as shown through a region of constant magnetic field of magnitude 0.55 T. What emf is induced in the top and bottom wires of the loop? What emf is induced in the side wires? If the loop has a resistance of 0.015 Ω, how much current flows through it?

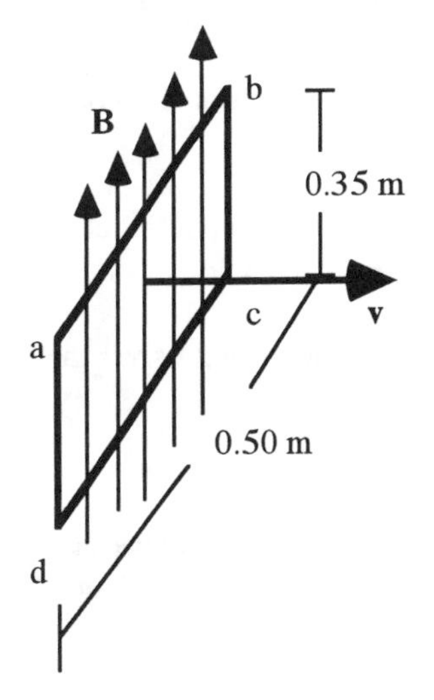

The top (ab) and bottom (cd) wires are moving perpendicular to the magnetic field, so equation (22.1) gives the emf across each of them to be

$$\text{emf} = vBL$$
$$\text{emf} = (22 \text{ m/s})(0.55 \text{ T})(0.50 \text{ m}) = 6.1 \text{ V}.$$

The end wires are oriented along the direction of the magnetic field so no emf will be produced in them. The current in the loop is driven by the NET emf induced in it. The polarity of the emf in the top wire is (a = +) the same as the polarity of the emf induced in the bottom wire (d = +). The net emf is the sum of the emfs around the loop.

$$\text{net emf} = +6.0\text{ V} - 6.0\text{ V} + 0\text{ V} + 0\text{ V}$$
$$\text{net emf} = 0\text{ V}.$$

Since the net emf is zero, there can be no current in the loop.

22.3 Magnetic Flux

As indicated previously, an emf can be produced in a coil by changing the magnetic field through the coil, the area of the coil and/or the orientation of the coil. These three methods of producing an emf are summarized by introducing the concept of magnetic flux Φ = BA cos , where B is the magnitude of the magnetic field penetrating the coil, A is the area of the coil and is the angle between the magnetic field vector and the normal to the surface. An emf can be produced in a coil by changing the magnetic flux through the coil. (Faraday's Law).

Example 3
A long solenoid of 150 turns per cm and radius 1.5 cm carries a dc current of 0.50 A. What is the total magnetic flux through a 0.25 m length of the solenoid due to its OWN magnetic field?

Previously, it was found that the magnetic field produced inside a long current-carrying solenoid is constant in magnitude and directed along the axis of the solenoid; $B = \mu_0 nI$.

Each turn of the solenoid has a flux $\Phi = BA\cos 0° = BA$ through it. The total flux through the solenoid is then

$$\Phi_T = NBA = N\mu_0 nIA$$
$$\Phi_T = \mu_0 n^2 LIA$$
$$\Phi_T = (4\pi \text{ x } 10^{-7}\text{ T m/A})(150 \text{ X } 10^2\text{ turns/m})^2(0.25\text{ m})(0.50\text{ A})(\pi)(1.5 \text{ X } 10^{-2}\text{ m})^2$$
$$\Phi_T = 2.5 \text{ x } 10^{-2}\text{ Wb}$$

22.4 Faraday's Law of Electromagnetic Induction

A change in the magnetic flux through a coil produces an induced emf in the coil given by (22.3).

Example 4
The current in the solenoid of example 3 changes from 0.50 A to 0 A in 1.5 ms. What is the magnitude of the average emf induced in the solenoid?

Faraday's law applied to the solenoid gives the magnitude of the emf induced in N turns.

$$\text{emf} = N(\Delta\Phi/\Delta t)$$

The flux changes through the solenoid because the decreasing current causes the magnetic field inside the solenoid to decrease to zero. The flux change through one turn is $\Delta\Phi = \mu_0 nIA$.

The magnitude of the emf induced in the solenoid is then

$$\text{emf} = (N\mu_0 nIA)/\Delta t = (2.5 \text{ x } 10^{-2}\text{ Wb})/(1.5 \text{ X } 10^{-3}\text{ s}) = 17\text{ V}$$

22.5 Lenz's Law

Faraday's Law gives the emf induced in a coil by a changing magnetic flux. Lenz's Law gives the polarity of the induced emf. The induced emf ALWAYS has a polarity which WOULD produce a current whose magnetic field WOULD oppose the change in flux that caused the induced emf in the first place. A key word here is WOULD. It is not necessary that a current actually flow in a closed circuit. The emf will be produced and its polarity will be given by Lenz's Law even if no closed circuit exists.

Example 5
An electromagnet generates a magnetic field which "cuts" through a coil as shown. What is the polarity of the emf generated in the coil if the applied field, B (a) points to the right and is increasing? (b) points to the right and is decreasing? (c) is pointing to the left and increasing? (d) is pointing to the left and is decreasing?

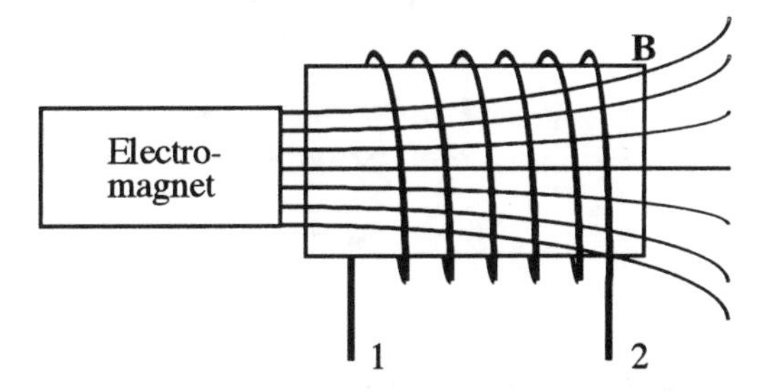

(a) If the magnetic field through the coil points to the right and is increasing, then the flux through the coil is increasing. Lenz's Law demands that if a current could flow then it would produce an "induced magnetic field", $\mathbf{B_I}$ which would oppose the increase in flux. The action of $\mathbf{B_I}$ is to produce a new flux through the coil to attempt to increase the total flux. $\mathbf{B_I}$ must point to the left, also. Now imagine that a resistor has been placed in series with the coil to provide a closed loop. What current direction through the resistor is needed to produce $\mathbf{B_I}$ in the coil? RHR-2 gives the current to be in the direction shown in the drawing. The resistor "sees" the coil as a source of emf much like a battery. Current leaves a source of emf at the + terminal and enters at the - terminal, so point 1 is POSITIVE while point 2 is NEGATIVE.

(b) If **B** points to the right and is decreasing, then $\mathbf{B_I}$ will attempt to produce a flux to increase the total flux through the coil. In this case $\mathbf{B_I}$ must be directed to the right and will be produced by a current flowing to the left through the resistor. This corresponds to point 1 NEGATIVE and point 2 POSITIVE.

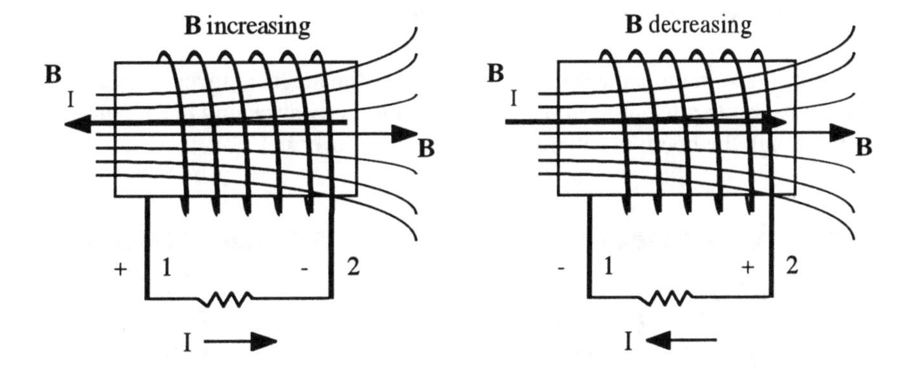

(c) and (d) The analysis of the situation where the applied field, **B**, points to the left is very similar to the analysis in parts (a) and (b). The drawing below gives the results.

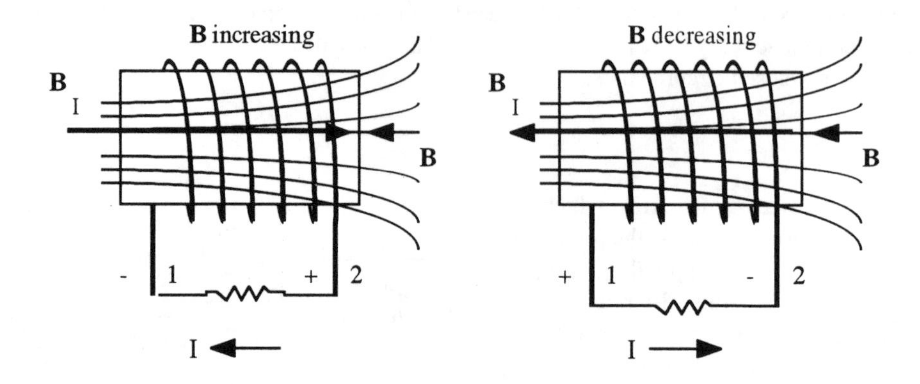

Sometimes Lenz's Law can be interpreted in a much more general way which may be useful. The following example illustrates one such interpretation.

Example 6
A metal hoop is suspended from a support by a string and a magnet is moved relative to the hoop as shown in the drawing. What is the direction of the induced current in the hoop if the magnet is moved (a) north end first toward the hoop? (b) south end first away from the hoop (c) south end first toward the hoop? (d) north end first away from the hoop?

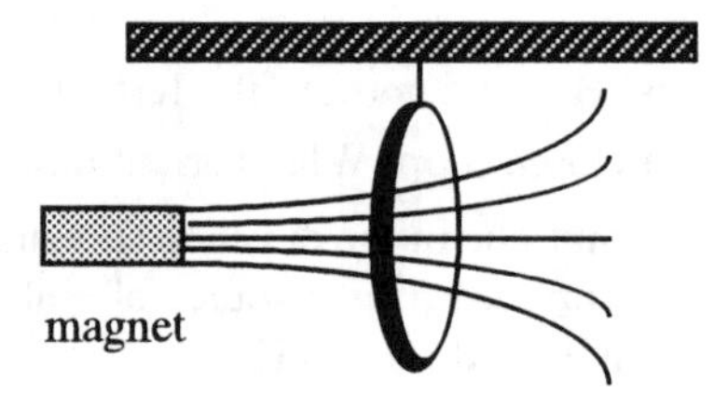

(a) Refer to the drawing shown below. If the north end of the magnet is moved toward the hoop, then the action of the induced current is to attempt to decrease the flux through the hoop. The flux through the hoop is increasing ultimately because the magnet is moving. The induced current will produce a north pole on the left of the loop to repel the magnet in an attempt to stop the relative motion. The left side of the coil is a north pole, hence the induced magnetic field, $\mathbf{B}_I$, is directed from right to left through the hoop. The current direction needed to produce this magnetic field is, according to RHR-2, CW as seen by the "eye".

(b) the situation here is analyzed in the same way as in part (a). As the magnet moves away from the hoop, the hoop develops a south pole on its left to attempt to stop the relative motion by attracting the magnet. The induced field is then from left to right through the hoop. The current as seen by the "eye" is CCW.

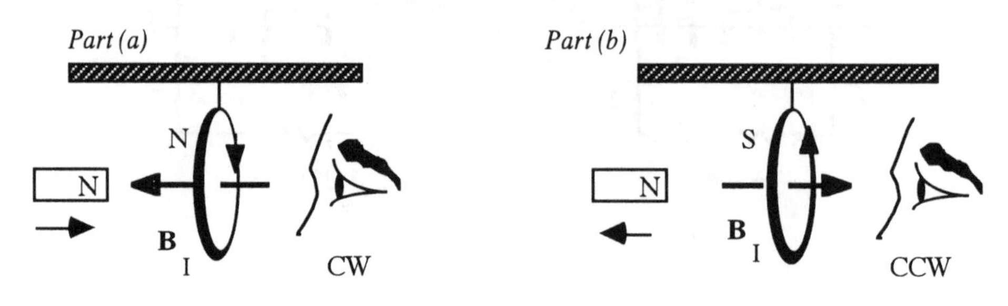

(c) Repeating of the above analysis for the magnet moving south end first toward the hoop results in a current flowing CCW. See the drawing below.

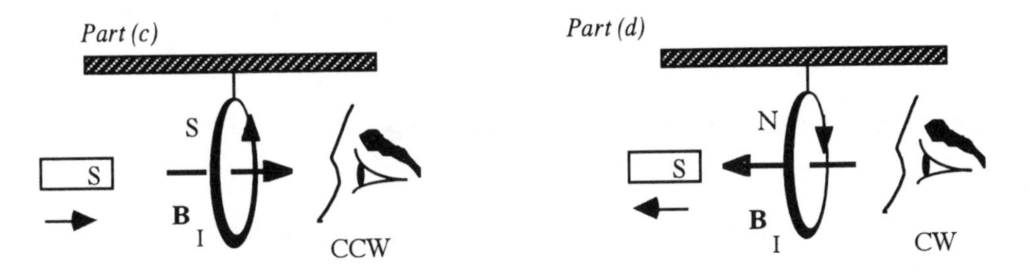

(d) Similarly, if the magnet moves away from the hoop with its south end toward the hoop a CW current will result as shown in the drawing.

22.7 The Electric Generator

Electromagnetic induction is used to produce an emf in a generator which can be used to power external devices such as motors. The simplest generator is described extensively in the text. It produces an emf given by (22.4).

Example 7
The maximum strength of the earth's magnetic field is about 7.0×10^{-5} T near the south magnetic pole. In principle, this field could be used with a rotating coil to generate 60.0 Hz ac electricity. What is the minimum number of turns (area per unit turn = 0.016 m^2) that the coil must have so as to produce an rms voltage of 120 V?

The peak emf produced by the generator is given by (22.4) to be $NAB\omega$. The rms voltage is then $V = (NAB\omega)/\sqrt{2}$. The number of turns needed is

$$N = \frac{\sqrt{2}\,V}{BA\omega} = \frac{\sqrt{2}\ (120\ \text{V})}{(7.0 \times 10^{-5}\ \text{T})(0.016\ \text{m}^2)(2\pi)(60.0\ \text{Hz})} = 4.0 \times 10^5$$

The very desirable induced emf which appears in a generator becomes a "back emf" in a motor. The back emf retards the motion of the armature in a motor by acting with a polarity opposite that of the voltage which is applied to run the motor. The net current "drawn" by an electric motor is given by (22.5).

Example 8
A motor is designed to operate on 117 V and draws a current of 37.5 A when it first starts up. At its normal operating speed, the motor draws a current of 4.10 A. Obtain (a) the resistance of the armature coil, (b) the back emf developed at normal speed, (c) the back emf at one third normal speed.

(a) When the motor first starts up, its back emf is zero since its armature is not yet turning, hence no induced emf can appear in it. Under these conditions (22.5) gives the resistance of the armature coil to be

$$R = V/I = (117\ \text{V})/(37.5\ \text{A}) = 3.12\ \Omega.$$

(b) At normal operating speed the rms back emf is given by (22.5) to be

$$\text{emf} = V - IR = 117\ \text{V} - (4.10\ \text{A})(3.12\ \Omega) = 104\ \text{V}.$$

(c) The peak emf at any speed is given by (22.4). Applying this equation to normal speed and one third speed and dividing the resulting equations gives the rms back emf at one third speed.

$$\text{emf} = (104 \text{ V})(1/3) = 34.7 \text{ V}$$

22.8 Mutual Inductance and Self-Inductance

Mutual Inductance

When two coils are placed close together and one of the coils (coil 1) carries a changing current, the second coil (coil 2) will develop an induced emf as the magnetic flux produced by coil 1 changes through it. In many cases, the flux through coil 2 is proportional to the current flowing in coil 1. The constant of proportionality is called the mutual inductance of the coils, M. The emf produced in coil 2 by a change in the current in coil 1 is given by (22.7). The mutual inductance of two coils depends on the geometry of the arrangement and any magnetic material which may be placed in the coils.

Example 9

A current in one coil changes uniformly from 2.0 A to 0.50 A in 45 ms. A voltage whose magnitude is 5.5 V is seen to develop in a second coil located close to the first. What is the mutual inductance of the two coils?

The magnitude of the voltage is given by (22.7) without the minus sign. Therefore, the mutual inductance is

$$M = \frac{(\text{emf})_2}{\frac{\Delta I_1}{\Delta t}} = \frac{5.5 \text{ V}}{\frac{1.5 \text{ A}}{(45 \text{ X } 10^{-3} \text{ s})}} = 0.17 \text{ H}$$

Self-Inductance

Self-induction occurs when a coil carrying a changing current induces an emf in itself. Of course, this comes about because the time changing current produces a changing magnetic field which "cuts" through the coil. The associated changing flux through the coil produces a self-induced emf in the coil. Often, the flux through a coil is proportional to the current through the coil. The constant of proportionality in this case is called the self-inductance of the coil. The emf developed in a coil due to self-induction is given by (22.9).

Example 10

A solenoid of 550 turns has a self- inductance of 5.2 mH. (a) What is the flux per turn through the solenoid when it carries a current of 25 mA? (b) What is the magnitude of the emf induced in the solenoid if the current changes uniformly to 0 A in 2.0 ms?

(a) The magnetic flux per turn through the solenoid is given by (22.8).

$$\Phi = LI/N = (5.2 \text{ X } 10^{-3} \text{ H})(25 \text{ X } 10^{-3} \text{ A})/(550 \text{ turns}) = 2.4 \text{ X } 10^{-7} \text{ Wb/turn}$$

(b) The magnitude of the induced emf is given by (22.9) without the minus sign.

$$\text{emf} = L\frac{\Delta I}{\Delta t} = (5.2 \text{ mH})\frac{25 \text{ mA}}{2.0 \text{ ms}} = 65 \text{ mV}.$$

Current-carrying inductors store energy in their magnetic fields much like charged capacitors store energy in their electric fields. As a matter of fact, the mathematical expression for the energy stored in inductors is similar in form to the expression for energy stored in capacitors. The storage of energy in these devices is of great importance in ac circuits which are studied in the next chapter.

22.9 Transformers

The transformer is probably the single most important application of mutual inductance in our technological world. It allows the "stepping up" and "stepping down" of time changing voltages. This is very important in the transmission of ac electricity. The relationship between the primary and secondary voltages and the "turns ratio" of the ideal transformer is given by (22.12) while the relationships between the primary and secondary currents and the turns ratio is given by (22.13).

Example 11

A particular "ac adapter" for a calculator is a transformer designed to be plugged into a wall socket and to operate on 120 V. It produces an output secondary voltage of 5.7 V and an output current of 240 mA. (a) What is the turns ratio of the adapter? (b) If the adapter were an ideal transformer, how much current would it draw from the wall socket? (c) If the adapter really draws 50.0 mA from the wall socket, what percentage of the power is lost in the transformer?

(a) The turns ratio can be found from (22.12).

$$\frac{N_s}{N_p} = \frac{V_s}{V_p} = \frac{5.7 \text{ V}}{120 \text{ V}} = 0.048$$

This can also be written as 1:21.

(b) The current drawn from the wall socket is the primary current in the transformer. Equation (22.13) gives

$$I_p = \left(\frac{V_s}{V_p}\right) I_s = \left(\frac{5.7 \text{ V}}{120 \text{ V}}\right)(240 \text{ mA}) = 11 \text{ mA}$$

(c) The power input to the primary coil is $P_p = I_p V_p = (50.0 \text{ X } 10^{-3} \text{ A})(120 \text{ V}) = 6.0 \text{ W}$. The power available at the secondary is $P_s = I_s V_s = (240 \text{ X } 10^{-3} \text{ A})(5.7 \text{ V}) = 1.4 \text{ W}$.

The percentage of power lost is

$$\% \text{ power lost} = \frac{6.0 \text{ W} - 1.4 \text{ W}}{6.0 \text{ W}} \text{ X } 100 \% = 77 \%$$

PRACTICE PROBLEMS

1. A 1.0 m long conducting rod is moving perpendicular to a 0.35 T magnetic field. How fast must the rod move in order to produce an emf of 12 V across its ends?

2. A conducting bar such as that in Figure 22.4b in the text moves perpendicular to a 1.5 T magnetic field with a constant speed of 150 m/s. The conducting bar delivers a current of 5.5 A to a 6.0 Ω load. How long is the bar? How much force is necessary to keep the bar moving at constant speed?

3. A cube, 0.50 m on a side, is placed in a uniform magnetic field of strength 0.45 T oriented in the direction of a diagonal of the front face of the cube. Find the magnetic flux through each of the faces of the cube. What is the total magnetic flux through the cube?

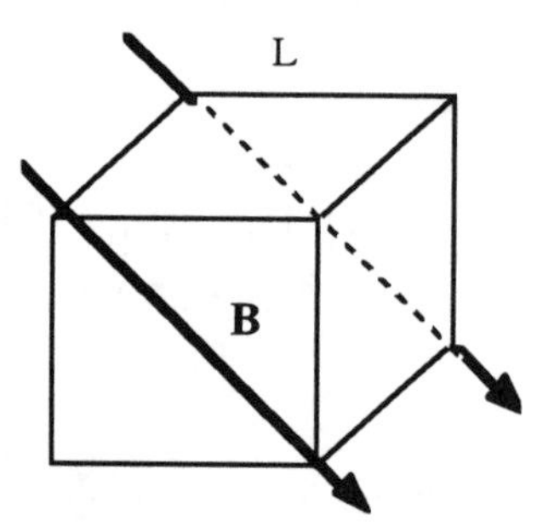

4. Use Faraday's Law to show that an emf = vBL is induced across a conducting bar of length, L, moving with a constant speed, v, perpendicular to a magnetic field of magnitude, B.

5. A 0.25 m long coil consists of 560 square turns 6.5 cm on a side. The coil is placed in a uniform magnetic field of strength 1.2 T. Initially the coil is oriented so that its axis coincides with the magnetic field direction. The coil is then rotated 90 ° about an axis perpendicular to the magnetic field direction in 0.20 ms. What is the magnitude of the emf induced in the coil?

6. The speed sensor for an automotive cruise control consists of a magnet mounted on the drive shaft along with a coil of wire mounted on the chassis. As the magnet passes the coil, an emf is induced for a short period of time. The number of induced emfs occurring per second allows a computer to determine the drive shaft speed and ultimately the speed of the car. What is the polarity of the emf when the magnet is at the three positions A, B, and C shown in the drawing? Assume the north pole of the magnet is closest to the coil.

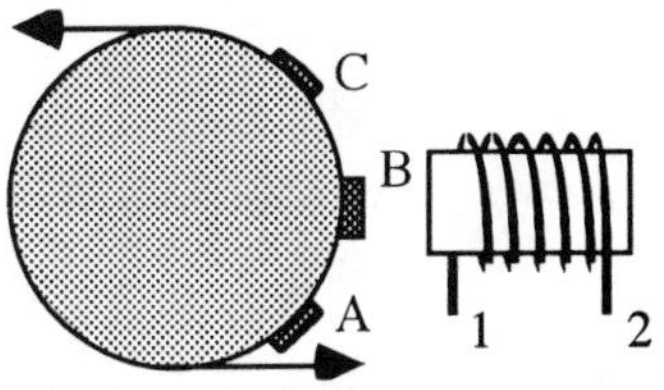

C
B
A
1
2

7. **CHALLENGE PROBLEM** A ring of aluminum is placed over the iron core of an electromagnet as shown in the drawing. The current in the coil of the electromagnet is 60.0 Hz ac. Use Lenz's Law to show that the flux change due only to the time changing magnetic field through the ring is insufficient to explain why the ring is repelled from the electromagnet.

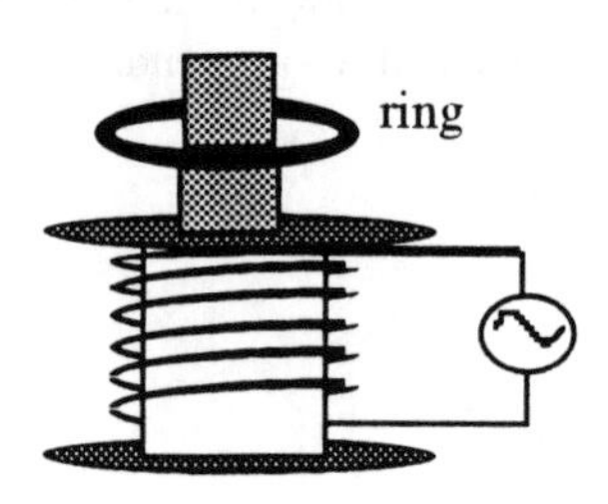

8. A generator consists of 150 rectangular turns of wire. Each turn has an area of 0.050 m^2. The generator is rotated in a 1.6 T magnetic field and produces an rms emf of 25 V. What is the frequency of the emf produced by the generator?

9. An experimenter needs an inductor with a self-inductance of 0.45 mH. The coil must be no longer than 2.5 cm and no more than 0.50 cm in diameter. How many turns of wire will the experimenter need if the coil can be assumed to be a perfect solenoid?

10. A 1:150 transformer has a primary voltage of 4500 V. (a) What is the voltage in the secondary coil? (b) If the transformer consumes 45 W, what are the currents in the primary and secondary coils?

HELPFUL SUGGESTIONS

1. It is only the component of the magnetic field in the direction of the normal that contributes to the magnetic flux through the surface. An analogy with water flow can be helpful in thinking about magnetic flux. Imagine that the lines of **B** correspond to the lines of flow for water. The magnetic flux through a surface can be thought of as the amount of water flowing through the surface in a given time. Only water flowing along the normal to the surface actually makes it through. Any water flowing perpendicular to the normal never really flows through the surface at all. This analogy can be used to find the net magnetic flux through a closed surface such as a cube. If the cube were made of screen material and immersed in flowing water, there would be no net flow into or out of the cube. What goes in must come out. In a similar way, a closed surface immersed in a magnetic field will have no net flux through it. Again, what goes in must come out.

2. The negative sign in Faraday's Law simply tells us that the polarity of the induced emf is opposite the change in flux which produced it. In actual application Faraday's law is used to give the magnitude of the emf. Lenz's law is then used to determine the actual polarity of the emf.

3. Faraday's Law applies to an emf developed around a closed loop when the magnetic flux through the loop is changing. The loop does NOT have to be a real conducting loop. Even an imaginary loop will do.

4. When you try to determine the polarity of an emf using Lenz's Law, it is often helpful to imagine a complete circuit in which current flows even if your problem does not include a complete circuit. Mentally connect a resistor between the points where you want to find the polarity of the emf. An imagined induced current flowing through the resistor will always flow from high to low potential.

5. The induced emf in a coil can be used to drive a current in an external circuit. In this way power can be supplied to the external circuit. The power comes from whatever agent is responsible for changing the flux through the coil; that is, whatever agent produces the rotation of the coil, the change in the area of the coil or the change in the magnetic field through the coil. This is why a generator handle is hard to turn when it is used to power a light bulb. You must supply the power by exerting a force on the crank handle.

6. Real transformers lose electrical power due to resistive heating and vibration of the coils. It is only for ideal transformers that the power into the primary equals the power out of the secondary.

EVERYDAY PHYSICS

1. Many of the concepts in this chapter can be demonstrated with the aid of a fairly sensitive galvanometer, a quantity of wire and a reasonably strong magnet. Galvanometers can be purchased or borrowed, wire is easy to obtain and strong magnets can be found in old speakers.
   Wind your own coil, attach it to the galvanometer and experiment with the emf induced by relative motion between the coil and magnet. (See Figure 22.1a in your text.) Try few windings, then many windings. Change the speed at which you move the magnet or coil. Switch the ends of the magnet. Do you get the expected results?

2. Interesting things can be done with an ac electromagnet with a soft iron core. Cut a ring from a large metal pipe or make a ring by soldering the ends of a piece of heavy wire together. Experiment with this as suggested by Practice Problem 7. Try thick rings and thin rings. Hold the rings over the iron core and note the heating of the rings. Cut the ring and experiment again. What happens? Why? You may want to ask your instructor to demonstrate these things if you cannot find a suitable electromagnet.

3. Use a continuous ring of wire suspended from a support along with a strong magnet to experiment with the situations discussed in Example 6. Are the results of Example 6 verified?

CHAPTER QUIZ

1. A 1.0 m long rod is moving with a constant speed of 21 m/s perpendicular to a 0.15 T magnetic field. What emf appears across the ends of the rod?
   a. 3.2 V b. 7.0 mV c. 140 V d. 0 V

2. A 1.0 m long rod is moving at a constant speed of 21 m/s perpendicular to a 0.15 T magnetic field. The rod is connected to an 5.0 Ω external resistor. How much force is required to keep the rod moving at constant speed?
   a. 6.4 N b. 0.38 N c. 8.0 N d. 0.094 N

3. A uniform magnetic field of magnitude 2.0 T acts at an angle of 22 ° to the plane of a 0.50 m X 0.25 m rectangular loop. What is the magnetic flux through the loop?
   a. 0.094 Wb b. 0.23 Wb c. 0.10 Wb d. 0.75 Wb

4. The north end of a magnet is moved toward a single loop of wire along the normal to the plane of the loop. What is the direction of the induced current, if any, in the loop as viewed from the magnet?
   a. No induced current flows
   b. counterclockwise
   c. clockwise
   d. Toward the magnet.

5. A magnet is moved quickly away from 0.065 m^2 loop of wire in such a way that the uniform magnetic field perpendicular to the coil changes from 1.5 T to 0.30 T in 65 ms. What is the magnitude of the average emf induced in the coil?
   a. 0.60 V b. 1.5 V c. 0.30 V d. 1.2 V

6. A counterclockwise torque is required to turn the coil of a generator which is supplying current to a light bulb. The magnetic field produced by the induced current in the coil generates________torque on the coil.
   a. no b. a CW c. an infinite d. a CCW

7. How many turns must a 0.10 m X 0.25 m rectangular generator coil turning in a 0.15 T magnetic field have in order to produce 120 V, rms, of 60.0 Hz electricity?
   a. 85 b. 120 c. 250 d. 640

8. The magnitude of the instantaneous emf induced in a generator coil is the greatest when the_________.
   a. flux through the coil is greatest
   b. magnetic field is the greatest
   c. magnetic field strength is the smallest
   d. the magnetic flux through the coil is zero

9. The self-inductance of a coil depends on________.
   a. the current through the coil
   b. the geometry of the coil
   c. the flux through the coil
   d. the voltage applied to the coil

10. The current through an inductor is increased by a factor of 4. How much does the energy stored in the inductor change?
    a. It increases by a factor of 4.
    b. It increases by a factor of 2.
    c. It decreases by a factor of 16.
    d. It increases by a factor of 16.

11. An ideal transformer receives 35 W of electrical power at 15 V from a power supply. It can deliver 5.0 A to a circuit connected to its secondary coil. What is the turns ratio of the transformer?
    a. 0.47 b. 2.1 c. 7.0 d. 3.0

SOLUTIONS AND ANSWERS

Practice Problems

---

1. Equation (22.1) applies to a rod moving through a magnetic field.

$$v = (\text{emf})/(BL) = (12\text{ V})/[(0.35\text{ T})(1.0\text{ m})] = \mathbf{34\ m/s}.$$

---

2. The emf produced by the moving bar is given by equation (22.1). This emf provides the current in the load. According to Ohm's law emf = IR. The length of the rod is

$$L = IR/vB = [(5.5\text{ A})(6.0\ \Omega)]/[(150\text{ m/s})(1.5\text{ T})] = \mathbf{0.15\ m}.$$

The force which must be exerted to keep the rod moving at a constant speed is

$$F = ILB = (5.5\text{ A})(1.5\text{ m})(0.15\text{ T}) = \mathbf{1.2\ N}.$$

---

3. The flux through an area is given by equation (22.2). For the:

front face: $\Phi_1 = BL^2 \cos 90° = \mathbf{0\ Wb}$.

back face: $\Phi_2 = BL^2 \cos 90° = \mathbf{0\ Wb}$.

top face: $\Phi_3 = BL^2 \cos 135° = \mathbf{-0.080\ Wb}$.

bottom face: $\Phi_4 = BL^2 \cos 45° = \mathbf{0.080\ Wb}$.

right face: $\Phi_5 = BL^2 \cos 135° = \mathbf{-\ 0.080\ Wb}$.

left face: $\Phi_6 = BL^2 \cos 45° = \mathbf{0.080\ Wb}$.

The total flux through the cube is the sum of the fluxes through the faces, so $\mathbf{\Phi = 0\ Wb}$.

---

4. Faraday's law can be used with real or imaginary loops. Construct an imaginary loop containing the moving bar as shown in the drawing.

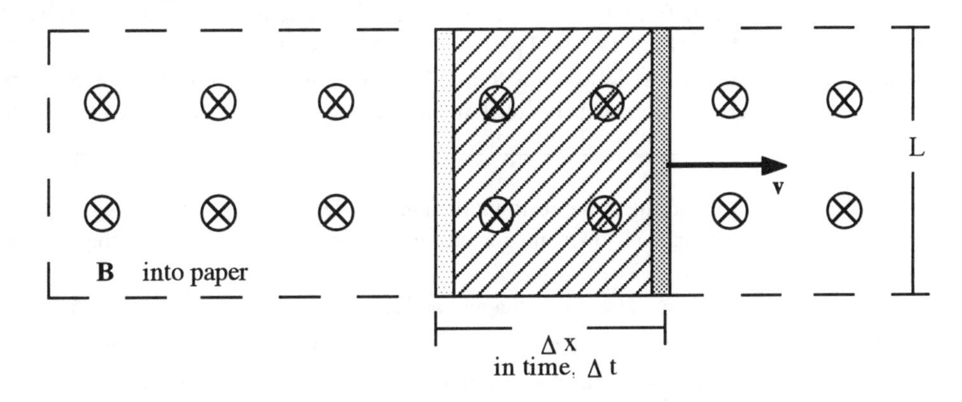

In time, Δt, the bar moves a distance, Δx, through the magnetic field. The flux change through the loop is entirely due to the change in area of the loop which is $\Delta A = L\,\Delta x$. flux change through the loop is

$$\Delta\Phi = B\,\Delta A = BL\,\Delta x.$$

Faraday's law gives the magnitude of the induced emf to be

$$\text{emf} = \Delta\Phi/\Delta t = BL\,(\Delta x/\Delta t).$$

Now $\Delta x/\Delta t$ is the speed of the bar, so

$$\text{emf} = vBL.$$

---

5. The flux change through the coil is due to a change in the orientation of the coil. The flux through the coil before it is rotated is

$$\Phi_0 = BNA \cos 0° = BNA.$$

The flux through the coil after it is rotated is

$$\Phi = BNA \cos 90° = 0 \text{ Wb}.$$

The change in flux during the 0.20 s time period is then

$$\Delta\Phi = BNA$$

and the magnitude of the emf induced in the coil is

$$\text{emf} = BNA/\Delta t = (1.2 \text{ T})(560)(6.5 \times 10^{-2} \text{ m})^2/(0.20 \times 10^{-3} \text{ s})$$
$$\text{emf} = \mathbf{1.4 \times 10^4 \text{ V}}.$$

---

6. Mentally connect a resistor between points 1 and 2 of the coil.

When the magnet is at **position A**, it is moving toward the coil and causing a change in the magnetic flux through the coil. A generalized interpretation of Lenz's law requires that the coil react to try to stop this motion. Therefore, the left end of the coil will become a north pole to repel the north pole of the magnet. The current in the coil needed to produce a north pole at the left end flows through the coil from 2 to 1. The current flows from the coil at 1, through the imaginary resistor, and back into the coil at 2. The **polarity of 1 is** then **POSITIVE** and **the polarity of 2 is NEGATIVE.**

When the magnet is at **position C**, then it is moving away from the coil. The coil will produce a south pole at its left end to attract the magnet in an attempt to stop its motion, hence the flux change. The current in the coil must flow through the coil from 1 to 2. The current in the imaginary resistor then flows from 2 to 1 which indicates that **point 2 is POSITIVE** and **point 1 is NEGATIVE**.
When the magnet is at **position B**, the left end of the coil is changing from a north to a south pole. At this position the flux through the coil is momentarily constant and **no emf is induced in the coil** at all.

---

7. An ac current in the electromagnet produces the following magnetic fields: B is up and increasing, B is up and decreasing, B is down and increasing, B is down and decreasing. The induced current in the ring produces the following induced fields, B_I.
B up and increasing $\Rightarrow$ B_I is down $\Rightarrow$ The electromagnet and ring repel since they are north to north.
B is up and decreasing $\Rightarrow$ B_I is up $\Rightarrow$ The electromagnet and ring attract since they are north to south..
B is down and increasing $\Rightarrow$ B_I is up $\Rightarrow$ The electromagnet and the ring are south to south and repel.
B is down and decreasing $\Rightarrow$ B_I is down $\Rightarrow$ The electromagnet and ring are south to north and attract.
The net force on the ring over one cycle of the ac current is obviously zero. The force of gravity over one cycle of the ac current is not zero, so if the time changing magnetic field through the ring is the only way the flux can change, the net force on the ring could not be zero and the ring will not float.

In the actual case, the flux through the ring changes as the ring moves up and down in the magnetic field since the magnetic field is not uniform. This flux change gives rise to an additional induced magnetic field which generates an average repulsive force on the ring.

---

8. The peak emf produced by the generator is given by equation (22.4) to be NABω, so the rms emf produced is NAB$\omega/\sqrt{2}$. Now the frequency of the emf is

$$f = \frac{\omega}{2\pi} = \frac{\sqrt{2}\ (\text{rms})}{2\pi NAB} = \frac{\sqrt{2}\ (25\ \text{V})}{2\pi(150)(0.050\ \text{m}^2)(1.6\ \text{T})} = \mathbf{0.47\ Hz}$$

---

9. The self-inductance of a solenoid is given in Example 10 in the text to be $L = \mu_0 n^2 Ax$. Here x is the length of the solenoid. Now

$$n = \sqrt{\frac{0.45 \times 10^{-3}\ \text{H}}{(4\pi \times 10^{-7}\ \text{T m/A})(\pi)(0.25 \times 10^{-2}\ \text{m})^2(2.5 \times 10^{-2}\ \text{m})}} = 27\,000\ \text{m}^{-1}$$

The total number of turns needed is then N = nx = **680**.

---

10. (a) Equation (22.12) gives

$$V_s = (N_s/N_p)V_p = (1/150)(4500\ \text{V}) = \mathbf{3.0 \times 10^1\ V}.$$

(b) The current delivered to the primary is

$$I_p = P/V_p = (45\ \text{W})/(4500\ \text{V}) = \mathbf{1.0 \times 10^{-2}\ A}.$$

If no power is lost in the transformer, the current in the secondary will be

$$I_s = P/V_s = (45\ \text{W})/(3.0 \times 10^1\ \text{V}) = \mathbf{1.5\ A}.$$

Quiz answers

1. a	3. b	5. d	7. b	9. b	11. a
2. d	4. b	6. b	8. d	10. d	

MCAT REVIEW PROBLEMS

In this passage, a "loop" refers to any closed segment of wire, such as the metallic wire rectangle in figure 1. If the plane of the loop is perpendicular to the magnetic field, and if the magnetic field is uniform over a discrete region of space (as in figure 1), then the magnetic flux through the loop is simply

$$\Phi = BA_{\text{in}}$$

where B is the magnetic field strength, and A_{in} is the area of the loop *inside* the magnetic field. For instance, if one third of the loop is "immersed" in the magnetic field, then A_{in} is one third the area of the whole rectangle.

Faraday's law of induction implies that, whenever the magnetic flux through a metal loop *changes*, current flows around the loop. No current flows when the magnetic flux stays constant.

In figure 1, the magnetic field points into the page with strength B_0, inside the dashed lines. Outside the dashed lines, the field vanishes.

Starting from the position drawn in figure 1, the loop is pushed rightward at constant speed, all the way through the magnetic field.

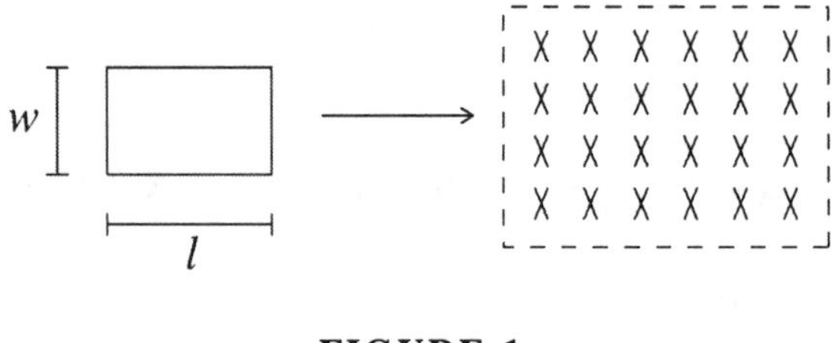

FIGURE 1

1. Which graph best shows the magnetic flux through the rectangle as a function of time?

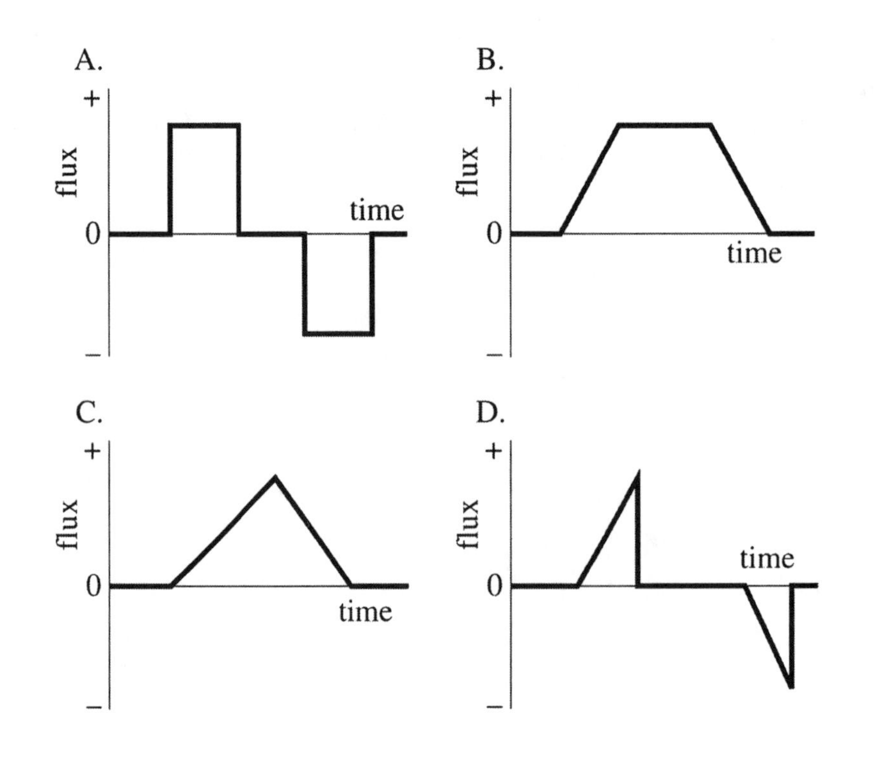

2. Current flows around the rectangle when the rectangle is:

   I. Entering the magnetic field
   II. Fully inside the magnetic field
   III. Leaving the magnetic field

   A. I only **B.** I and II **C.** I and III **D.** I, II, and III

---

3. Which diagram best represents the magnetic field lines generated by *one* side of the rectangle, when current flows through it?

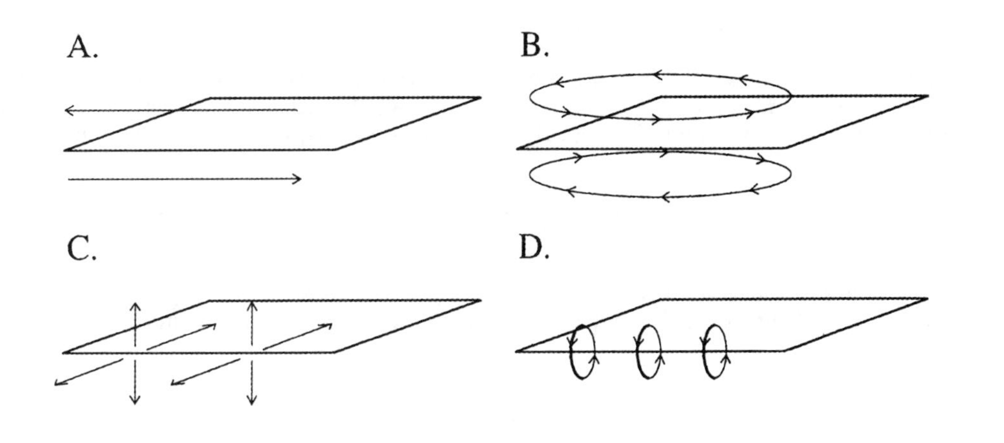

---

4. Which of the following could NOT cause current to flow around this metal rectangle?

   A. Hold the rectangle motionless near a power line through which the electric current is decreasing.
   B. Move the rectangle towards a bar magnet.
   C. Move a bar magnet towards the rectangle.
   D. Place a motionless bar magnet inside the rectangle.

ANSWERS TO MCAT REVIEW PROBLEMS

---

1. **B.** While the rectangle enters the magnetic field, the percentage of its area inside the field steadily increases, from 0 to 100%. So, A_{in} gradually increases from 0 to $A_{rect} = lw$, the rectangle's area. Therefore, the flux steadily increases.

   Once fully inside the field, the magnetic flux neither increases nor decreases. It stays constant, at $\Phi = B_0 lw$. So, although the rate of change of the flux drops to zero, the flux itself stays large.

   Finally, as the rectangle exits the field, the percentage of its area inside the field steadily drops, from 100% to 0. So, the magnetic flux steadily decreases from $B_0 lw$ to 0. Crucially, the flux never becomes negative. It merely becomes less and less positive, until reaching zero.

---

2. **C.** According to the passage, current flows around the rectangle only when the magnetic flux through it *changes*. As discussed in question 1, the magnetic flux through the rectangle changes only while the rectangle enters or leaves the field. When fully immersed, the rectangle experiences a constant flux, $\Phi = B_0 lw$.

---

3. **D.** The field lines generated by a current-carrying straight wire trace circles around the wire. Diagram C represents the *electric* field produced by a positively charged rod. Unfortunately, magnetic fields behave less intuitively.

---

4. **D.** Although the formula $\Phi = BA_n$ applies only to uniform magnetic fields, we can infer that the magnetic flux depends on the strength of the field passing through the rectangle, and on the area of the rectangle inside the field. So, in all four cases, the rectangle experiences a magnetic flux. But current flows around the rectangle only when the flux *changes*. And the flux changes only if the strength of the field piercing through the rectangle changes, or if the percentage of the rectangle inside the field changes.

   In scenario A, the decreasing current in the power line generates a decreasing field through the rectangle. Therefore, the flux through the rectangle changes, causing current to flow around the rectangle. In scenarios B and C, the rectangle gets closer to the bar magnet. Consequently, the field through the rectangle increases. So once again, the flux changes. By contrast, in scenario D, the magnetic field through the rectangle does not increase or decrease with time. The flux might be huge, but it's not *changing*. Therefore, no current flows around the rectangle.

Chapter 23
Alternating Current Circuits

PREVIEW

In this chapter you will learn the basics of alternating current (ac) circuits. You will learn how capacitors, resistors, and inductors are combined in series with ac voltage sources to produce alternating currents. Phasor diagrams are used to illustrate the relationships between current and voltage in ac circuits. Resonance circuits and their applications are discussed. An introduction to semiconductor devices is included as well.

QUICK REFERENCE

Important Terms

Capacitive reactance
A quantity which determines how much rms-current exists in a capacitor in response to a given rms-voltage across the capacitor. The units of reactance are ohms (Ω).

Inductive reactance
A quantity which determines how much rms-current exists in an inductor in response to a given rms-voltage across the inductor. The units of reactance are ohms (Ω).

Impedance
The total opposition to the flow of charge in a series RCL-circuit. The resistance, capacitive reactance, and inductive reactance all contribute to the impedance.

Resonance
A condition which occurs when the frequency of the oscillating voltage supplied by a generator matches the natural frequency of the components in the circuit. At resonance the impedance of the series RCL-circuit has a minimum value equal to the resistance R, and the rms-current has a maximum value.

n-type semiconductors
Materials such as silicon and germanium that are "doped" by adding impurity atoms which contribute mobile electrons to the material.

p-type semiconductors
Materials such as silicon and germanium that are "doped" by adding impurity atoms which contribute mobile positive holes to the material.

p-n junction diode
A device formed from an n-type and a p-type semiconductor. When placed in a circuit it will allow current to flow in only one direction in that circuit.

Equations

The **rms-voltage across a capacitor** can be expressed as,

$$V_{rms} = I_{rms} X_C \tag{23.1}$$

where X_C is known as the **capacitive reactance** and is given by,

$$X_C = \frac{1}{2\pi fC} \tag{23.2}$$

Here, f is the frequency in Hz, C is the capacitance in farads, and X_C is measured in ohms.

The **rms-voltage across an inductor** can be expressed as,

$$V_{rms} = I_{rms} X_L \tag{23.3}$$

where X_L is known as the **inductive reactance** and is given by,

$$X_L = 2\pi fL \tag{23.4}$$

The **total rms-voltage in a series RCL-circuit** is given by,

$$V_{rms}^2 = V_R^2 + (V_L - V_C)^2 \tag{23.5}$$

The **total rms-voltage** is **related** to **the rms-current** through the following expressions,

$$V_{rms} = I_{rms} Z \tag{23.6}$$

$$Z = \sqrt{R^2 + (X_L - X_C)^2} \tag{23.7}$$

where Z is known as the **impedance** of the RCL-circuit. The **phase angle** of an RCL-circuit is defined as

$$\tan \phi = \frac{X_L - X_C}{R} \tag{23.8}$$

The **average power dissipated in an RCL-circuit** can be written as,

$$\overline{P} = I_{rms} V_{rms} \cos \phi \tag{23.9}$$

where $\cos \phi$ is called the **power factor** of the circuit, defined as $\cos \phi = R/Z$. When the impedance of a circuit is minimized, i.e., when $X_L = X_C$ in equation (23.7) and thus $Z = R$, the circuit will **resonate** with

$$f_0 = \frac{1}{2\pi\sqrt{LC}} \tag{23.10}$$

where f_0 is known as the **resonance frequency** of the circuit.

DISCUSSION OF SELECTED SECTIONS

23.1 Capacitors and Capacitive Reactance

The rms-voltage across a resistor is $V_{rms} = I_{rms}R$. For a capacitor, notice that the term X_C replaces the resistance in equation (23.1). The **capacitive reactance** X_C, like resistance, is measured in ohms and determines how much rms current exists in a capacitor in response to a given rms-voltage across the capacitor.

The capacitive reactance depends on the capacitance of the capacitor and the frequency of the ac voltage applied to the capacitor according to

$$X_C = \frac{1}{2\pi fC} \qquad (23.2)$$

Thus, according to equation (23.2), the capacitive reactance is large for small frequencies f. When X_C is large, the current is small since $I_{rms} = V_{rms}/X_C$. When the frequency is high, X_C is small, and I_{rms} is large.

Example 1
A 5.00 μF capacitor is in series with an ac voltage source which supplies an rms voltage of 32.0 V. Determine the rms-current in the circuit when the frequency of the source is (a) 60.0 Hz and (b) 6.00×10^3 Hz.

(a) At 60.0 Hz,

$$X_C = \frac{1}{2\pi fC} = \frac{1}{2\pi(60.0\ \text{Hz})(5.00 \times 10^{-6}\ \text{F})} = 531\ \Omega$$

$$I_{rms} = \frac{V_{rms}}{X_C} = \frac{32.0\ \text{V}}{531\ \Omega} = 0.0603\ \text{A}$$

(b) For 6.00×10^3 Hz,

$$X_C = \frac{1}{2\pi fC} = \frac{1}{2\pi(6.00 \times 10^3\ \text{Hz})(5.00 \times 10^{-6}\ \text{F})} = 5.31\ \Omega$$

$$I_{rms} = \frac{V_{rms}}{X_C} = \frac{32.0\ \text{V}}{5.31\ \Omega} = 6.03\ \text{A}$$

Let's compare the effects that resistors and capacitors have on circuits. Consider the behavior of the **instantaneous** (not rms) voltage or current for a resistor as a function of time, as in the following diagram.

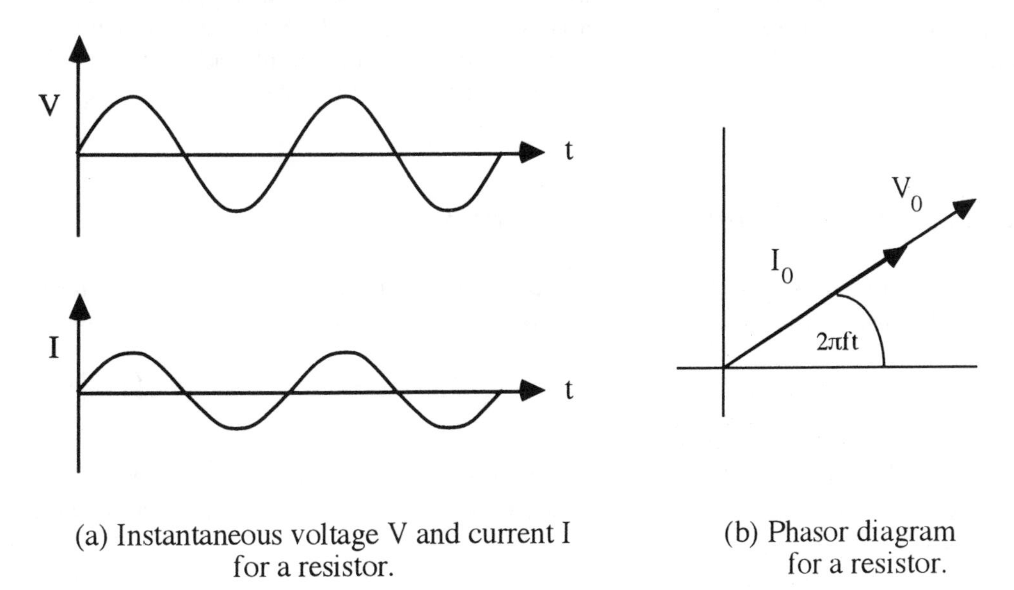

(a) Instantaneous voltage V and current I for a resistor.

(b) Phasor diagram for a resistor.

The graphs in figure (a) shown above indicate that, when only resistance is present, the voltage and current are proportional to each other at every moment. As the voltage across the resistor increases, the current increases proportionately. As the voltage drops, the current follows suit. For this reason, the current in a resistance R is said to be **in phase** with the voltage across the resistor. This aspect is depicted in figure (b) above through the use of a **phasor diagram**. In this model, voltage and current are represented by rotating arrows, called phasors, whose lengths correspond to the maximum voltage V_0 and the maximum current I_0. The **phasors rotate counterclockwise** at a frequency f. For a resistor, the phasors are colinear as they rotate, because the voltage and current are in phase. Now consider the situation for capacitors, as shown in the following diagrams.

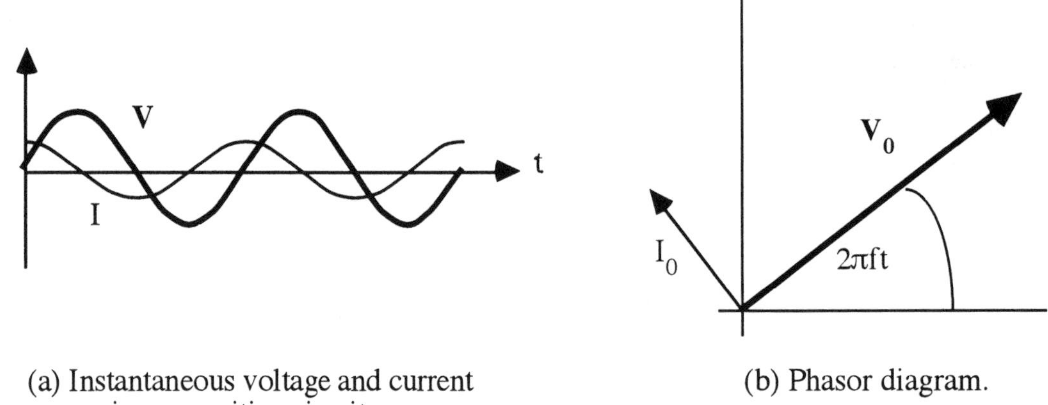

(a) Instantaneous voltage and current in a capacitive circuit.

(b) Phasor diagram.

For a capacitor, the current and voltage are **not** in phase. As the voltage increases at the beginning of the cycle, the charge on the capacitor begins to build up. The current, or rate of flow of charge, has a maximum positive value at the start of the charging process, when there is no charge on the capacitor. As charge builds up, and the voltage across the capacitor increases, the flow of charge is impeded, and hence the current drops. When the capacitor is fully charged, no further charge can flow, so the instantaneous current is zero. The capacitor then begins to discharge, and current begins to flow in the opposite (negative) direction. When all the charge is gone from the capacitor, the voltage across the capacitor goes to zero, and the current has its maximum negative value. The voltage and current are therefore one-quarter cycle out of step, or 90° out of phase. From the figure it is seen that the current through a capacitor **leads** the voltage across the capacitor by 90°. In figure (b) note that the phasors remain perpendicular while rotating, because the phase angle between the current and voltage is 90°. Since the current leads the voltage for a capacitor, the current phasor is ahead of the voltage phasor in the direction of rotation (counterclockwise). The instantaneous voltage and current are given by the vertical components of the phasors.

23.2 Inductors and Inductive Reactance

An inductor, as discussed in Section 22.8, is usually a coil of wire which develops a voltage that opposes a change in the current. The rms-voltage across an inductor is given by the expression,

$$V_{rms} = I_{rms} X_L \qquad (23.3)$$

which is analogous to $V_{rms} = I_{rms} R$ for a resistor. The term that appears in place of R in equation (23.3), X_L, is called the **inductive reactance**. It depends on the inductance of the inductor and the frequency of the generator according to the relation

$$X_L = 2\pi f L \qquad (23.4)$$

Equation (23.4) indicates that the larger the frequency of the voltage generator, the larger the inductive reactance, and hence the smaller the rms-current. Low frequencies mean small inductive reactances and large currents.

Example 2
A 7.50-mH inductor is in a circuit with a generator which provides an rms-voltage of 32.0 V. Find X_L and the rms current at frequencies of (a) 60.0 Hz and (b) 6.00 x 10^3 Hz.

(a) At 60.0 Hz,

$$X_L = 2\pi fL = 2\pi(60.0 \text{ Hz})(7.50 \times 10^{-3} \text{ H}) = 2.83 \ \Omega$$

$$I_{rms} = \frac{V_{rms}}{X_L} = \frac{32.0 \text{ V}}{2.83 \ \Omega} = 11.3 \text{ A}$$

(b) At 6.00 x 10^3 Hz,

$$X_L = 2\pi fL = 2\pi(6.00 \times 10^3 \text{ Hz})(7.50 \times 10^{-3} \text{ H}) = 283 \ \Omega$$

$$I_{rms} = \frac{V_{rms}}{X_L} = \frac{32.0 \text{ V}}{283 \ \Omega} = 0.113 \text{ A}$$

By virtue of its inductive reactance, an inductor affects the amount of current that exists in an ac circuit. Consider the following figure (a) in which we plot a graph of the instantaneous voltage and current versus time for a circuit containing only an inductor.

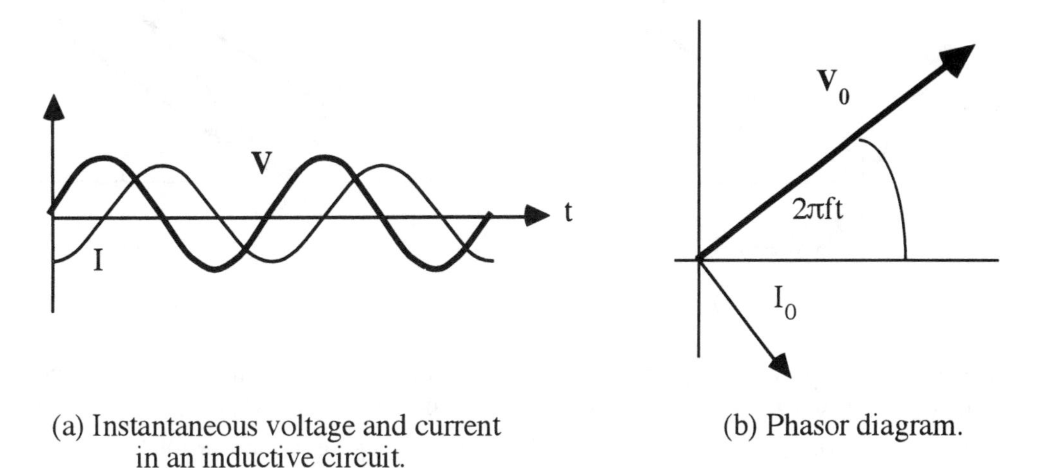

(a) Instantaneous voltage and current in an inductive circuit.

(b) Phasor diagram.

At the points where the current is a maximum or a minimum, the current doesn't change much with time, and hence the voltage generated by the inductor to oppose the ac current is zero. At points where the current is zero, the graph is at its steepest, and the current has its largest rate of change. Correspondingly, the voltage generated by the inductor to oppose the ac current has its largest value (positive or negative). Since the current reaches its maximum after the voltage does, it is said that the current **lags** behind the voltage by a phase angle of 90°.

In figure (b) above, the current and voltage phasors remain perpendicular as they rotate, for there is a 90° phase angle between them. The current phasor lags the voltage phasor, relative to the direction of rotation (counterclockwise), in contrast to the situation for the capacitor. Once again, the instantaneous voltage and current are given by the vertical components of the phasors.

23.3 Circuits Containing Resistance, Capacitance, and Inductance

Capacitors, inductors, and resistors can be combined in a single circuit, as shown in the following figure.

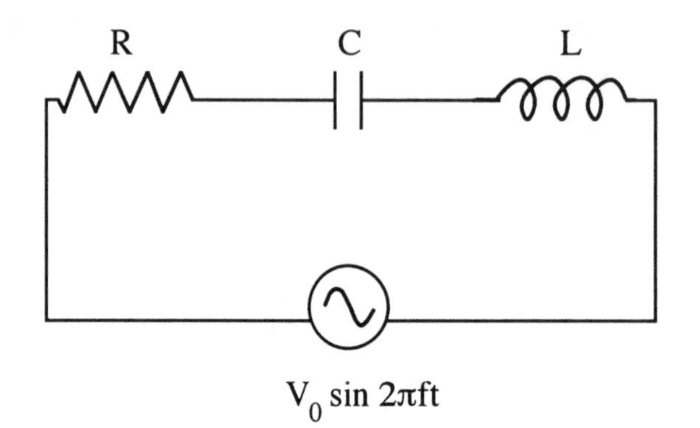

In a series RCL-circuit the total opposition to the flow of charge is called the **impedance** of the circuit. In order to determine how the maximum voltages across the resistor, V_R, the capacitor, V_C, and the inductor, V_L, combine to yield the maximum voltage, V_0, we need to study the phasor diagram for the combination. Kirchhoff's loop rule indicates that the phasors add together to give the total voltage V_0 that is supplied to the circuit by the generator. The addition, however, must be a vector addition to take into account the different directions. Consider the phasor diagram for the RCL-circuit shown above.

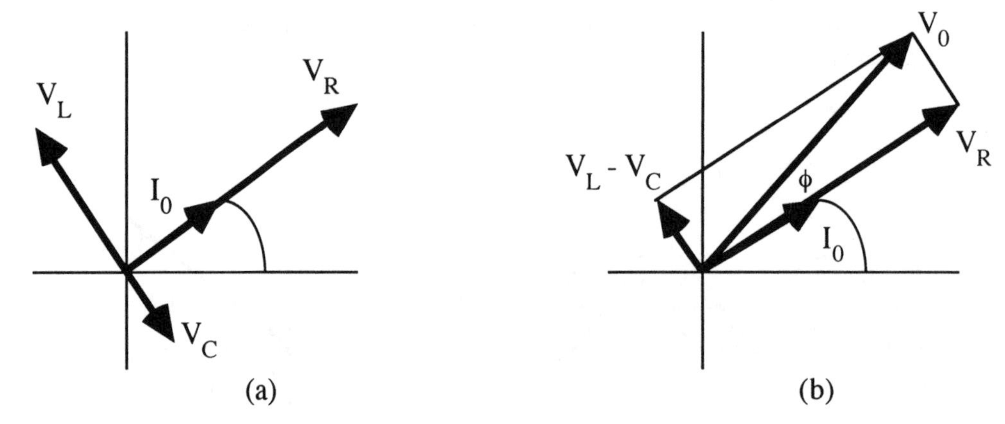

In figure (a) the phasors for V_R, V_C, and V_L are shown relative to the maximum current I_0. Notice that the current is in phase with V_R, leads V_C by 90°, and lags behind V_L by 90°. Figure (b) shows how to add the voltages together vectorially to produce V_0, the resultant voltage. From the drawing we see that the total voltage is given by:

$$V_0^2 = V_R^2 + (V_L - V_C)^2$$

The rms-voltage of the circuit is obtained by dividing this equation by $(\sqrt{2})^2$ so that

$$V_{rms}^2 = V_R^2 + (V_L - V_C)^2 \qquad (23.5)$$

where the symbols on the right now refer to rms-quantities.

Finally, since $V_R = I_{rms}R$, $V_C = I_{rms}X_C$, and $V_L = I_{rms}X_L$, for the entire RCL-circuit it follows that

$$V_{rms} = I_{rms} Z \tag{23.6}$$

$$Z = \sqrt{R^2 + (X_L - X_C)^2} \tag{23.7}$$

where the quantity Z is known as the **impedance** of the circuit. The **phase angle** ϕ between the current phasor I_0 and the voltage phasor V_0, shown in part (b) of the above diagram, is given by

$$\tan \phi = \frac{X_L - X_C}{R} \tag{23.8}$$

The average power dissipated by the circuit is given by

$$\overline{P} = I_{rms} V_{rms} \cos \phi \tag{23.9}$$

where the term $\cos \phi = R/Z$ is called the **power factor** of the circuit.

Example 3

A series RCL-circuit consists of a 65.0-Ω resistor, a 3.00-μF capacitor, and a 65.0-mH inductor. The generator has a frequency of 444 Hz and an rms-voltage of 18.0 V. Obtain (a) the rms-voltage across and current through each element and, (b) the electrical power consumed by the circuit.

(a) First determine the impedance of the circuit,

$$X_C = \frac{1}{2\pi fC} = \frac{1}{2\pi(444\text{ Hz})(3.00 \times 10^{-6}\text{ F})} = 119\ \Omega$$

$$X_L = 2\pi fL = 2\pi(444\text{ Hz})(6.50 \times 10^{-2}\text{ H}) = 181\ \Omega$$

$$Z = \sqrt{R^2 + (X_L - X_C)^2} = \sqrt{(65.0\ \Omega)^2 + (181\ \Omega - 119\ \Omega)^2} = 9.0 \times 10^1\ \Omega$$

The rms-current in the circuit is therefore,

$$I_{rms} = \frac{V_{rms}}{Z} = \frac{18.0\text{ V}}{9.0 \times 10^1\ \Omega} = 0.20\text{ A}$$

The rms-voltages across each device are therefore

$$V_R = I_{rms} R = (0.20\text{ A})(65.0\ \Omega) = 13\text{ V}$$

$$V_C = I_{rms} X_C = (0.20\text{ A})(119\ \Omega) = 24\text{ V}$$

$$V_L = I_{rms} X_L = (0.20\text{ A})(181\ \Omega) = 36\text{ V}$$

(b) To find the power dissipated by the circuit we need to first find the power factor. We know that

$$\cos \phi = \frac{R}{Z} = \frac{65.0\ \Omega}{9.0 \times 10^{1}\ \Omega} = 0.72$$

Therefore, the power dissipated by the circuit is

$$\overline{P} = I_{rms} V_{rms} \cos \phi = (0.20\ A)(18.0\ V)(0.72) = 2.6\ W$$

23.4 Resonance in Electric Circuits

Under certain conditions in a series RCL-circuit a situation known as **resonance** can occur. Resonance occurs when the frequency of the vibrating force exactly matches the natural frequency of the object to which the force is applied. In the electric case the vibrating force is provided by the oscillating electric field that is related to the voltage supplied by the generator. In a resonant series RCL-circuit the current reaches its largest possible value. The largest value of I_{rms} occurs when the impedance is at its minimum value. From equation (23.7) we see that this occurs when $X_L = X_C$. Using the definitions for reactance, equations (23.2) and (23.4), the **resonant frequency** f_0 occurs when $2\pi f_0 L = 1/(2\pi f_0 C)$, that is, when

$$f_0 = \frac{1}{2\pi\sqrt{LC}} \tag{23.10}$$

Note that the resonant frequency depends on the inductance and capacitance, but not the resistance.

Example 4
A series RCL-circuit consists of a 150-Ω resistor, a 5.5-nF capacitor, and a 48-mH inductor. The circuit is connected to the terminals of an ac oscillator whose output is 75 V and whose frequency is adjustable over a wide range. (a) What is the resonant frequency of the circuit? (b) How much current flows in the circuit at resonance?
(c) What are the voltages across each of the circuit components?

(a) The resonant frequency is,

$$f_0 = \frac{1}{2\pi\sqrt{LC}} = \frac{1}{2\pi\sqrt{(48 \times 10^{-3}\ H)(5.5 \times 10^{-9}\ F)}} = 9800\ Hz$$

(b) At resonance, we see from equation (23.7) that $Z = R$. Therefore,

$$I_{rms} = \frac{V_{rms}}{R} = \frac{75\ V}{150\ \Omega} = 0.50\ A$$

(c) The voltages are: $V_R = I_{rms} R = (0.50\ A)(150\ \Omega) = 75\ V$

$$X_C = 1/(2\pi fC) = 1/[2\pi(9800\ Hz)(5.5 \times 10^{-9})] = 3.0 \times 10^3\ \Omega,$$

$$V_C = (0.50\ A)(3.0 \times 10^3\ W) = 1500\ V$$

$$X_L = 2\pi fL = 2\pi(9800\ Hz)(48 \times 10^{-3}\ H) = 3.0 \times 10^3\ \Omega;\ V_L = (0.50\ A)(3.0 \times 10^3\ \Omega) = 1500\ V$$

PRACTICE PROBLEMS

1. A capacitor has a capacitance of 1.90 μF. What is the reactance of this capacitor at frequencies of (a) 333 Hz, (b) 165 Hz, and (c) 66.6 Hz?

2. An inductor has an inductance of 121 mH. What is the reactance of this inductor at frequencies of (a) 333 Hz, (b) 165 Hz, and (c) 66.6 Hz?

3. A 5.0-mH inductor is connected to an ac voltage source whose rms-voltage is 24 V. The rms-current flowing in the circuit is 1.2 A. What is the frequency of the voltage source?

4. A 2700-Ω resistor and a 1.1 μF capacitor are connected in series across a generator (60.0 Hz, 120 V). Determine the power dissipated in the circuit.

5. A circuit consists of a 215-Ω resistor and a 0.200-H inductor. These two elements are connected in series across a generator that has a frequency of 106 Hz and a voltage of 234 V. (a) What is the current in the circuit? (b) Determine the phase angle between the current and the voltage.

6. A 66 Hz ac generator operates in a series RCL-circuit which has an rms-current of 1.0 A. The following rms voltages are found across the resistor, capacitor, and inductor, respectively:

$$V_R = 55 \text{ V}, \quad V_C = 32 \text{ V}, \quad V_L = 48 \text{ V}.$$

(a) Determine the source voltage and the phase angle between the current and the voltage. (b) Determine the values of R, C, and L.

7. The elements in a series RCL-circuit are a 106-Ω resistor, a 3.30-μF capacitor, and a 31.0-mH inductor. What is the impedance of the circuit and the phase angle between the current and the voltage when the frequency is (a) 215 Hz and (b) 609 Hz? In each case interpret the algebraic sign of the phase angle, remembering that a positive sign means the voltage "leads" the current and the circuit is more inductive than capacitive.

8. A series RCL-circuit has a resonance frequency of 55 kHz. If the value of the inductance is 3.5 mH, what is the value of the capacitance?

9. A series RCL-circuit contains a 240-V, 120-Hz ac voltage, a 2200-Ω resistor, a coil of 3.0-H inductance, and a capacitance C that can be varied. (a) What value should C have in order for the voltage to lead the current by 45°? (b) At this value of C, find the rms-current in the circuit. (c) What is the voltage across the coil?

10. A series RCL-circuit has R = 25 Ω, C = 1.8 μF, and L = 6.3 mH. (a) At what frequency of an ac generator will the circuit resonate? (b) What is the rms-current at resonance if the ac voltage is 25 V?

HELPFUL SUGGESTIONS

1. Resistance, capacitive reactance, and inductive reactance all have units of ohms (Ω). Remember, however, that while the resistance R is fixed, X_C and X_L vary with frequency. So as the frequency varies, the values of X_C and X_L vary as well, and must be re-computed for each value of f. This is not true of the resistance.

2. Capacitors offer low reactance, and hence large currents, at high frequencies in an ac circuit. Inductors offer low reactance, and hence large currents, at low frequencies in an ac circuit. In terms of the impedance of an ac circuit, therefore, capacitors dominate the behavior of the circuit at low frequencies while inductors dominate the behavior at high frequencies.

3. The rms-voltages of the elements in a series RCL-circuit do NOT in general add up to the rms-voltage of the source. This is because the voltages across the individual elements in the circuit are not in phase with one another. In order to add voltages, it must be done vectorially, as illustrated in the text.

4. Remember that the current LEADS the voltage by 90° in a purely CAPACITIVE circuit. On the other hand, the current LAGS behind the voltage by 90° in a purely INDUCTIVE circuit. Since the current flows in the same direction through all the elements of the circuit, this means that the voltage across the capacitor and the inductor are 180° out of phase with one another. Thus, the reactances of these two elements act to oppose one another, completely canceling each other out at the resonant frequency.

5. The power used by an ac circuit depends on the phase angle ϕ, or the power factor cos ϕ, for the circuit. Since the average power varies as $\cos \phi = R/Z$, no power is dissipated when $R = 0$. However, the maximum power is dissipated when $Z = R$, which occurs when X_L and X_C are equal.

6. The resonance frequency of an RCL-circuit depends only on L and C, and is independent of the resistance. Varying the resistance in a resonant circuit acts only to change the "sharpness" of the resonance peak. When the resistance is small, the current-vs-frequency graph falls off sharply on either side of the maximum current. When R is large, the falloff is more gradual, and the graph isn't as "sharp".

EVERYDAY PHYSICS

1. The current in houses and buildings is always ac current. The commercial ac circuits in the United States involve 60 Hz frequencies, although other frequencies are used in other parts of the world. Alternating currents are desirable for commercial use because it is easier to produce ac current than it is to produce dc current. In addition, it is easy to "step up" ac currents to high voltages using a transformer. This allows for the transmission of electric power across great distances without substantial power loss.

2. An example of a resonance circuit is a heterodyne metal detector, which utilizes two capacitor/inductor oscillator circuits. Each circuit normally produces the same resonance frequency. When the device comes near a piece of metal, the inductance in the probe of the device decreases, causing the resonance frequency to change in that circuit. The difference in frequency between the two circuits is then detectable by the presence of a beat frequency which can be heard through a set of headphones.

CHAPTER QUIZ

1. A capacitor is connected across a source of ac voltage. If the frequency of the voltage source is increased while the rms-voltage is kept fixed,
   a. the rms-current remains fixed.
   b. the rms-current increases.
   c. the rms-current decreases.
   d. no current flows in the circuit.

2. An inductor is connected across a source of ac voltage. If the frequency of the voltage source is increased while the rms-voltage is kept fixed,
   a. the rms-current remains fixed.
   b. the rms-current increases.
   c. the rms-current decreases.
   d. no current flows in the circuit.

3. A resistor is connected to an ac voltage source. In this circuit, the current
   a. leads the voltage by 90°.
   b. lags the voltage by 90°.
   c. is in phase with the voltage.
   d. none of the above.

4. A capacitor is connected to an ac voltage source. In this circuit, the current
   a. leads the voltage by 90°.
   b. lags the voltage by 90°.
   c. is in phase with the voltage.
   d. none of the above.

5. An inductor is connected to an ac voltage source. In this circuit, the current
   a. leads the voltage by 90°.
   b. lags the voltage by 90°.
   c. is in phase with the voltage.
   d. none of the above.

6. For a series RCL-circuit
   a. the power dissipated by the circuit is a minimum at resonance.
   b. the power dissipated by the circuit is not a maximum at resonance.
   c. the power dissipated by the circuit is I^2R regardless of the frequency of the circuit.
   d. none of the above statements is correct.

7. In a series RCL-circuit, resonance occurs when
   a. X_C is greater than X_L.
   b. X_C is less than X_L.
   c. X_C and X_L are zero.
   d. X_C is equal to X_L.

8. A series RCL-circuit has a capacitance of 2.0 μF and an inductance of 3.9 mH. What is the resonance frequency of the circuit?
   a. 49 Hz
   b. 780 Hz
   c. 1800 Hz
   d. 2.0×10^7 Hz

9. What is the capacitive reactance of a 4.0-μF capacitor at a frequency of 220 Hz?
   a. 180 Ω
   b. 880 Ω
   c. 5.5 Ω
   d. 55 Ω

10. What is the inductive reactance of a 4.0-mH inductor at a frequency of 220 Hz?
    a. 180 Ω
    b. 880 Ω
    c. 5.5 Ω
    d. 55 Ω

11. A 4.0-μF capacitor, 4.0-mH inductor, and 180-Ω resistor are in series with an ac voltage source whose frequency is 220 Hz. The impedance of the circuit is
    a. 175 Ω
    b. 180 Ω
    d. 2.3 Ω
    d. 250 Ω

SOLUTIONS AND ANSWERS

Practice Problems

---

1. The capacitive reactance is calculated from equation (23.2), $X_C = 1/(2\pi fC)$. Therefore,

(a) $X_C = 1/[2\pi(333\text{ Hz})(1.90 \times 10^{-6}\text{ F})] = \mathbf{252\ \Omega}$.

(b) $X_C = 1/[2\pi(165\text{ Hz})(1.90 \times 10^{-6}\text{ F})] = \mathbf{508\ \Omega}$.

(c) $X_C = 1/[2\pi(66.6\text{ Hz})(1.90 \times 10^{-6}\text{ F})] = \mathbf{1260\ \Omega}$.

---

2. (a) $X_L = 2\pi fL = 2\pi(333\text{ Hz})(0.121\text{ H}) = \mathbf{253\ \Omega}$.
(b) $X_L = 2\pi(165\text{ Hz})(0.121\text{ H}) = \mathbf{125\ \Omega}$.
(c) $X_L = 2\pi(66.6\text{ Hz})(0.121\text{ H}) = \mathbf{50.6\ \Omega}$.

---

3. We know

$$V_{rms} = I_{rms}X_L = I_{rms}(2\pi fL) \Rightarrow f = V_{rms}/(2\pi I_{rms}L) = (24\text{ V})/[2\pi(1.2\text{ A})(5.0 \times 10^{-3}\text{ H})] = \mathbf{640\ Hz}.$$

---

4. Find the impedance, $Z^2 = R^2 + X_C{}^2$ where

$$X_C = 1/(2\pi fC) = 1/[2\pi(60.0\text{ Hz})(1.1 \times 10^{-6}\text{ F})] = 2400\ \Omega.$$

Thus,

$$Z = \sqrt{(2700\ \Omega)^2 + (2400\ \Omega)^2} = 3600\ \Omega.$$

The power factor is

$$\cos\phi = R/Z = (2700\ \Omega)/(3600\ \Omega) = 0.75.$$

The power is

$$P = I_{rms}V_{rms}\cos\phi = (V_{rms}{}^2/Z)\cos\phi = (120\text{ V})^2(0.75)/(3600\ \Omega) = \mathbf{3.0\ W}.$$

---

5. Find the impedance of the circuit, $Z^2 = R^2 + X_L{}^2$ where

$$X_L = 2\pi fL = 2\pi(106\text{ Hz})(0.200\text{ H}) = 133\ \Omega.$$

Thus,

$$Z = \sqrt{(215\ \Omega)^2 + (133\ \Omega)^2} = 253\ \Omega.$$

We can now find the current and phase angle:

(a) The rms-current is therefore,

$$I_{rms} = V_{rms}/Z = (234\text{ V})/(253\ \Omega) = \mathbf{0.925\ A}.$$

(b) Since $\cos\phi = R/Z$,

$$\phi = \cos^{-1}[(215\ \Omega)/(253\ \Omega)] = \mathbf{31.7^\circ}.$$

---

6. Find the reactances using $V_{rms} = I_{rms}X$. So,

$$R = V_R/I = (55\ V)/(1.0\ A) = 55\ \Omega;$$

$$X_C = V_C/I = (32\ V)/(1.0\ A) = 32\ \Omega;\ X_L = V_L/I = (48\ V)/(1.0\ A) = 48\ \Omega.$$

(a) Equation (23.5),

$$V_{rms}^2 = V_R^2 + (V_L - V_C)^2 = (55\ V)^2 + (48\ V - 32\ V)^2 \Rightarrow V_{rms} = \mathbf{57\ V}.$$

Equation (23.8),

$$\tan\phi = (X_L - X_C)/R = (16\ \Omega)/(55\ \Omega) = 0.29 \Rightarrow \phi = \tan^{-1}(0.29) = \mathbf{16°}.$$

(b) We know

$$X_C = 1/(2\pi fC) \Rightarrow C = 1/(2\pi fX_C) = 1/[2\pi(66\ Hz)(32\ \Omega)] = \mathbf{75\ \mu F}.$$

$$X_L = 2\pi fL \Rightarrow L = X_L/(2\pi f) = (48\ \Omega)/[2\pi(66\ Hz)] = \mathbf{0.12\ H}.\text{ Of course, } R = \mathbf{55\ \Omega}.$$

---

7. (a) At 215 Hz; $X_L = 2\pi fL = 41.9\ \Omega$, $X_C = 1/(2\pi fC) = 224\ \Omega$, $R = 106\ \Omega$, and

$$Z^2 = R^2 + (X_L - X_C)^2 = (106\ \Omega)^2 + (41.9\ \Omega - 224\ \Omega)^2 \Rightarrow Z = \mathbf{211\ \Omega}.$$

We also know,

$$\tan\phi = (X_L - X_C)/R \Rightarrow \phi = \tan^{-1}(-182\ \Omega/106\ \Omega) = \mathbf{-59.8°}.$$

Since the voltage LAGS the current by 59.8°, the circuit is **capacitive**.

(b) At 609 Hz; $X_L = 119\ \Omega$, $X_C = 79.2\ \Omega$, so that

$$Z^2 = (106\ \Omega)^2 + (119\ \Omega - 79.2\ \Omega)^2 \Rightarrow Z = \mathbf{113\ \Omega}.$$

$$\phi = \tan^{-1}(39.8\ \Omega/106\ \Omega) = \mathbf{+20.6°}.$$

So the voltage leads the current and the circuit is **inductive**.

---

8. Using equation (23.10),

$$f_0 = 1/[2\pi\sqrt{(LC)}] \Rightarrow C = 1/(4\pi^2 f_0^2 L) = 1/[4\pi^2(55 \times 10^3\ Hz)^2\ (3.5 \times 10^{-3}\ H)] = \mathbf{2.4\ nF}.$$

---

9. (a) Since $\tan\phi = (X_L - X_C)/R$ and since $\phi = 45°$, we have

$$X_C = X_L - R\tan 45° \Rightarrow 1/(2\pi fC) = 2\pi fL - R\tan 45°.$$

Solving,

$$C = 1/(4\pi^2 f^2 L - 2\pi fR\tan 45°) = 1/[4\pi^2(120\ Hz)^2(3.0\ H) - 2\pi(120\ Hz)(2200\ \Omega)\tan 45°] = \mathbf{21\ \mu F}.$$

(b) The impedance of the circuit is

$$Z^2 = (2200\ \Omega)^2 + (2200\ \Omega)^2 \Rightarrow Z = 3100\ \Omega.$$

So that,

$$I_{rms} = V_{rms}/Z = (240\ V)/(3100\ \Omega) = \mathbf{0.077\ A}.$$

(c) We know

$$V_L = I_{rms}X_L = (0.077\ A)(2300\ \Omega) = \mathbf{180\ V}.$$

---

10. (a) The resonance frequency is

$$f_0^2 = 1/(4\pi^2 LC) = 1/[4\pi^2(7.5 \times 10^{-3}\ H)(1.8 \times 10^{-6}\ F)] \Rightarrow f_0 = \mathbf{1500\ Hz}.$$

(b) At resonance, $X_L = X_C$ so that Z = R. Therefore, the rms-current is given by,

$$I_{rms} = V_{rms}/Z = (25\ V)/(25\ \Omega) = \mathbf{1.0\ A}.$$

Quiz answers

1. b
2. c
3. c
4. a
5. b
6. c
7. d
8. c
9. a
10. c
11. d

MCAT REVIEW PROBLEMS

A resistanceless LC circuit consists of a capacitor and an inductor, as drawn in figure 1. When the current flowing through an inductor *changes*, a voltage (emf) gets created in the inductor that opposes the change in the current. So, inductors have an "inertia" with respect to changes in current, just like masses have inertia with respect to changes in velocity.

In an LC circuit, charge oscillates back and forth between the two capacitor plates, just like a pendulum's position oscillates back and forth between the two endpoints of its swing. For instance, in figure 1, suppose the bottom plate initially carries negative charge. Excess electrons flow off the bottom plate, through the inductor, and onto the top plate. So, the top plate now carries the negative charge. But then, the negative charge "sloshes" back onto the bottom plate. In this manner, the charge oscillates.

At any given moment, the energy stored in the capacitor is $U_C = \frac{1}{2}\frac{Q^2}{C}$, where Q denotes the net charge on the top plate, and C denotes the capacitance. The energy stored in the inductor is $U_L = \frac{1}{2}LI^2$, where I denotes the current through the circuit, and L is a constant called the inductance. Since the circuit is essentially resistanceless, no heat dissipates as current oscillates back and forth.

In this circuit, $C = 0.10$ farads and $L = 2.0$ henrys. (Farads and henrys are both SI units.) Initially, at time $t = 0$, the bottom plate has charge -1.0 coulombs, and no current is flowing.

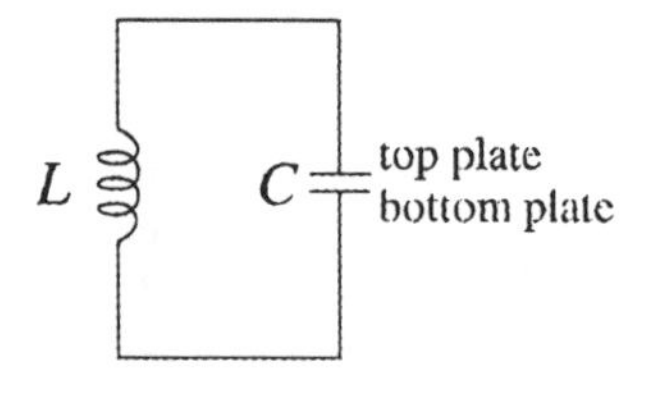

FIGURE 1

1. Which graph best represents the charge on the top plate as a function of time?

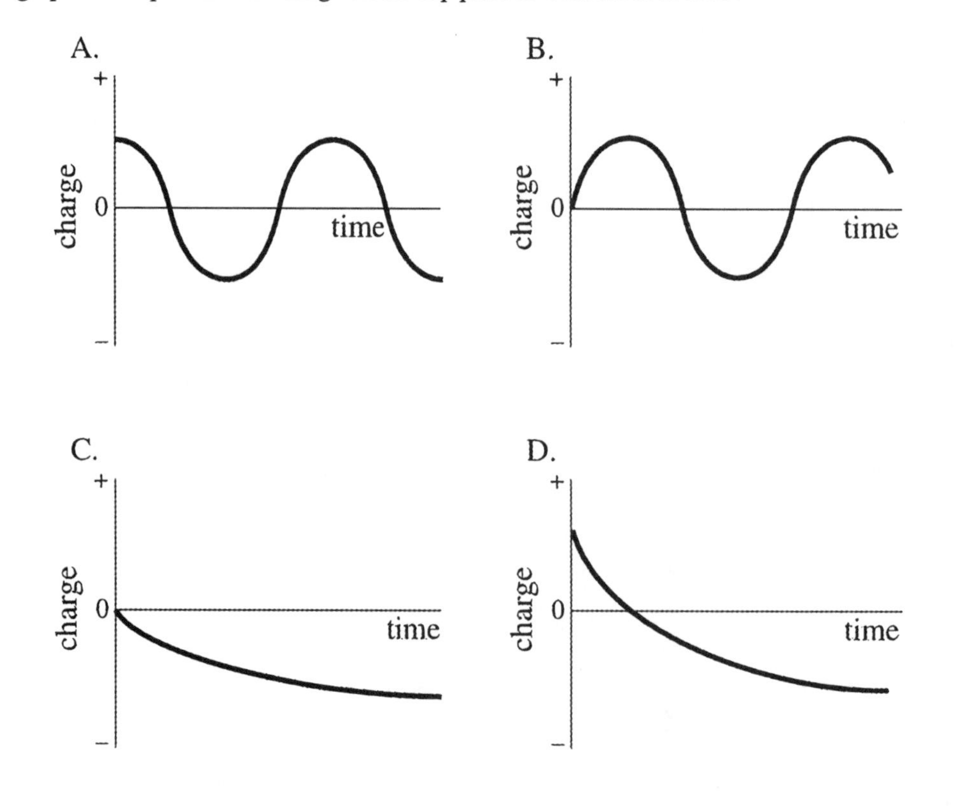

2. At the moment the top plate has charge -1.0 coulombs, the current through the circuit is most nearly:

A. 0 amperes **B.** 1 ampere **C.** 2 amperes **D.** 3 amperes

---

3. At the moment the top plate has no net charge, the current through the circuit is most nearly:

A. 0 amperes **B.** 1 ampere **C.** 2 amperes **D.** 3 amperes

---

ANSWERS TO MCAT REVIEW PROBLEMS

---

1. **A.** According to the passage, the charge oscillates between the plates just like a pendulum oscillates between the endpoints of its swing. So, like a pendulum's position, the charge on the top plate (and the bottom plate) oscillates sinusoidally. This narrows down the answer to A or B.

   Fortunately, the passage provides additional information. Since a circuit carries zero net charge, and since the bottom plate initially carries negative charge, the top plate initially carries positive charge, at $t = 0$. Graph B incorrectly shows the initial charge on the top plate as zero.

---

2. **A.** You can address this problem using either intuitive reasoning about oscillatory motion, or conservation of energy.

   Energy conservation. As charge sloshes back and forth, the total energy of the circuit stays the same. Therefore, the initial energy equals the energy at any later time.

   Since the top plate initially has charge 1 coulomb, the capacitor initially carries energy

$$U_{C\,\text{initial}} = \frac{1}{2}\frac{Q_{\text{initial}}^2}{C} = \frac{1}{2}\frac{(1\text{ coulomb})^2}{.1\text{ F}} = 5\text{ J}.$$

By contrast, because no current flows at $t = 0$, the inductor carries initial energy

$$U_{L\,\text{initial}} = \frac{1}{2} L\, I_{\text{initial}}^2 = 0.$$

Hence, the circuit initially carries a total energy of 5 joules. By setting this initial energy equal to the energy at a later time, we can finish the problem. . .

When the top plate has charge -1 coulomb, it carries energy $U_C = \frac{1}{2}\frac{(-1\text{ coulomb})^2}{.1\text{ F}} = 5$ joules. So, at that moment, the capacitor contains *all* of the circuit's energy. Therefore, the inductor carries no energy. Since $U_L = \frac{1}{2}LI^2$, it follows that $I = 0$.

Oscillation reasoning. This conclusion makes intuitive sense if you think of a pendulum. When displaced by $x = 1$ cm, a pendulum swings until reaching $x = -1$ cm, at which point it momentarily comes to rest, while "turning around" to swing the other way. Similarly, the -1 coulomb of excess charge on the bottom plate "swings" (flows) to the top plate, until that plate carries -1 coulomb. At this point, the charge stops flowing for a moment, to "turn around" and flow the other way.

---

3. **C.** Again, use conservation of energy. As shown above, the circuit has a total energy of 5 joules. So, at all times,

$$U_C + U_L = \frac{1}{2}\frac{Q^2}{C} + \frac{1}{2}LI^2 = 5\text{ joules}.$$

When $Q = 0$, the capacitor carries none of that energy. Therefore, the inductor must carry all 5 joules:

$$U_L = \frac{1}{2}LI^2 = 5\text{ joules}.$$

Solve this equation for I to get

$$I = \sqrt{\frac{2(5\text{ joules})}{L}} = \sqrt{\frac{2(5\text{ joules})}{2\text{ henrys}}} = \sqrt{5\frac{\text{joules}}{\text{henry}}},$$

which is about 2 amperes.

CHAPTER 24
Electromagnetic Waves

PREVIEW

In this chapter you are introduced to electromagnetic waves, of which radio waves, infrared radiation, visible light, ultraviolet radiation, x-rays and gamma rays are familiar examples. Electromagnetic waves as oscillating electric and magnetic fields moving through space are investigated. Evidence that electromagnetic waves are transverse waves is demonstrated by an analysis of the polarization of the waves.

QUICK REFERENCE

Important Terms

Electromagnetic wave
Fluctuating electric and magnetic fields which propagate together.

Near field
The electric and magnetic fields which exit in the region close to an antenna.

Radiation field
The electric and magnetic fields formed far from an antenna which propagate as an electromagnetic wave.

Electromagnetic spectrum
Electromagnetic waves arranged in order of their frequency.

Intensity
The intensity of an electromagnetic wave is the total energy per unit time per unit area carried by an electromagnetic wave.

Unpolarized wave
A wave in which the vibrations can occur randomly in any direction.

Linearly polarized wave
A wave in which the vibrations always occur along one direction.

Polarizing material
A material which produces a linearly polarized wave.

Transmission axis
The transmission axis of a polarizing material is the preferred transmission direction of an electric field passing through the material.

Equations

The **speed** of an **electromagnetic wave**

$$c = \frac{1}{\sqrt{\varepsilon_0 \mu_0}} \qquad (24.1)$$

The **total energy density** of an **electromagnetic wave:**

$$u = \frac{1}{2}\varepsilon_0 E^2 + \frac{1}{2\mu_0}B^2 \qquad (24.2a)$$

$$u = \varepsilon_0 E^2 \qquad (24.2b)$$

$$u = \frac{1}{\mu_0}B^2 \qquad (24.2c)$$

The **relationship** between the magnitudes of the **electric field** and the **magnetic field** in an electromagnetic wave:

$$E = cB \qquad (24.3)$$

The **intensity** of an **electromagnetic wave:**

$$S = cu \qquad (24.4)$$

$$S = cu = \frac{1}{2}c\varepsilon_0 E^2 + \frac{c}{2\mu_0}B^2 \qquad (24.5a)$$

$$S = c\varepsilon_0 E^2 \qquad (24.5b)$$

$$S = \frac{c}{\mu_0}B^2 \qquad (24.5c)$$

Malus' Law gives the **average intensity** of light **through an analyzer:**

$$\bar{S} = \overline{S_0}\cos^2\theta \qquad (24.7)$$

DISCUSSION OF SELECTED SECTIONS

24.1 The Nature of Electromagnetic Waves

Oscillating currents in wires produce oscillating magnetic fields in the region about the wire. Faraday's Law requires that a changing magnetic flux through a surface, not necessarily a "real" surface, induce an emf around a loop enclosing the surface. An emf is a potential difference and is associated with an electric field. Hence, an oscillating current results in an oscillating magnetic field which produces an oscillating electric field. Oscillating electric fields can also produce oscillating magnetic fields. Once the process is started the oscillating magnetic and electric fields produce each other and can become detached from the original source (the current carrying wire) and travel through space as an independent wave - an electromagnetic wave.

24.3 The Speed of Light

Example 1
Radio waves are a form of electromagnetic radiation. If a spacecraft leaves the earth and travels a great distance, some time is required for radio transmissions from the earth to reach the spacecraft and for a reply by the spacecraft to

return to the earth. What is the minimum delay in the reply to a message sent from earth if the spacecraft is close to Mars and is 3.78×10^{11} m from the earth?

The time that the radio waves take to travel from the earth to the spacecraft is

$$t = s/c = (3.78 \times 10^{11}\ \text{m})/(3.00 \times 10^{8}\ \text{m/s}) = 1.26 \times 10^{3}\ \text{s}.$$

The time taken for the reply to travel from the spacecraft to the earth is the same. The total time that the radio waves spend in transit is then 2.52×10^{3} s which is **42 min**. This is the total time required for the reply to reach the earth even if the spacecraft broadcasts the reply with no delay.

Example 2
A distance is to be measured by reflecting a pulse of laser light from a mirror and measuring the total time taken for the pulse to return to a detector at the site of the laser. How much time elapses for a distance of 152 m?

The total time required for the pulse of laser light to return to the detector is

$$t = 2(152\ \text{m})/3.00 \times 10^{8}\ \text{m/s}) = 1.01 \times 10^{-6}\ \text{s}.$$

This time, though short, is well within the measurement capabilities of modern electronics.

24.4 The Energy Carried by Electromagnetic Waves

The energy density in an electromagnetic wave is just the sum of the energy densities of the electric and magnetic fields. The magnitude of the electric and magnetic fields are related by $E = cB$. These give equations (24.2) which can be used to calculate the energy density of an electromagnetic wave. Note that E and B in these equations are sinusoidal functions of time, hence u is also a function of time. The equations can be interpreted as giving the average energy density if the rms values of E and B are used.

Example 3
What is the average energy density in an electromagnetic wave whose maximum magnetic field is 9.33×10^{-7} T? We must use the rms values of E and B to get the average energy density in the wave.
$B_{rms} = B_0/\sqrt{2} = 6.60 \times 10^{-7}$ T. Then equation (24.2c) gives the

$$\begin{aligned}\text{average energy density} &= B^2_{rms}/\mu_0 = (6.60 \times 10^{-7}\ \text{T})^2/(4\pi \times 10^{-7}\ \text{T m/A}) \\ &= 3.47 \times 10^{-7}\ \text{J/m}^3.\end{aligned}$$

The intensity of an electromagnetic wave is the energy that it carries per second per square meter. Equations (24.5) give the time varying intensity. Again these equations may be used to find average intensities if the rms values of E and B are used.

Example 4
Sunlight falling on the ground on a certain day has an average intensity of 1100 W/m^2. What are the maximum values of the electric and magnetic fields in the sunlight?

Equation (24.5b) can be used to find the rms value of E if S is interpreted as being the average intensity of the light.

$$E_{rms} = \sqrt{\frac{\bar{S}}{c\varepsilon_0}} = \sqrt{\frac{1100\ \text{W/m}^2}{(3.00 \times 10^{8}\ \text{m/s})(8.85 \times 10^{-12}\ \text{C}^2/\text{N m}^2)}} = 640\ \text{N/C}$$

The maximum value of the electric field is then

$$E_{max} = \sqrt{2}\, E_{rms} = 910 \text{ N/C}$$

Similarly, equation (24.5c) is used to find the rms value of the magnetic field.

$$B_{rms} = \sqrt{\frac{\mu_0 \bar{S}}{c}} = \sqrt{\frac{(4\pi \text{ X } 10^{-7} \text{ T m/A})(1100 \text{ W/m}^2)}{3.00 \text{ X } 10^8 \text{ m/s}}} = 2.1 \text{ X } 10^{-6} \text{ T}$$

The maximum magnetic field is then

$$B_{max} = \sqrt{2}\, B_{rms} = 3.0 \text{ X } 10^{-6} \text{ T}$$

24.6 Polarization

Only transverse waves may be polarized. Many sources of electromagnetic waves produce a random polarization. That is, the direction of the electric field vector may change randomly. If all possible electric field directions are filtered out except for one, then the electromagnetic wave is linearly polarized. A polarizing material in essence does this filtering. It limits the electric field oscillations to a single direction by selectively transmitting it and reflecting or absorbing all others.

Example 5
A device made from polarizing material is often called a polarizer. The intensity of unpolarized light reaching a polarizer is S. Show that the intensity of the light transmitted through the polarizer is 1/2 S.
The randomly polarized light has an electric field vector which oscillates in all possible directions. At any given time the electric field vector of the light wave can be resolved into a component along the transmission axis of the polarizer and a component perpendicular to the transmission axis. Since the light is randomly polarized, these two components will, on the average be equal. The component parallel to the transmission axis is

$$E \cos 45° = \frac{1}{\sqrt{2}} E.$$

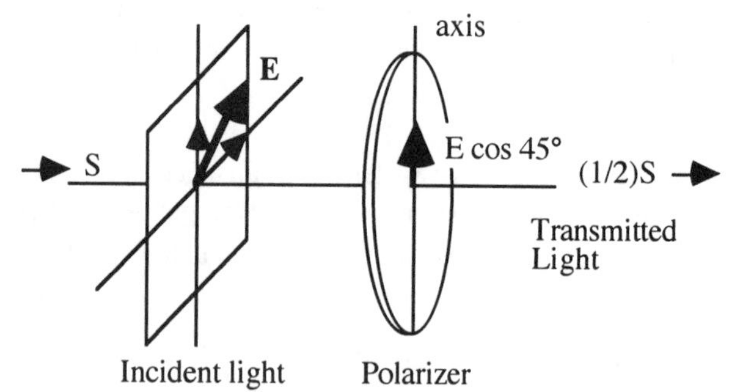

This is the only component transmitted and corresponds to an intensity given by (24.5b) of $(1/2)\, c\varepsilon_0 E^2$. The intensity of the light incident on the polarizer is $S = c\varepsilon_0 E^2$. Comparing, we see that the intensity of the light transmitted through the polarizer is (1/2) S.

Malus' Law, equation (24.7), gives the average intensity of previously polarized light transmitted through a sheet of polarizing material whose transmission axis is an angle, θ, to the electric field direction of the light. If the light were originally polarized by a sheet of polarizer, the second sheet of polarizing material is referred to as an analyzer.

Example 6
When the axes of two sheets of polarizing material are perpendicular, or crossed (see drawing), no light is transmitted. However, when a third sheet is placed between the other two and has its axis tilted with respect to both of them, then light is transmitted. Explain.

When two sheets of polarizing materials are crossed, the first transmits only the electric field component along its transmission axis, hence no component of electric field exists along the axis of the second sheet. No electric field is transmitted through the pair, so the average intensity of the transmitted light is zero.

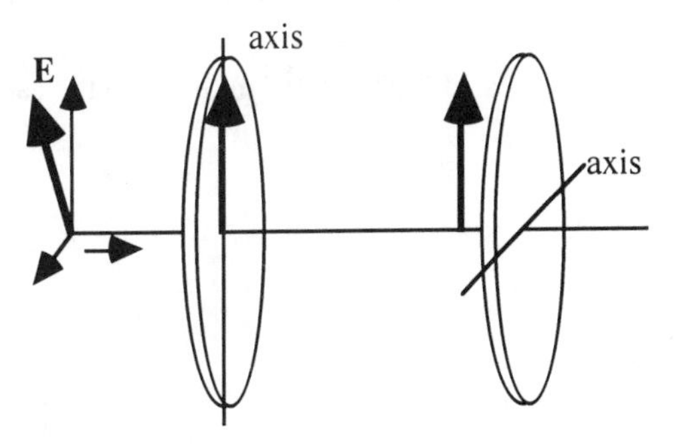

When the third sheet is introduced between the crossed pair, it transmits the component of the electric field along its axis. This electric field now has a component along the axis of the last sheet and some light will be transmitted.

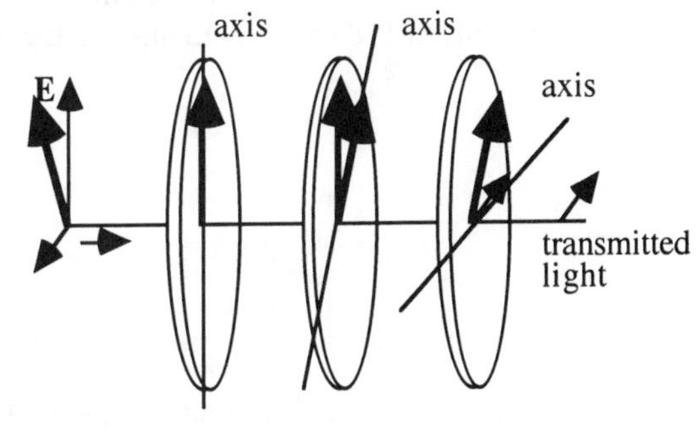

PRACTICE PROBLEMS

1. A certain star is located 2.93×10^{15} miles from the earth. How long ago did the light reaching the earth leave the star?

2. Television transmissions leaving the earth could possibly be received by inhabitants of another planet in a galaxy far far away. Imagine that somewhere there are intelligent creatures watching an episode of *I Love Lucy* which was broadcast in 1952. In 1988 how far in miles is the planet from the earth? Ignore the extra days in leap years.

3. A certain laser produces a cylindrical beam of light with a diameter of 1.5 mm and an average power of 1.5 W. The laser beam can be focused by a lens to a circular spot 2.0 μm in radius. (a) What is the average intensity of the focused beam? (b) What are the maximum values of the electric and magnetic fields in the focused beam?

4. Unpolarized light of average intensity, I_0, is incident on polarizing sheets with crossed axes. A third sheet is placed between the other two and has its transmission axis oriented at an angle, θ, to the transmission axis of the first polarizer. Find an expression for the average intensity of light, I, transmitted through the system as a function of the angle, θ.

5. The arrangement in Figure 24.6 can be used as a simple radio transmitter if an energy source is connected across the transformer in place of the amplifier circuit. The energy source excites the LC circuit to produce a current in the coil which oscillates at the natural frequency of the circuit. This oscillating current produces an electromagnetic wave which travels away from the antenna. What frequency of radio wave can be produced by a circuit in which L = 0.15 mH and C = 4.7 pF?

6. Linearly polarized light falls on a sheet of polarizing material whose transmission axis is oriented at an angle, θ, to the direction of polarization of the light. What fraction of the incident intensity is transmitted through the sheet for $\theta = 15.0°$, $30.0°$, $60.0°$ and $75.0°$?

7. A vertical radio antenna broadcasts vertically polarized radio waves. What fraction of the radio wave intensity arriving at a receiving antenna is available to produce a current if the receiving antenna is oriented 35° to the vertical?

8. A flat coil of wire is used with an LC-tuned circuit as a receiving antenna. The coil has a radius of 25 cm and consists of 450 turns. The transmitted radio wave has a frequency of 1.2 MHz. The magnetic field of the wave is parallel to the normal of the coil and has a maximum value of 2.0×10^{-13} T. Using Faraday's law of induction estimate the magnitude of the average emf induced in the antenna during one quarter of a wave period.

HELPFUL SUGGESTIONS

1. Much of what you have learned about waves on strings and sound waves is directly applicable to electromagnetic waves. The major differences between the types of waves are in their speeds, frequency ranges and media through which they travel. Electromagnetic waves, unlike other wave types, require NO medium. They travel quite well through empty space.

2. Equations (24.2) and equations (24.5) as stated refer to the instantaneous values of energy density and intensity, respectively. They can be modified to give the average values of energy density and intensity by replacing E and B with their rms values.

3. Only transverse waves can be polarized. Demonstrating that a wave can be polarized is good evidence that the wave is transverse.

4. Malus' law, equation (24.7), applies to light which passes through an analyzer after being linearly polarized by a polarizer or some other means. Reflection is an additional means of linearly polarizing light. The angle, θ, in all cases is the angle between the polarization direction of the incident light and the transmission axis of the analyzer.

EVERYDAY PHYSICS

1. The polarization of a transverse wave on a rope can be investigated using the procedure suggested in Figure 24.17. Two boards held about an inch apart can be substituted for the board with a slit shown in the figure. Tie a rope or heavy cord to an immovable object and have a friend hold the boards vertically on each side of the rope. A circular wave can be generated by rapidly moving the free end of the rope in a circle while pulling to apply some tension. The waves on the rope should have a circular polarization until they arrive at the boards. The boards should limit the horizontal motion of the rope and allow it to move only vertically. The resulting wave is vertically polarized.

2. The polarization of light upon reflection can be seen by holding a piece of window glass so that it reflects light from a bulb at a low angle into your eyes. Put on a pair of polaroid sunglasses and observe how the intensity of the light reaching your eye diminishes. Try different orientations of the glass relative to the bulb. How does the amount of horizontally polarized light change?

3. If you wear polaroid sunglasses, you may have noticed an interesting effect when looking at the back window of an automobile when it is illuminated by strong sunlight. It looks "blotchy" or "checkered" when you have the sunglasses on, but not when you look at the window without the glasses. Can you explain this?

4. Old time 3-D movies were made by projecting a red and a green image separated by a small angle on the screen. The audience wore glasses with one red lens and one green lens. Each eye saw only one image since only the red image could be seen through the red lens and only the green image could be seen through the green lens. The brain interpreted the slightly displaced images seen by each eye as depth in the scene, hence 3-D. How can polarized light be used to produce a 3-D effect? What advantages would this technique have?

CHAPTER QUIZ

1. The lowest frequency of visible light corresponds to the color______.
   a. blue   b. green   c. red   d. yellow

2. Which of the following cannot travel through a vacuum.
   a. microwaves   b. visible light   c. x-rays   d. sound

3. The electric and magnetic fields in the radiation field of an oscillating current are_________.
   a. perpendicular to each other
   b. unrelated to each other
   c. Parallel to each other.
   d. both along the direction of propagation of the wave.

4. A current is produced in the vertical metal antenna of an automobile by the radio wave's______.
   a. electric field
   b. magnetic field
   c. electric and magnetic fields
   d. intensity

5. A popular radio station broadcasts on a frequency of 101.1 MHz. What is the vacuum wavelength of the radio waves broadcast by the station?
   a. 2.97 m   b. 0.337 m   c. 16 500 nm   d. 97.2 m

6. The electric field in an electromagnetic wave has an rms strength of 150 N/C. What is the rms strength of the magnetic field in the wave?
   a. 5.0×10^{-7} T   b. 4.5×10^{10} T   c. 7.07×10^{-7} T   d. 6.63×10^{10} T

7. A vertically polarized laser beam with an average intensity of 1.19×10^5 W/m^2 is incident on a sheet of polarizing material whose transmission axis is oriented at 30° to the vertical. What is the intensity of the light transmitted by the sheet?
   a. 1.03×10^5 W/m^2 b. 8.93×10^4 W/m^2 c. 5.95×10^4 W/m^2 d. 1.19×10^5 W/m^2

8. An electromagnetic wave has an average intensity of 2.0×10^4 W/m^2. What is the energy density in the wave?
   a. 2.0×10^4 J/m^3 b. 6.0×10^{12} J/m^3 c. 6.7×10^{-5} J/m^3 d. zero

9. A beam of laser light spreads as it travels. Hence its energy density decreases. If its energy density decreases by a factor of 4, how much has the electric field of the wave changed?
   a. It doubles.
   b. It increases by a factor of 4.
   c. It decreases by a factor of two.
   d. It decreases by a factor of four.

10. A sheet of polarizing material has it axis oriented at 60.0° to the vertical. Another sheet has its axis oriented 60.0° clockwise from that of the first sheet. A third sheet has its axis oriented 60.0° CW from that of the second sheet. What fraction of the intensity of unpolarized light incident on the first polarizer will be transmitted through the last sheet?
    a. 1/4   b. 1/8   c. 1/16   d. 1/32

SOLUTIONS AND ANSWERS

Practice Problems

---

1. The time taken for the light to travel from the star to the earth is

$$t = s/c = (2.93 \times 10^{15}\ \text{mi})(1.609 \times 10^{3}\ \text{m/1 mi})/(3.00 \times 10^{8}\ \text{m/s})$$
$$t = \mathbf{1.57 \times 10^{10}\ s = 498\ y}.$$

---

2. The distance traveled by the light in its trip from the earth to the planet is

$$s = ct = (3.00 \times 10^{8}\ \text{m/s})(6.22 \times 10^{-4}\ \text{mi/1 m})(36\ \text{y})(365\ \text{d/1 y})(24\ \text{h/1 d})(3600\ \text{s/1 h})$$
$$s = \mathbf{2.12 \times 10^{14}\ mi}.$$

---

3. (a) The average intensity of the concentrated beam is

$$\bar{S} = \frac{P}{A} = \frac{1.5\ \text{W}}{\pi(2.0 \times 10^{-6}\ \text{m})^2} = 1.2 \times 10^{11}\ \text{W/m}^2$$

(b) Equation (24.5b) written for the average intensity gives

$$E_m = \sqrt{2}\ E_{rms} = \sqrt{\frac{2\bar{S}}{c\varepsilon_0}} = \sqrt{\frac{2(1.2 \times 10^{11}\ \text{W/m}^2)}{(3.00 \times 10^{8}\ \text{m/s})(8.85 \times 10^{-12}\ \text{C}^2/\text{N m}^2)}} = 9.5 \times 10^{6}\ \text{N/C}$$

Equation (24.3) then gives

$$B_m = E_m/c = \mathbf{0.032\ T}.$$

---

4. The unpolarized incident light has an electric field E_0. Only 1/2 S_0 of the light is transmitted through the first sheet and its electric field is $1/\sqrt{2}\ E_0$. Only the component of the electric field along the second sheet's axis is transmitted, that is, $1/\sqrt{2}\ E_0 \cos\theta$. The transmission axis of the last sheet is at an angle of $90° - \theta$ to the electric field, so the component of the field which is transmitted through the last sheet is $1/\sqrt{2}\ E_0 \cos\theta \sin\theta$. The intensity of the transmitted light is then $S = cE^2$ or

$$S = 1/2\ c_0 E_0^2 \cos^2\theta \sin^2\theta = \mathbf{\tfrac{1}{2} I_0 \cos^2\theta \sin^2\theta}.$$

---

5. The frequency of the radio wave produced by the oscillating current has the same frequency as the current.

$$f = \frac{\omega}{2\pi} = \frac{1}{2\pi}\sqrt{\frac{1}{LC}} = \frac{1}{2\pi}\sqrt{\frac{1}{(0.15 \times 10^{-3}\ \text{H})(4.7 \times 10^{-12}\ \text{F})}} = 6.0 \times 10^{6}\ \text{Hz}$$

---

6. Malus' law, equation (24.7) gives

$$\frac{\bar{S}}{\bar{S}_0} = \cos^2\theta$$

which gives for: $\theta = \mathbf{15.0°}$, **0.933**; $\theta = \mathbf{30.0°}$, **0.750**; $\theta = \mathbf{60.0°}$, **0.250**; $\theta = \mathbf{75.0°}$, **0.0670**.

---

7. Malus's law applies since the radio waves are already linearly polarized. The fraction of the radio wave intensity available to produce a current is $\cos^2 35° =$ **0.67**.

---

8. The flux through the coil changes from 0 Wb to $B_m NA$ in 1/4 of a wave period, T. The magnitude of the average emf induced in the coil during this time is, according to Faraday's law,

$$\text{emf} = B_m NA/(1/4\ T)$$

or

$$\text{emf} = 4B_m NAf = 4(2.0 \times 10^{-13}\ T)(450)(\pi)(25 \times 10^{-2}\ m)^2\ (1.2 \times 10^6\ Hz)$$

$$\text{emf} = \mathbf{8.5 \times 10^{-5}\ V}.$$

Quiz answers

1. c	3. a	5. a	7. b	9. c
2. d	4. a	6. a	8. c	10. d

MCAT REVIEW PROBLEMS

A scientist performs an experiment to measure the speed of light, using a laser and the rotating opaque wheel pictured in figure 1. The laser light passes through the wheel only if the small hole in the wheel is aligned directly in front of the laser. When this alignment happens, a pulse of laser light travels to the distant mirror—which is 10 miles behind the wheel—and reflects back towards the wheel. The reflected light passes through the wheel and triggers the light detector only if the small hole is displaced 180° from its initial position in front of the laser.

When the scientist performs the experiment, she finds that the smallest rate of rotation of the wheel needed to ensure that the light detector gets triggered is 5000 revolutions per second (rev/s).

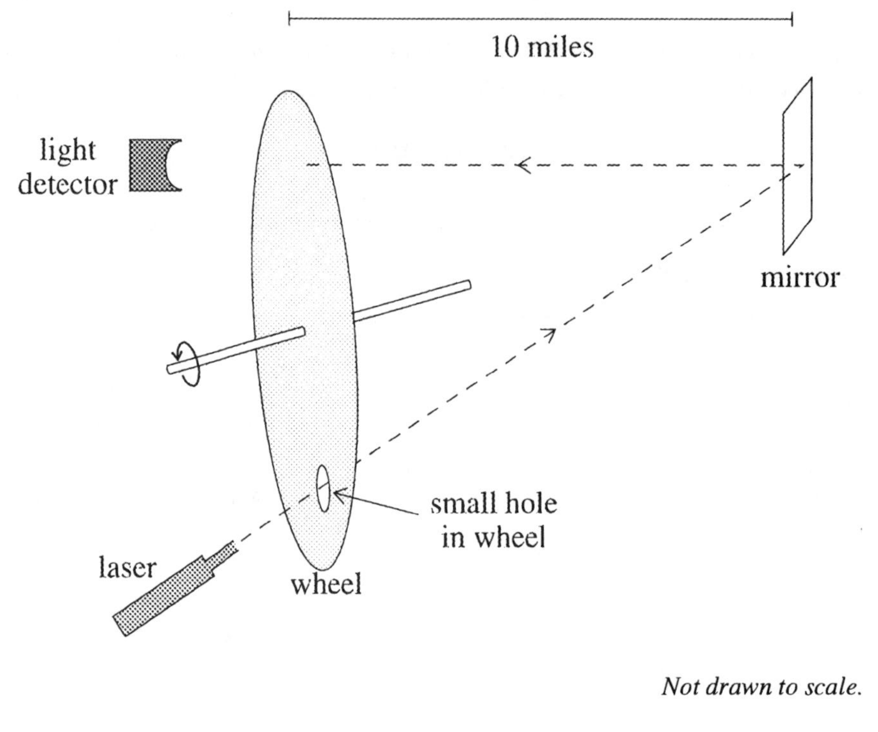

FIGURE 1

1. What other angular speeds of the wheel will also ensure that the light detector gets triggered?

   A. 10,000 rev/s, but not 15,000 rev/s
   B. 15,000 rev/s, but not 10,000 rev/s
   C. Both 10,000 rev/s and 15,000 rev/s
   D. Neither 10,000 rev/s nor 15,000 rev/s

2. When spinning at 5000 rev/s, the wheel completes one revolution in 2.0×10^{-4} seconds. According to this experiment, the speed of light is approximately:

   A. 0.5×10^5 miles per second
   B. 1.0×10^5 miles per second
   C. 2.0×10^5 miles per second
   D. 4.0×10^5 miles per second

3. In a vacuum, which of the following does NOT travel at the speed of light?

   A. X-rays from hospital equipment
   B. Infrared radiation from a hot spoon
   C. Polarized radio waves
   D. Beta radiation from a decaying atom

---

4. Unpolarized laser light passes through a linearly polarizing material and emerges back into the air. As compared to the light originally emitted by the laser, the light that passed through the polarizing material has a different:

   A. Wavelength
   B. Speed
   C. Maximum amplitude
   D. None of the above

ANSWERS TO MCAT REVIEW PROBLEMS

---

1. **B.** Suppose the small hole in the wheel is aligned directly in front of the laser at time $t = 0$. So, a pulse of laser light passes through the wheel at that moment.

   Let t denote the pulse's round trip "travel time" from the wheel to the mirror and back to the wheel. At time t, the small hole must be half a revolution (180°) displaced from its initial position in front of the laser; otherwise, the reflected pulse cannot pass through the wheel and trigger the light detector. When the wheel rotates at 5000 rev/s, the detector does indeed get triggered; no smaller angular speed "works." Therefore, at 5000 rev/s, the wheel must complete half a revolution in time t.

   If the wheel rotates twice as fast, at 10,000 rev/s, then the small hole completes a *full* revolution in time t. Therefore, when the reflected pulse reaches the wheel, the small hole again finds itself in front of the laser, instead of 180° displaced from that point. Consequently, the reflected pulse slams into "solid wheel," instead of passing through the hole. The light detector does not get triggered. So, 10,000 rev/s does not "work."

   By contrast, at 15,000 rev/s, the wheel completes one-and-a-half revolutions in time t, So, when the reflected pulse reaches the wheel, the small hole is in the "right" place, half a revolution displaced from its initial position in front of the laser.

   In general, the light detector gets triggered if, in time t, the wheel completes half a revolution, or one-and-a-half revolutions, or two-and-a-half revolutions, or n-and-a-half revolutions, where n is any integer.

---

2. **C.** As emphasized in question 1, when the wheel rotates at 5000 rev/s, it completes half a revolution in time t, where t denotes the round-trip travel time of light from the wheel to the mirror and back to the wheel. At this angular speed, the wheel takes 2.0×10^{-4} seconds to complete a full revolution. Therefore, it takes $t = 1.0 \times 10^{-4}$ seconds to complete half a revolution. In this time, the light travels 20 miles: 10 miles from the wheel to the mirror, and another 10 miles from the mirror back to the wheel.

   In summary, the light covers $x = 20$ miles in $t = 1.0 \times 10^{-4}$ seconds. Therefore, it travels at speed

$$v = \frac{x}{t} = \frac{20 \text{ miles}}{1.0 \times 10^{-4} \text{ s}} = 2.0 \times 10^{5} \frac{\text{miles}}{\text{s}}.$$

   A common mistake is to say the light travels 20 miles in 2.0×10^{-4} seconds. But in 2.0×10^{-4} seconds, the wheel completes a *full* revolution, and hence, the small hole ends up in front of the laser. Therefore, if the pulse completes its round trip in 2.0×10^{-4} seconds, it reaches the wheel with the hole in the "wrong" place. In order to trigger the detector, the reflected pulse must reach the wheel when the small hole is *half* a revolution displaced from its initial position. In other words, the pulse must complete its 20-mile round trip in the time it takes the wheel to finish *half* a revolution—1.0×10^{-4} seconds, not 2.0×10^{-4} seconds.

---

3. **D.** You can solve by elimination, even if you know nothing about beta radiation. X-rays, infrared radiation, and radio waves are all electromagnetic waves. In a vacuum, all electromagnetic waves travel at the speed of light.

---

4. **D.** The linearly polarizing material transmits only those light waves for which the electric field is aligned in a certain direction. Roughly speaking, it absorbs the "mis-aligned" light, thereby reducing the intensity of the laser beam. But the light waves that *do* get transmitted are unchanged; they have the same speed, wavelength, and maximum amplitude as before.

CHAPTER 25
The Reflection of Light: Mirrors

PREVIEW

In this chapter the reflection of light from mirrors is discussed. You will see how reflections are formed by plane and spherical mirrors, how to construct ray diagrams for reflecting surfaces, and how to determine magnification.

QUICK REFERENCE

Important Terms

Wave fronts
Surfaces on which all points of a wave are in the same phase of motion.

Rays
Lines that are perpendicular to the wave fronts and point in the direction of the velocity of the wave.

Law of reflection
For light that reflects off a smooth surface, (a) the incident ray, the reflected ray, and the normal to the surface all lie in the same plane, and (b) the angle of reflection equals the angle of incidence.

Virtual image
An image from which rays of light do not actually come, but only appear to do so.

Real image
An image from which rays of light actually emanate.

Plane mirror
A flat reflecting surface which forms an upright, virtual image that is located as far behind the mirror as the object is in front of the mirror. In addition, the heights of the image and the object are equal.

Spherical mirror
A reflecting surface that has the shape of a section from the surface of a sphere.

Principal axis
The straight line drawn through the center of curvature and the middle of the mirror's surface.

Paraxial rays
Rays that lie close to the principal axis.

Radius of curvature
The distance from the center of curvature to the mirror.

Focal point
A point on the principal axis where paraxial rays that are parallel to the principal axis converge after being reflected from a concave mirror. For a convex mirror, it is the point on the principal axis from which such rays appear to emanate.

Focal length
The distance from the focal point to the middle of the mirror.

Magnification
The ratio of the image height to the object height.

Equations

The **focal length** of a **concave mirror**:

$$f = \frac{1}{2}R \tag{25.1}$$

The **focal length** of a **convex mirror**:

$$f = -\frac{1}{2}R \tag{25.2}$$

The **mirror equation**:

$$\frac{1}{d_o} + \frac{1}{d_i} = \frac{1}{f} \tag{25.3}$$

The magnification equation:

$$m = \frac{\text{Image height}}{\text{Object height}} = -\frac{d_i}{d_o} \tag{25.4}$$

DISCUSSION OF SELECTED SECTIONS

25.2 The Reflection of Light

Suppose a ray of light is incident on a flat, reflecting surface, such as the mirror shown in the diagram. The angle of incidence θ_i is the angle that the incident ray makes with respect to the normal to the surface. The angle of reflection θ_r is the angle that the reflected ray makes with the normal. The **law of reflection** states that the incident ray, the reflected ray, and the normal to the surface all lie in the same plane, and the angle of reflection equals the angle of incidence.

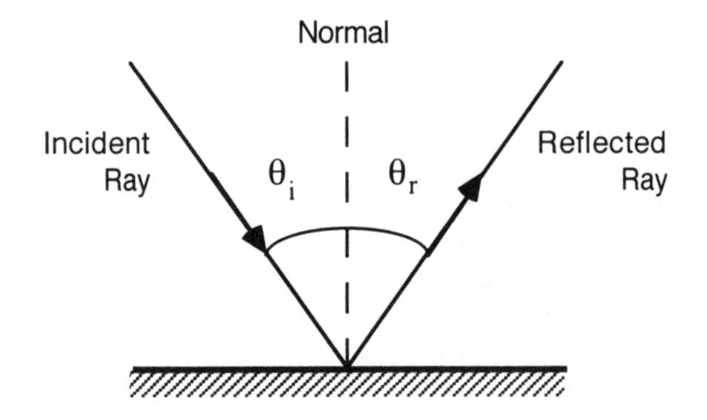

Example 1
Two plane mirrors are separated by 120°, as the drawing illustrates. If a ray strikes mirror M_1, at a 65° angle of incidence, at what angle θ does it leave mirror M_2?

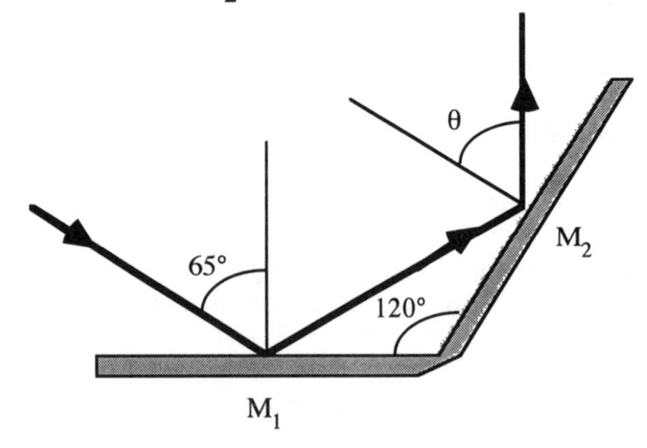

The ray leaves M_1 at an angle of 65° with respect to the normal, or 90° - 65° = 25° with respect to the plane of M_1. The first reflected ray therefore strikes M_2 at an angle of 180° - (120° + 25°) = 35° with respect to the plane of M_2. The second angle of incidence is 90° - 35° = 55°. Finally, the angle of reflection from M_2, θ, is equal to the angle of incidence for mirror M_2, which is just θ = 55°.

25.3 The Formation of Images by a Plane Mirror

When you look into a plane mirror, you see an image of yourself which is upright, has left-right reversal, is located as far behind the mirror as you are in front of it, and is the same size as you are. To illustrate how this image is formed, consider the following **ray diagram**.

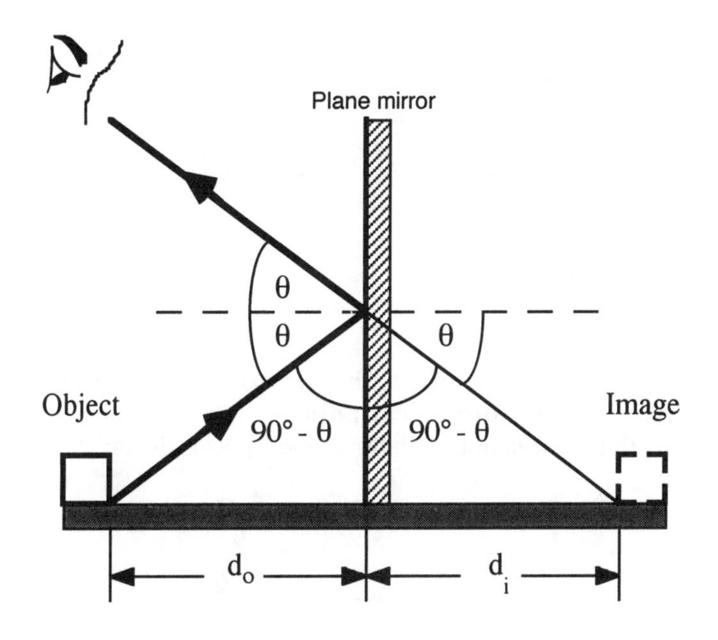

A ray of light leaving the bottom of the object reflects from the mirror and enters the eye. To the eye, it seems that the ray originates from behind the mirror, as shown. However, rays of light do not actually emanate from behind the mirror, and we refer to this as a **virtual image**. All plane mirrors produce virtual images.

From the diagram, it can be demonstrated that the image is located as far behind a plane mirror as the object is in front of it. That is, the object distance d_o equals the image distance d_i. In addition, by drawing a ray of light from the top of the object, rather than from the bottom, we find that the height of the image equals the height of the object.

25.4 Spherical Mirrors

A spherical mirror has the shape of a section from the surface of a sphere. If the inside or concave surface of the mirror is polished, we have a **concave mirror**, shown below.

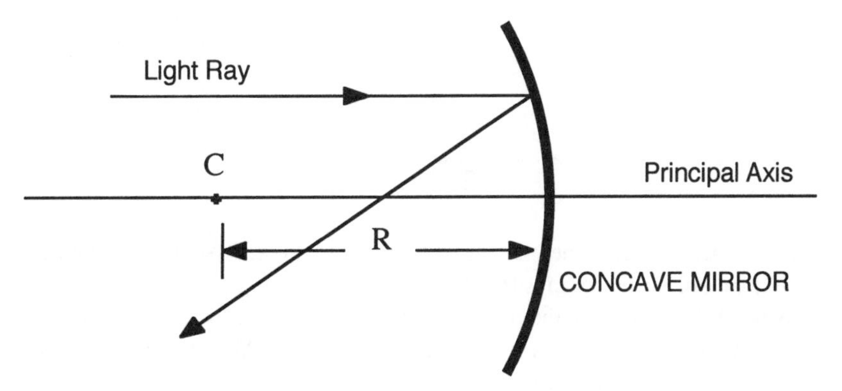

If the outside or convex surface is polished, we have a **convex mirror**, as in the figure. Each type of mirror has a radius of curvature R, where the center of curvature is located at point C. The **principal axis** of the mirror is a straight line drawn through C and the midpoint of the mirror.

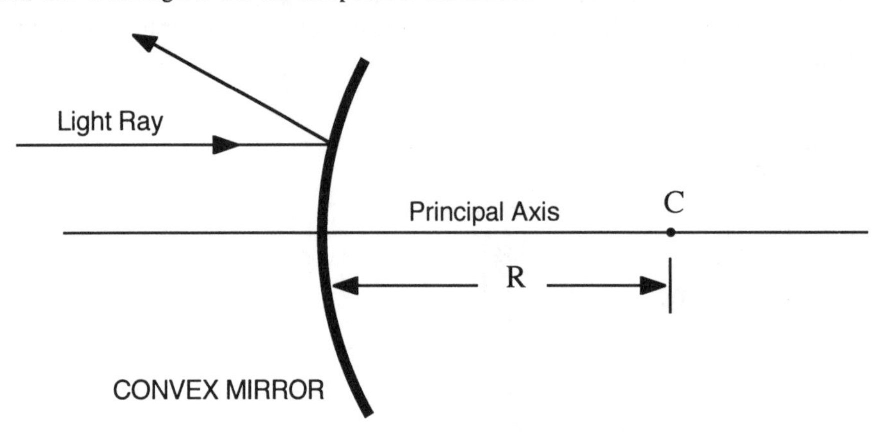

If light rays incident on a spherical mirror are parallel to the principal axis, and lie close to this axis, the reflected rays all pass through a common point, called the **focal point** F. The distance between this focal point and the mirror is called the **focal length** f of the mirror, as shown in the following figure.

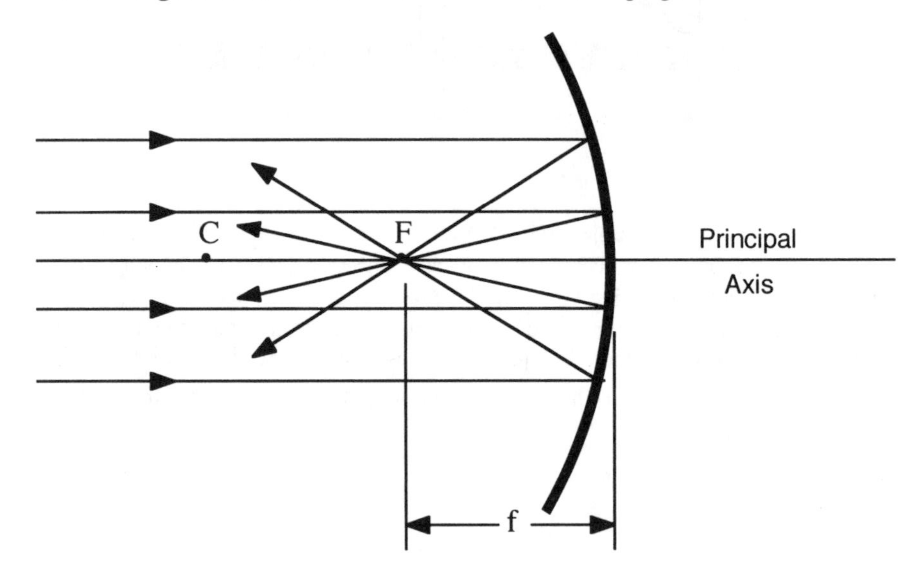

The focal length f and radius of curvature R for spherical mirrors are related. The relationships are as follows:
For a concave mirror:

$$f = \frac{1}{2}R \tag{25.1}$$

For a convex mirror:

$$f = -\frac{1}{2}R \tag{25.2}$$

Thus, the magnitude of the focal length is one-half the radius of curvature. It should be pointed out, however, that equations (25.1) and (25.2) are only valid for rays that lie close to the principal axis, that is, for **paraxial rays**. For spherical mirrors, rays which lie too far from the principal axis will not be reflected through the focal point, and this produces blurred images. This effect is known as **spherical aberration**. Also note that the focal length for convex mirrors is taken by convention to be a negative number.

25.5 The Formation of Images by Spherical Mirrors

When locating the position and size of an image formed by a **concave** mirror, three paraxial rays are useful.

Ray 1. A ray initially parallel to the principal axis, which will pass through the focal point F upon reflection.

Ray 2. A ray passing through the focal point, which will be reflected parallel to the principal axis, according to the principle of reversibility.

Ray 3. A ray traveling through the center of curvature C, which strikes the mirror perpendicularly, and is therefore reflected back on itself.

The procedure used to locate the image is as follows: Draw each of the three rays discussed above from a single point on the object (for example the top of the object). Upon reflection, these rays will converge at a point. This point will then represent the position of the top of the image, as shown in the following figure.

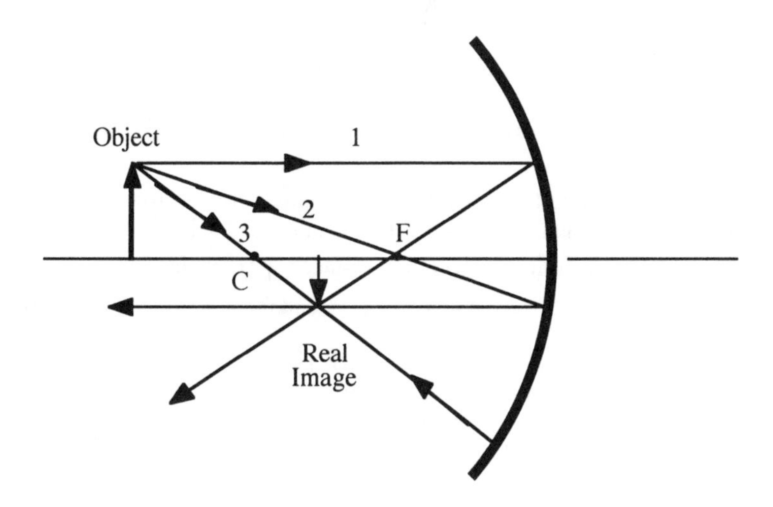

Notice that the image formed by the concave mirror above is a real image, is inverted, and smaller than the object. The image is real because the light rays actually pass through the image. It is possible, under certain conditions, to obtain a virtual image with a concave mirror. This occurs when the object is placed between the focal point F and the mirror.

The procedure for determining the location and size of an image for a **convex** mirror uses the same three paraxial rays discussed above. However, the focal point and center of curvature lie behind the mirror. Therefore, we have:

Ray 1. A ray initially parallel to the principal axis, which appears to originate from the focal point F after reflection from the mirror.

Ray 2. This ray heads toward F, emerging parallel to the principal axis after reflection.

Ray 3. A ray initially traveling toward the center of curvature C, which strikes the mirror perpendicularly and reflects back on itself.

The three rays discussed above appear to come from a point on a virtual image that is behind the mirror. The virtual image is upright, and smaller that the object, as shown in the following ray diagram.

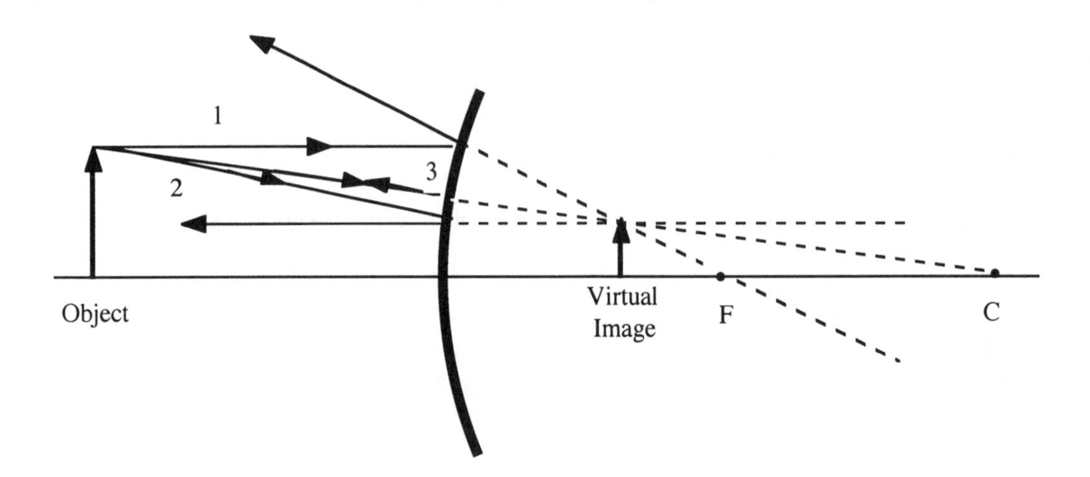

25.6 The Mirror Equation and the Magnification Equation

Ray diagrams are useful in determining the properties of the image formed by spherical mirrors. However, a more precise description of the image is available. Consider an object located a distance d_o from a spherical mirror, having a height h_o. The image formed by the mirror is located a distance d_i from the mirror and has a height h_i. If f is the focal length of the mirror, the object distance, image distance, and focal length are related by the **mirror equation**:

$$\frac{1}{d_o} + \frac{1}{d_i} = \frac{1}{f} \tag{25.3}$$

The **magnification** m of a mirror is defined as the ratio of the image height to the object height: $m = h_i/h_o$. It can be shown (see proof in text) that the ratio of the image and object heights is equal to the ratio of the image and object distances. That is, $h_i/h_o = d_i/d_o$. The **magnification equation** can therefore be written as

$$m = \frac{\text{Image height}}{\text{Object height}} = -\frac{d_i}{d_o} \tag{25.4}$$

By convention, the minus sign is inserted in equation (25.4) to help determine whether the object is upright or inverted. The magnification is positive if the image is upright, and negative if the image is inverted with respect to the object.

Example 2
An object is placed 40.0 cm in front of a concave mirror whose radius of curvature is 30.0 cm. Find the position of the image and the magnification.

First find the focal length of the concave mirror using equation (25.1):

$$f = \frac{1}{2}R = \frac{1}{2}(30.0\text{ cm}) = 15.0\text{ cm}$$

We can now determine the position of the image using the mirror equation,

$$\frac{1}{d_i} = \frac{1}{f} - \frac{1}{d_o} = \frac{1}{15.0\text{ cm}} - \frac{1}{40.0\text{ cm}} = 0.0417\text{ cm}^{-1} \Rightarrow d_i = 24.0\text{ cm}$$

Since d_i is a positive quantity, the image is **real**. To find the magnification use equation (25.4),

$$m = -\frac{d_i}{d_o} = -\frac{24.0\text{ cm}}{40.0\text{ cm}} = -0.600$$

So the image is only 60.0% as large as the object. The minus sign indicates that the image is **inverted** with respect to the object.

Example 3
An object is placed 10.0 cm in front of a concave mirror of focal length 15.0 cm. Find the image distance and magnification.

The mirror equation gives us:

$$\frac{1}{d_i} = \frac{1}{f} - \frac{1}{d_o} = \frac{1}{15.0\text{ cm}} - \frac{1}{10.0\text{ cm}} = -0.033\text{ cm}^{-1} \Rightarrow d_i = -3.0 \times 10^1\text{ cm}$$

So the image is a **virtual image** as the minus sign for d_i indicates. The magnification is

$$m = -\frac{d_i}{d_o} = -\frac{(-3.0 \times 10^1\text{ cm})}{10.0\text{ cm}} = 3.0$$

So the image is 3.0 times larger than the object. The fact that the magnification is positive in this case means the image is upright with respect to the object.

Example 4
An object is placed 75 cm in front of a convex mirror of focal length - 55 cm. Find the properties of the image.

The image distance is:

$$\frac{1}{d_i} = \frac{1}{f} - \frac{1}{d_o} = \frac{1}{-55\text{ cm}} - \frac{1}{75\text{ cm}} = -0.032\text{ cm}^{-1} \Rightarrow d_i = -31\text{ cm}$$

The image is of course virtual, and the magnification is

$$m = -\frac{d_i}{d_o} = -\frac{(-31\text{ cm})}{75\text{ cm}} = 0.41.$$

So the image is upright (m is positive) and smaller than the object (m is less than unity).

PRACTICE PROBLEMS

1. A 179-cm tall boy wants to buy a mirror that is tall enough so he can see himself full length in the mirror. What is the minimum height of such a mirror?

2. You hold a plane mirror 1.0 m in front of your eyes and are able to see a 15 m high tree behind you. If the mirror is 20.0 cm high, and the tree image completely fills the mirror, how far are you standing from the tree?

3. The image of an object viewed in a concave mirror of focal length 25.0 cm appears 75.0 cm in front of the mirror. Find the location of the object and the magnification.

4. An object is placed 35 cm in front of a convex mirror of focal length - 25 cm. Find the location and magnification of the image.

5. A concave makeup mirror is designed so the virtual image is twice the size of the object, when the distance between the object and mirror is 15 cm. (a) Determine the radius of curvature of the mirror. (b) Draw a ray diagram to scale, showing this situation.

6. Complete the following table. As an example, the first row of the table shows a completed set of data.

Type of Mirror	Radius of Curvature	Focal Length	Object Distance	Image			Magnif.
				Distance	Real?	Inverted?	
Concave	20.0 cm	+10.0 cm	+6.67 cm	-20.0 cm	NO	NO	+ 3.0
Plane	——	—	+45 cm				
	50.0 cm		+5.0 cm				+ 0.84
				+75 cm			- 3.0
		+20.0 cm		-40.0 cm			
Convex				-10.0 cm			0.33 (sign?)
Concave	30.0 cm			+150 cm			

HELPFUL SUGGESTIONS

1. It is important to remember the sign conventions that are used with the mirror equation and the magnification equation. These conventions apply to both concave and convex mirrors:

Object distance

d_o is + if the object is in front of the mirror (real object).

d_o is - if the object is behind the mirror (virtual object).

Image distance

d_i is + if the image is in front of the mirror (real image).

d_i is - if the image is behind the mirror (virtual image).

Focal length

f is + for a concave mirror.

f is - for a convex mirror.

Magnification

m is + for an image that is upright with respect to the object.

m is - for an image that is inverted with respect to the object.

EVERYDAY PHYSICS

1. The side view mirror on the passenger side of most cars reads "Caution, objects are closer than they appear". What does this tell you about the type of mirror being used for these side view mirrors?

2. Set up two pocket mirrors at right angles to one another and place a small object between them. You'll see three objects in the mirrors. Change the angle of the mirrors and see how many images you can obtain. Notice what happens when the mirrors are nearly facing one another.

3. Convex and concave mirrors are quite commonly found in everyday situations. Makeup mirrors use concave mirrors to provide magnified images of one's face. Store security systems usually employ large convex mirrors which allow large areas of the store to be seen as a single compact image. As you will see in chapter 26, concave mirrors are used in a type of telescope known as a reflecting telescope.

CHAPTER QUIZ

1. A concave spherical mirror has a radius of curvature of 100 cm. What is its focal length?
   a. 50 cm b. 100 cm c. 200 cm d. 300 cm

2. If a real object is placed inside the focal point of a concave mirror, the image is
   a. real and upright b. real and inverted c. virtual and upright d. virtual and inverted

3. The image formed by a convex spherical mirror is
   a. sometimes real, sometimes virtual
   b. sometimes erect, sometimes inverted
   c. always real and inverted
   d. always virtual and upright.

4. Convex spherical mirrors produce images which
   a. are always larger than the actual object
   b. are always smaller than the actual object
   c. are always the same size as the actual object
   d. are sometimes larger, sometimes smaller.

5. A man is 6.0 ft tall. What is the smallest size plane mirror he can use to see his entire image?
   a. 3.0 ft b. 6.0 ft c. 12 ft d. 24 ft

6. An object is placed at the radius of curvature of a concave spherical mirror. The image formed by the mirror is
   a. located at the focal point of the mirror.
   b. located between the focal point and the radius of curvature of the mirror.
   c. located at the center of curvature of the mirror.
   d. located out beyond the center of curvature of the mirror.

7. An object is placed 60 cm in front of a concave mirror. The real image formed by the mirror is located 30 cm in front of the mirror. What is the object's magnification?
   a. + 2 b. - 2 c. + 0.5 d. - 0.5

8. An object is placed 60 cm in front of a convex mirror. The virtual image formed by the mirror is located 30 cm behind the mirror. What is the object's magnification?
   a. + 2 b. - 2 c. + 0.5 d. - 0.5

9. An object is placed 20.0 cm in front of a concave mirror whose focal length is 25.0 cm. Where is the image located?
   a. 1.0×10^2 cm in front of the mirror
   b. 1.0×10^2 cm behind the mirror
   c. 5.0×10^1 cm in front of the mirror
   d. 5.0×10^1 cm behind the mirror

10. An object is placed 20.0 cm in front of a concave mirror whose focal length is 25.0 cm. What is the magnification of the object?
    a. + 5.0 b. - 5.0 c. + 0.20 d. - 0.20

11. An object is placed 40.0 cm in front of a convex mirror. The image appears 15 cm behind the mirror. What is the focal length of the mirror?
    a. + 24 cm b. + 11 cm c. - 11 cm d. - 24 cm

12. An object is placed in front of a concave mirror of focal length 50.0 cm and a real image is formed 75 cm in front of the mirror. How far is the object from the mirror?
    a. 25 cm b. 30 cm d. 150 cm d. - 150 cm

SOLUTIONS AND ANSWERS

Practice Problems

---

1. Referring to example 1 in the textbook it is shown that a mirror needs to be one-half the height of the person in order to view his entire height in the mirror. Therefore, the height of the mirror is **89.5 cm**.

---

2. The ray diagram can be drawn in the following way:

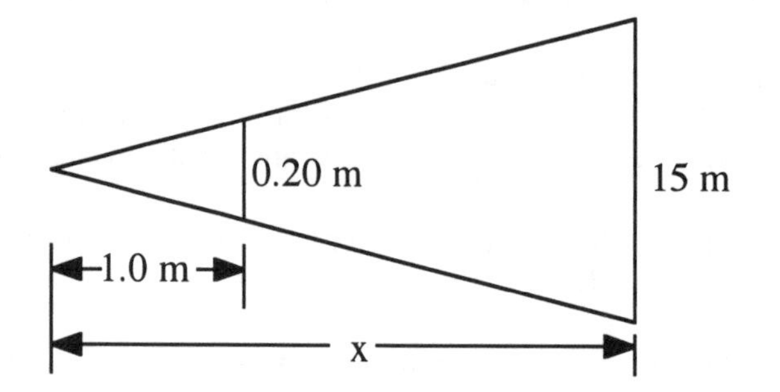

To find the distance x, use the two similar triangles to obtain the following ratio:

$$\frac{x}{1.0\text{ m}} = \frac{15\text{ m}}{0.20\text{ m}} \Rightarrow x = \frac{(15\text{ m})(1.0\text{ m})}{0.20\text{ m}} = 75\text{ m}$$

Therefore, the tree is 75 - 1.0 m = **74 m** behind you.

---

3. We have

$$1/d_o = 1/f - 1/d_i = 1/(25.0\text{ cm}) - 1/(75.0\text{ cm}) = 0.0267\text{ cm}^{-1} \Rightarrow d_o = \mathbf{37.5\text{ cm}}.$$

The magnification is

$$m = -d_i/d_o = -(75.0\text{ cm})/(37.5\text{ cm}) = \mathbf{-2.00}.$$

---

4. We have

$$1/d_i = 1/f - 1/d_o = 1/(-25\text{ cm}) - 1/(35\text{ cm}) = -0.069\text{ cm}^{-1} \Rightarrow d_i = \mathbf{-15\text{ cm}}.$$

The magnification is

$$m = -d_i/d_o = -(-15\text{ cm})/(35\text{ cm}) = \mathbf{+0.43}.$$

---

5. The magnification is $m = -d_i/d_o = +2$ where d_o is 15 cm. Therefore, $d_i = -2d_o = -3.0 \times 10^1$ cm. Therefore, the focal length is found from $1/f = 1/d_o + 1/d_i$ which yields f = + 30 cm. Therefore, R = 2f = **60 cm**.

---

6. See next page for the completed table.

Quiz answers

1. a	5. a	9. b
2. c	6. c	10. a
3. d	7. d	11. d
4. b	8. c	12. d

Practice Problem 6

Type of Mirror	Radius of Curvature	Focal Length	Object Distance	Image			Magnif.
				Distance	Real?	Inverted?	
Concave	20.0 cm	+10.0 cm	+6.67 cm	-20.0 cm	NO	NO	+ 3.0
Plane	——	——	+45 cm	-45 cm	NO	NO	+ 1.0
Convex	50.0 cm	- 25.0 cm	+5.0 cm	-4.2 cm	NO	NO	+ 0.84
Concave	38 cm	+19 cm	+25 cm	+75 cm	YES	YES	- 3.0
Concave	40.0 cm	+20.0 cm	+13.3 cm	-40.0 cm	NO	NO	+ 3.0
Convex	30.0 cm	-15.0 cm	+30.0 cm	-10.0 cm	NO	NO	+0.33
Concave	30.0 cm	+15.0 cm	+16.7 cm	+150 cm	YES	YES	- 9.0

MCAT REVIEW PROBLEMS

At the microscopic level, the law of reflection always holds true: The angle of incidence of an incoming light ray equals the angle at which it reflects off the surface. But at a macroscopic level, light bounces off some surfaces at all different angles. This phenomenon is called "diffuse reflection." It happens because all but the smoothest surfaces contain microscopic nooks and crevices, as shown in figure 1.

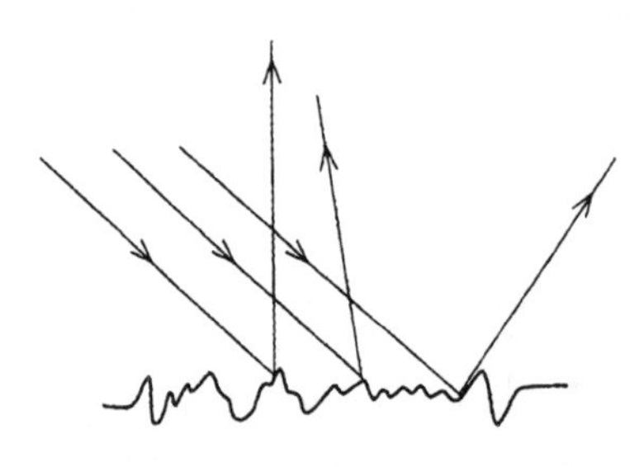

FIGURE 1

Consider the metal coin in figure 2. The right half is carefully polished, while the left half is unpolished.

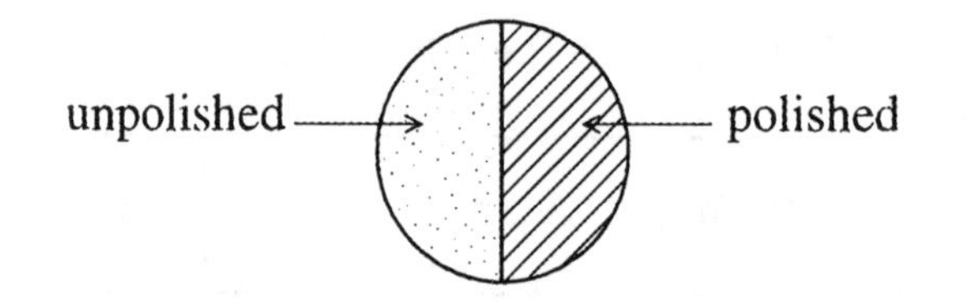

FIGURE 2

In a dark room, this coin is placed face-up on the floor and illuminated with a thin-beamed flashlight, as shown in figure 3. Ann and Bob look at the coin from the positions indicated in figure 3.

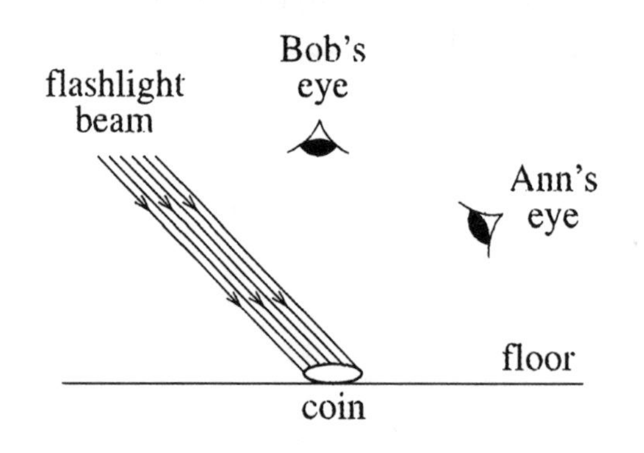

FIGURE 3

1. According to Ann, which half of the coin (if either) is brighter?

   A. The polished half
   B. The unpolished half
   C. Both halves look equally bright.
   D. Both halves look completely dark.

---

2. According to Bob, which half of the coin (if either) is brighter?

   A. The polished half
   B. The unpolished half
   C. Both halves look equally bright.
   D. Both halves look completely dark.

---

3. Light striking a painted concrete wall reflects in all different directions. But a "beam" of sound hitting the same wall reflects at a single angle, equal to the angle of incidence. Why?

   A. Diffuse reflection happens only to light waves, not to other kinds of waves.
   B. Sound waves are slower than light waves.
   C. Sound waves have a longer wavelength than light waves.
   D. Sound waves are longitudinal "pressure" waves, not transverse waves.

---

4. According to the law of reflection, which graph best shows the relationship between the angle of incidence and the angle of reflection?

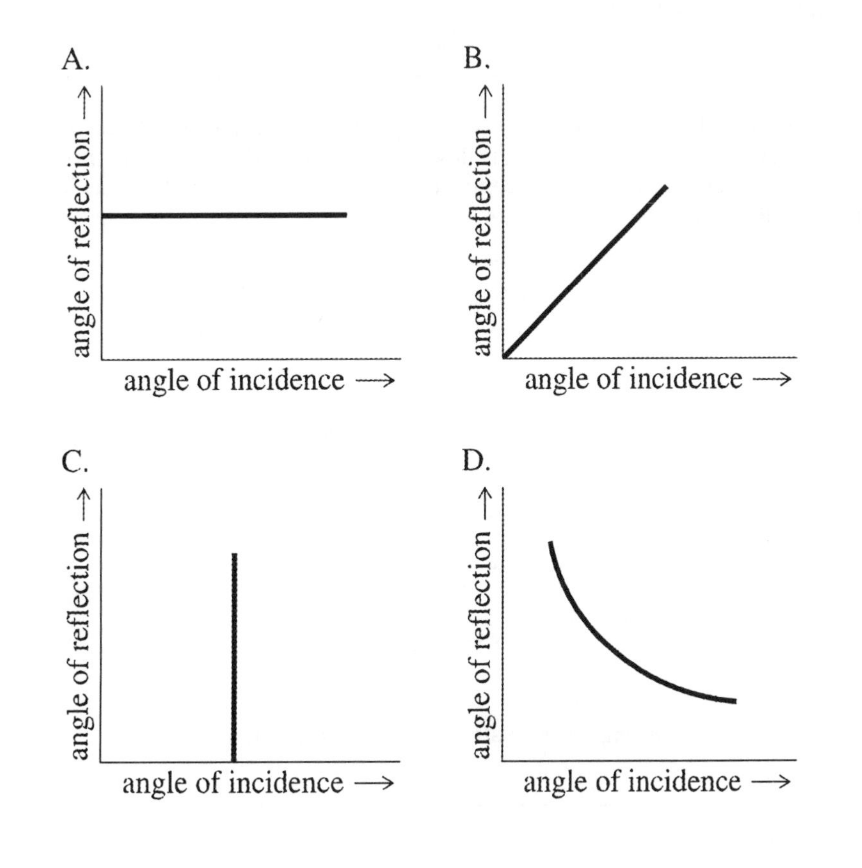

ANSWERS TO MCAT REVIEW PROBLEMS

1. **A.** Light reflecting off the coin at the angle of incidence reaches Ann's eye.

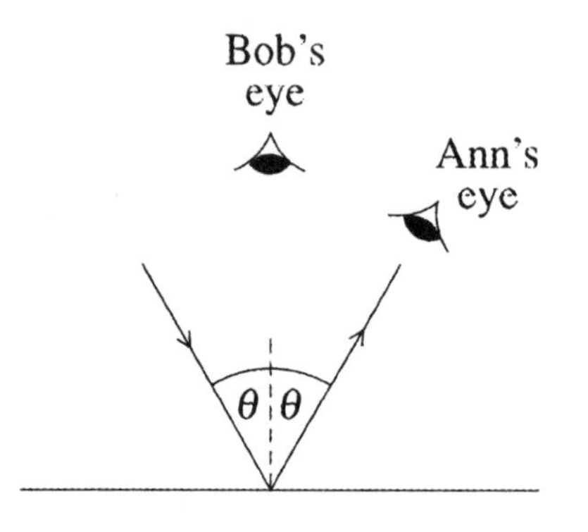

FIGURE 4

Since the polished half is smoother than the unpolished half, the polished half behaves more a like mirror. In other words, most of the light reaching the polished half reflects at the angle of incidence, instead of undergoing diffuse reflection. Therefore, Ann sees a high percentage of the light that hits the polished part of the coin.

By contrast, light hitting the unpolished half undergoes more diffuse reflection, bouncing off the coin in all different directions. So, a smaller percentage of that light reaches Ann's eye. To her, the unpolished half looks less bright than the polished half.

---

2. **B.** As shown in figure 4, most of the light reflecting off the polished part of the coin travels in the same direction, towards Ann's eye. Consequently, Bob sees very little of that light. To him, the polished half of the coin looks dim or completely dark. By contrast, since the unpolished half reflects light diffusely, some of that light reaches Bob's eye.

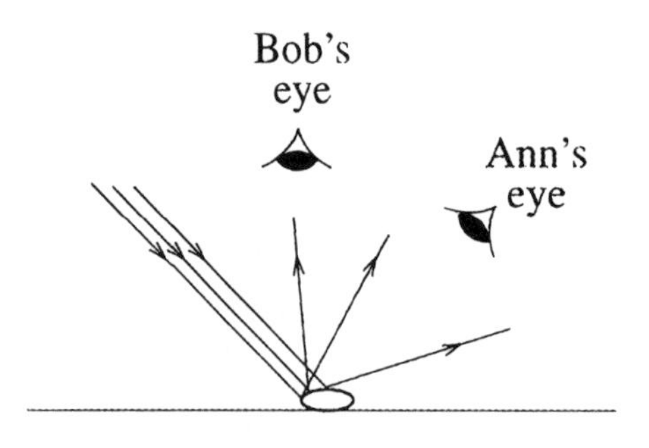

FIGURE 5

So, he sees more light from the unpolished half than he sees from the polished half.

---

3. **C.** All kinds of waves can reflect diffusely, if the surface contains nooks and crevices. But whether something counts as a "nook" or "crevice" depends on the wavelength.

   For instance, suppose the nooks on the painted concrete wall are about a hundredth of a millimeter (10^{-5} meters) wide and deep. Since light waves are between 10^{-7} and 10^{-6} meters long, those nooks seem large from the standpoint of a light wave. Specifically, since light waves can "get inside" those nooks and crevices, the waves bounce off at all different angles. By contrast, even the shortest sound waves are about a centimeter (10^{-2} meters) long. So, the nooks are 1000 times smaller than a sound wave. From the standpoint of a sound wave, the painted wall "looks" smooth, because those waves cannot fit inside the nooks.

   To see why this matters, visualize a tennis ball bouncing off the wall. It bounces at a predictable angle, because the wall "seems" smooth to the tennis ball. More precisely, the ball cannot "get inside" the tiny nooks. In order to make the tennis ball "reflect diffusely," we must chisel some tennis-ball-sized nooks into the wall. The ball would bounce off these large nooks at all different angles. Similarly, to make sound reflect diffusely, we must chisel some sound-wave-sized nooks into the wall.

---

4. **B.** According to the law of reflection, the angle of incidence equals the angle of reflection. Therefore, when one of those angles increases, so does the other. By contrast, graph A "says" that the angle of reflection stays constant as the angle of incidence increases. Graph C "says" that the angle of incidence cannot vary. And graph D "says" that the angle of reflection is inversely proportional to the angle of incidence.

CHAPTER 26
The Refraction of Light: Lenses And Optical Instruments

PREVIEW

In this chapter you will be introduced to one of the most useful properties of light: refraction at an interface between two transparent materials. Snell's law governs how much refraction occurs at an interface. You will learn to apply Snell's law to find how much the path of a light ray is bent in passing through an interface and how much the path of a light ray is displaced when passing through a pane of glass.

A special case of Snell's law when the angle of incidence is sufficiently large leads to the phenomenon of total internal reflection at the interface. This phenomenon has importance in the transmission of light through optical fibers and is responsible for the brilliance of the light reflected from a diamond.

The polarization of light by reflection from a boundary is discussed as well as the dispersion of light into its rainbow of colors when it is refracted through a prism.

Many applications of refraction involve lenses. You will learn how lenses refract light to form images of the object from which the light originated. In particular you will learn to calculate where the images will be formed, what kind of image is formed and the size of the image for the special case of thin lenses. After treating thin lenses separately, you will study thin lenses in combinations. Two familiar applications of lenses in combination are the microscope and the telescope which are also discussed along with the camera, the human eye, and the magnifying glass.

QUICK REFERENCE

Important Terms

Interface
The boundary between two materials.

Refraction
The change in the direction of travel of light as it passes from one material to another.

Index of refraction
The ratio of the speed of light in a vacuum to the speed of light in a particular material.

Apparent depth
The depth of a submerged object as seen by someone not under the water. The apparent depth is smaller than the actual depth of the object due to the refraction of light as it travels from the water into the air.

Critical angle
The angle of incidence at which the angle of refraction is 90°.

Total internal reflection
The phenomenon in which a light ray incident on an interface is totally reflected and no refraction occurs at an interface.

Brewster angle
The angle of incidence of a light ray at which the reflected ray is completely polarized. At the Brewster angle the reflected and refracted rays are 90° to each other.

Dispersion
The spreading of light into its color components.

Focal point
The point where a lens focuses light rays that are near the principle axis and parallel to it.

Focal length
The distance between a lens and its focal point.

Converging lens
A lens which causes incident parallel rays to converge at the focal point.

Diverging lens
A lens which causes incident parallel rays to diverge after passing through the lens, the rays then appearing to come from the focal point.

Camera
A device which uses a converging lens system to produce real, inverted images on film.

Human eye
It is very much like a camera. It forms a real, inverted image on a light-sensitive surface, called the retina.

Accommodation
The process by which the focal length of the eye is automatically adjusted, so that objects at different distances can be made to produce focused images on the retina.

Near point
The point nearest the eye at which an object can be placed and still produce a sharp image on the retina.

Far point
The location of the farthest object on which the fully relaxed eye can focus.

Nearsightedness (myopia)
A condition in which the eye can focus on nearby objects, but not on distant ones. The condition can be corrected through the use of glasses made from diverging lenses.

Farsightedness (hyperopia)
A condition in which the eye can focus on distant objects, but not on nearby ones. The condition can be corrected through the use of glasses made from converging lenses.

Refractive power
A figure of merit of a lens measured in diopters and given by 1/f, where f is the focal length of the lens in meters.

Angular size
The angle an object subtends at the eye of the viewer.

Angular magnification
The ratio of the final angular size produced by an optical instrument to the reference angular size of the object.

Magnifying glass
A single converging lens that forms an enlarged, upright, and virtual image of an object which is placed at or inside the focal point of the lens.

Compound microscope
A device consisting of an objective lens and an eyepiece lens which produces an enlarged, inverted, and virtual image of an object.

Astronomical telescope
A device which magnifies distant objects with the aid of objective and eyepiece lenses. It produces a final image which is inverted and virtual.

Spherical aberration
An effect in which rays from the outer edge of a spherical lens are not focused at the same point as rays that pass through the center of the lens, thus causing images to be blurred.

Chromatic aberration
An undesirable lens effect arising when a lens focuses different colors at different points.

Equations

The **index of refraction of a material** is

$$n = \frac{\text{Speed of light in a vacuum}}{\text{Speed of light in the material}} = \frac{c}{v} \qquad (26.1)$$

Snell's law of refraction

$$n_1 \sin \theta_1 = n_2 \sin \theta_2 \qquad (26.2)$$

The **apparent depth** of an object with the **observer directly above the object** is

$$d' = d\left(\frac{n_2}{n_1}\right) \tag{26.3}$$

The **critical angle for total internal reflection** is given by

$$\sin\theta_c = \frac{n_2 \sin 90°}{n_1} = \frac{n_2}{n_1} \qquad (n_1 > n_2) \tag{26.4}$$

Brewster's Law

$$\tan\theta_B = \frac{n_2}{n_1} \tag{26.5}$$

The **thin lens equation**

$$\frac{1}{d_o} + \frac{1}{d_i} = \frac{1}{f} \tag{26.6}$$

The **magnification equation**

$$m = \frac{\text{Image height}}{\text{Object height}} = \frac{h_i}{h_o} = -\frac{d_i}{d_o} \tag{26.7}$$

The **refractive power** of a lens is given by:

$$\text{Refractive power (in diopters)} = \frac{1}{\text{f (in meters)}} \tag{26.8}$$

The **angular magnification of an optical instrument** is:

$$M = \frac{\text{Angular size produced by instrument}}{\text{Reference angular size}} = \frac{\theta'}{\theta} \tag{26.9}$$

The **angular magnification of a magnifying glass** is:

$$M = \frac{\theta'}{\theta} \cong \left(\frac{1}{f} - \frac{1}{d_i}\right) N \tag{26.10}$$

The **angular magnification of a compound microscope** is given by:

$$M \cong -\frac{(L - f_e)N}{f_o f_e} \tag{26.11}$$

The **angular magnification of an astronomical telescope** is:

$$M = \frac{\theta'}{\theta} \cong -\frac{f_o}{f_e} \qquad (26.12)$$

DISCUSSION OF SELECTED SECTIONS

26.2 Snell's Law and the Refraction of Light

When light travels across an interface from one medium to another, it undergoes a change in its speed as indicated by the different indices of refraction of the materials. This speed change results in a bending (refraction) of the path of the light unless it strikes the interface at normal incidence (See figure 26.2 in the text.). Snell's laws (equation 26.2) can be used to calculate the amount of bending.

Example 1
Light strikes an interface between two materials of refractive indices , n_1 and n_2, at an angle θ_1 to the normal to the surface. Show that a ray of the light is bent towards the normal if $n_1 < n_2$ and that a ray is bent away from the normal if $n_1 > n_2$.
Snell's law gives for any light ray

$$\sin\theta_2 = \frac{n_1}{n_2}\sin\theta_1$$

If $n_1 < n_2$ then $\sin\theta_2 < \sin\theta_1$ and $\theta_2 < \theta_1$. This means that the ray is bent toward the normal.

If $n_1 > n_2$ then $\sin\theta_2 > \sin\theta_1$ and $\theta_2 > \theta_1$. This means that the ray is bent away from the normal.

Example 2
A corner reflector is placed on the bottom of a lake. Light is shined from a boat at an angle of 35° to the surface of the water strikes the corner reflector, and passes through the water and back into the air. At what angle to the surface does the light leave the water. Note: It is a property of a corner reflector that it reflects light back along a path that is parallel to its original path.

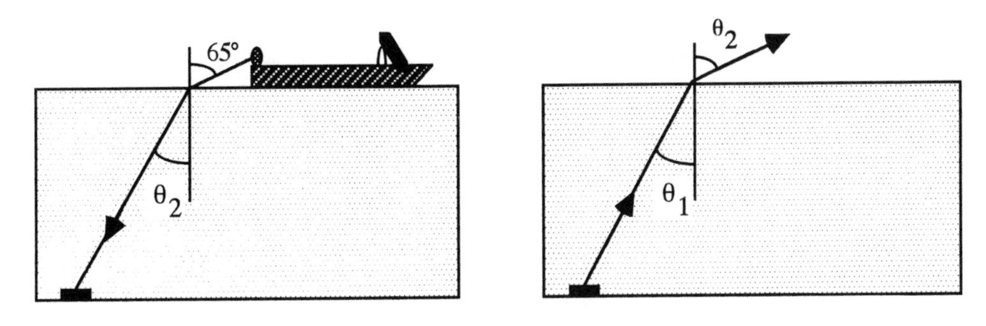

Original ray

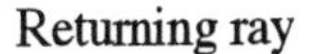

Returning ray

Snell's law applied to the light as it travels from the air to the water gives the angle of refraction into the water

$$\theta_2 = \sin^{-1}\left(\frac{n_1}{n_2}\sin\theta_1\right) = \sin^{-1}\left(\frac{1.00}{1.33}\sin 65^\circ\right) = 43^\circ$$

Similarly, the angle of refraction into the air for the returning light is

$$\theta_2 = \sin^{-1}\left(\frac{n_1}{n_2}\sin\theta_1\right) = \sin^{-1}\left(\frac{1.33}{1.00}\sin 43°\right) = 65^c$$

The angle of refraction from the water to the air is the same as the original incident angle. This illustrates a useful property of light rays. A correct ray diagram is still valid if the ray reverses direction.

Example 3
An airborne pelican looks directly down and sees a fish which is actually 0.50 m below the surface of the water. At what apparent depth does the pelican see the fish?

The light leaving the fish and arriving at the pelican's eye travels along the normal to the water-air interface, so equation (26.3) may be used to find the apparent depth of the fish.

$$d' = d\left(\frac{n_2}{n_1}\right) = (0.50\text{ m})\left(\frac{1.00}{1.33}\right) = 0.38\text{ m}$$

Example 4
The windshields on old automobiles were flat instead of curved. What is the displacement, as seen by the driver, of objects whose light strikes a 0.81 cm thick flat windshield of index of refraction 1.50 at an angle of 45°?
Do you think that the displacement of the objects as seen by the driver will hinder his driving ability?

Refer to the drawing. A light ray from the object is displaced by an amount d as it travels through the windshield. The ray strikes the air-glass interface of the windshield at an angle of $\theta_1 = 45°$ to the normal. Snell's law requires that the angle of refraction be

$$\sin\theta_2 = (n_1/n_2)\sin\theta_1 = (1.00/1.33)\sin 45°$$
$$\theta_2 = 32°$$

The right triangles (123 and 134) shown in the drawing have a common hypotenuses, hence

$$\frac{t}{\cos\theta_2} = \frac{d}{\sin(\theta_1 - \theta_2)}$$

$$d = \left[\frac{\sin(\theta_1 - \theta_2)}{\cos\theta_2}\right] t = \left[\frac{\sin(45° - 32°)}{\cos 32°}\right](0.81\text{ cm})$$

$d = 0.21$ cm.

This represents a very small displacement of the objects as seen by the driver and should not hinder his driving ability.

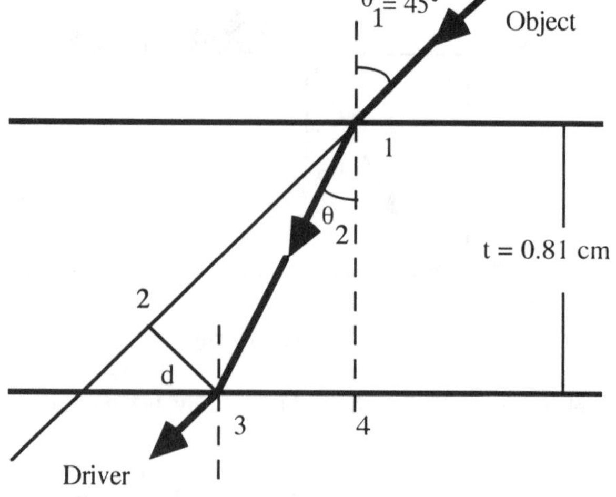

26.4 Polarization and the Reflection and Refraction of Light

When light strikes a nonmetallic surface at an angle of incidence equal to the Brewster angle, then the reflected light will be totally 100 % polarized parallel to the surface. The light transmitted through the surface, that is the refracted light, is then partially polarized perpendicular to the surface.

Example 5
The light produced inside of a laser tube is unpolarized. The light emitted by many lasers is polarized, however, because special windows, called Brewster windows are used at the ends of the tube. The windows are inclined at the Brewster angle to the axis of the tube.(a) Explain how these windows can completely polarize the laser light and determine the direction of polarization of the emitted light. (b) What is the necessary angle of inclination of the windows if the gas inside the tube has n = 1.00 and the windows are made of fused quartz?

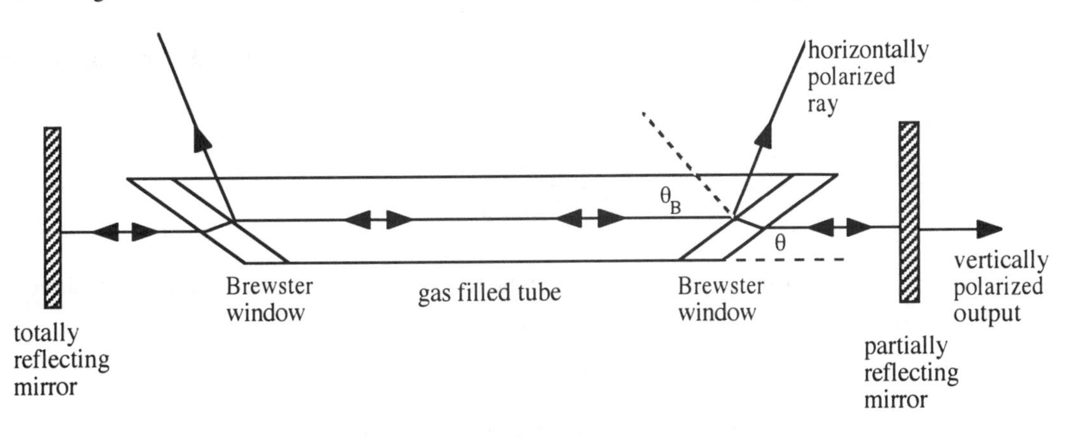

(a) Refer to the drawing. Not all of the laser light produced inside the tube is emitted as output light. Some is reflected back into the tube by a partially reflecting output mirror. This light has already passed through the Brewster window and is partially polarized in the vertical direction. It is this light that is amplified in the laser tube and passes through another window before being reflected back by a totally reflecting mirror. After several passes through the tube, the light becomes totally polarized in the vertical direction since the horizontal component is reflected out of the tube, hence the output beam is vertically polarized.

(b) The angle of inclination of the window to the axis of the tube, θ, is $90° + \theta_B$. Now

$$\theta_B = \tan^{-1}(1.458/1.00) = 55.6°$$

so $\theta = 34.4°$.

26.7 The Formation of Images by Lenses
28.8 The Thin Lens Equation and the Magnification Equation

The same equation applies to image formation by THIN lenses as applied to image formation by mirrors.

$$\frac{1}{d_o} + \frac{1}{d_i} = \frac{1}{f}$$

For a single **lens** the following conventions are used. These are the same conventions as used in the text, but are stated in a somewhat different manner.

1. The object is placed to the left of the lens.
2. Real images fall to the right of the lens.

3. Virtual images fall to the left of the lens.
4. The object distance, d_0, is always positive.
5. The image distance, d_i, is positive for real images and negative for virtual images.
6. The focal length, f, is positive for a converging lens and negative for a diverging lens.
7. The magnification, m, is positive for an upright image and negative for an inverted image.

Example 6

A converging lens can form both real and virtual images depending on where the object is placed relative to the focal point of the lens. (a) For what object distances will it produce a real image? Virtual image? (b) For what object distances will it produce an upright image? Inverted image? (c) For what object distances will the image be the same size as the object? Larger than? Smaller than?

(a) Whether the image is real or virtual is determined by the sign of the image distance, d_i. The thin lens equation gives the image distance to be

$$d_i = \frac{d_0 f}{d_0 - f}$$

Now the object distance, d_0, is positive for real objects and the focal length of a converging lens is positive, so the sign of d_i is determined by whether d_0 or f has the larger value.

d_i is positive and the image is real if $d_0 > f$.
d_i is negative and the image is virtual if $d_0 < f$.

(b) Whether or not the image is up right or inverted is determined by the sign of the magnification,

$$m = -\frac{d_i}{d_0}$$

If $d_0 < f$, then d_i is negative and m is positive so the image is upright.
If $d_0 > f$, then d_i is positive and m is negative so the image is inverted.

(c) Solving the thin lens equation for d_i and substituting into the magnification equation yields

$$m = -\frac{f}{d_0 - f}$$

Remember that both d_0 (real object)and f are positive for a converging lens.

The magnitude of the magnification is equal to one if $f - d_0 = f$, that is, $d_0 = 2f$.

The magnitude of the magnification is greater than one if $d_0 - f < f$, that is, $d_0 < 2f$.

The magnitude of the magnification is less than one if $d_0 - f > f$, that is $d_0 > 2f$.

The magnitude of the magnification is INFINITE if $d_0 = f$. This is the case where the image rays are parallel.

The above results are summarized in the following diagram.

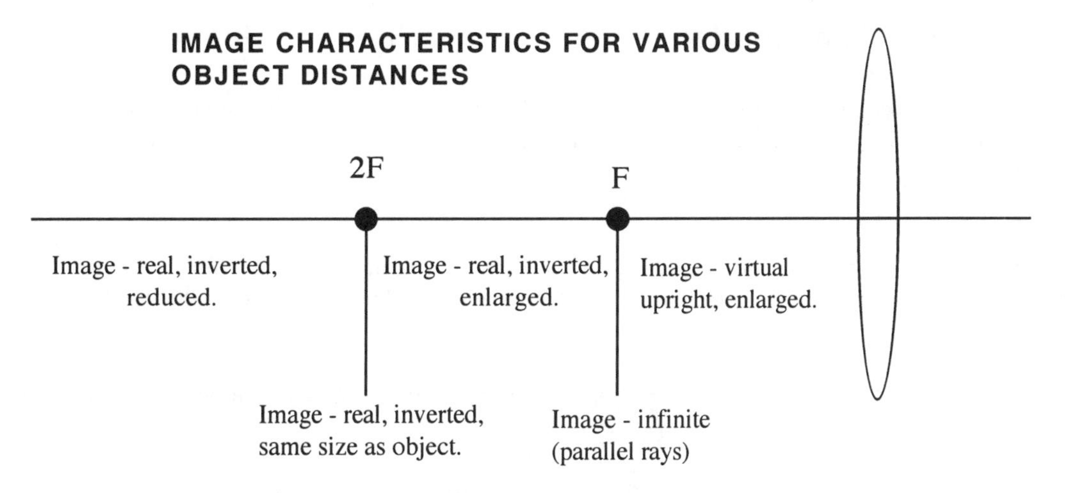

Example 7
Repeat example 6 for a diverging lens.

(a) The thin lens equation gives

$$d_i = \frac{d_0 f}{d_0 - f}$$

This is the same as for the converging lens except that f is negative for a diverging lens. This requires that d_i be NEGATIVE for ALL object distances, hence the image formed by a diverging lens is ALWAYS VIRTUAL for real objects.

(b) The magnification is again

$$m = -\frac{d_i}{d_0}$$

In part (a) we found that d_i is always negative for a diverging lens, so the magnification is ALWAYS POSITIVE for real objects and corresponds to an UPRIGHT image.

(c) The magnification is also

$$m = -\frac{f}{d_0 - f}$$

where f is negative for real objects. The magnitude of the magnification is ALWAYS LESS THAN ONE, since the denominator is always larger than the numerator in the above fraction. These results are summarized in the following diagram.

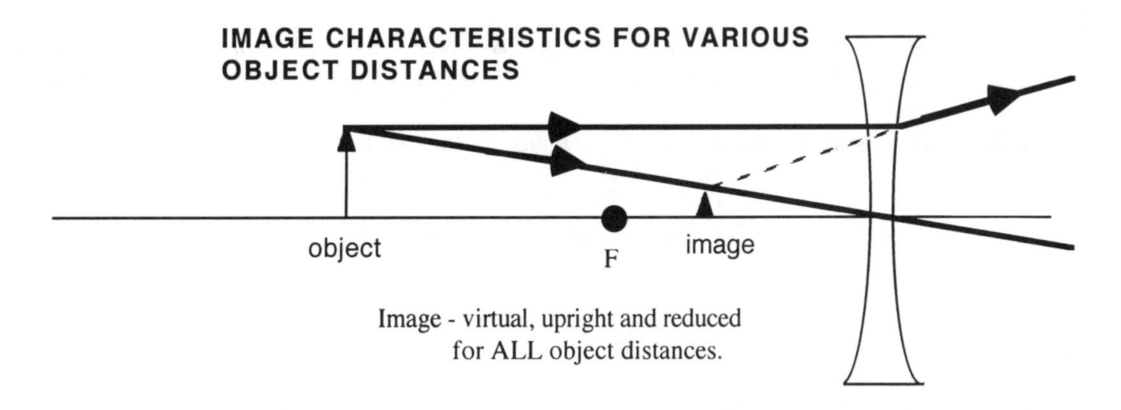

26.9 Lenses in Combination

Finding the final image location for a combination of lenses involves successive applications of the thin lens equation. Find the image location for each lens always using the image of the previous lens as the object for the following lens.

Example 8

An object is placed 15.0 cm to the left of a diverging lens of focal length, - 15.0 cm, which is 25 cm to the left of a converging lens with a focal length of 10.0 cm. Find the location of the final image of the combination. Describe the image.

First, ignore the converging lens and find the image location of the diverging lens alone. The thin lens equation gives

$$\frac{1}{d_{i1}} = \frac{1}{f} - \frac{1}{d_{01}} = \frac{1}{-15.0\text{ cm}} - \frac{1}{15.0\text{ cm}}$$

$$d_{i1} = -7.50\text{ cm}$$

This image serves as the object for the second lens located a distance $d_{02} = 25.0\text{ cm} + 7.50\text{ cm} = 32.5\text{ cm}$ away. The thin lens equation gives for the second lens

$$\frac{1}{d_{i2}} = \frac{1}{f} - \frac{1}{d_{02}} = \frac{1}{10.0\text{ cm}} - \frac{1}{32.5\text{ cm}}$$

$$d_{i2} = 14.4\text{ cm}$$

The final image distance is POSITIVE, so the image is REAL. The overall magnification of the combination is

$$m' = \left(-\frac{d_{i1}}{d_{01}}\right)\left(-\frac{d_{i2}}{d_{02}}\right) = \left(\frac{7.50\text{ cm}}{15.0\text{ cm}}\right)\left(\frac{-14.4\text{ cm}}{32.5\text{ cm}}\right) = -0.222$$

The final image is SMALLER than the object and is INVERTED since the magnification is negative.

In the above example, the image of the first lens was located on the left of the second lens, hence the object was real and the object distance was positive. If the image of the first lens happens to fall to the right of the second lens, then it forms a virtual object for the second lens. The object distance in this case is considered to be negative.

Example 9

Consider an object lying 15.0 cm to the left of a converging lens of focal length, 10.0 cm, which is located 12.0 cm to the left of a diverging lens of focal length, - 15.0 cm. Where is the final image located? Describe the image.

Again, ignore the second lens and find the location of the image for the first lens.

$$\frac{1}{d_{i1}} = \frac{1}{f} - \frac{1}{d_{01}} = \frac{1}{10.0\text{ cm}} - \frac{1}{15.0\text{ cm}}$$

$$d_{i1} = 3.0 \times 10^{1}\text{ cm}$$

This image serves as the object for the second lens, but it lies to the right of the lens.

Now, with d_{o2} = 12.0 cm - 3.0 X 10^1 cm = - 18 cm,

$$\frac{1}{d_{i2}} = \frac{1}{f} - \frac{1}{d_{02}} = \frac{1}{-15.0\text{ cm}} - \frac{1}{-18\text{ cm}}$$

$$d_{i2} = -9.0 \text{ X } 10^1 \text{ cm}$$

The final image lies 9.0 X 10^1 cm to the left of the second lens and is a VIRTUAL image. The overall magnification is

$$m' = \left(-\frac{d_{i1}}{d_{01}}\right)\left(-\frac{d_{i2}}{d_{02}}\right) = \left(\frac{-30.0\text{ cm}}{15.0\text{ cm}}\right)\left(\frac{9.0 \text{ X } 10^1\text{ cm}}{-18\text{ cm}}\right) = +1.0 \text{ X } 10^1$$

The final image is ten times larger than the object and is upright as indicated by the positive magnification.

26.10 The Human Eye

The anatomy of the human eye is discussed in the textbook, so we will concentrate our discussion on the optics of the eye. The human eye is very similar to a camera; both have a lens system and a variable-aperture diaphragm (the iris and pupil fulfill this role in the eye). In the eye, the image formed on the retina is real, inverted, and smaller than the object, as is the case for the camera.

Unlike the camera, however, the eye has a fixed image distance, that is, the distance between the lens and the retina is constant. In order to focus on objects at different distances, therefore, the focal length of the lens must be adjustable. This adjustment is produced by physically changing the shape of the lens, thus changing the focal length. The ciliary muscle is responsible for "tensing" or "relaxing" the eye's lens. As an object moves closer to the eye, the ciliary muscle automatically tenses, literally squeezing the lens, increasing its curvature, and thus shortening its focal length. This permits a sharp image to be formed on the retina of this nearby object. For distant objects, the ciliary muscle relaxes, the curvature of the lens decreases, and the focal length increases to provide a sharp focus on the retina. The process in which the focal length of the lens changes for different object distances is referred to as **accommodation**. The eye cannot accommodate perfectly for all object distances, however. If an object is placed too close to the eye, a clear focus cannot be attained. The point nearest the eye at which an object can be placed and still produce a sharp image on the retina is called the **near point**. The near point for the normal eye is about 25 cm. Similarly, the **far point** of the eye is the location of the farthest object on which the fully relaxed eye can focus. For normal vision, the far point is located nearly at infinity.

A **nearsighted (myopic)** person can only focus clearly on nearby objects but not on distant ones. For such a person, the far point of the fully relaxed eye is not at infinity, but may be only a few meters away. The image of a distant object falls in front of the retina, as shown in the figure below, thus causing the image to be out of focus. To correct this, a diverging lens is used to refocus the light back on the retina.

Nearsightedness (myopia) can be corrected using a diverging lens as shown in the figure at the top of the next page. The diverging lens acts to produce a virtual image at the far point of the unaided myopic eye. Thus, instead of the eye trying to focus at infinity, it will focus on a virtual image located at the far point of the eye.

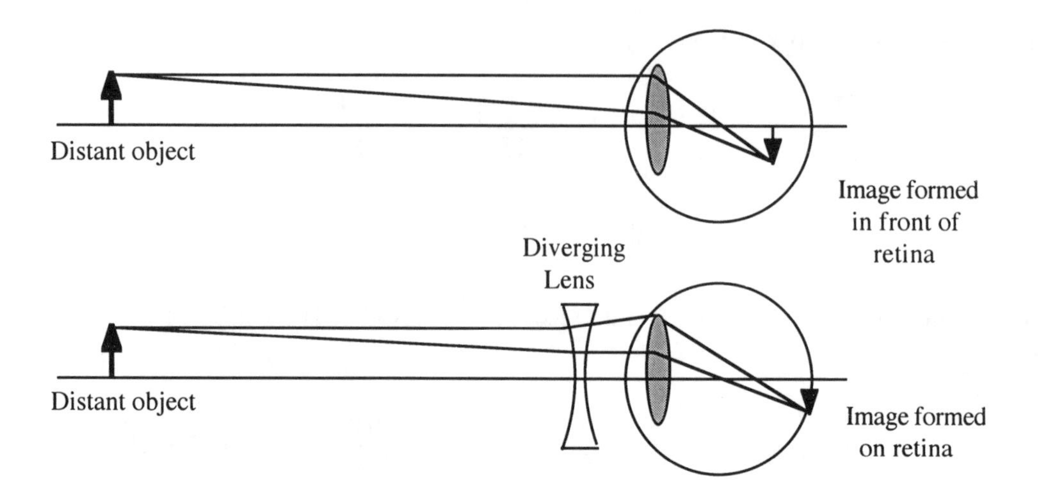

Example 10 *Use of a diverging lens to correct nearsightedness*
A nearsighted person is diagnosed to have a far point located only 350 cm from his eyes. Determine the focal length of contact lenses that will enable him to see distant objects clearly.

The virtual image formed by the diverging lens is located at the far point of the eye. Since contact lenses are placed right against the eyes, d_i = - 350 cm, the negative sign indicating a virtual image. Since the object is far away, the object distance is taken to be infinity, that is, $d_o = \infty$. The thin-lens equation (26.6) then gives

$$\frac{1}{f} = \frac{1}{d_o} + \frac{1}{d_i} = \frac{1}{\infty} + \frac{1}{-350 \text{ cm}} \quad \Rightarrow f = -350 \text{ cm}.$$

Note that the focal length is negative since it is a diverging lens.

A **farsighted (hyperopic)** person can usually see distant objects clearly, but cannot focus on those nearby. Whereas the near point of a "normal" eye is located about 25 cm from the eye, the near point of the farsighted eye may be much farther away. As shown in the figure below, an object located inside the near point of a hyperopic eye will be focused behind the retina, thus producing a blurred image. To correct for this effect, a converging lens is used to refocus the light onto the retina, so that a clear image may be formed. The converging lens forms a virtual image located at the near point of the hyperopic eye. Thus, instead of the eye trying to focus on the nearby object, it will focus on the virtual image located at the near point of the unaided eye. As shown in the figure below, farsightedness (hyperopia) can be corrected using a converging lens.

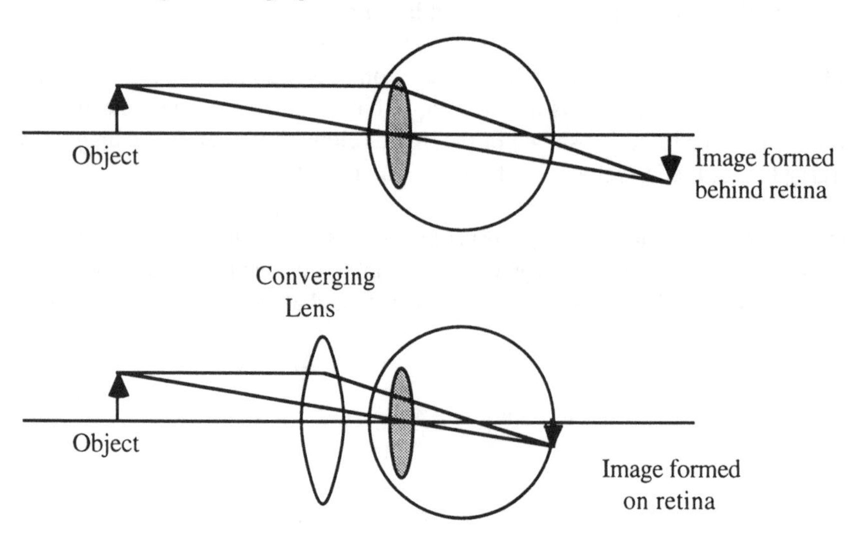

Example 11 *Use of a converging lens to correct farsightedness*
A farsighted person has a near point located 350 cm from the eyes. Obtain the focal length of the converging lens used in contact lenses so that this person can read a book held 25 cm from the eyes.

Since contact lenses are worn directly against the eyes, the object distance is d_o = 25 cm. Likewise, the lens forms an image at the near point, so d_i = - 350 cm. Therefore,

$$\frac{1}{f} = \frac{1}{d_o} + \frac{1}{d_i} = \frac{1}{25\text{ cm}} + \frac{1}{-350\text{ cm}} = 0.037\text{ cm}^{-1} \Rightarrow f = 27\text{ cm}.$$

Notice the focal length is positive since it is a converging lens.

The **refractive power** of a lens is used to describe the degree to which a lens refracts light. It is defined as:

$$\text{Refractive power (in diopters)} = \frac{1}{\text{f (in meters)}} \tag{26.8}$$

The unit of refractive power is the diopter. One diopter is 1 m^{-1}.

Example 12
What is the refractive power of the lens described in example 11?

The focal length (in meters) of the lens in example 11 is f = 27 cm = 0.27 m. Therefore, the refractive power is

$$\text{Refractive power (in diopters)} = \frac{1}{\text{f (meters)}} = \frac{1}{0.27\text{ m}} = 3.7\text{ diopters.}$$

Note that the refractive power of a converging lens is positive, while the refractive power of a diverging lens would be negative.

26.11 Angular Magnification and the Magnifying Glass

As an object is moved farther and farther away from the eye, it appears to get smaller. As far as the brain is concerned, the size of the image on the retina determines how large the object appears to be. Usually, however, the image size is not discussed directly, but rather, the angle θ subtended by the image can be used as a measure of the size. The angle θ is called the **angular size** of the object or image. This angle can be expressed in terms of the height h_o and distance d_o of the object according to the relation

$$\theta \text{ (in radians)} = \text{Angular size} \cong \frac{h_o}{d_o}$$

The approximation for angular size is demonstrated in the following figure.

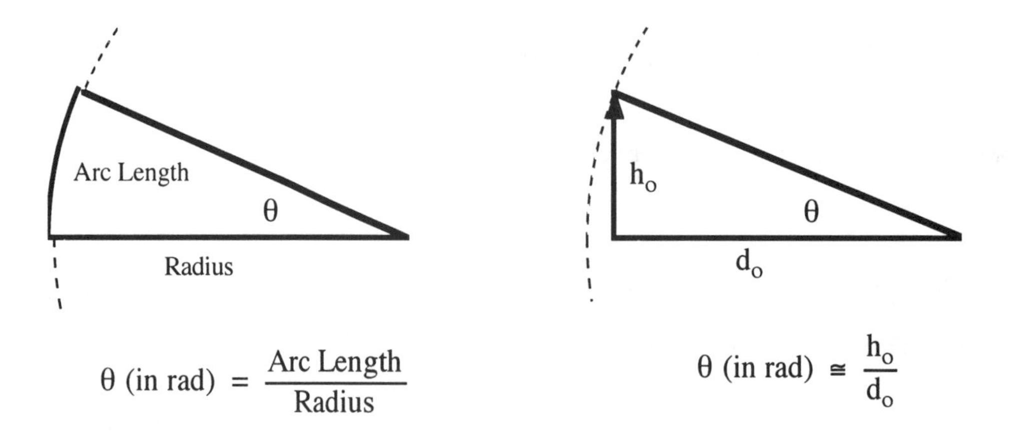

Example 13
Determine the angular size of the sun as viewed from the earth. The sun has a diameter of 1.39×10^6 km and is at a distance of 1.50×10^8 km from the earth.

The angle subtended by the sun is

$$\theta \cong \frac{h_o}{d_o} = \frac{1.39 \times 10^6 \text{ km}}{1.50 \times 10^8 \text{ km}} = 9.27 \times 10^{-3} \text{ rad} = 0.531°$$

An optical instrument, such as a magnifying glass, allows us to view small or distant objects, because it produces a larger image on the retina than would be possible otherwise. If θ' is the angular size of the final image produced by the instrument, the **angular magnification** (or **magnifying power**) M is defined as

$$M = \frac{\text{Angular size produced by instrument}}{\text{Reference angular size}} = \frac{\theta'}{\theta} \qquad (26.9)$$

The simplest type of magnifying device is a magnifying glass. This is usually a single converging lens that forms an upright, virtual, enlarged image of an object placed at or inside the focal point of the lens.

For a magnifying glass held close to the eye, the angular magnification is approximately

$$M = \frac{\theta'}{\theta} \cong \left(\frac{1}{f} - \frac{1}{d_i}\right) N \qquad (26.10)$$

where N is the distance from the eye to the near point.

Example 14
A converging lens whose focal length is 11.0 cm is being used to examine a small object by a person whose near point is located 45 cm from his eyes. What is (a) the maximum angular magnification of the lens and (b) the angular magnification when the lens is positioned so the eye is fully relaxed?

(a) The maximum magnification occurs when the image is placed at the near point, so d_i = - N and thus,

$$M \cong \frac{N}{f} + 1 = \frac{45 \text{ cm}}{11.0 \text{ cm}} + 1 = 5.1$$

(b) When the eye is relaxed, $d_i = -\infty$, so equation (26.10) becomes

$$M \cong \frac{N}{f} = \frac{45 \text{ cm}}{11.0 \text{ cm}} = 4.1$$

26.12 The Compound Microscope

To increase the angular magnification beyond that possible with a magnifying glass, an additional converging lens can be included to "premagnify" the object. The resulting optical instrument is called a **compound microscope**. The magnifying glass is called the **eyepiece** of the microscope, and the additional lens is called the **objective**. If f_o is the focal length of the objective, f_e the focal length of the eyepiece, and L the distance between the objective and eyepiece, the angular magnification of the compound microscope is given by

$$M \cong -\frac{(L - f_e)N}{f_o f_e} \qquad (26.11)$$

Example 15
The objective of a compound microscope has a focal length of 5.0 mm, while that of the eyepiece is 1.8 cm. If the two lenses are separated by 25.0 cm, determine the angular magnification if a person with normal vision is using the microscope (that is, N = 25 cm).

$$M \cong -\frac{(L - f_e)\,N}{f_o f_e} = -\frac{(25.0 \text{ cm} - 1.8 \text{ cm})(25 \text{ cm})}{(0.50 \text{ cm})(1.8 \text{ cm})} = -640$$

Note that when the eyepiece is used alone, the angular magnification would be $M = (N/f_e) + 1 = 15$. The compound microscope therefore represents an improvement in angular magnification by a factor of 640/15 = 43 over that of the magnifying glass alone.

26.13 The Telescope

A telescope is an instrument for magnifying distant objects, such as stars and planets. A telescope consists of an objective lens and an eyepiece. The "first image" is formed by the objective lens and is real, inverted, and smaller than the object. If this first image falls inside the focal point of the eyepiece, the eyepiece acts like a magnifying glass. It forms a final image that is virtual, enlarged, and located near infinity. This final image can then be viewed with a fully relaxed eye.

The angular magnification of a telescope can be written as:

$$M = \frac{\theta'}{\theta} \cong -\frac{f_0}{f_e} \tag{26.12}$$

Example 16
A refracting telescope has an objective and an eyepiece that have refractive powers of 1.25 diopters and 250 diopters, respectively. Find the angular magnification of the telescope.

To find the angular magnification we first need the focal lengths of the objective and eyepiece. Using equation (26.8) we can find the focal lengths. We have

$$\text{Refractive Power (diopters)} = \frac{1}{\text{f (in meters)}}$$

therefore,

$$f_o = \frac{1}{1.25\text{ diopters}} = 0.80\text{ m} \qquad f_e = \frac{1}{250\text{ diopters}} = 0.0040\text{ m}$$

Finally, the angular magnification is given by

$$M = -\frac{f_o}{f_e} = -\frac{0.80\text{ m}}{0.0040\text{ m}}$$

$$M = -2.0 \times 10^2.$$

PRACTICE PROBLEMS

1. Light reflected from a fish strikes the surface of the water at an angle of 38° to the normal. What is the angle of refraction of the light into the air?

2. A clear spring has white shells on its bottom which appear to the occupants of a glass bottom boat to be only 45.0 ft below. What is the actual depth of the spring?

3. A laser beam is directed at an angle, θ, to the normal of one side of a rectangular block of fused quartz. Part of the light striking the side of the block passes into the block and strikes an adjacent face. For what angle θ will the laser beam be totally reflected in the block?

4. The light reflected from a solid of unknown refractive index is 100 % horizontally polarized when the light strikes the solid at an angle of 65° to the normal. What is the index of refraction of the solid?

5. Fill-in the following table for a single lens. Include the appropriate signs on ALL numbers.

Lens Type	focal length	object distance	image distance	magnification	real/virtual	inverted/upright
converging	25 cm	15 cm				
converging		22 cm	+22 cm			
diverging			12 cm	0.50		
diverging	11 cm		5 cm			
	8 cm		18 cm			inv

6. Two converging lenses with focal lengths 15 cm and 25 cm are placed 18 cm apart. An object is located 8.0 cm to the left of the 15 cm focal length lens. Where is the final image formed?

7. A converging lens is used to read the small print in a contract. The lens is held 9.0 cm from the print and produces a magnification of + 2.5. What is the focal length of the lens?

8. A camera is supplied with a 35.0 mm focal length lens. A 2.00 m tall man stands 16.0 m from the camera and has his picture taken. What is the size of the image of the man that the lens produces on the film?

9. A farsighted person cannot focus clearly on objects that are less than 145 cm from his eyes. To correct this problem, the person wears eyeglasses that are located 2.0 cm in front of his eyes. Determine the focal length that will permit this person to read a newspaper at a distance of 32.0 cm from his eyes.

10. A certain eye has a near point of 11.0 cm and a far point of 15.0 cm. (a) What is the refractive power of the lens that is required to place the far point at infinity? (b) What is the near point distance when the person uses the lens found in part (a)? (c) What is the refractive power of the lens required to yield a near-point distance of 25 cm?

11. A jeweler whose near point is 65 cm from his eye uses a small magnifying glass to examine a watch held 5.0 cm from the lens. Find the angular magnification of the magnifying glass.

12. A microscope has an objective of 4.0-mm focal length and an eyepiece of 2.0-cm focal length. The distance between the two lenses, fixed by the microscope barrel, is 20.0 cm. Determine the magnification of this instrument for a person whose near point distance is 25 cm.

13. A microscope with a 22.0 cm barrel has an eyepiece whose focal length is 1.60 cm. What focal length of the objective lens will produce an angular magnification of - 450? Take the near point distance to be 25 cm.

14. A reflecting telescope uses a concave mirror whose radius of curvature is 4.00 m. If the angular magnification of the telescope is - 160.0, what is the focal length of the eyepiece?

15. A refracting telescope has an angular magnification of - 240. The length of the telescope barrel is 2.60 m. What are the focal lengths of the objective and eyepiece lenses?

16. The refracting telescope at the Yerkes Observatory has an objective whose focal length is 19.4 m. Its eyepiece has a focal length of 10.0 cm. (a) What is the angular magnification of the telescope? (b) If the telescope is used to look at a lunar crater (diameter 1500 m), what is the size of the first image, assuming the moon is 3.85×10^8 m from the earth?

HELPFUL SUGGESTIONS

1. You will simply have to memorize the sign conventions to be used in the thin lens equation. An application of this equation will yield signed numbers where the signs tell you whether the image is real, virtual, etc. An easy way to determine whether an image is virtual or real from the ray diagram is to imagine that you place a small piece of paper at the image site. If light rays would actually fall on the paper and produce an image, then the image is real. If no light rays could fall on paper, then the image is virtual.

2. Keep in mind that diverging lenses can form ONLY virtual images of real objects, while converging lenses may form either real or virtual images depending on where the object is placed.

3. Confusion sometimes results when students think about virtual images formed by lenses. A virtual image has little meaning unless there is an additional converging lens in the system to form a final real image by using the virtual image as its object. In many cases the additional converging lens in the system is YOUR eye. You can "see" a virtual image produced by a lens because the converging lens in your eye forms a real image on the retina.

4. When dealing with combinations of lenses, mirrors, or both, analyze each lens or mirror separately. The image of the previous element forms the object for the next element and so on. The final image position can thus be found. The overall magnification of a combination is the product of the individual magnifications.

5. The thin lens equation involves reciprocals of the object distance, image distance, and focal length. When calculating numerical values, use the 1/x function on your calculator like you did for parallel resistors.

6. Remember that a diverging lens is needed to correct for nearsightedness. If the far point for the myopic eye is known, the correcting lens produces a virtual image at this point. The far point then allows us to determine the image distance (which is **negative** for a virtual image). The object distance is taken to be infinity and the focal length of the correcting lens can then be found using the thin-lens equation (26.6). Recall that the focal length is negative for this diverging lens.

7. A converging lens is needed to correct for farsightedness. In this case, the near point of the hyperopic eye is usually given. The converging lens produces a virtual image at this near point, so the image distance (negative sign) is known. For normal vision, the near point is usually taken to be 25 cm, so this becomes the object distance. The focal length of the converging lens (positive in this case) can then be determined using equation (26.6).

8. The refractive power (measured in diopters) is defined as one divided by the focal length f, where f MUST be expressed in meters.

9. The angular magnification equations for various optical instruments, equations (26.10), (26.11), and (26.12), are approximations to the true angular magnification. This is because we used the fact that $\theta \approx h_o/d_o$ and $\theta' \approx h_i/d_i$ as depicted in Figure 26.40. However, for small angles, this approximation is reasonably accurate and can therefore be employed in almost all cases.

EVERYDAY PHYSICS

1. Glass or even plastic lenses are commonly available in toy and department stores. Generally these are converging lenses intended to be used as magnifying glasses. Obtain some and use them to verify the properties of lenses presented in this chapter. Try focusing the light from a bulb onto a piece of paper. Can you focus an image of the sun on a piece of paper and determine the focal length of the lens? BE CAREFUL not to direct the image into your eye.

2. One of the earliest known lenses was simply a drop of clear liquid. You can demonstrate this by rubbing a small film of clear wax on a newspaper and placing a drop of water over a picture or print. If you use a "slick" magazine there is no need for the wax since the water will not soak in readily.

3. The phenomenon of apparent depth can be readily seen by looking into an aquarium either from the top or from the side. The fish and objects in the aquarium appear to be closer to you and larger than they really are. How does this relate to lenses?

4. If you partially submerge a stick in water at an angle, it appears bent. Why? Try it.

5. The atmosphere produces lensing effects because layers of air at different altitudes have different indices of refraction. The moon appears larger when it is rising than when it is directly overhead. Can you explain this?

6. Have you ever produced a rainbow on a sunny day in the fine mist from a garden hose. If not, try it. If you have try it again and relate what you see to the contents of this chapter.

7. Obtain a converging lens (from a pair of reading glasses, a camera, or film projector) and view an object with it. Determine where you should place the object and your eye to produce the largest image. Determine the magnification of the lens by observing a piece of lined paper directly and through the lens and comparing the spacing between the lines.

8. Make a "pinhole camera" by removing one of the ends of a box and covering the opening with a piece of tissue paper, wax paper, or onion-skin paper. Make a small, clean-cut pin hole at the other end of the box. Aim the camera at a bright object in a darkened room and you will see the inverted image on the tissue paper. Notice what happens if you change the size of the pinhole.

CHAPTER QUIZ

1. Light is incident on an air-water interface at an angle of 25° to the normal. What angle does the refracted ray make with the normal?
   a. 19° b. 34° c. 25° d. 90°

2. You are under the water in a clear lake looking at the surface and see the image of a fish due to total internal reflection. What is the minimum angle that the light leaving the fish makes with the normal to the surface of the lake?
   a. 42° b. 53° c. 49° d. 37°

3. White light is incident at an angle to the surface of a triangular piece of glass. Which color of light deviates most from its original path after leaving the glass?
   a. red b. orange c. green d. blue

4. Light reflected from a boundary between an unknown substance and air is seen to become 100% polarized when the angle of incidence is 62.0°. What is the index of refraction of the unknown substance?
   a. 1.88 b. 1.13 c. 2.14 d. 0.532

5. A converging lens has a focal length of 15 cm. An object is placed 9.0 cm from the lens. Describe the image formed.
   a. real, upright, enlarged
   b. real, inverted, reduced in size
   c. virtual, inverted, reduced in size
   d. virtual, upright, enlarged

6. An object is placed 10.0 cm from a diverging lens which forms an image 6.5 cm from the lens. What is the focal length of the lens? Include the sign.
   a. + 3.9 cm b. - 16.5 cm c. - 21.2 cm d. - 18.6 cm

7. Under what conditions does a diverging lens form a virtual image of a real object?
   a. Only if $d_0 > f$.
   b. Only if $d_0 < f$.
   c. Only if $d_0 = f$
   d. A diverging lens *always* forms a virtual image of a real object.

8. A convex lens of focal length 25 cm receives light from the sun. A diverging lens of focal length - 12 cm is placed 37 cm to the right of the converging lens. Where is the final image located relative to the diverging lens?
   a. 6 cm to the left b. 25 cm to the left c. At infinity d. 12 cm to the right

9. A lens produces a enlarged, virtual image. What kind of lens is it?
   a. converging b. diverging c. It could be either diverging or converging.

10. A camera lens focuses light from a 12.0 m tall building located 35.0 m away on film 50.0 mm behind the lens. How tall is the image of the building on the film?
    a. 17.1 mm b. 7.00 mm c. 2.50 cm d. 1.25 mm

11. The pupil of the eye changes in size to adjust for
    a. objects at different distances
    b. objects of different sizes
    c. different colors
    d. different amounts of light

12. To use a magnifying glass, the object should be placed
    a. as close to the lens as possible
    b. just within the lens' focal point
    c. just beyond the focal point
    d. some distance beyond the focal point

13. A camera employs a _______________ lens to form _______________ images.
    a. diverging ... real b. diverging ... virtual c. converging ... real d. converging ... virtual

14. In the eye, the position of the image on the retina is adjusted by changing the
    a. position of the lens
    b. focal length of the lens
    c. diameter of the pupil
    d. length of the eyeball

15. What power lens is needed to correct for nearsightedness where the uncorrected far point is 250 cm?
    a. +2.5 diopters b. - 2.5 diopters c. + 0.4 diopters d. - 0.4 diopters

16. What power lens is needed to correct for farsightedness where the uncorrected near point is 250 cm?
    a. +3.6 diopters b. - 3.6 diopters c. + 0.28 diopters d. - 0.28 diopters

17. The objective of a telescope has a focal length of 220 cm and is used with an eyepiece of focal length 25 mm. What is the angular magnification of this telescope?
    a. - 8.8 b. - 88 c. - 880 d. - 550

18. Spherical lenses suffer from
    a. chromatic aberration, but not spherical aberration.
    b. spherical aberration, but not chromatic aberration.
    c. both spherical and chromatic aberration.
    d. neither spherical nor chromatic aberration.

19. You want a lens to use as a simple magnifying glass which gives a maximum magnification of 6. If the near point is taken to be 25 cm, what is the focal length of this magnifying glass?
    a. + 4 cm b. - 4 cm c. + 5 cm d. - 5 cm

20. A dime (diameter 1.8 cm) is placed a distance of 30.0 meters from your eye. What is the angle subtended by the dime at this distance?
    a. 6.0×10^{-4} rad b. 0.54 rad c. 6.0×10^{-2} rad d. 54 rad

SOLUTIONS AND ANSWERS

Practice Problems

---

1. Snell's law gives

$$n_w \sin \theta_1 = n_a \sin \theta_2$$
$$\sin \theta_2 = (1.33) \sin 38°$$
$$\theta_2 = \mathbf{55°}$$

---

2. Equation (26.3) gives the actual depth to be

$$d = (n_1/n_2)\, d' = (1.33/1.00)(45.0 \text{ ft}) = \mathbf{59.9 \text{ ft}}.$$

---

3. Snell's law gives for the refraction into the block

$$(1.00) \sin \theta = n_2 \sin \theta_2$$

If the beam is totally internally reflected at the adjacent face, the incident angle on that face must be at least

$$\theta_c = \sin^{-1} (1.000/1.458) = 43.30°.$$

Since the block is rectangular, $\theta_2 + \theta_c = 90°$ so θ_2 is less than 46.70° for the internal reflection to occur.

$$\sin \theta = (1.458) \sin 46.70° = 1.06.$$

The sine function cannot be greater than one, so the angle, θ, which will produce the internal reflection must be less than or equal to 90°. That is, ALL incident angles will cause the light to be totally internally reflected at the adjacent face as long as the light strikes that face.

---

4. The incident angle of 65° must be Brewster's angle for the reflected light to be 100 % polarized. Equation (26.5)

$$n_2 = n_1 \tan \theta_B = (1.00) \tan 65° = \mathbf{2.1}.$$

---

5. See table that follows "Quiz Answers".

---

6. Find the image location for the first lens using the thin lens equation.

$$1/d_{i1} = 1/f_1 - 1/d_{o1} = 1/(15 \text{ cm}) - 1/(8.0 \text{ cm})$$

which gives $d_{i1} = -17$ cm.

This image forms the object of the second lens located 17 cm + 18 cm = 35 cm away. The thin lens equation gives for the second lens

$$1/d_{i2} = 1/f_2 - 1/d_{o2} = 1/(25 \text{ cm}) - 1/(35 \text{ cm})$$

which gives $d_{i2} = 87.5$ cm.

The final image lies **87.5 cm to the right of the second lens** (25 cm) lens.

---

7. The fine print serves as the object for the lens, $d_0 = 9.0$ cm. The image distance is then

$$d_i = - md_0 = - (2.5 \text{ cm})(9.0 \text{ cm}) = - 22 \text{ cm}$$

The thin lens equation gives

$$1/f = 1/d_0 + 1/d_i = 1/(9.0 \text{ cm}) + 1/(-22 \text{ cm})$$

so f = **+ 15 cm.**

---

8. The angular size remains fixed so $\theta = h_i/d_i = h_o/d_o$. Solving for the image size h_i yields:

$$h_i = h_o(d_i/d_o) = (2.00 \text{ m})(35.0 \text{ mm})/(16.0 \text{ m}) = \mathbf{4.38 \text{ mm}}.$$

---

9. The near point is 145 cm and eyeglasses are 2.0 cm in front of the eyes. Therefore, $d_i = - 143$ cm. The object is placed 32.0 cm from the eyes so $d_o = + 30.0$ cm. The focal length is obtained from equation (26.6),

$$1/f = 1/d_o + 1/d_i = 1/(30.0 \text{ cm}) + 1/(- 143 \text{ cm}) = 0.026 \text{ cm}^{-1}$$

hence, f = **38 cm**.

---

10. (a) We want to move the far point to infinity, so

$$1/f = 1/\infty + 1/(- 0.15 \text{ m}) = \mathbf{- 6.7 \text{ diopters}}.$$

(b) We have

$$1/d_o = 1/f - 1/d_i = - 6.7 \text{ m}^{-1} - 1/(- 0.11 \text{ m}) = 2.4 \text{ m}^{-1}.$$

So that d_o = **42 cm**.

(c) For the near point at 25 cm:

$$1/f = 1/(0.25 \text{ m}) + 1/(- 0.11 \text{ m}) = \mathbf{- 5.1 \text{ diopters}}.$$

---

11. The angular magnification is

$$M = \theta'/\theta = (1/f - 1/d_i)N = N/d_o = (65 \text{ cm})/(5.0 \text{ cm}) = \mathbf{13}.$$

---

12. Use equation (26.11):

$$M = - (L - f_e)N/f_o f_e = - (20.0 \text{ cm} - 2.0 \text{ cm})(25 \text{ cm})/[(0.40 \text{ cm})(2.0 \text{ cm})] = \mathbf{- 560}.$$

---

13. Use equation (26.11) and solve for the objective focal length, f_o. We obtain,

$$f_o = - (L - f_e)N/f_e M = - (22.0 \text{ cm} - 1.60 \text{ cm})(25 \text{ cm})/[(1.60 \text{ cm})(- 450)] = \mathbf{0.71 \text{ cm}}.$$

---

14. The radius of curvature is 4.00 m, so the focal length is f = R/2 = 2.00 m. Using equation (26.12),

$$M = - f_o/f_e.$$

Solving for the eyepiece focal length yields:

$$f_e = - f_o/M = - (2.00 \text{ m})/(- 160.0) = \mathbf{1.25 \times 10^{-2} \text{ m}}.$$

---

15. The length of the barrel of a refracting telescope equals the sum of the focal lengths of the objective and eyepiece. Therefore, $f_o + f_e = 2.60$ m. From equation (26.12),

$$M = - f_o/f_e = - 240.$$

Thus, $f_o = 240\ f_e$. Combining equations,

$$240\ f_e + f_e = 2.60 \text{ m}.$$

The eyepiece focal length is $f_e = 0.011$ m = **11 mm**. The focal length of the objective is therefore,

$$f_o = 2.60 - 0.011 = \mathbf{2.59 \text{ m}}.$$

---

16. (a) The magnification is:

$$M = - f_o/f_e = - (19.4 \text{ m})/(0.100 \text{ m}) = \mathbf{- 194}.$$

(b) The angular size of the crater is

$$\theta = h_o/d_o = (1500 \text{ m})/(3.85 \times 10^8 \text{ m}) = 3.9 \times 10^{-6} \text{ rad}.$$

The angular size of the first image is $\theta' = M\theta = (194)(3.9 \times 10^{-6} \text{ rad}) = 7.6 \times 10^{-4}$ rad. The size of the first image is therefore,

$$h = \theta' f_e = (7.6 \times 10^{-4} \text{ rad})(10.0 \text{ cm}) = \mathbf{7.6 \times 10^{-3} \text{ cm}}.$$

Quiz answers

1. a	6. d	11. d	16. a
2. c	7. d	12. b	17. b
3. d	8. a	13. c	18. c
4. a	9. a	14. b	19. c
5. d	10. a	15. d	20. a

Practice Problem 5

Lens Type	focal length	object distance	image distance	magnification	real / virtual	inverted / upright
converging	+ 25 cm	+ 15 cm	**-38 cm**	**+2.5**	**V**	**up**
converging	**+11 cm**	+ 22 cm	+22 cm	**-1.0**	**R**	**inv**
diverging	**-24 cm**	**+24 cm**	- 12 cm	+ 0.50	**V**	**up**
diverging	- 11 cm	**+9.2cm**	-5 cm	**+0.54**	**V**	**up**
converging	+ 8 cm	**+14 cm**	+18 cm	**-1.3**	**R**	inv

MCAT REVIEW PROBLEMS

Inside a substance such as glass or water, light travels more slowly than it does in a vacuum. If c denotes the speed of light in a vacuum and v denotes its speed through some other substance, then

$$v = \frac{c}{n},$$

where n is a constant called the index of refraction.

To good approximation, a substance's index of refraction does not depend on the wavelength of light. For instance, when red and blue light waves enter water, they both slow down by about the same amount. More precise measurements, however, reveal that n varies with wavelength. Table 1 presents some indices of refraction of Cutson glass, for different wavelengths of visible light. A nanometer (nm) is 10^{-9} meters. In a vacuum, light travels at $c = 3.0 \times 10^8$ m/s.

TABLE 1:
Indices of refraction of Cutson glass

approximate color	wavelength in vacuum (nm)	n
yellow	580	1.500
yellow-orange	600	1.498
orange	620	1.496
orange-red	640	1.494

1. Inside Cutson glass:

   A. Orange light travels faster than yellow light.
   B. Yellow light travels faster than orange light.
   C. Orange and yellow light travel equally fast.
   D. We cannot determine which color of light travels faster.

2. For blue-green light of wavelength 520 nm, the index of refraction of Cutson glass is probably closest to:

   A. 1.49 **B.** 1.50 **C.** 1.51 **D.** 1.52

3. For visible light, which graph best expresses the index of refraction of Cutson glass as a function of the frequency of the light?

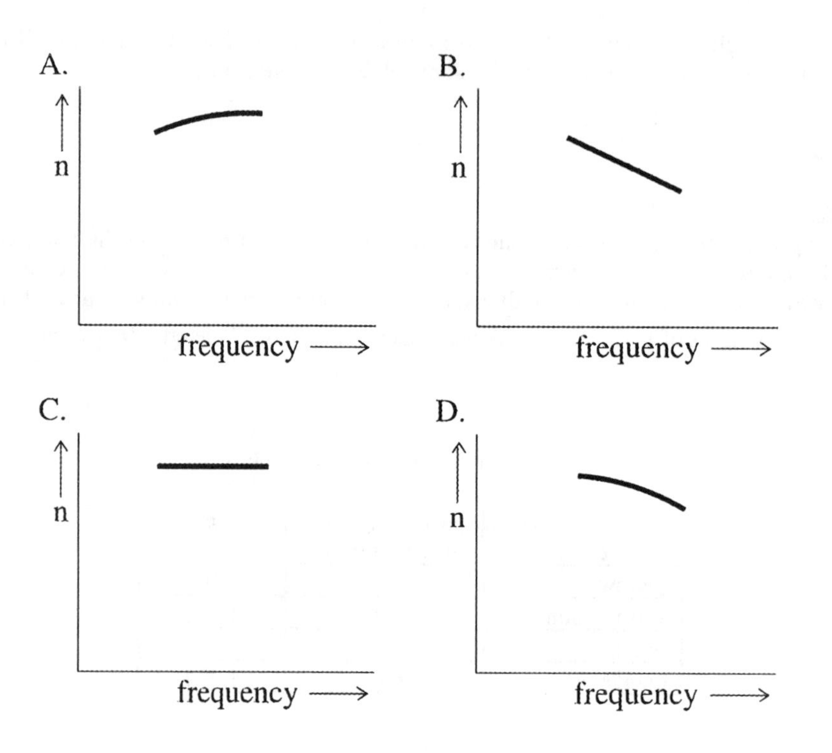

---

4. Which of the following phenomena happens BECAUSE n varies with wavelength?

- **A.** A lens focuses light.
- **B.** A prism breaks sunlight into different colors.
- **C.** Total internal reflection ensures that light travels down a fiber-optic cable.
- **D.** Light rays entering a pond change direction at the pond's surface.

ANSWERS TO MCAT REVIEW PROBLEMS

---

1. **A.** The velocity of light is inversely proportional to the index of refraction: $v = \frac{c}{n}$. So, the lower the n, the higher the v. According to table 1, Cutson glass has a slightly lower n for orange light than it has for yellow light. Therefore, inside Cutson glass, orange light travels slightly faster.

---

2. **C.** Table 1 suggests a linear relationship between wavelength and index of refraction. Starting with yellow light, we can extrapolate this linear relationship to blue-green light.

As compared to yellow light waves, blue-green light waves are 60 nm shorter (520 nm vs. 580 nm). By reading table 1 from bottom to top, we see that each 20 nm *decrease* in wavelength corresponds to a .002 *increase* in n. So, since yellow and blue-green light differ in wavelength by 60 nm, the corresponding n's should differ by .006:

$$n_{\text{blue-green}} = n_{\text{yellow}} + .006 = 1.500 + .006 = 1.506,$$

which is closer to 1.51 than it is to 1.50.

You can also address this question by extrapolating table 1:

blue-green	***520***	***1.506***
	540	*1.504*
	560	*1.502*
yellow	580	1.500
yellow-orange	600	1.498
orange	620	1.496
orange-red	640	1.494

---

3. **A.** As just discussed, index of refraction decreases linearly with *wavelength*. Graph B shows a linear relationship. But the problem wants the relationship between *frequency* and n, not between wavelength and n. Since higher wavelengths correspond to lower frequencies, and vice versa, n increases with frequency.

To clarify this reasoning, we should review the relationship between wavelength and frequency. For *all* waves, not just light waves,

$$v = \lambda f,$$

where v denotes velocity, λ denotes wavelength, and f denotes frequency. In a vacuum, all light waves travel at speed c. So, frequency and wavelength are inversely proportional: $f = \frac{c}{\lambda}$. Higher wavelengths correspond to lower frequencies, and vice versa. Therefore, since n decreases at higher wavelengths, it increases at higher frequencies.

Again, extending table 1 can help us spot this relationship. Using $f = \frac{c}{\lambda}$, we can find the frequency of different colors of light. For instance,

$$f_{\text{yellow}} = \frac{c}{\lambda_{\text{yel.}}} = \frac{3.0 \times 10^8 \text{ m/s}}{580 \times 10^{-9} \text{ m}} = 5.17 \times 10^{14} \text{ Hz},$$

$$f_{\text{yellow-orange}} = \frac{3.0 \times 10^8 \text{ m/s}}{600 \times 10^{-9} \text{ m}} = 5.0 \times 10^{14} \text{ Hz},$$

and so on. By inserting these frequencies into table 1, we see that n increases with frequency; higher frequencies correspond to higher n's.

color	vacuum λ (nm)	frequency (Hz)	n
yellow	580	5.17×10^{14}	1.500
yellow-orange	600	5.00×10^{14}	1.498
orange	620	4.84×10^{14}	1.496
orange-red	640	4.69×10^{14}	1.494

---

4. **B.** According to Snell's law, the angle through which light bends upon entering glass from air (or vice versa) depends on the glass' index of refraction. Since different colors correspond to different n's, those different colors bend differently when they enter and leave the prism. See figure 1. In this way, the prism "pulls apart" sunlight into its component colors.

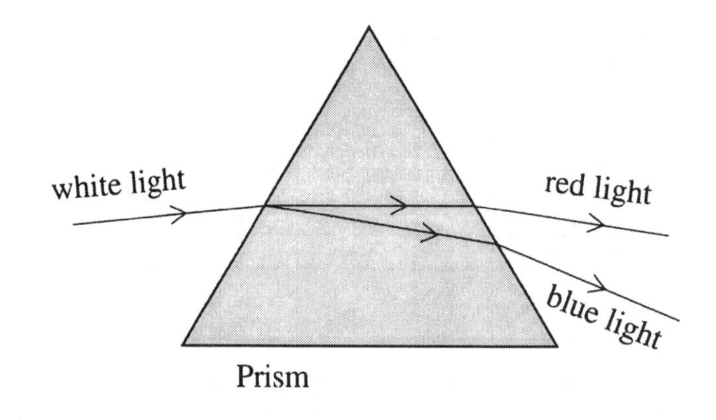

FIGURE 1

When light passes through a lens, this same phenomenon occurs; different colors get bent by different angles. Therefore, different colors converge upon slightly different focal points. But even without this "aberration," a lens would still focus light. In fact, lenses would focus light more sharply if different colors got bent equally!

Similarly, light would still refract at the pond's surface if n_{water} were the same for all wavelengths.

CHAPTER 27
Interference and the Wave Nature of Light

PREVIEW

Light has wave properties. The geometrical optics presented up to now has not fully addressed this fact. In this chapter you will be introduced to wave optics which is the study of the wave nature of light. The wave properties of light include several familiar wave phenomena, such as superposition, interference, and diffraction.

QUICK REFERENCE

Important Terms

Geometrical optics
The study of the straight line motion of light and the reflection and refraction of light.

Wave (Physical) Optics
The study of the wave nature of light including interference and diffraction.

Superposition of light waves
Adding the electric fields of two or more light waves to produce the electric field of a new wave.

Constructive interference
The superposition of two or more light waves which results in a reinforcement of the waves.

Destructive interference
The superposition of two or more light waves which results in a cancellation of the waves.

Coherent sources
Sources of light whose waves maintain a constant phase relationship.

Diffraction of light
The bending of light waves around obstacles or the edges of an opening.

Resolving power
The ability of an optical instrument to distinguish between two closely spaced objects.

Diffraction grating
A device with a large number of closely spaced parallel slits which is used to separate white light into its various colors.

Principal maxima of a grating
The bright fringes produced by a diffraction grating.

Equations

Young's Double Slit Interference

For **bright fringes**

$$\sin\theta = m\frac{\lambda}{d} \qquad m = 0, 1, 2, 3, \ldots \qquad (27.1)$$

For **dark fringes**

$$\sin\theta = (m + \tfrac{1}{2})\frac{\lambda}{d} \qquad m = 0, 1, 2, 3, \ldots \qquad (27.2)$$

Thin Film Interference

The **wavelength which is important** for thin film interference is the wavelength of light in the film.

$$\lambda_{film} = \frac{\lambda_{vacuum}}{n} \qquad (27.3)$$

Diffraction

Single-slit dark fringes

$$\sin\theta = m\frac{\lambda}{W} \qquad m = 1, 2, 3, \ldots \qquad (27.4)$$

Circular opening, first dark fringe

$$\sin\theta = 1.22\frac{\lambda}{D} \qquad (27.5)$$

Rayleigh criterion for resolution of two objects

$$\theta_{min} \cong 1.22\frac{\lambda}{D} \qquad (27.6)$$

Principal maxima of a diffraction grating of spacing, d

$$\sin\theta = m\frac{\lambda}{d} \qquad m = 0, 1, 2, 3, \ldots \qquad (27.7)$$

Principles

Principle of **linear superposition of light** - when two or more light waves are at the same place at the same time the electric field of the waves combine to give the total electric field of the resulting wave.

Huygen's principle - every point on a wavefront acts as a source of wavelets which move forward with the same speed as the wave; the waveform at the later instant is the surface tangent to the wavelets.

DISCUSSION OF SELECTED SECTIONS

27.2 Young's Double-Slit Experiment

Coherent light falling on two closely spaced narrow slits is observed to produce a pattern of alternating bright and dark images of a slit (fringes) on a screen. The central fringe is the brightest with the brightness of the fringes decreasing on either side of the central fringe.

If the light falling on the slits consists of several different frequencies (colors), then overlapping interference patterns will be produced and the fringes will be less visible. If the light of each frequency is not coherent, then the pattern due to each frequency will change randomly in time. The average pattern seen by an observer is then "washed out". Most common sources of light produce such "washed out" patterns, hence much of the interference of light is not obvious in everyday life.

A good source of light for showing interference patterns is a laser, since the laser produces light that is coherent and practically of a single frequency. Young used sunlight which was passed through a small slit. Since a small sample of sunlight falling on the earth came from virtually one point on the sun, it is largely coherent.

Example 1
A helium-neon laser beam (λ = 633 nm) illuminates two narrow slits. The third bright fringe of the resulting pattern is 15° from the central maximum. How far apart are the slits?

Equation (27.1) applies to the position of a bright fringe in a double slit pattern. Solving for the separation of the slits, d, gives

$$d = \frac{m\lambda}{\sin\theta} = \frac{3(633 \times 10^{-9}\ \text{m})}{\sin 15°} = 7.3 \times 10^{-6}\ \text{m} = 7.3\ \mu\text{m}.$$

Example 2
The interference pattern of a double-slit (d = 55.1 μm, λ = 551 nm) falls on a screen located 1.00 m from the slits. How far apart are the third and fourth bright fringes?

The location of a bright fringe is given by equation (27.1). The third bright fringe is located at an angle of

$$\theta = \sin^{-1}\left(\frac{m\lambda}{d}\right) = \sin^{-1}\left(\frac{3(551 \times 10^{-9}\ \text{m})}{55.1 \times 10^{-6}\ \text{m}}\right) = 1.72°$$

and a distance from the central fringe of

$$y_3 = L \tan\theta = (1.00\ \text{m}) \tan(1.72°) = 3.00 \times 10^{-2}\ \text{m} = 3.00\ \text{cm}.$$

The fourth bright fringe is at an angle of

$$\theta = \sin^{-1}\left(\frac{m\lambda}{d}\right) = \sin^{-1}\left(\frac{4(551 \times 10^{-9}\ \text{m})}{55.1 \times 10^{-6}\ \text{m}}\right) = 2.29°$$

The position of the fourth bright fringe relative to the central fringe is

$$y_4 = L \tan\theta = (1.00\ \text{m}) \tan(2.29°) = 4.00 \times 10^{-2}\ \text{m} = 4.00\ \text{cm}.$$

The distance between the fringes is then

$$y_4 - y_3 = 4.00\ \text{cm} - 3.00\ \text{cm} = 1.00\ \text{cm}.$$

An alternative way to find the distance between adjacent fringes if they correspond to small angular positions is to note that tan θ and the sin θ are approximately the same for small angles (< 10°). The above analysis gives

$$\sin\theta = m\lambda/d \text{ and } \tan\theta = y_m/L$$

for the mth fringe. The position of the mth fringe is then

$$y_m = m\lambda L/d.$$

A similar expression for the $(m + 1)^{th}$ fringe gives

$$y_{m+1} = (m + 1)\lambda L/d$$

so that the distance between the adjacent fringes is

$$y_{m+1} - y_m = \lambda L/d.$$

Note that the distance between adjacent fringes is constant as long as the angle is small, that is, the fringes are close to the central fringe.

Using the numbers in example 2 we obtain

$$y_4 - y_3 = (551 \times 10^{-9}\ m)(1.00\ m)/(55.1 \times 10^{-6}\ m) = 1.00 \times 10^{-2}\ m = 1.00\ cm.$$

This is the same result as before, so we can conclude that our approximation is valid.

Example 3
What is the largest number of dark fringes that could appear on a screen in a double slit arrangement (d = 0.100 mm) when the slits are illuminated with light of wavelength 455 nm?

If the screen were infinitely large, then all fringes with angles from 0° to 90° could fall on the screen. Equation (27.2) gives the m for the last fringe to be

$$m = (d/\lambda) \sin\theta - 1/2 = (0.100 \times 10^{-3}\ m/455 \times 10^{-9}\ m) \sin 90.0° - 1/2 = 219$$

There is a dark fringe for each m from 0 to 219, so there are 220 fringes which could appear on a sufficiently large screen.

27.3 Thin-Film Interference

All thin film problems involve at least three layers of materials as shown in the drawing. The region with n_2 is the actual film and the wavelength of the light in this material. λ_{film}, is the ONLY wavelength that is important in the analysis.

Any ray which reflects from a boundary of increasing index of refraction will suffer a 180° phase change upon reflection. A ray which reflects from a boundary of decreasing index of refraction undergoes no phase shift.

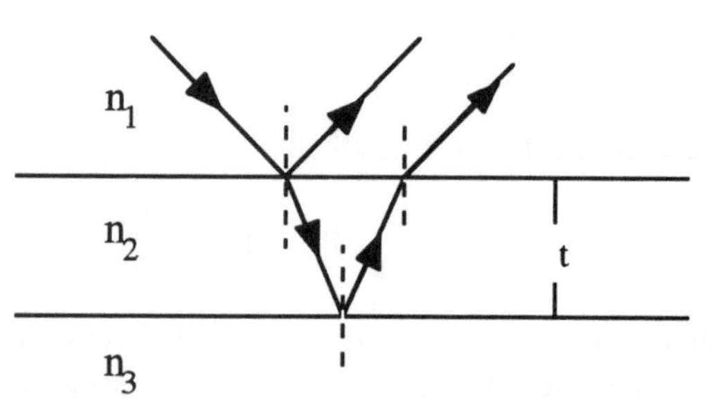

The following cases can be analyzed.

1. If $n_1 > n_2 > n_3$, then no phase shift occurs at either boundary. The only phase shift that can occur is due to the difference in distance traveled between the two reflected waves.

   Bright fringes occur for

$$2t = m\lambda_{film}$$

   Dark fringes occur for

$$2t = (m + 1/2)\lambda_{film}$$

2. If $n_1 < n_2 < n_3$, then a phase shift of 180° occurs at EACH boundary and the net phase shift of the reflected waves is again due solely to their path difference.

Bright fringes occur for

$$2t = m\lambda_{film}$$

Dark fringes occur for

$$2t = (m + 1/2)\lambda_{film}$$

3. If n_1 and n_3 are either both larger or smaller than n_2, then a 180° phase shift will occur at only one boundary and the reflected waves will be shifted $\lambda/2$ by the reflection.

Bright fringes occur for

$$2t = (m + 1/2)\lambda_{film}$$

Dark fringes occur for

$$2t = m\lambda_{film}$$

Example 4
A certain region of a soap bubble reflects red light of vacuum wavelength λ = 650 nm. What is the minimum thickness that this region of the soap bubble could have? Take the index of refraction of the soap film to be 1.41.

The soap film has air on both sides, so the reflections of the light produce a net 180 ° phase shift. The condition for bright fringes is $2t = (m + 1/2)\lambda_{film}$.

$$t = \frac{(m + 1/2)\lambda_{film}}{2} = \frac{(m + 1/2)\lambda}{2\,n} = \frac{(1/2)(650 \text{ X } 10^{-9} \text{ m})}{2(1.41)} = 1.2 \text{ X } 10^{-7} \text{ m}$$

Example 5
A physics professor wants to find the diameter of a human hair by placing it between two flat glass plates, illuminating the plates with light of vacuum wavelength λ = 552 nm and counting the number of bright fringes produced along the plates. The Prof finds 125 bright fringes between the edge of the plates and the hair. What is the diameter of the hair?

The reflections from the boundaries will cause a net 180° phase shift. The condition for bright fringes is $2t = (m + 1/2)\lambda_{film}$. Now m = 124 since there is a bright fringe for m = 0 and $\lambda_{film} = \lambda/n$.

$$t = \frac{(m + \frac{1}{2})\lambda_{film}}{2} = \frac{(m + \frac{1}{2})\lambda}{2n} = \frac{(124 + \frac{1}{2})(552 \text{ X } 10^{-9} \text{ m})}{2(1.00)} = 3.44 \text{ X } 10^{-5} \text{ m}$$

27.5 Diffraction

Light diffracts through an opening such as a slit and subsequently spreads. If a screen is placed in the path of the light, a point on the screen will receive light which has come from each point in the opening and interference of essentially an infinite number of waves occurs at the point. This interference is totally destructive at some points on the screen and gives dark fringes while the interference is totally constructive at other points and gives rise to bright fringes. Of course, intermediate interferences occur between the bright and dark fringes. The central fringe of the resulting diffraction pattern is wide and intense while the fringes to either side of the central fringe fall off in intensity and are narrower. Equation (27.4) locates the positions of the dark fringes. Note that this is the same equation which governed the diffraction of sound (see Equation 17.1).

27.6 Resolving Power

The diffraction of light through an opening places limits on our, or an instrument's, ability to distinguish between closely spaced objects. Since the pupil of our eye is a circular opening, we see diffraction patterns which fall on the retina. This is the reason that stars which are so far away that they should be point sources of light appear to us to be circles of light. We see the diffraction pattern of a point source of light. If two such sources of light are close together, their individual diffraction patterns may overlap on the retina so much that they look like one source. When do two sources look like two sources, that is, when are they resolved? The Rayleigh criterion, equation (27.6), gives a useful, if not unique, definition. The assumption is that two objects are "just" resolved when the first minimum of the diffraction pattern of one falls on the central maximum of the other.

Example 6
The light from a red star which is essentially a point object is viewed by a child on a dark night when the child's pupil has a diameter of 7.00 mm. How wide does the star appear on the retina of the child? Assume that the index of refraction of the vitreous humor in the eye is 1.36, that light seen by the child has a wavelength of 625 nm, and that the distance from the pupil to the retina is 3.5 cm.

The child will see the star as the central fringe of the diffraction pattern which falls on the retina. The angular position of the first minimum of the pattern is given by

$$\sin\theta = 1.22\ (\lambda/D)$$

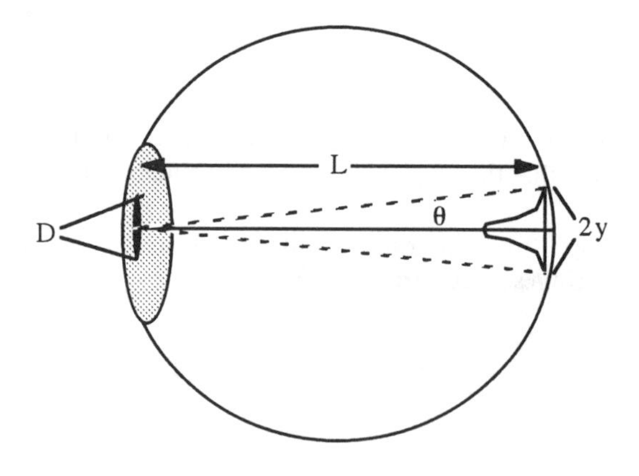

where λ is the wavelength of light within the eye, $\lambda = (625\ \text{nm})/(1.36) = 4.60 \times 10^{-7}$ m and $d = 7.00 \times 10^{-3}$ m.

$$\theta = 4.59 \times 10^{-3\,\circ}.$$

The width of the central maximum is

$$2y = 2L\tan\theta = 2(3.5 \times 10^{-2}\ \text{m})\tan(4.59 \times 10^{-3\,\circ}) = 5.61 \times 10^{-6}\ \text{m}.$$

Example 7
The headlights of a certain automobile are 1.50 m apart. How far away could the car be and the headlights just be resolved by a person's eye? Use the data for the eye given in the previous example and assume the wavelength of the light is 625 nm.

The headlines will be just resolved according to the Rayleigh criterion when the angle subtended by the headlights at the pupil is $\theta = 1.22\ \lambda/D$. From the diagram (next page) we see that $\theta = d/x$ if the angle is small. Then, since $\lambda = (625\ \text{nm})/(1.36) = 4.60 \times 10^{-7}$ m,

$$x = \frac{dD}{1.22\ \lambda} = \frac{(1.50\ \text{m})(7.00 \times 10^{-3}\ \text{m})}{1.22(4.60 \times 10^{-7}\ \text{m})} = 1.87 \times 10^{4}\ \text{m}$$

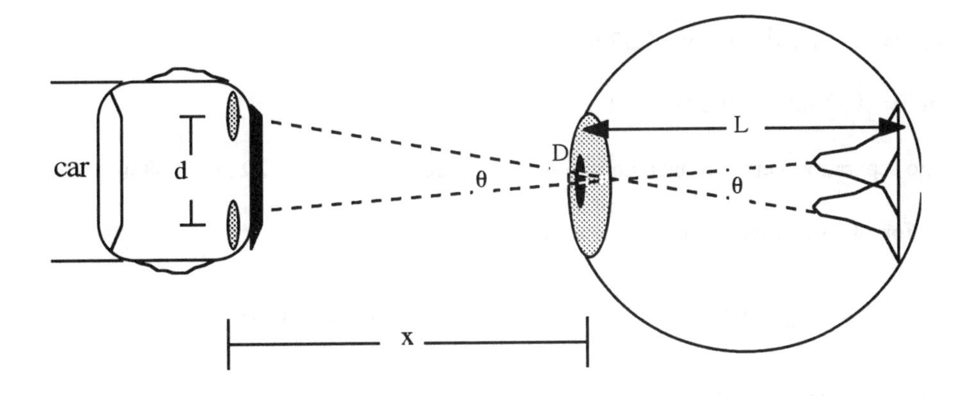

27.7 The Diffraction Grating

A set of closely spaced parallel slits is referred to as a diffraction grating. A grating with a large number of very closely spaced slits is capable of producing bright fringes which are much narrower and sharper than the bright fringes of a double slit. The bright fringes of a grating are often referred to as the principal maxima or orders. The positions of the principal maxima of a grating is given by equation (27.7). Here, d is the spacing between the slits of the grating. Usually a grating is specified by the number of slits or lines per inch or per cm. To find the spacing, d, simply take the reciprocal.

Example 8
How many orders (principal maxima) can be produced by a 1500 lines/cm grating when illuminated with green light whose wavelength is $\lambda = 510$ nm?

The largest angle that an order could possibly have is 90°. The order which is closest to having this angle is, according to equation (27.7)

$$m = \frac{d}{\lambda} = \frac{(1/1500) \text{ X } 10^{-2} \text{ m}}{510 \text{ X } 10^{-9} \text{ m}} = 13$$

The grating produces 13 orders along with the zero order, so it has a total of 14 orders.

The diffraction grating has a very important application in spectroscopes due to its ability to separate light into its various wavelengths (colors). Light of different wavelengths are diffracted at different angles, so the grating disperses light into its colors as a prism does. The grating produces a "rainbow" spectrum in each of its orders except the zero order. The grating in the above example can produce 13 rainbow spectra. The separation of the colors is larger in the higher orders which makes them especially useful for detailed analysis of the frequencies of light.

Example 9
A mixture of red (630 nm) and blue (490 nm) light passes through a 5500 slits/cm grating and falls on a screen located 1.0 m away. What are the separations of the colors on the screen in the first order and in the second order?

The angular position of a wavelength in the first order is

$$\theta = \sin^{-1}\frac{\lambda}{d} = \sin^{-1}[(5500 \text{ X } 10^{2} \text{ m}^{-1})\lambda]$$

The position of this wavelength on the screen is $y = L \tan \theta$.

For λ = 630 nm, $\theta = 2.0 \times 10^{1}$ ° and y = 0.37 m.

For λ = 490 nm, θ = 16 ° and y = 0.29 m.

The separation between the colors in the first order is 0.37 m - 0.29 m = 0.08 m = 8 cm.

The angular position of a wavelength in the second order is

$$\theta = \sin^{-1}\frac{2\lambda}{d} = \sin^{-1}\left[2(5500 \times 10^{2}\ \mathrm{m}^{-1})\lambda\right]$$

For λ = 630 nm, θ = 44° and y = 0.97 m.

For λ = 490 nm, θ = 33° and y = 0.65 m.

The separation of the colors in the second order is 0.97 m - 0.65 m = 0.32 m = 32 cm.

PRACTICE PROBLEMS

1. Relative to the central fringe, what is the position on a screen located 2.0 m away of the third order fringe in a double slit arrangement ($d = 0.058$ mm) when the slits are illuminated with $\lambda = 480$ nm light?

2. Two parallel slits are illuminated by light composed of two wavelengths, one of which is 486 nm. On viewing the screen the light whose wavelength is known produces its third dark fringe at the same place where the light whose wavelength is unknown produces its fourth-order bright fringe. The fringes are counted relative to the central or zeroth-order bright fringe. What is the unknown wavelength?

3. A certain region of a soap bubble appears red ($\lambda = 630$ nm) due to reflection. What are the three smallest possible thicknesses of this region of the bubble? Take $n = 1.36$ for the soap film.

4. The minima of the diffraction pattern of light passing by a human hair lie in the same places as the minima of the diffraction pattern of a slit the same width as the hair. The distance between the two minima on either side of the central fringe is measured to be 2.2 cm on a screen located 2.5 m from a hair when $\lambda = 550$ nm light is used.. What is the diameter of the hair?

5. Two stars, each located 580 light years from earth, are 2.4 light years apart. What is the smallest diameter telescope which could just resolve them using 630 nm light?

6. There are 5620 lines per centimeter in a grating that is used with light whose wavelength is 471 nm. A flat observation screen is located a distance of 0.750 m from the grating. What is the minimum width that the screen must have so the centers of all the principal maxima formed on either side of the central maximum fall on the screen?

7. A hologram is a photographic exposure of the interference between laser light reflected from an object and light coming directly from the laser. When the film is developed, there are ridges and valleys in the emulsion on the film which can act like a diffraction grating when light is shined through it. The light striking the hologram is diffracted into a three dimensional image of the object that was originally used. Assume that laser light $\lambda = 630$ nm is incident on a small spot of the hologram is diffracted in the first order through an angle of 22° to form an image. What is the effective lines per cm on this spot of the hologram?

8. You have a 4500 lines per cm grating available and need to separate light of 535 nm and 585 nm by a distance of at least 15 cm on a screen 0.75 m away. What is the smallest order of diffraction you could use?

9. Light of wavelength 630 nm is incident on a small circular hole. The light travels to a wall located 2.0 m from the hole. What must be the diameter of the hole so that the first dark ring has a diameter of 2.5 cm?

HELPFUL SUGGESTIONS

1. The wave nature of light is mysterious only because we rarely see (notice?) its manifestations in everyday life. You can help yourself with understanding by relating what you know about the interference and diffraction of other types of waves: sound, water surface waves, etc.

2. Many of the equations presented in this chapter are similar in form, and this can lead to confusion. Pay particular attention to the circumstances under which each equation is applicable instead of just memorizing them.

3. When you are required to find the separations between two low order fringes on a screen, you can sometimes make use of the small angle approximation. That is, sin θ is approximately equal to tan θ and also to θ in radian measure. This is a useful approximation ONLY if θ is less than about 10°. If in doubt---DON'T USE IT!

4. The only effect that the indices of refraction of the materials above and below a film have on the interference is in determining the phase shifts of the light upon reflection at the two boundaries. This phase shift on reflection is either 0° or 180°. It is the index of refraction OF THE FILM which determines the wavelength of the light in the film.

EVERYDAY PHYSICS

1. The low cost, low power laser has made the investigation of interference and diffraction quite easy and a lot of fun. If you have access to a low power laser such as a helium-neon laser, you can perform double slit interference, single and double slit diffraction, diffraction through a hole, and thin film interference.

   You can make your own slits, holes etc. in aluminum foil using simple tools such as razor blades and needles. Perhaps your instructor or a friendly physics major will demonstrate these if you cannot locate a laser. If you can use a laser , please observe the safety precautions on the warning label. Never direct a laser beam into anyone's eyes.

2. Examples of thin film interference abound in everyday life. Just stop to notice them and take a little time to play.

   One of the easiest and most fascinating thin film projects you can do is to "play" with soap films made in your own kitchen or from purchased bottles of bubbles. Blow some bubbles and observe the colors reflected from the films. Why do the colors appear to change position? Are there any black areas on the films? Can you explain?

   Often merchandise are "shrink wrapped" in clear plastic film to protect them before purchase. Air is often trapped in thin layers between the plastic film and a surface of the merchandise. Light reflected from the air layer produces interference fringes. This effect is especially common on TV screens which have been covered with a protective plastic film.

   Oil films on water also show thin film interference quite nicely and are easy to produce. Place light oil a little at a time on the surface of water in a glass and shine a flashlight on the surface. You can change the thickness of the film by adding more oil.

3. As mentioned in the text CDs have light diffracting properties due to their closely spaced tracks. If you have a CD, look at it. Also, you may want to take a look at an ordinary 33 1/3 rpm record. The grooves are so close together on some records that you see a rainbow dispersion of reflected light. A good magnifying glass will allow you to see the grooves in more detail.

4. The ultimate in diffraction effects is, of course, the hologram which produces a three dimensional image by acting like a very complex diffraction grating. Holograms are becoming quite common. They are being used on credit cards since they are very difficult to fake. Some magazines have published holograms on their covers. Obtain some holograms and investigate them. You might take a close look at them with a magnifying glass or a microscope. Can you see the lines?

CHAPTER QUIZ

1. The term coherent sources refers to sources whose light has
   a. the same frequency.
   b. the same amplitude.
   c. the same frequency and a constant phase relationship.
   d. different frequencies and a constant phase relationship.

2. Two thin slits are spaced 152 μm apart. How many bright fringes lie on one side of the central fringe in the interference pattern formed when the slits are illuminated by 515 nm light?
   a. one b. 132 c. 295 d. 315

3. Which of the following colors of light would produce a double slit interference pattern with the largest separation between the bright fringes?
   a. red b. orange c. green d. blue

4. The condition for constructive interference of light of vacuum wavelength, λ, being reflected from an oil film of thickness, t, and index of refraction , $n > 1.33$, floating on water is
   a. $2t = m\lambda$ b. $2t = m(\lambda/n)$ c. $2t = (m + 1/2)\lambda$ d. $2t = (m + 1/2)(\lambda/n)$

5. Light of wavelength 480 nm passes through a 0.85 mm wide slit and strikes a screen located 1.0 m away. What is the width of the central bright fringe formed on the screen?
   a. 5.5×10^{-5} m b. 2.2×10^{-4} m c. 33 mm d. 1.1 mm

6. A particular region on a freshly produced soap film changes in color from red to green to blue. What can you say about the thickness of this region of the film?
   a. The thickness is increasing.
   b. The thickness is decreasing.
   c. It is uniform in thickness
   d. No conclusion can be reached.

7. A pin hole camera is simply a box with a small hole in one side. Photographic film is placed against the side opposite the hole. Light passing through the hole forms an image on the film. What is the smallest angular separation of two points on an object that can be resolved by a pinhole camera which has a 0.25 mm diameter hole? Use red light of $\lambda = 630$ nm for your calculation.
   a. 2.5×10^{-3} radians b. 3.1×10^{-3} radians c. 1.3×10^{-3} rad d. 0.23 °

8. Multicolored light in the wavelength range from 630 nm to 450 nm falls on a 9500 lines/cm diffraction grating. How many complete "rainbow spectra" will be produced by the grating?
   a. one b. two c. none d. an infinite number

9. A spy satellite is in orbit 35 000 m above the surface of the earth. A camera on board the satellite has an effective lens diameter of 0.50 m. If the camera uses red light of wavelength 630 nm, what is the smallest separation of objects on the surface of the earth which can be resolved by the camera?
   a. 8.2 m b. 4.4 cm c. 1.2 km d. 54 mm

SOLUTIONS AND ANSWERS

Practice Problems

---

1. The angular position of the third order fringe is

$$\theta = \sin^{-1}[3(480 \times 10^{-9}\ m)/(0.058 \times 10^{-3}\ m)] = 1.4°$$

The position of the fringe measured from the center of the screen is

$$y = L \tan\theta = (2.0\ m) \tan 1.4° = \mathbf{5.0\ cm.}$$

---

2. Equation (27.2) written for the third dark fringe of the known $\lambda = 486$ nm is:

$$\sin\theta = (2 + 1/2)\lambda/d$$

and, equation (27.1) written for the fourth order bright fringe of the unknown wavelength, λ' gives

$$\sin\theta = 4\lambda'/d.$$

The fringes are in the same position, so

$$\lambda' = (5/8)\lambda = (5/8)(486\ nm) = \mathbf{3.04 \times 10^2\ nm.}$$

---

3. There is a 180° phase shift only at the soap-air interface, so the constructive interferences occur for

$$2t = (m + 1/2)\lambda_{film}$$

Thus, the three smallest thicknesses are:

$$2t_1 = (1/2)\lambda_{film}$$
$$2t_2 = (3/2)\lambda_{film}$$
$$2t_3 = (5/2)\lambda_{film}$$

$\lambda_{film} = \lambda/n$, so

$$t_1 = \mathbf{115\ nm}$$
$$t_2 = \mathbf{350\ nm}$$
$$t_3 = \mathbf{580\ nm.}$$

---

4. The first minima of the diffraction pattern of a slit, hence a hair of width, W, is found from equation (27.4). The position of the first minimum on the screen is $y = L \tan\theta$, so

$$\theta = \tan^{-1}(1.1 \times 10^{-2}\ m/2.5\ m) = 0.25°.$$

Now $$W = \lambda/\sin\theta = (550 \times 10^{-9}\ m)/\sin 0.25° = 1.2 \times 10^{-4}\ m = \mathbf{0.12\ mm.}$$

---

5. The Rayleigh criterion for the resolution of two objects gives the needed diameter: $D = 1.22\,\lambda/\theta_{min}$.

The angle subtended by the stars is

$$\theta = (2.4\ ly)/(580\ ly) = 4.1 \times 10^{-3}\ rad.$$

Then, from the Rayleigh criterion:

$$D = (1.22)(630 \times 10^{-9}\ m)/(4.1 \times 10^{-3}\ rad) = \mathbf{1.9 \times 10^{-4}\ m.}$$

---

6. The screen must have a total width,

$$2y = 2L \tan\theta \text{ where } \theta = \sin^{-1} m\lambda/d.$$

The order of the last maximum which can fall on one side of the screen is

$$m = (d/\lambda) \sin 90° = 3.77 \text{ or } 3$$

The angle locating the last maximum on the screen is

$$\theta = \sin^{-1} (3\lambda/d) = 52.7°$$

Now half the width of the screen is

$$y = (0.750) \tan 52.7° = 0.985\ m.$$

The required screen width is then **1.97 m.**

---

7. The distance between the "slits" is

$$d = \lambda/\sin\theta = (630 \times 10^{-9}\ m)/\sin 22° = 1.7 \times 10^{-6}\ m = 1.7 \times 10^{-4}\ cm$$

Then

$$1/d = \mathbf{5900\ lines/cm}$$

---

8. The angular position of the wavelengths on the screen is $\theta = y/L$ and $\theta' = y'/L$ where a small angle approximation has been used. Now $\theta = m\lambda/d$ and $\theta' = m\lambda'/d$, so

$$m = (y' - y)d/(\lambda' - \lambda)L = 8.9$$

The smallest order which could be used to see both lines is the **9th**.

---

9. The angular width of the first dark ring is

$$\theta \approx 1.22\ \lambda/D.$$

The diameter of the first dark ring is

$$d \approx 2L\theta \text{ so}$$

$$D \approx 2(1.22)(2.0\ m)(630 \times 10^{-9}\ m)/(2.5 \times 10^{-2}\ m) = \mathbf{1.2 \times 10^{-4}\ m.}$$

Quiz answers

1. c	3. a	5. d	7. b	9. d
2. c	4. d	6. b	8. a	

MCAT REVIEW PROBLEMS

Figure 1 shows an experiment designed to explore some of the properties of light. The laser emits a coherent, monochromatic beam of intensity 10^6 W/m^2. These beam splitters transmit 50% of the light reaching them, and reflect the other 50%, no matter which side of the beam splitter the light hits. For instance, after reaching beam splitter 1, half the laser light travels along path A and the other half travels along path B. The cross-sectional area of each of these two beams equals the cross-sectional area of the original "unsplit" beam. These cross-sectional areas stay constant as the beams travel along their respective paths.

Unlike the beam splitters, the mirrors reflect essentially 100% of the light reaching them. The light detectors register the intensity of the light reaching them.

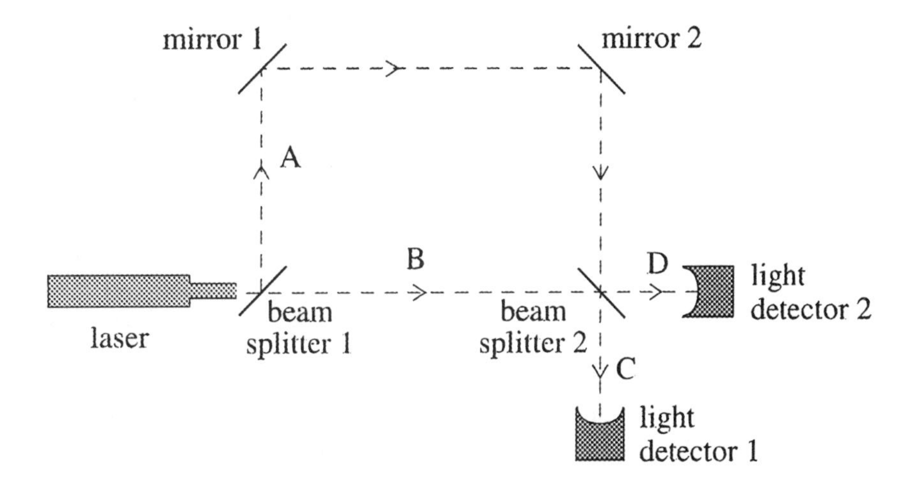

FIGURE 1

1. A student holds a block in path A, midway between beam splitter 2 and mirror 2. The intensity of laser light reaching the block is:

   A. 2×10^6 W/m^2 **B.** 10^6 W/m^2
   C. 5×10^5 W/m^2 **D.** 2.5×10^5 W/m^2

2. While the student blocks the light traveling along path A (as described in question 1), what percentage of the light emitted by the laser reaches detector 1?

   A. 100% **B.** 50% **C.** 25% **D.** 0%

3. Now the student removes the block, allowing light to travel unobstructed along both paths. Approximately what intensity does detector 1 register?

   A. 10^6 W/m^2 **B.** 5×10^5 W/m^2
   C. 2.5×10^5 W/m^2 **D.** We cannot determine the answer without more information.

4. By varying the distance between the mirrors and beam splitters, a scientist could use this experimental set-up to determine which properties of the laser light?

A. Speed, but not wavelength
B. Wavelength, but not speed
C. Both speed and wavelength
D. Neither speed nor wavelength

ANSWERS TO MCAT REVIEW PROBLEMS

---

1. **C.** Half the laser light travels along path A, while the other half takes path B. More precisely, path A "gets" half the total power emitted by the laser. Since the cross-sectional area of the path A beam equals the cross-sectional area of the original unsplit beam, and since intensity is power per cross-sectional area, the path A beam must be half as intense as the original unsplit beam. This reasoning applies no matter where along path A we determine the intensity.

---

2. **C.** Because of the block, no light from path A reaches either detector. Of the total light emitted by the laser, half travels along path B. And when the path B light reaches beam splitter 2, half of it transmits through to detector 2, while the other half reflects down to detector 1. So, detector 1 receives half the intensity of the path B beam, which has half the intensity of the original unsplit laser beam.

---

3. **D.** If light waves did not interfere, then each detector would receive 50% of the total intensity. But the light reaching detector 1 consists of an overlapping combination of light from paths A and B. If the path A beam and path B beam are "out of phase" when they reach path C, then detector 1 registers *zero* intensity. In other words, if "crests" from path A overlap with "troughs" from path B, as illustrated in figure 2, then those two light beams cancel each other out; the intensity vanishes, due to destructive interference. A human eye at detector 1 would see nothing, because there is nothing to see!

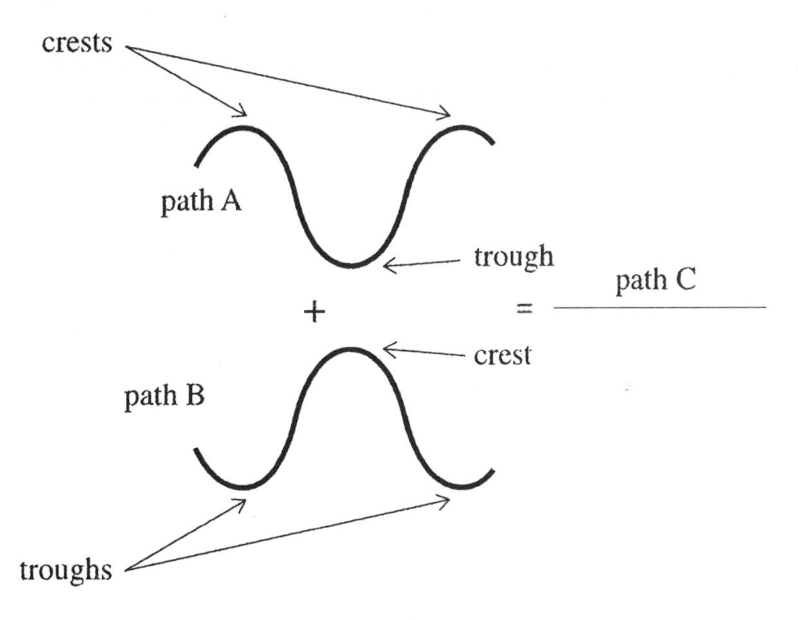

FIGURE 2

Alternatively, if the path A and path B beams are in phase when they reach path C, then detector 1 registers a higher intensity than would otherwise be expected, because the two beams constructively interfere.

Without more information, we cannot determine whether beams A and B interfere constructively, destructively, or at some intermediate level when they reach path C. Hence, we cannot guess the intensity registered by detector 1.

---

4. **B.** As figure 3 shows, lengthening a "light path" by half a wavelength *flips the phase* at the end of that path. The phase at point X corresponds to a trough. But if we lengthen that light path by half a wavelength, to point Y, then the phase corresponds to a crest.

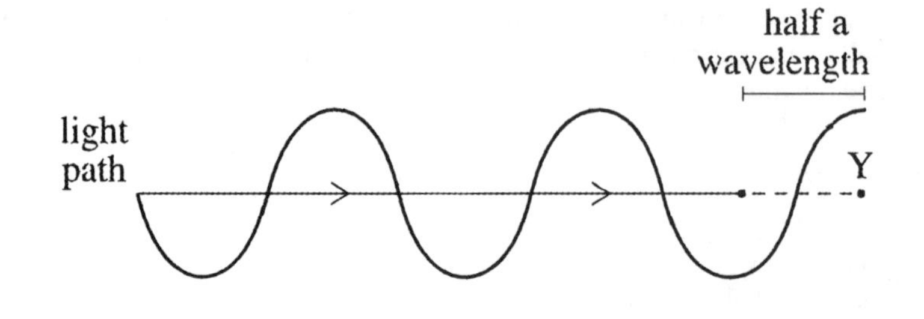

FIGURE 3

Using this insight, we can determine the laser light's wavelength. First, adjust the mirrors and beam splitters so that detector 1 registers zero intensity. This destructive interference implies that crests from path B meet *troughs* from path A. Next, gradually lengthen path A by moving the mirrors away from the beam splitters, until detector 1 registers the highest intensity reached during the experiment. This "peak" intensity indicates perfectly constructive interference; crests from path B now meet *crests* from path A.

In summary, before we lengthened path A, crests from B met *troughs* from A. After we lengthened path A, crests from B met *crests* from A. Therefore, as explained in figure 3, we must have "flipped the phase" at the end of path A. In other words, we must have lengthened path A by half a wavelength.

So, by measuring how far we just moved the mirrors, we can calculate how far we lengthened path A. That distance equals half a wavelength. In this way, we just measured the wavelength of the laser light.

By contrast, this experiment cannot tell us the speed of light, because we cannot determine the "travel time" of light along path A or path B.

CHAPTER 28
Special Relativity

PREVIEW

In this chapter you will be introduced to some of the fascinating ideas contained in Einstein's theory of special relativity. You will learn how the theory has changed the way we view space and time, with the ideas of time dilation and length contraction. The relation between mass and energy will also be examined as well as some of the other seemingly strange predictions made by this famous theory.

QUICK REFERENCE

Important Terms

Event
A physical "happening" that occurs at a certain place and time.

Inertial reference frame
A frame of reference in which Newton's law of inertia is valid. That is, if the net force acting on a body is zero, the body either remains at rest or moves with constant velocity. In other words, the acceleration of a body is zero when measured in an inertial frame if there is no net force.

Relativity postulate
The laws of physics are the same in every inertial reference frame.

Speed of light postulate
The speed of light in a vacuum, measured in any inertial reference frame, always has the same value of c, no matter how fast the source of light and the observer are moving relative to each other.

Proper time interval
The time interval between two events measured by an observer who is at rest relative to the events and who views these events as occurring at the same place.

Time dilation
The lengthening of the time interval between two events, as measured by an observer who is in motion with respect to the events. It implies that "moving clocks run slow".

Proper length
The length of an object measured by an observer who is at rest relative to the object.

Length contraction
The shortening of distances as measured by a moving observer.

Mass-energy equivalence
Energy and mass are related through the equation $E_0 = mc^2$.

Equations

The **time dilation** equation is:

$$\Delta t = \frac{\Delta t_0}{\sqrt{1 - \frac{v^2}{c^2}}} \qquad (28.1)$$

The equation for **length contraction** is:

$$L = L_0\sqrt{1 - \frac{v^2}{c^2}} \quad (28.2)$$

The equation for the **relativistic momentum** is

$$p = \frac{mv}{\sqrt{1 - \frac{v^2}{c^2}}} \quad (28.3)$$

The **total energy** of an object is:

$$E = \frac{mc^2}{\sqrt{1 - \frac{v^2}{c^2}}} \quad (28.4)$$

The **rest energy** of an object is

$$E_0 = mc^2 \quad (28.5)$$

The **relativistic kinetic energy** is:

$$KE = E - E_0 = mc^2\left(\frac{1}{\sqrt{1 - \frac{v^2}{c^2}}} - 1\right) \quad (28.6)$$

The **velocity-addition** formula is:

$$u = \frac{u' + v}{1 + \frac{u'v}{c^2}} \quad (28.7)$$

DISCUSSION OF SELECTED SECTIONS

28.2 The Postulates of Special Relativity

The theory of special relativity deals with **inertial reference frames**. These are frames of reference in which Newton's law of inertia is valid. That is, if the net force acting on a body is zero, the body either remains at rest or moves at a constant velocity. The acceleration of such a body is zero when measured in an inertial reference frame. Rotating and otherwise accelerating reference frames are not inertial frames.

Einstein based his theory of special relativity on two fundamental assumptions or postulates.

1. **The Relativity Postulate**. The laws of physics are the same in every inertial reference frame.
2. **The Speed of Light Postulate**. The speed of light in a vacuum, measured in any inertial reference frame, always has the same value of c, no matter how fast the source of light and the observer are moving relative to each other.

We will find some very unusual predictions made by the theory of special relativity, most of which will not appeal to our "common sense". These seemingly strange predictions occur as a direct result of the application of the two postulates just stated. So keep in mind that these ideas, which may seem strange to our intuitive notions of space and time, are actually experimentally verified and accepted theories of physics.

28.3 The Relativity of Time: Time Dilation

Consider an observer on the earth and one on a spacecraft moving with respect to the earth. The theory of special relativity reveals that the earth-based observer sees time passing more slowly for the astronaut on the spacecraft than for himself. For example, suppose the astronaut has a clock on his spaceship. He measures the time interval between successive "ticks" on his clock (two **events**) to be Δt_0, which we will call the **proper time interval**. However, the earth-based observer measures a time interval Δt between these two events that is GREATER than the time interval Δt_0 measured by the astronaut. This effect is known as **time dilation** because the time interval Δt is dilated or "expanded" relative to Δt_0. In essence, the moving clock runs more slowly from the point of view of the earth-based observer. The relationship between the two time intervals is given by:

$$\Delta t = \frac{\Delta t_0}{\sqrt{1 - \frac{v^2}{c^2}}} \qquad (28.1)$$

where v is the relative speed between the two observers, and c is the speed of light in a vacuum.

Example 1
An astronaut travels away from the earth at v = 0.98c. If one hour ticks off on the astronaut's clock, what is the time interval that an earth observer measures on the astronaut's clock?

In equation (28.1) the proper time interval Δt_0 is taken to be 1.0 h, v = 0.98c, and we have

$$\Delta t = \frac{\Delta t_0}{\sqrt{1 - \frac{v^2}{c^2}}} = \frac{1.0\text{ h}}{\sqrt{1 - \frac{(0.98c)^2}{c^2}}} = 5.0\text{ h}$$

So, according to the earth observer, it takes 5.0 earth-hours for one hour to "tick off" on the astronaut's clock!

28.4 The Relativity of Length: Length Contraction

Because of time dilation, observers moving at a constant velocity relative to each other measure different time intervals between two events. Let's calculate the distance traveled by the astronaut in example 1, as measured by the earth observer and the astronaut. Suppose the spacecraft travels from point A to point B, the trip taking 5.0 h as measured by the earth observer. The astronaut measures a distance $L = v\,\Delta t_0 = (0.98c)(1.0\text{ h}) = 0.98$ light-hours. (NOTE: a light-hour is the distance that light travels in one hour, i.e., $(3.0 \times 10^8\text{ m/s})(3600\text{ s/1 h}) = 1.1 \times 10^{12}$ m.) The earth observer measures the distance to be $L_0 = v\,\Delta t = (0.98c)(5.0\text{ h}) = 4.9$ light-hours. For the astronaut, the distance has shortened, as compared to the earth observer's distance. This is an example of the phenomenon known as **length contraction**.

If L_0 is the length between two points as measured by an observer at rest with respect to these points (L_0 is referred to as the **proper length**), and L is the length measured by an observer moving at a constant speed v with respect to these points, the equation relating L and L_0 is given by:

$$L = L_0\sqrt{1 - \frac{v^2}{c^2}} \qquad (28.2)$$

It is important to note that this length contraction occurs only along the direction of motion. Those dimensions that are perpendicular to the motion are not shortened.

Example 2
An astronaut carries a meter stick in his spaceship. The meter stick is aligned in the direction of motion of his spaceship, which is moving at a speed of 0.8c relative to the earth. What is the length of the meter stick, as measured by an observer back on earth?

The proper length of the meter stick is $L_0 = 1$ m. The speed of the spaceship is $v = 0.8c$. The length of the meter stick, as measured by the earth observer, is therefore

$$L = L_0\sqrt{1 - \frac{v^2}{c^2}} = (1\ \text{m})\sqrt{1 - \frac{(0.8c)^2}{c^2}} = 0.6\ \text{m}$$

So an observer on earth measures the astronaut's meter stick to be only 60 cm long!

28.5 Relativistic Momentum

When an object is moving at a speed v relative to an observer, the observer measures a momentum, p, called the **relativistic momentum** given by

$$p = \frac{mv}{\sqrt{1 - \frac{v^2}{c^2}}} \qquad (28.3)$$

As the velocity approaches the speed of light, the term in the denominator of the equation approaches zero and the relativistic momentum approaches infinity. When $v = c$, the relativistic momentum becomes infinite, and to accelerate the mass up to the speed of light would require an infinite amount of energy. An object with finite mass can therefore never have a speed equal to or greater than the speed of light. The speed of light therefore represents the "ultimate" speed for any such object.

Example 3
A proton whose mass is 1.67×10^{-27} kg is traveling at a speed $v = 0.999c$ relative to the earth. What is the relativistic momentum of the proton as measured by an earth based observer?

$$p = \frac{mv}{\sqrt{1 - \frac{v^2}{c^2}}} = \frac{(1.67 \times 10^{-27}\ \text{kg})(0.999\ c)}{\sqrt{1 - (0.999)^2}}$$

$$p = \mathbf{1.12 \times 10^{-17}\ kg\ m/s}.$$

28.6 The Equivalence of Mass and Energy

An important result of special relativity is that mass and energy are equivalent. The equivalence between energy and mass is stated in

$$E = \frac{mc^2}{\sqrt{1 - \frac{v^2}{c^2}}}$$

where E represents the total energy of an object that is moving at speed v relative to an observer.

When an object is at rest relative to an observer, its energy is $E_0 = mc^2$, where E_0 represents the **rest energy** of the object. If the object is in motion, it has kinetic energy as well, so its total energy is a sum of its rest energy and its kinetic energy, $E = E_0 + KE$. Using equation (28.4), we can now express the kinetic energy of the object as

$$KE = E - E_0 = mc^2\left(\frac{1}{\sqrt{1 - \frac{v^2}{c^2}}} - 1\right) \qquad (28.6)$$

Example 4
For the proton of example 3 calculate (a) its rest energy, (b) its total energy, and (c) its kinetic energy.

(a) The rest energy is:

$$E_0 = mc^2 = (1.67 \times 10^{-27}\text{ kg})(3.00 \times 10^8\text{ m/s})^2 = 1.50 \times 10^{-10}\text{ J}$$

(b) The total energy is:

$$E = \frac{mc^2}{\sqrt{1 - \frac{v^2}{c^2}}} = \frac{(1.67 \text{ X } 10^{-27}\text{ kg})(3.00 \text{ X } 10^8\text{ m/s})^2}{\sqrt{1 - \frac{(0.999\text{ c})^2}{c^2}}} = 3.37 \text{ X } 10^{-9}\text{ J.}$$

(c) The kinetic energy is:

$$KE = E - E_0 = 3.37 \times 10^{-9}\text{ J} - 1.50 \times 10^{-10}\text{ kg} = 3.22 \times 10^{-9}\text{ J.}$$

28.7 The Relativistic Addition of Velocities

Consider the following situation shown in the drawing. A rocket is moving with a velocity v (in the +x direction) with respect to the earth. An earth observer and an astronaut on the rocket both see a flying saucer moving in the +x direction. However, each measures a different value for the speed of the saucer. The earth observer measures a speed u, while the astronaut measures a speed u' for the saucer.

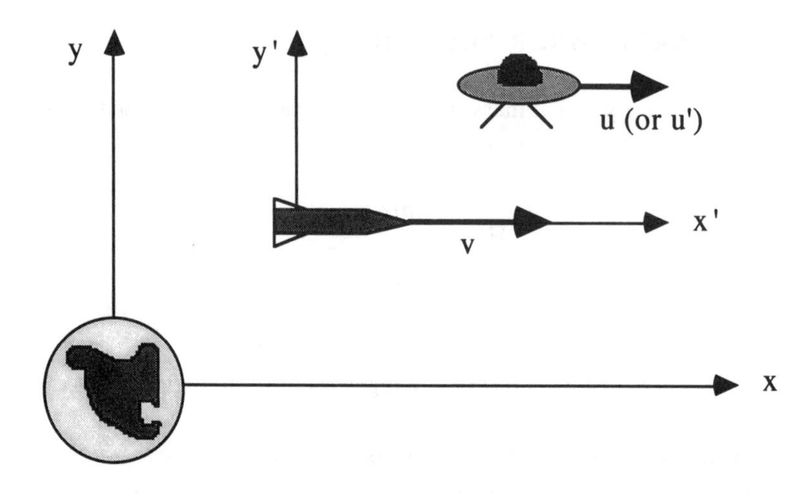

The theory of special relativity states that the velocities u and u' are related according to the **velocity-addition formula**:

$$u = \frac{u' + v}{1 + \frac{u'v}{c^2}}$$

When the velocities are one dimensional, they can have either positive or negative values, depending on whether they are directed along the positive or negative direction.

Example 5

An astronaut travels in her spaceship in the +x direction at a speed of 0.60c relative to earth. She looks out her window and sees a flying saucer speeding past her at 0.90c in the +x' direction. What is the speed of the flying saucer as measured by an observer on earth?

Using the velocity-addition formula with v = 0.60c and u' = 0.90c we obtain:

$$u = \frac{u' + v}{1 + \frac{u'v}{c^2}} = \frac{0.90c + 0.60c}{1 + \frac{(0.90c)(0.60c)}{c^2}} = 0.97c$$

The earth observer measures the speed of the flying saucer to be u = 0.97c, rather than the value of 1.5c which we would have obtained if we didn't know about relativity. Equation (28.7) ensures that relative velocities will never be greater than the speed of light.

Example 6

The spaceship in example 5 fires a photon torpedo (a burst of electromagnetic energy traveling at the speed of light) at the flying saucer. (a) How fast is the photon torpedo traveling with respect to an observer on earth? (b) How fast is the photon torpedo traveling with respect to an observer on the flying saucer?

(a) Using v = 0.60c and u' = c in equation (28.7) we obtain:

$$u = \frac{u' + v}{1 + \frac{u'v}{c^2}} = \frac{c + 0.60c}{1 + \frac{c(0.60c)}{c^2}} = \frac{1.60c}{1.60} = c$$

(b) Let the rest system (unprimed) be the flying saucer and the spaceship be the primed system. We therefore have, $v = -0.90c$, and $u' = c$. The velocity of the photon torpedo as viewed by the saucer is u, which is

$$u = \frac{u' + v}{1 + \frac{u'v}{c^2}} = \frac{c + (-0.90c)}{1 + \frac{c(-0.90c)}{c^2}} = \frac{0.10c}{0.10} = c$$

So all observers measure the same value for the speed of the photon torpedo, namely, the speed of light. This is, or course, in keeping with the second postulate of special relativity.

PRACTICE PROBLEMS

1. An unstable nuclear particle has a lifetime of 3.00×10^{-6} s when at rest on earth. If the particle is moving at 99.9% of the speed of light relative to earth, what is its lifetime as measured by an earth based observer?

2. If the lifetime of pions at rest in the laboratory is 2.6×10^{-8} s, at what speed must the pions travel with respect to the laboratory so that their lifetime is 7.8×10^{-8} s as measured by a laboratory observer?

3. As seen by an observer on earth, the Tau Ceti star system is 11.3 light years distant. A spaceship is sent to Tau Ceti at a speed of 0.95c. (a) What is the distance to Tau Ceti as measured by an astronaut on the spaceship? (b) How much does the astronaut age according to the astronaut's clock?

4. At what rate must matter be converted to energy in order to supply the United States with the 1×10^{20} J of energy it consumes each year?

5. A proton has a rest energy of 938 MeV. If the proton is moving at a speed of 0.999 999c, what is its kinetic energy in MeV?

6. Electrons are accelerated in a linear accelerator so that their total energy is 3.0×10^4 times their rest energy. Traveling at this speed, what do the electrons measure for the length of the accelerator which is 2.0 km long in the laboratory?

7. An observer on the earth sees a spaceship approaching at a velocity of 0.50c. The spacecraft then launches an exploration vehicle that, according to the earth observer, approaches at 0.70c. What is the velocity of the exploration vehicle relative to the spaceship?

8. Consider the three blocks labeled 1, 2, and 3 in the figure below. Block 1 slides at a speed of 0.50c relative to the ground; block 2 slides at speed 0.50c relative to block 1; and block 3 slides at speed 0.50c relative to block 2. Find the speed of (a) block 2 relative to the ground; (b) block 3 relative to the ground.

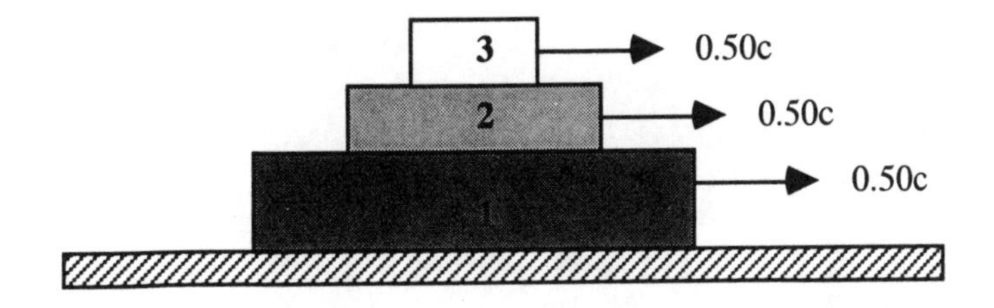

HELPFUL SUGGESTIONS

1. Confusion often occurs regarding the use of the time-dilation equation. The time intervals Δt and Δt_0 are easily reversed. Just remember that the person measuring the time interval between events occurring in his own reference frame measures the proper time Δt_0. If the observer is measuring an event occurring in a reference frame moving with respect to the observer, he measures the relativistic time Δt. Note that the proper time interval is always shorter than the relativistic time interval.

2. The same confusion sometimes arises regarding the use of the length-contraction equation. Again, an observer at rest with respect to an object always measures the proper length L_0 for that object, and this length is always greater than the length measured by an observer in motion with respect to the object.

3. When adding relativistic velocities using equation (28.7) it is important to correctly define the various relative velocities. You should try to visualize each inertial reference frame before trying to "plug in" numbers. The velocity of one reference frame with respect to another is v. An observer on earth, for example, sees an object moving in space with a velocity u. An observer in a spaceship moving at speed v relative to earth will measure the speed of the same object to be u'. Care should be taken to include the proper sign for each of the relative velocities. For example, if an object is moving at a speed $u = +\ 0.5c$ relative to the earth, and the spaceship is moving at $v = +\ 0.8c$ relative to earth (all in the same direction), the speed of the object as measured by the spaceship, u', will be negative since it is moving faster than the object, relative to earth.

EVERYDAY PHYSICS

There are not many applications of special relativity in everyday physics. You should realize by now that the effects of special relativity discussed here are only noticeable when speeds approach the speed of light. In fact, even at speeds of 20% the speed of light, the effects of time dilation, and length contraction only amount to about 2%. Currently, the greatest speeds we've attained (for space vehicles such as Pioneer 10 which is racing out of the solar system) are only about 5×10^{-5} c, so no one in our lifetime is ever likely to experience the effects of special relativity. However, we shouldn't let this deter us from fantasizing about what a trip at near light speeds might be like. For example, a trip to Alpha Centuri on a spaceship traveling at 0.98c would take about 4.4 years as seen by an earth based observer. However, the trip would take a little less than a year as measured by an astronaut on the spaceship! If we could travel toward the center of our Milky Way galaxy at 0.999 9997c, we would reach the center in about 40 000 years, according to earth based observers. The astronaut on this vehicle would make the trip in a mere 30 years!

In a sense, time travel is possible with special relativity. An astronaut making a round trip to the center of the galaxy would age 60 years during his entire trip. However, 80 000 years of earth history have passed, so he has literally traveled into the future. But alas, all that he knew is gone and all his friends are now long dead. Furthermore, according to relativity, there is no way to reverse the process. Time travel is a one way trip, ever forward in time, without hope of going back!

CHAPTER QUIZ

1. If you were able to travel in a rocket at speeds close to the speed of light, **you** would notice that, at these speeds:
   a. your shape had undergone a distortion.
   b. your pulse rate had decreased.
   c. your shape had undergone a distortion and your pulse rate had decreased.
   d. your shape and pulse rate are the same as they were when the rocket was at rest.

2. According to a stationary observer, a person traveling at a relativistic speed appears to:
   a. have a shorter life.
   b. contract in the direction of motion.
   c. expand in the direction of motion.
   d. contract in the direction of motion and have a shorter life.

3. Consider two spaceships, each traveling in a straight line at 0.5c. Ship X travels directly toward the sun and ship Y travels directly away from the sun. Astronauts on each vehicle measure the speed of the light coming from the sun. What do they measure for this speed?
   a. Ship X measures the speed to be greater than c, while ship Y measures the speed to be less than c.
   b. Ship Y measures the speed to be greater than c, while ship X measures the speed to be less than c.
   c. They both measure the speed to be less than c.
   d. They both measure the speed to be exactly c.

Questions 4 - 6 are based on the following: An astronaut travels at a speed of 0.87c relative to earth. He carries a clock, a meter stick aligned in the direction of motion of his ship, and has a mass of 85 kg.

4. If the astronaut sees that 1 hour ticks off on his clock, how much time passes according to an earth observer?
   a. 0.5 h   b. 1 h   c. 2 h   d. 4 h

5. What is the length of the meter stick as measured by the earth observer?
   a. 0.50 m   b. 0.87 m   c. 1.0 m   d. 2.0 m

6. What is the mass of the astronaut as viewed from the earth?
   a. 42 kg   b. 85 kg   c. 170 kg   d. 340 kg

7. Two spaceships travel through space toward one another. As seen from earth, each is moving at 0.50c. What is the speed of one ship, as measured by an observer on the other?
   a. 0.50c   b. 1.0c   c. 0.80c   d. 1.2c

8. If we were able to convert 1 g of matter into energy, with 100% efficiency, how much energy would result?
   a. 9×10^{13} J   b. 9×10^{19} J   c. 9×10^{20} J   d. 9×10^{25} J

9. A spaceship visits Alpha Centuri, which is 4.3 light years from earth, and returns. If the spaceship travels at 0.60c for virtually the entire trip, how long was the ship gone, according to an earth observer?
   a. 7.2 years   b. 14 years   c. 11 years   d. 21 years

10. A spaceship visits Alpha Centuri, which is 4.3 light years from earth, and returns. If the spaceship travels at 0.60c for virtually the entire trip, how long was the ship gone, according to the spaceship observer?
    a. 7.2 years   b. 14 years   c. 11 years   d. 21 years

SOLUTIONS AND ANSWERS

Practice Problems

---

1. The proper time is $\Delta t_0 = 3.00 \times 10^{-6}$ s. Using v = 0.999c,

$$\Delta t = \frac{\Delta t_0}{\sqrt{1 - v^2/c^2}} = 22.4\ \Delta t_0$$

$$\Delta t = \mathbf{6.72 \times 10^{-5}\ s}.$$

---

2. Using $\Delta t_0 = 2.6 \times 10^{-8}$ s and $\Delta t = 7.8 \times 10^{-8}$ s,

$$\Delta t_0/\Delta t = 0.33 = \sqrt{1 - v^2/c^2}.$$

Squaring this expression and solving for v yields,

$$1 - v^2/c^2 = 0.11$$

so that $v^2 = 0.89c^2$ and therefore, v = **0.94c**.

---

3. (a) The proper length is L_0 =11.3 light years. The distance to Tau Ceti as measured by the astronaut is

$$L = L_0\sqrt{1 - v^2/c^2}.$$

Therefore,

$$L = (11.3\ \text{ly})\sqrt{1 - (0.95\ c)^2} = \mathbf{3.5\ light\ years}.$$

(b) Since Δt = 11.3 years,

$$\Delta t_0 = \Delta t\sqrt{1 - v^2/c^2} = (11.3\ \text{y})\sqrt{1 - (0.95\ c)^2}\,] = \mathbf{3.5\ years}.$$

---

4. The power required to produce 1×10^{20} J in one year is

$$P = (1 \times 10^{20}\ \text{J/yr})(1\ \text{yr}/365\ \text{d})(1\ \text{d}/86\,400\ \text{s}) = 3 \times 10^{12}\ \text{J/s}.$$

So,

$$P = E_0/t = mc^2/t$$

which leads to

$$m/t = P/c^2 = (3 \times 10^{12}\ \text{J/s})/(3.00 \times 10^8\ \text{m/s})^2 = \mathbf{3 \times 10^{-5}\ kg/s}.$$

---

5. According to equation (28.6) the kinetic energy of the proton is

$$KE = (938\ \text{MeV})\left[\frac{1}{\sqrt{1 - \frac{(0.999\,999\ c)^2}{c^2}}} - 1\right] = \mathbf{6.62 \times 10^5\ MeV}.$$

---

6. According to equation (28.4) the energy is

$$E = 3.0 \times 10^4 E_0 = E_0/\sqrt{(1 - v^2/c^2)}$$

so that

$$\sqrt{1 - v^2/c^2} = 1/(3.0 \times 10^4).$$

Now, $L = L_0\sqrt{1 - v^2/c^2}$ and using $L_0 = 2.0$ km we obtain,

$$L = (2.0 \text{ km})/(3.0 \times 10^4) = 6.7 \times 10^{-5} \text{ km} = \mathbf{6.7\ cm}.$$

---

7. Using equation (28.7) we can solve for the velocity of the exploration vehicle as seen by the spaceship, u'.

$$u' = (u - v)/[1 - (uv/c^2)] = (0.70c - 0.50c)/[1 - (0.70c)(0.50c)/c^2]$$
$$u' = 0.20c/0.65 = \mathbf{0.31c}.$$

---

8. (a) Choose the ground to be the rest system and block 1 to be the moving system, so v = 0.50c. Since block 2 is moving at 0.50c relative to block 1, u' = 0.50c. Therefore,

$$u = (u' + v)/(1 + u'v/c^2) = \mathbf{0.80c}.$$

(b) Now choose block 2 to be the moving system (the ground is still the rest system). Since block 2 is moving at 0.80c relative to the ground, v = 0.80c. Block 3 moves at a speed of 0.50c relative to block 2 so u' = 0.50c. Therefore,

$$u = (0.50c + 0.80c)/[1 + (0.50c)(0.80c)/c^2] = \mathbf{0.93\ c}.$$

Quiz answers

1. d
2. b
3. d
4. c
5. a
6. b
7. c
8. a
9. b
10. c

MCAT REVIEW PROBLEMS

When subatomic particles undergo reactions, energy is conserved, but mass is not necessarily conserved. However, a particle's mass "contributes" to its total energy, in accordance with Einstein's famous equation,

$$E = mc^2.$$

In this equation, E denotes the energy a particle carries *because* of its mass. The particle can also have additional energy due to its motion and its interactions with other particles.

Consider a neutron at rest, and well separated from other particles. It decays into a proton, an electron, and an undetected third particle:

$$\text{neutron} \rightarrow \text{proton} + \text{electron} + ???.$$

Table 1 summarizes some data from a single neutron decay. An MeV (mega electron volt) is a unit of energy. Column 2 shows the rest mass of the particle *times* the speed of light squared.

TABLE 1:
Data from a single neutron decay

particle	mass × c^2 (MeV)	kinetic energy (MeV)
neutron	940.97	0.00
proton	939.67	0.01
electron	0.51	0.39

1. Assuming table 1 contains no major errors, what can we conclude about the mass × c^2 of the undetected third particle?

   A. It is 0.79 MeV.
   B. It is 0.39 MeV.
   C. It is less than or equal to 0.79 MeV; but we cannot be more precise.
   D. It is less than or equal to 0.39 MeV; but we cannot be more precise.

2. Given table 1, which properties of the undetected third particle can we calculate?

   A. Total energy, but not kinetic energy
   B. Kinetic energy, but not total energy
   C. Both total energy and kinetic energy
   D. Neither total energy nor kinetic energy

3. Consider an ensemble of particles, all of which have the same positive kinetic energy but different masses. For this ensemble, which graph best represents the relationship between the particle's mass and its total energy?

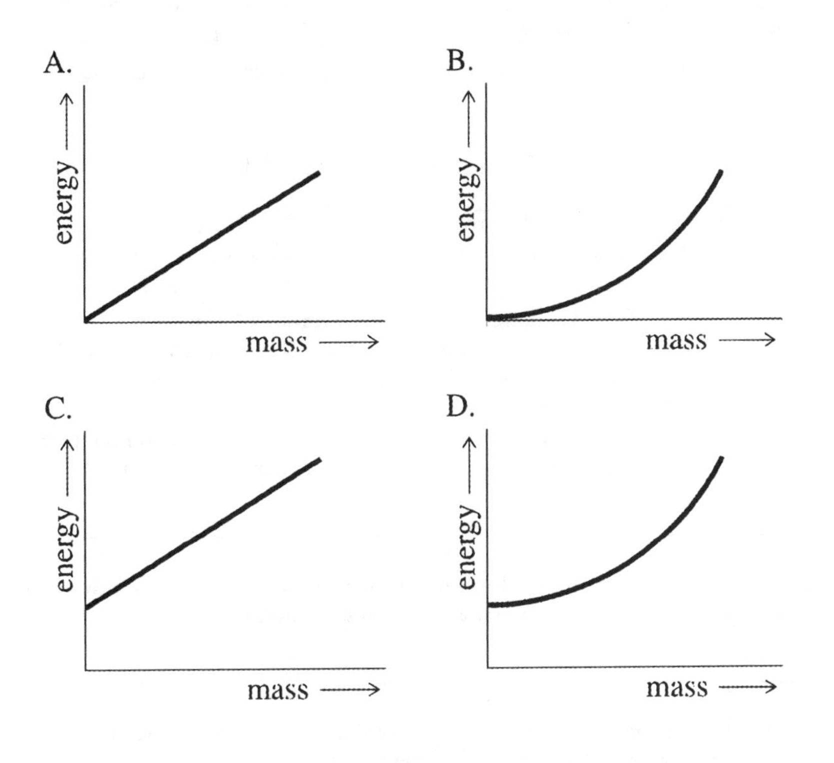

---

4. Could this reaction occur?

$$\text{proton} \rightarrow \text{neutron} + \text{other particles}$$

A. Yes, if the other particles have much more kinetic energy than mass energy.
B. Yes, but only if the proton has potential energy (due to interactions with other particles).
C. No, because a neutron is more massive than a proton.
D. No, because a proton is positively charged while a neutron is electrically neutral.

ANSWERS TO MCAT REVIEW PROBLEMS

---

1. **D.** According to the passage, subatomic reactions do *not* conserve mass. So, we cannot find the third particle's mass by setting m_{neutron} equal to $m_{\text{proton}} + m_{\text{electron}} + m_{\text{third particle}}$.

   By contrast, the *total* energy—in this case, the sum of "mass energy" and kinetic energy—*is* conserved. If E denotes total energy, then

$$E_{\text{neutron}} = E_{\text{proton}} + E_{\text{electron}} + E_{\text{third particle}}.$$

   The neutron has energy 940.97 MeV. The proton has energy 939.67 MeV + 0.01 MeV = 939.68 MeV. The electron has energy 0.51 MeV + 0.39 MeV = 0.90 MeV. Therefore, the third particle has energy

$$\begin{aligned} E_{\text{third particle}} &= E_{\text{neutron}} - E_{\text{proton}} - E_{\text{electron}} \\ &= 940.97 \text{ MeV} - 939.68 \text{ MeV} - 0.90 \text{ MeV} \\ &= 0.39 \text{ MeV}. \end{aligned}$$

   We just found the third particle's *total* energy, the sum of its mass energy and kinetic energy. Without more information, we cannot figure out how much of that total energy is mass energy.

---

2. **A.** As just shown, energy conservation allows us to calculate the third particle's total energy. But we don't know what percentage of that total is mass energy vs. kinetic energy.

---

3. **C.** The "c^2" in $E = mc^2$ can mislead you into thinking that E varies with the square of mass. But E varies linearly with m. Indeed, we can write Einstein's equation as

$$E = (\text{constant})m,$$

   where the constant happens to equal c^2. By contrast, graphs B and D represent an equation such as $E = (\text{constant})m^2$.

   Since the particles in this ensemble all have positive kinetic energy, the total energy is positive even when the mass is zero. (Relativistic physics permits this, because kinetic energy no longer equals $\frac{1}{2}mv^2$.) So, the graph does not begin at the origin.

---

4. **B.** A proton with *only* mass energy ($E_{\text{total}} = mc^2$) could never decay into a neutron and other particles, because a neutron has more mass energy than a proton does. Therefore, the reaction products would have more energy than the original proton, in violation of energy conservation.

   But suppose the proton has *additional* energy, such as potential energy from its interactions with other particles in a nucleus. Then, the total energy of the proton—mc^2 *plus* potential energy—can equal the total energy of the products. In this case, the products (neutron and other particles) have more mass and more kinetic energy than the proton did. Therefore, during the reaction, the increase in mc^2 and the increase in kinetic energy must be offset by an equally big decrease in the system's potential energy. This happens when a nucleus undergoes β^+ decay, which you will study in chapter 31.

CHAPTER 29
Particles and Waves

PREVIEW

This chapter begins the study of a branch of what is often called modern physics. Modern physics was developed mainly during this century and shows a new way of looking at nature which is both intriguing and often strange. One branch of modern physics called quantum mechanics views light as having both wave and particle properties and views matter as also having both particle and wave properties. Relativity is also part and parcel to modern physics and was discussed in an earlier chapter.

The main ideas of quantum mechanics grew out of the failure of classical physics to explain the results of two experiments: blackbody radiation and the photoelectric effect. Albert Einstein's explanation of the photoelectric effect led to the concept of light as being packets of energy called photons. Louis de Broglie, noticing that light appeared to behave both as a wave and a particle, suggested that electrons, protons, etc, which were normally thought of as particles may also behave like waves.

One of the foundation principles of quantum mechanics is the Heisenberg uncertainty principle which asserts that we cannot know some quantities such as the position and the momentum of an object accurately at the same time. In fact, the best we can do is to know the probability of the object having a certain momentum and a certain position.

QUICK REFERENCE

Important Terms

Wave-particle duality
The idea that waves can exhibit particle-like characteristics and particles can exhibit wave-like characteristics.

Blackbody radiation
The light emitted by a perfect emitter and absorber of light (a blackbody) due to its temperature.

Photoelectric effect
The phenomenon in which light incident on a surface such as metal ejects electrons from the surface.

Photon
A discrete packet of light energy postulated to exist by Albert Einstein to explain the photoelectric effect.

Work function
The minimum energy needed to eject an electron from a metal surface.

Compton effect
The phenomenon in which an x-ray photon is scattered from an electron with a change in frequency of the photon.

Compton wavelength of an electron
Half the maximum wavelength change of a photon in a Compton scattering with an electron.

de Broglie wavelength
The wavelength of the wave associated with a material particle. The de Broglie wavelength of a particle is inversely proportional to the momentum of the particle.

Wave function (Ψ)
A function which contains all of the information which can be known about a particle or system of particles. The square of the wavefunction $(\Psi)^2$ is interpreted as being related to the probability of finding the particle at a certain position.

Quantum mechanics
The theoretical framework for determining the wavefunction of a particle or system of particles.

Uncertainty principle
A principle of quantum mechanics which asserts that it is impossible to know with 100% certainty the position and momentum of a particle at the same time.

Equations

Blackbody Radiation

The possible **energies** of the **atomic vibrators** which produce **blackbody radiation.**

$$E = nhf \qquad n = 0, 1, 2, 3, \ldots \tag{29.1}$$

Photoelectric Effect

Energy of a photon

$$E = hf \tag{29.2}$$

The **photoelectric** equation

$$hf = KE_{max} + W_0 \tag{29.3}$$

Compton Effect

Conservation of energy for the **Compton effect**

$$hf = hf' + KE \tag{29.4}$$

The **momentum of a photon**

$$p = \frac{hf}{c} = \frac{h}{\lambda} \tag{29.6}$$

The **wavelength change** of the **scattered photon**

$$\lambda' - \lambda = \frac{h}{mc}(1 - \cos\theta) \tag{29.7}$$

The **deBroglie wavelength** of a particle

$$\lambda = \frac{h}{p} \tag{29.8}$$

The **Heisenberg uncertainty principle**

$$(\Delta p_y)(\Delta y) \geq \frac{h}{2\pi} \tag{29.10}$$

$$(\Delta E)(\Delta t) \geq \frac{h}{2\pi} \tag{29.11}$$

Planck's constant

$$h = 6.63 \text{ X } 10^{-34} \text{ J s}$$

$$h = 4.14 \text{ X } 10^{-15} \text{ (eV) s}$$

DISCUSSION OF SELECTED SECTIONS

29.3 Photons and the Photoelectric Effect

The explanation of the photoelectric effect by Albert Einstein led to the concept of the photon as a "particle" or quantum of light. The main results of the photoelectric effect experiment are:

1. Light incident on a clean metal surface can cause electrons to be ejected from the surface.
2. The time between the introduction of the light and the ejection of electrons is short.
3. Whether or not electrons were actually ejected from the metal depends on the frequency of the light. Above a certain frequency electrons are ejected, below this frequency no electrons are ejected.
4. If electrons are ejected, a higher frequency of light causes the electrons to have more kinetic energy (higher speeds).

Classical physics was useless in explaining the details of the photoelectric experiment because electromagnetic waves should:

1. give electrons in the metal energy a little at a time since the energy of a wave is spread out. Indeed it should take a long time (even days) for an electron to get enough energy to be ejected from the metal.
2. impart energy to the electron independent of the frequency of the wave. The energy carried by an electromagnetic wave depends on the square of its electric field.

Einstein assumed that light consisted of photons each of energy $E = hf$. This frequency dependence of the energy carried by light allowed Einstein to completely explain the photoelectric effect by an application of the conservation of energy (equation 29.3).

Example 1
What is the frequency of light which will cause electrons to be emitted from a magnesium surface whose work function is $W_0 = 3.68$ eV with a kinetic energy of 2.52 eV?
Equation (29.3) gives

$$f = \frac{KE_{max} + W_0}{h} = \frac{(1.60 \text{ X } 10^{-19} \text{ J/ 1 eV})(2.52 \text{ eV} + 3.68 \text{ eV})}{6.63 \text{ X } 10^{-34} \text{ J s}}$$

$$f = 1.50 \text{ X } 10^{15} \text{ Hz}$$

Example 2
A laser beam has an average intensity of 1.5 W/m^2 and a wavelength of 515 nm. What is the density of photons in the beam?

The energy density of the beam is

$$u = \frac{\overline{S}}{c} = \frac{1.5 \text{ W/m}^2}{3.00 \text{ X } 10^8 \text{ m/s}} = 5.0 \text{ X } 10^{-9} \text{ J/m}^3$$

A single photon has energy

$$E = hf = hc/\lambda = (6.63 \text{ X } 10^{-34} \text{ J s})(3.00 \text{ X } 10^8 \text{ m/s})/(515 \text{ X } 10^{-9} \text{ m}) = 3.86 \text{ X } 10^{-19} \text{ J}.$$

The density of photons in the beam is then

$$\text{density} = u/E = (5.0 \text{ X } 10^{-9} \text{ J/m}^3)/(3.86 \text{ X } 10^{-19} \text{ J}) = 1.3 \text{ X } 10^{10} \text{ photons/m}^3.$$

29.4 The Momentum of a Photon and the Compton Effect

Evidence that light has particle properties comes from the collision of a photon with an electron. In this collision it can be demonstrated that BOTH energy and momentum are conserved if the momentum of the photon can be considered to be $p = h/\lambda$. The change in the wavelength of the scattered photons is given by equation (29.7).

Example 3
What is the momentum of a 0.055 nm photon? How fast must an electron travel to have this momentum (Assume the speed is much less than the speed of light)?

The momentum of the photon is

$$p = h/\lambda = (6.63 \times 10^{-34}\ \text{J s})/(0.055 \times 10^{-9}\ \text{m}) = 1.21 \times 10^{-23}\ \text{kg m/s}.$$

An electron would have to travel with a speed of

$$v = p/m = (1.21 \times 10^{-23}\ \text{kg m/s})/(9.11 \times 10^{-31}\ \text{kg}) = 1.33 \times 10^{7}\ \text{m/s}$$

to have the same momentum.

Example 4
A 3.10 MeV photon collides with a stationary electron and is scattered backwards. How much energy does the photon impart to the electron?

The photon gives the electron the same amount of energy that it loses since the collision is elastic. The change in the photon energy is

$$\Delta E = hf' - hf = hc(1/\lambda' - 1/\lambda)$$

The initial wavelength of the electron is

$$\lambda = c/f = hc/E$$
$$= (4.14 \times 10^{-15}\ \text{eV s})(3.00 \times 10^{8}\ \text{m/s})/(3.10 \times 10^{6}\ \text{eV}) = 4.01 \times 10^{-13}\ \text{m}.$$

The final wavelength of the photon is given by (29.7)

$$\lambda' = \lambda + (h/mc)(1 - \cos\theta) = 4.01 \times 10^{-13}\ \text{m} + (2.43 \times 10^{-12}\ \text{m})(1 - \cos 180°)$$

$$= 4.01 \times 10^{-13}\ \text{m} + 4.86 \times 10^{-12}\ \text{m} = 5.26 \times 10^{-12}\ \text{m}.$$

Now

$$\Delta E = hc\left(\frac{1}{\lambda'} - \frac{1}{\lambda}\right) = (6.63 \times 10^{-34}\ \text{J s})(3.00 \times 10^{8}\ \text{m/s})\left(\frac{1}{5.26 \times 10^{-12}\ \text{m}} - \frac{1}{4.01 \times 10^{-13}\ \text{m}}\right)$$

$$\Delta E = -4.58 \times 10^{-13}\ \text{J}$$

The energy imparted to the electron is then 4.58×10^{-13} J.

29.5 The de Broglie Wavelength and the Wave Nature of Matter

Much of what you have already learned about waves applies to the wave nature of particles. The connection between the wave and particle nature of matter is in the de Broglie relationship (equation 29.8) which gives the wavelength of a particle. In principle, all matter has a wave nature, hence a de Broglie wavelength.

Example 5
A 75 kg person is running with a speed of 2.0 m/s. What is the de Broglie wavelength of the person? Comment on the potential importance of the wave nature of common objects. Equation (29.8) gives

$$\lambda = \frac{h}{mv} = \frac{6.63 \text{ X } 10^{-34} \text{ J s}}{(75 \text{ kg})(2.0 \text{ m/s})} = 4.4 \text{ X } 10^{-36} \text{ m}$$

This is an incredibly short wavelength compared to the size of objects in our everyday world. It's little wonder that we do not see the wave nature of people.

Example 6
An electron has been accelerated from rest through a potential difference of 2.5 X 10^3 V. What is the deBroglie wavelength of the electron? Comment on the potential importance of the wave nature of electrons.

The speed of the electron can be found by applying the conservation of mechanical energy

$$v = \sqrt{\frac{2eV}{m}} = \sqrt{\frac{2(1.6 \text{ X } 10^{-19} \text{ C})((2.5 \text{ X } 10^{3} \text{ V})}{9.11 \text{ X } 10^{-31} \text{ kg}}} = 3.0 \text{ X } 10^{7} \text{ m/s}$$

The de Broglie wavelength is then

$$\lambda = \frac{h}{mv} = \frac{6.63 \text{ X } 10^{-34} \text{ J s}}{(9.11 \text{ X } 10^{-31} \text{ kg})(3.0 \text{ X } 10^{7} \text{ m/s})} = 2.4 \text{ X } 10^{-11} \text{ m}$$

The wavelength of the electron is small but comparable to the size of atoms and molecules. The wave nature of electrons is important, and we should expect them to interfere, diffract and generally behave like the other kinds of waves that you have studied.

29.6 The Heisenberg Uncertainty Principle

Since particles can behave like waves, it becomes difficult to tell exactly where a particle is located. It is very much like generating a wave on a pond and asking "Where is it?" Since the wave is spread out on the surface of the pond, its position is ambiguous. How much a water wave is spread out depends on the wavelength of the wave.

The Heisenberg uncertainty principle (equation 29.10) addresses this ambiguity for matter waves. The uncertainty of the knowledge of a particle's position is related to the uncertainty in the simultaneous knowledge of the particle's momentum. The more accurately you know the position, the less accurately you can know the momentum, and vice versa. Uncertainty also applies to the knowledge of the particle's energy and the time available for measuring the energy (equation 29.11). If you try to measure the energy in a short time the measurement will be very uncertain. However, if your measurement of the energy is made over a long time interval it can be quite accurate.

Example 7
A measurement of the position of an electron shows it to be within a region which is 0.15 nm wide. How much uncertainty is there in the electron's momentum? Compare this uncertainty with the momentum of an electron traveling at 5.5×10^6 m/s.

$$\Delta p \geq \frac{\frac{h}{2\pi}}{\Delta x} = \frac{\frac{6.63 \times 10^{-34}\ \text{J s}}{2\pi}}{0.15 \times 10^{-9}\ \text{m}} = 7.0 \times 10^{-25}\ \text{kg m/s}$$

The momentum of an electron traveling at 5.5×10^6 m/s is

$$p = mv = (9.11 \times 10^{-31}\ \text{kg})(5.5 \times 10^6\ \text{m/s}) = 5.0 \times 10^{-24}\ \text{kg m/s}.$$

$$\Delta p/p = (7.0 \times 10^{-25}\ \text{kg m/s})/(5.0 \times 10^{-24}\ \text{kg m/s}) = 0.14.$$

The uncertainty is 14 % of the momentum.

Example 8
An atom loses energy by radiating away a photon. The process takes 1.5×10^{-8} s. What is the uncertainty in the energy of the photon? In its frequency?

Equation (29.11) gives the smallest uncertainty in the energy when the time interval is $\Delta t = 1.5 \times 10^{-8}$ s.

$$\Delta E = \frac{\frac{h}{2\pi}}{\Delta t} = \frac{\frac{6.63 \times 10^{-34}\ \text{J s}}{2\pi}}{1.5 \times 10^{-8}\ \text{s}} = 7.0 \times 10^{-27}\ \text{J}$$

Now the frequency is $f = E/h$. If the energy of the photon is uncertain by at least ΔE, then the frequency of the photon is uncertain by at least

$$\Delta f = \Delta E/h = (7.0 \times 10^{-27}\ \text{J})/(6.63 \times 10^{-34}\ \text{J s}) = 1.1 \times 10^7\ \text{Hz}.$$

PRACTICE PROBLEMS

1. At night approximately 530 photons per second must enter an unaided human eye for an object to be seen, assuming the light is green. The light bulb in Example 1 in the text emits green light uniformly in all directions, and the diameter of the pupil of the eye is 7.0 mm. What is the minimum distance from which the bulb could be seen?

2. A photon scattered from an electron suffers a decrease in its wavelength of 25 % of the Compton wavelength of the electron. Through what angle was the electron scattered?

3. The repeating crystal structure of a solid can act like a diffraction grating for electrons. The atoms in the crystal correspond to the lines on a grating. Electrons traveling at a speed of 4.6×10^7 m/s are diffracted by a crystal whose atomic spacing is 150 nm. About how many orders of electron diffraction are produced?

4. Electrons traveling at a speed of 2.4×10^5 m/s are diffracted through a single slit and fall on a screen located 25 cm away. How wide must the slit be to produce a central fringe 0.50 mm wide?

5. Can any object ever be considered to be truly at rest? Explain why or why not on the basis of the Heisenberg uncertainty principle.

6. A vacuum in classical physics is considered to be empty space devoid of energy and matter. Is there such a thing as empty space devoid of energy or matter according to the uncertainty principle? HINT: Consider what could happen in a short time period, say 10^{-20} s.

7. In "Mr. Thompkins in Wonderland", by George Gamow, Mr. Thompkins, while listening to a stimulating lecture on quantum mechanics, doses off. He dreams that he is in a land where Planck's constant is a large value, say $h = 100.0$ J s. In this land a herd of giraffes move through a row of evenly spaced trees with a speed of 0.12 m/s. If each giraffe has a mass of 360 kg, what must be the spacing of the trees for the giraffes to produce 3 complete orders of diffraction?

HELPFUL SUGGESTIONS

1. The wave particle duality means that matter and light can have both wave and particle properties. Matter usually shows us the particle side of its nature when the objects in question are large and slow (baseballs, cars, etc.). In contrast light shows its particle side when it interacts with small objects like electrons. Matter manifests its wave characteristics when the objects are small and fast (electrons, protons, etc.), while light behaves as waves when it interacts with objects such as slits.

2. Often you are given the wavelength of light and need to calculate the energy of a photon of the light in electron volts. The procedure is $E = hf = h(c/\lambda) = hc/\lambda$.
   The constant $hc = (6.63 \times 10^{-34}\text{ J s})(3.00 \times 10^{8}\text{ m/s}) = 1.99 \times 10^{-25}\text{ J m}$. Now $1\text{ J} = 1/(1.60 \times 10^{-19})\text{ eV}$, so
   $$E = (1.24 \times 10^{-6}\text{ eV m})/\lambda.$$
   This is a handy expression to keep in mind so that you don't have to do the calculation repeatedly.

3. Remember that the uncertainty principle places limits on what we can know. This is fundamental in nature and does not depend on how well we design our experiments. Even a "perfect" experiment will yield an uncertainty in the position and momentum of an object.

EVERYDAY PHYSICS

1. The concepts presented in this chapter do not lend themselves to direct observation in everyday life. Many devices, principally electronic devices, have been designed using quantum mechanical principles and work as planned. This is indirect evidence of the validity of quantum mechanics. Two of these devices are:

a. *Tunnel diodes*
A tunnel diode is an electronic device which is useful in various electronic circuits. The tunnel diode operates only because of the wave/probability effect called "tunneling" of electrons. Tunneling can be understood by thinking about throwing a ball against a building. It is quite obvious that if you do not give the ball enough energy to rise at least to the top of the building, it will not appear on the other side. Instead, it will hit the wall and bounce off. Imagine that the ball is a wave and has a wavelength about the size of the building. The wave may extend to the top of the building even if the main part of it does not go that high. Since the wave represents a probability of finding the ball at a particular location, there is a small probability that the ball will be at the top of the building. Hence it may make it over even though the ball was not given the required amount of energy.
The tunnel diode conducts a current through an electrical potential barrier even though the electrons do not have enough energy to make it over the barrier. This is due solely to the wave nature of the electron.

b. *Photocells*
Photocells convert light energy into electrical energy. They work on the same principles as the photoelectric emission studied in this chapter, except that they use semiconductor material instead of metals. Photons are absorbed by electrons which gain enough energy to move to another region of the semiconductor causing a charge separation, hence an emf. The emf can be used to produce currents in an external circuit as long as light shines on the photocell to replenish the energy lost by the currents. Photocells would not work at all if light did not have its energy concentrated as photons.

2. The wave function which represents an object, such as yourself, usually has a value everywhere in space although its value may be quite small. Hence, Ψ^2 has a value everywhere in space. Since Ψ^2 represents a probability of locating you, there is a some probability of you being anywhere in space. When I try to find you, there is some probability, albeit small, that you are in another solar system. Can you relate this to the uncertainty principle? Could this be used to explain some missing persons reports?

CHAPTER QUIZ

1. A photon has an energy of 5.42 eV. What is the frequency of the photon?
   a. 1.31×10^{15} Hz b. 6.63×10^{14} Hz c. 5.42×10^{10} Hz d. 8.17 X 1033 Hz

2. Light falling on a metal whose work function is 4.08 eV ejects electrons from the surface with a kinetic energy of 1.29 eV. What is the frequency of the light?
   a. 4.45×10^{19} Hz b. 6.73×10^{14} Hz c. 1.30×10^{15} Hz d. 8.95×10^{-19} H

3. The Compton wavelength of an electron is 2.43×10^{-12} m. A 0.056 nm photon is scattered from an electron at an angle of 60.0°. What is the wavelength of the scattered photon?
   a. 2.43×10^{-12} m b. 1.21×10^{-12} m c. 5.72×10^{-11} m d. 5.47×10^{-11} m

4. What is the de Broglie wavelength of a 0.50 g bullet traveling with a speed of 320 m/s?
   a. 4.14×10^{-36} m b. 2.4×10^{33} m c. 4.1×10^{-33} m d. 0 m

5. An electron traveling with a speed of 6.5×10^{6} m/s passes through a slit of width, W. Which of the following widths would you expect to produce the most diffraction?
   a. 1.0 cm b. 1.0 mm c. 1.0 μm d. 1.0 nm

6. The Heisenberg uncertainty principle implies that the momentum and position cannot be accurately determined at the same time. What is the truth of this statement?
   a. It is true as stated.
   b. It is true only if your experiments are not perfect.
   c. It is true if particles are NOT waves.
   d. It is definitely not true.

7. A particle of mass 0.35 μg is trapped in a box of length 1.0 μm. What is the minimum uncertainty in the velocity of the particle along the length of the box ?
   a. zero b. 3.0×10^{-19} m/s c. 1.5 m/s d. 3.2×10^{-7} m/s

8. How much mass could possibly "appear" and" disappear" in a vacuum during the time period 1.0×10^{-15} s.
   a. 3.5×10^{-28} kg b. 1.1×10^{-19} kg c. 1.2×10^{-36} kg d. none

9. A wavefunction, Ψ, describes a particle. What describes the particle's probability?
   a. Ψ b. Ψ^2 c. Ψ^3 d. Δp

10. The *wave-particle duality* refers to the fact that ________ can exhibit both wave and particle characteristics.
    a. light b. matter c. both light and matter d. nothing

SOLUTIONS AND ANSWERS

Practice Problems

---

1. According to Example 1, 3.6×10^8 photons/s are emitted by the bulb. These photons spread out over the surface of a sphere as they leave the bulb. After traveling a distance, R, to reach the eye, the light has an intensity

$$I = (3.6 \times 10^8 \text{ photons/s})/(4\pi R^2).$$

In order to the eye to detect the light the intensity entering the eye must be

$$I = (530 \text{ photons/s})/\pi r^2$$

The maximum distance that the light can travel and still be seen is

$$R = r\sqrt{(3.6 \times 10^8 \text{photons / s}) / (2120 \text{photons / s})} = \mathbf{1.4 \times 10^5 \ m}$$

---

2. We must have $\lambda' - \lambda = (h/mc)(0.250)$. Comparing this with equation (29.7) reveals that

$$1 - \cos\theta = 0.250 \Rightarrow \theta = \mathbf{41.4\ °}.$$

---

3. The order number of the last order produced is

$$m = (d/\lambda) \sin 90° = d/\lambda.$$

The wavelength of the electron is

$$\lambda = h/p = (6.63 \times 10^{-34} \text{ J s})/[(9.11 \times 10^{-31} \text{ kg})(4.6 \times 10^7 \text{ m/s})] = 1.6 \times 10^{-11} \text{ m}.$$

The number of orders produced is then

$$m = \mathbf{9400}.$$

---

4. Using the small angle approximation and the same notation as used in the text, we find that the angle of diffraction is $\theta = \lambda/W$ and $\theta = y/L$ so

$$W = \lambda(L/y).$$

The de Broglie wavelength of the electrons is

$$\lambda = h/p = (6.63 \times 10^{-34} \text{ J s})/[(9.11 \times 10^{-31} \text{ kg})(2.4 \times 10^5 \text{ m/s})] = 3.0 \times 10^{-9} \text{ m}.$$

The slit width must be

$$W = (3.0 \times 10^{-9} \text{ m})[(2.5 \times 10^{-2} \text{ m})/(0.25 \times 10^{-3} \text{ m})] = \mathbf{3.0 \times 10^{-6}\ m}.$$

---

5. **No**. The Heisenberg uncertainty principle requires that the minimum uncertainty in the position of an object be

$$\Delta x = (h/2\pi)/\Delta p.$$

If an object is truly at rest, then Δp is zero and the object's position is completely uncertain. If an object is truly at rest though, it should be quite easy to find with great certainty. The contradiction leads us to the conclusion that the object cannot be at rest.

---

6. **No**. A measurement of the energy in the "vacuum" in a short time period will yield at least

$$\Delta E = (h/2\pi)/\Delta t.$$

For $\Delta t = 10^{-20}$ s the energy will be more than $\Delta E = 1.1 \times 10^{-14}$ J = 0.069 MeV.

---

7. The highest order of diffraction produced is m = 3. The separation needed between the trees is

$$d = (3\lambda/m) \sin 90° = 3\lambda.$$

The wavelength is

$$\lambda = h/p = (100.0 \text{ J s})/[(360 \text{ kg})(0.12 \text{ m/s})] = 2.3 \text{ m}.$$

Hence,

$$d = 3(2.3 \text{ m}) = \mathbf{6.9 \ m}.$$

Quiz answers

1. a	3. c	5. d	7. b	9. b
2. c	4. c	6. a	8. c	10. c

MCAT REVIEW PROBLEMS

Although in many respects light behaves like a wave, in certain situations the "particle" nature of light becomes more apparent. According to quantum theory, a "particle" of light—called a photon—has energy

$$E = hf,$$

where f is the frequency of the waves corresponding to that color of light, and $h = 6.6\times 10^{-34}$ J·s is a fundamental constant of nature called Planck's constant. Biophysics experiments reveal that the human eye can often "see" a single photon of visible light. Visible wavelengths range from 375 nanometers for violet light to 750 nanometers for red light, where a nanometer is 10^{-9} meters. Light travels at 3.0×10^{8} m/s.

---

1. Consider two 100-watt lasers, the first of which emits red light, the second of which emits violet light. Which laser, if either, emits more photons per second?

   A. The red-light laser
   B. The violet-light laser
   C. Both lasers emit the same number of photons per second.
   D. We cannot determine the answer without more information.

---

2. Which graph best shows the relationship between the wavelength of light and the energy of the photons comprising that light?

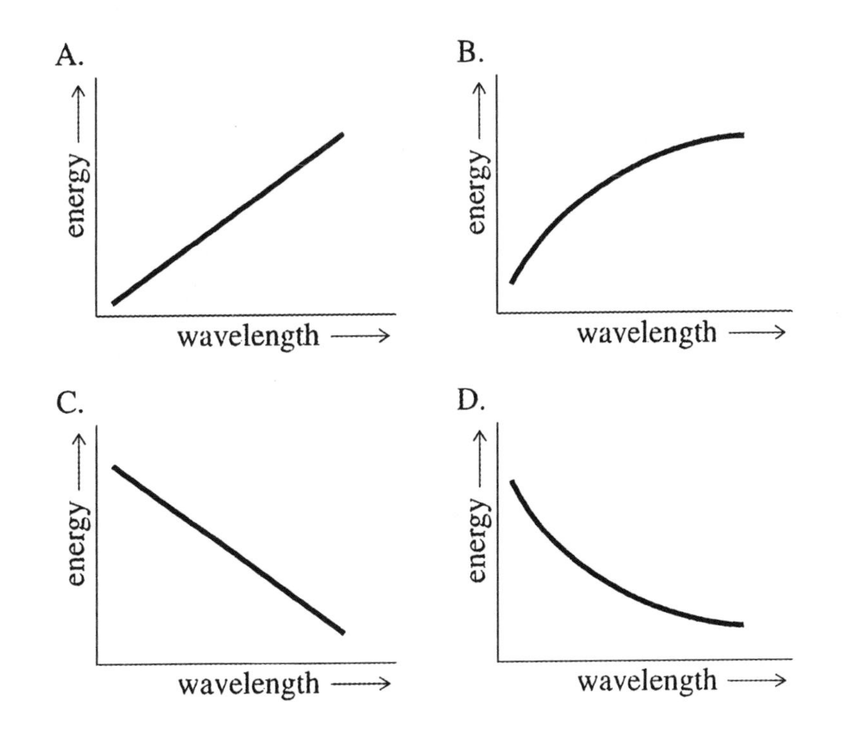

---

3. If someone "sees" a single photon of red light, about how much energy does her eye absorb?

   A. 10^{-19} joules **B.** 10^{-28} joules **C.** 10^{-31} joules **D.** 10^{-40} joules

---

4. Which of the following phenomena is explained much better by a wave theory of light than by a particle theory of light?

   A. Reflection
   B. Light travels in straight lines
   C. Interference
   D. Intensity drops with the square of distance.

---

5. As compared to the photons comprising red light, the photons comprising X-rays have higher:

   A. Wavelength
   B. Speed (in a vacuum)
   C. Mass
   D. Energy

ANSWERS TO MCAT REVIEW PROBLEMS

---

1. **A.** Since both lasers have the same power, they both emit the same energy per second. But each violet photon carries more energy than a red photon, as explained below. Therefore, the red-light laser must emit more photons per second, in order to match the energy emitted by the violet-light laser.

   To see why a red photon carries less energy than a violet photon, recall the relationship between speed, wavelength, and frequency:

$$v = \lambda f.$$

   This equation applies to all waves, not just electromagnetic waves. Solve for f to get $f = \frac{v}{\lambda}$. So, since all wavelengths of light travel equally fast in a vacuum, wavelength and frequency are inversely proportional. Higher wavelengths correspond to lower frequencies.

   Because red light waves are twice as long as violet light waves, red light has half the frequency of violet light. Since $E = hf$, each red photon carries half the energy of a violet photon. Therefore, as compared to the violet-light laser, the red-light laser must emit twice as many photons per second, in order to produce the same energy.

---

2. **D.** As just discussed, $f = \frac{v}{\lambda}$. Therefore, a photon carries energy

$$E = hf = h\frac{v}{\lambda}.$$

   The photon energy is *inversely proportional* to the wavelength, λ. For instance, doubling λ cuts the energy in half. Tripling λ cuts the energy in third. Only graph D captures this relationship. Graph C expresses a linear decrease, not an inverse proportionality. For an intuitive discussion of the difference, see MCAT review chapter 11, question 4.

---

3. **A.** Since red light has a frequency of

$$f = \frac{v}{\lambda} = \frac{3.0 \times 10^8 \text{ m/s}}{750 \times 10^{-9} \text{ m}} = 4.0 \times 10^{14} \text{ Hz},$$

   a red photon carries an energy of

$$\begin{aligned} E = hf &= (6.6 \times 10^{-34} \text{ J·s})(4.0 \times 10^{14} \text{ Hz}) \\ &= 2.6 \times 10^{-19} \text{ J}. \end{aligned}$$

   Even without a calculator, you can answer this question quickly. Since all four choices differ by several orders of magnitude (e.g., 10^{-28} joules is a billion times smaller than 10^{-19} joules), precise calculations are unnecessary. For instance, by using $v = 10^8$ m/s, $\lambda = 10^{-6}$ m, and $h = 10^{-34}$ J·s, you just need to add and subtract exponents. These numbers yield $E = 10^{-20}$ J, which unambiguously picks out option A.

---

4. **C.** Particles travel in straight line and "reflect" (bounce) off surfaces. Furthermore, by assuming that a light source spews out photons equally in all directions, we can explain why intensity decreases with the square of distance. For instance, your textbook's derivation of the intensity formula in section 16.7 "works" equally well for waves or particles. But classical particles cannot "interfere" with each other. For example, when two marbles collide, they cannot "cancel" each other out (destructive interference). Only waves can do that.

---

5. **D.** For the MCAT, you should know that X-rays are electromagnetic waves of much lower wavelength, and therefore much higher frequency, than visible light. Since $E = hf$, higher frequency implies higher energy.

Like all electromagnetic radiation, X-rays have no mass and travel at speed $c = 3.0 \times 10^8$ m/s through a vacuum.

CHAPTER 30
The Nature of the Atom

PREVIEW

In this chapter you will study the nature of the atom. You will learn about line spectra, the Bohr model of the hydrogen atom, energy level diagrams, the quantum mechanical picture of the hydrogen atom, the Pauli exclusion principle, X-rays, and lasers.

QUICK REFERENCE

Important Terms

Line spectrum
A series of discrete electromagnetic wavelengths emitted by the atoms of a low-pressure gas that is subjected to a sufficiently high potential difference. Certain groups of discrete wavelengths are referred to as a series.

Bohr model
The electron in a single electron atom exists in circular orbits called stationary orbits. A photon is emitted when an electron changes from a higher energy orbit to a lower energy orbit.

Ionization energy
The energy needed to remove an electron completely from an atom.

Principal quantum number
An integer number, n, which determines the total energy of the hydrogen atom.

Orbital quantum number
An integer number, l , which determines the angular momentum of the electron due to its orbital motion.

Magnetic quantum number
An integer number, m_l , which determines the component of the angular momentum along a specific direction.

Spin quantum number
A number, m_s, which is used to describe the intrinsic spin angular momentum of an electron.

Pauli exclusion principle
No two electrons in an atom can have the same set values for the four quantum numbers n, l , m_l , m_s. This determines the way in which the electrons in multiple-electron atoms are distributed into shells and subshells, thus explaining the periodic table.

X-rays
Electromagnetic waves emitted when high-energy electrons strike a metal target contained within an evacuated glass tube.

Laser
A device that generates electromagnetic waves via a process known as "stimulated emission". The process produces waves that are coherent and may be confined to a very narrow beam.

Equations

The wavelengths of the **spectral line series of hydrogen** are given by:

$$\text{Lyman series:} \quad \frac{1}{\lambda} = R\left(\frac{1}{1^2} - \frac{1}{n^2}\right) \qquad n = 2, 3, 4, \ldots \tag{30.1}$$

$$\text{Balmer series:} \quad \frac{1}{\lambda} = R\left(\frac{1}{2^2} - \frac{1}{n^2}\right) \qquad n = 3, 4, 5, \ldots \tag{30.2}$$

$$\text{Paschen series:} \quad \frac{1}{\lambda} = R\left(\frac{1}{3^2} - \frac{1}{n^2}\right) \qquad n = 4, 5, 6, \ldots \tag{30.3}$$

The **energy of the photon** emitted when an electron changes from a higher to a lower energy level is:

$$E_i - E_f = hf \tag{30.4}$$

The **total energy of the atom** (Bohr theory) is given by:

$$E = KE + EPE = \frac{1}{2}mv^2 - \frac{kZe^2}{r} \tag{30.5}$$

Equating the centripetal and Coulomb forces yields:

$$mv^2 = \frac{kZe^2}{r} \tag{30.6}$$

The total energy can therefore be written:

$$E = \frac{1}{2}\left(\frac{kZe^2}{r}\right) - \frac{kZe^2}{r} = -\frac{kZe^2}{2r} \tag{30.7}$$

According to Bohr, the **angular momentum L_n** is given by:

$$L_n = mv_n r_n = n\frac{h}{2\pi} \qquad n = 1, 2, 3, \ldots \tag{30.8}$$

The **radius of the nth Bohr orbit** is:

$$r_n = \left(\frac{h^2}{4\pi^2 mke^2}\right)\frac{n^2}{Z} \qquad n = 1, 2, 3, \ldots \tag{30.9}$$

Substituting the constants into equation (37.9) yields:

$$r_n = (5.29 \times 10^{-11}\ \text{m})\frac{n^2}{Z} \qquad n = 1, 2, 3, \ldots \tag{30.10}$$

The **energy of the nth Bohr orbit** is:

$$E_n = -\left(\frac{2\pi^2 mk^2 e^4}{h^2}\right)\frac{Z^2}{n^2} \qquad n = 1, 2, 3, \ldots \qquad (30.11)$$

The **Bohr energy levels** can then be written as:

$$E_n = -(2.18 \times 10^{-18}\,J)\frac{Z^2}{n^2} \qquad n = 1, 2, 3, \ldots \qquad (30.12)$$

The **Bohr energy levels** expressed **in electron volts**:

$$E_n = -(13.6\,eV)\frac{Z^2}{n^2} \qquad n = 1, 2, 3, \ldots \qquad (30.13)$$

The **wavelengths radiated by the hydrogen** are given by:

$$\frac{1}{\lambda} = \frac{2\pi^2 mk^2 e^4}{h^3 c}(Z^2)\left(\frac{1}{n_f^2} - \frac{1}{n_i^2}\right) \qquad (30.14)$$

$$n_i, n_f = 1, 2, 3, \ldots \text{ and } n_i > n_f$$

The **magnitude of the angular momentum of an electron** is:

$$L = \sqrt{l\,(l+1)}\,\frac{h}{2\pi} \qquad (30.15)$$

The **angular momentum in the z direction** is:

$$L_z = m_l \frac{h}{2\pi} \qquad (30.16)$$

The **minimum wavelength of the photon emitted by an X-ray tube** is:

$$\lambda_0 = \frac{hc}{eV} \qquad (30.17)$$

DISCUSSION OF SELECTED SECTIONS

30.2 Line Spectra

When a low-pressure gas contained within an evacuated tube is subjected to a sufficiently large potential difference, the gas will emit electromagnetic waves. With a grating spectrometer the individual wavelengths emitted by the gas can be separated and identified as a series of bright fringes or lines. The series of lines is called a **line spectrum**. The simplest line spectrum is that of the hydrogen atom. In schematic form, the following figure illustrates one series of lines for atomic hydrogen, the **Balmer series**.

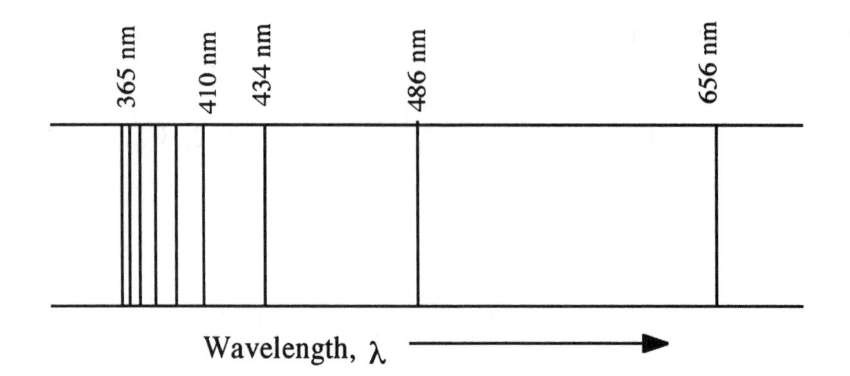

Balmer found an empirical equation that gave the values for the observed wavelengths. This equation is given below, along with similar equations that apply to the **Lyman** and **Paschen** series of hydrogen lines.

Lyman series: $$\frac{1}{\lambda} = R\left(\frac{1}{1^2} - \frac{1}{n^2}\right) \qquad n = 2, 3, 4, \ldots$$

Balmer series: $$\frac{1}{\lambda} = R\left(\frac{1}{2^2} - \frac{1}{n^2}\right) \qquad n = 3, 4, 5, \ldots$$

Paschen series: $$\frac{1}{\lambda} = R\left(\frac{1}{3^2} - \frac{1}{n^2}\right) \qquad n = 4, 5, 6, \ldots$$

In these equations, the constant term R, known as the **Rydberg constant**, has the value $R = 1.097 \times 10^7 \text{ m}^{-1}$.

Example 1
Find the wavelength of the n = 4 Balmer line. Using equation (30.2) we obtain,

$$\frac{1}{\lambda} = R\left(\frac{1}{2^2} - \frac{1}{n^2}\right) = (1.097 \times 10^7 \text{ m}^{-1})\left(\frac{1}{2^2} - \frac{1}{4^2}\right) = 2.057 \times 10^6 \text{ m}^{-1} \Rightarrow \lambda = 486 \text{ nm}$$

30.3 The Bohr Model of the Hydrogen Atom

Bohr's theory of the hydrogen atom has a positively charged nucleus and a negatively charged electron moving around it in a circular orbit. Bohr hypothesized that in a hydrogen atom there can only be certain values of the total energy (electron kinetic energy plus potential energy). These allowed energy levels correspond to different orbits for the electron as it moves around the nucleus, the larger orbits being associated with larger total energies. An electron in one of these orbits does not radiate electromagnetic waves. These orbits are sometimes referred to as **stationary orbits** or **stationary states**.

Bohr theorized that a photon is emitted only when the electron changes orbits from a larger one with higher energy to a smaller one with a lower energy. If E_i is the energy of the initial orbit (higher energy orbit), E_f the energy of the final orbit (lower energy), and the energy of the photon is hf, where f is the frequency and h is Planck's constant, then

$$E_i - E_f = hf$$

For an electron of mass m and speed v in an orbit of radius r, the total energy is the kinetic energy of the electron plus the electric potential energy. Therefore, we can write

$$E = KE + PE = \frac{1}{2}mv^2 - \frac{kZe^2}{r}$$

Equation (30.5) can be modified to yield the radii of the various orbits r_n and the energies of these orbits E_n. As detailed in the textbook, by using the Coulomb electrostatic force as the centripetal force (equation 30.6), and by assuming that the allowed energy levels have angular momenta that are integral multiples of Planck's constant (equation 30.8), the radii for the **Bohr orbits** are shown to be

$$r_n = (5.29 \times 10^{-11}\ m)\frac{n^2}{Z} \qquad n = 1, 2, 3, \ldots$$

For the hydrogen atom (Z = 1) and the smallest Bohr orbit (n = 1), $r_1 = 5.29 \times 10^{-11}$ m. This particular value is often referred to as the **Bohr radius**. Equation (30.10) can now be used in equation (30.7) to show that the energy of the nth Bohr orbit is given by

$$E_n = -(2.18 \times 10^{-18}\ J)\frac{Z^2}{n^2} \qquad n = 1, 2, 3, \ldots$$

This energy is often expressed in electron volts, where 1 eV = 1.60×10^{-19} J. Therefore, we can write

$$E_n = -(13.6\ eV)\frac{Z^2}{n^2} \qquad n = 1, 2, 3, \ldots$$

It is useful to represent the energy values given by equation (30.13) on an **energy level diagram**. For the hydrogen atom (Z = 1), the highest energy level corresponds to n = ∞ in equation (30.13) and has an energy of 0 eV. This is the energy of the atom when the electron is completely removed (r = ∞) from the nucleus and is at rest. The lowest energy level corresponds to n = 1 and has a value of - 13.6 eV. The lowest energy level is called the **ground state** while higher energy levels are called **excited levels**. To determine the wavelengths radiated by the atom, equation (30.13) for the energies can be used in equation (30.4) with f = c/λ to yield equation (30.14). The Lyman, Balmer, and Paschen series of lines in the hydrogen atom spectrum correspond to transitions that the electron makes between higher and lower energy levels, as shown in the diagram below.

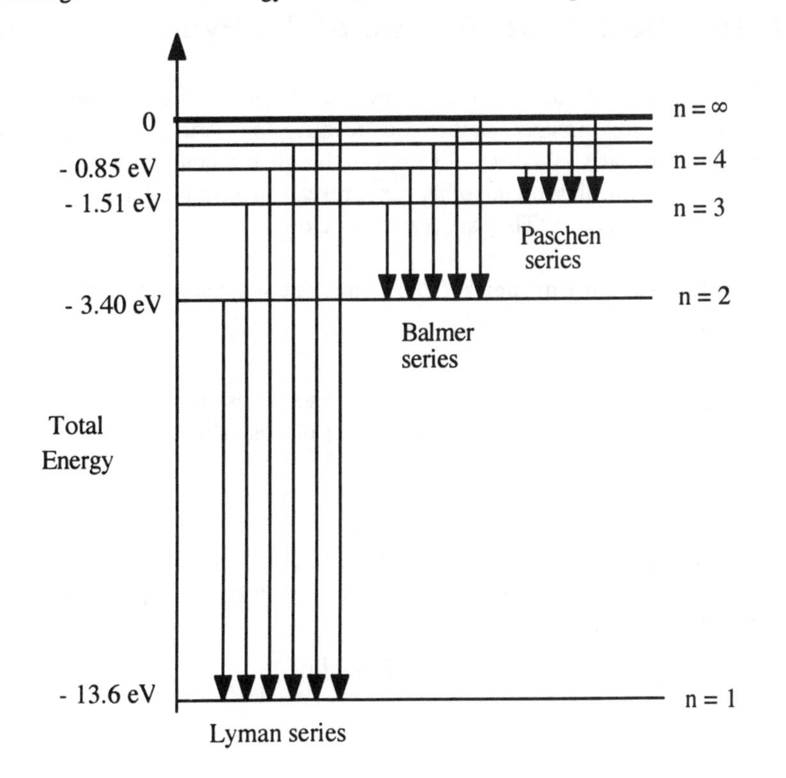

The wavelengths of the line spectra are given by:

$$\frac{1}{\lambda} = \frac{2\pi^2 mk^2 e^4}{h^3 c} (Z^2) \left(\frac{1}{n_f^2} - \frac{1}{n_i^2} \right)$$

$$n_i, n_f = 1, 2, 3, \ldots \text{ and } n_i > n_f$$

Example 2
Lines in the Paschen series are produced when electrons, excited to higher levels, make transitions to the n = 3 level. Determine (a) the longest wavelength in the series, (b) the wavelength that corresponds to the transition from $n_i = 7$ to $n_f = 3$. (c) What spectral region are these lines found in?

(a) The longest wavelength corresponds to the transition which has the smallest energy change. This occurs between levels $n_i = 4$ to $n_f = 3$. Using equation (30.14) with Z = 1, we find that

$$\frac{1}{\lambda} = (1.097 \times 10^7 \text{ m}^{-1}) \left(\frac{1}{3^2} - \frac{1}{4^2} \right) = 5.33 \times 10^5 \text{ m}^{-1} \Rightarrow \lambda = 1880 \text{ nm}$$

(b) For the transition from $n_i = 7$ to $n_f = 3$:

$$\frac{1}{\lambda} = (1.097 \times 10^7 \text{ m}^{-1}) \left(\frac{1}{3^2} - \frac{1}{7^2} \right) = 9.95 \times 10^5 \text{ m}^{-1} \Rightarrow \lambda = 1010 \text{ nm}$$

(c) According to Figure 24.9, these lines lie in the **infrared** region of the electromagnetic spectrum.

30.5 The Quantum Mechanical Picture of the Hydrogen Atom

The picture of the hydrogen atom that quantum mechanics provides differs in a number of ways from the Bohr model. The Bohr model uses a single integer number n to identify the various electron orbits and the associated energies. Because this number can have only discrete values, rather than a continuous range of values, n is called a **quantum number**. Quantum mechanics, on the other hand, reveals that four different quantum numbers are needed to describe each state of the hydrogen atom. They are described below.

1. **The principal quantum number n.** This is the same number used by Bohr to describe the total energy of the atom and can have only integer values: n = 1, 2, 3, . . . $E_n = -(13.6 \text{ eV}) Z^2/n^2$.

2. **The orbital quantum number l.** This number determines the angular momentum of the electron due to its orbital motion. Only the following integer values of l are allowed:

$$l = 0, 1, 2, \ldots, (n - 1)$$

For instance, if n = 1, the orbital quantum number can only have the value $l = 0$. For n = 5, possible values for l are $l = 0, 1, 2, 3,$ and 4. According to the Schrodinger equation, the magnitude L of the angular m omentum of the electron is

$$L = \sqrt{l\,(l + 1)}\, \frac{h}{2\pi}$$

3. **The magnetic quantum number m_l.** This number determines the component of the angular momentum along a specific direction, which is called the z direction by convention. The values that m_l can have depend on the value of l, with only the following positive and negative integers being permitted:

$$m_l = 0, \pm 1, \pm 2, \ldots, \pm l$$

For example, if the angular momentum quantum number is $l = 2$, then the magnetic quantum number can have values $m_l = -2, -1, 0, +1, +2$. The Schrodinger equations shows that the component L_z of the angular momentum in the z direction is

$$L_z = m_l \frac{h}{2\pi}$$

4. **The spin quantum number m_s.** This number is needed because the electron has an intrinsic property called spin angular momentum. There are only two possible values for the spin quantum number. They are:

$$m_s = +\frac{1}{2} \quad \text{(spin up)} \qquad \text{or} \qquad m_s = -\frac{1}{2} \quad \text{(spin down)}$$

Example 3
Write down the eighteen possible sets of four quantum numbers that exist when the principal quantum number is 3.

Using the rules listed above, for n = 3 we have:

n	l	m_l	m_s
3	0	0	1/2
3	0	0	-1/2
3	1	1	1/2
3	1	1	-1/2
3	1	0	1/2
3	1	0	-1/2
3	1	-1	1/2
3	1	-1	-1/2
3	2	2	1/2
3	2	2	-1/2
3	2	1	1/2
3	2	1	-1/2
3	2	0	1/2
3	2	0	-1/2
3	2	-1	1/2
3	2	-1	-1/2
3	2	-2	1/2
3	2	-2	-1/2

30.6 The Pauli Exclusion Principle and the Periodic Table of the Elements

In multiple-electron atoms, all electrons with the same value of n are said to be in the same **shell**. For example, electrons with n = 1 are in a single shell (sometimes called the K shell), those with n = 2 are in another shell (the L shell), those with n = 3 are in another (the M shell), and so on. Those electrons with the same values for both n and l are said to be in the same **subshell**. So the n = 1 shell has one subshell, $l = 0$. The n = 2 shell has two subshells, $l = 0$ and $l = 1$, and so forth.

In most atoms, the electron spends most of its time in the lowest energy level n = 1 called the **ground state**. However, when a multiple-electron atom is in the ground state, not every electron is crowded into the n = 1 shell. The reason for this is that electrons obey the **Pauli exclusion principle**. This principle states that no two electrons in an atom can have the same set of values for the four quantum numbers n, l, m_l, and m_s.

Because of the Pauli exclusion principle, there is a maximum number of electrons that can fit into an energy level or subshell. For example, the n = 1, l = 0 subshell can hold at most two electrons. The n = 2, l = 1 subshell, however, can hold six electrons, because with l = 1 there are three possibilities for m_l (-1, 0, 1), each of which have two values of m_s (+1/2, -1/2). The maximum number of electrons the l the subshell can hold is $2(2l + 1)$. So, for instance, the n = 5, l = 4 subshell can hold 2[2(4) + 1] = 18 electrons.

Example 4

How many electrons can be put into (a) the n = 4 shell and (b) the n = 5 shell?

(a) Since there are $2(2l + 1)$ possible electrons in each subshell, and since l can have values up to (n - 1), for n = 4, we have l = 0, 1, 2, and 3. So for l = 0 there are two electrons, for l = 1 there are six electrons, for l = 2 there are ten electrons and for l = 3 there are fourteen electrons. For the n = 4 shell, therefore, there are a maximum of N = 2 + 6 + 10 + 14 = **32 electrons.**

(b) For the n = 5 shell, the l = 4 subshell can have 18 possible electrons. Therefore, in the entire n = 5 shell there can be as many as N = 2 + 6 + 10 + 14 + 18 = **50 electrons.**

A shorthand notation is often used for the electronic configuration of the atom. For instance, the l = 0 subshell is called the s subshell. The l = 1 and 2 subshells are known as the p and d subshells, respectively. Higher values of l are referred to as f, g, h, etc. This convention of letters is used in a shorthand notation that indicates the principal quantum number n, the orbital quantum number l, and the number of electrons in the n, l, subshell. We have

$$2p^5$$

Number of electrons = 5

n = 2 ; l = 1

The ground state configurations for a few elements in the periodic table are listed below.

Element	*Number of Electrons*	*Electronic Configuration*
Hydrogen	1	$1s^1$
Helium	2	$1s^2$
Lithium	3	$1s^22s^1$
Boron	5	$1s^22s^22p^1$
Neon	10	$1s^22s^22p^6$
Sodium	11	$1s^22s^22p^63s^1$
Aluminum	13	$1s^22s^22p^63s^23p^1$
Argon	18	$1s^22s^22p^63s^23p^6$
Potassium	19	$1s^22s^22p^63s^23p^64s^1$

PRACTICE PROBLEMS

1. Determine the radius and energy of the hydrogen atom in the n = 3 state of the Bohr model.

2. What are the wavelengths of the longest and shortest wavelength lines in the Brackett series which has $n_f = 4$.

3. Compute the ionization energy of doubly ionized lithium Li^{2+} (Z = 3).

4. Light of 1090-nm wavelength is emitted from a hydrogen discharge tube. If the transition that produces this emission occurs to the $n_f = 3$ level, from which level n_i did it originate?

5. Assume that 1200 hydrogen atoms are initially in the n = 4 state. The electrons then make transitions to lower energy levels. (a) How many distinct spectral lines will be emitted? (b) Assume, for simplicity, that all possible transitions to lower levels are equally probable from a given excited state. Determine the total number of photons emitted.

6. Write down the fourteen sets of the four quantum numbers that correspond to the electrons in a completely filled 4f shell.

7. Write down the ground state electronic configuration for iron ($Z = 26$).

8. Suppose tungsten ($Z = 74$) has all but one of its electrons removed. (a) How much energy (in eV) is required to remove the last electron completely if the ionized atom is initially in the ground state? (b) What would be the wavelength of the photon emitted if the remaining electron were to make the transition $n = 2$ to $n = 1$? (c) In what region of the spectrum is such a photon found?

9. Suppose the X-ray machine in a dentist's office uses a potential difference of 25 kV to operate the X-ray tube. What is the shortest X-ray wavelength emitted by the machine?

10. Calculate the approximate minimum energy necessary to observe the $n = 2$ to $n = 1$ characteristic-X ray of gold.

HELPFUL SUGGESTIONS

1. Remember that equations (30.1), (30.2), and (30.3) give $1/\lambda$, not the wavelength itself. The numbers obtained using these equations must be inverted to get the wavelengths.

2. In the Bohr theory, the principal quantum number n is used to obtain the energy E_n (using equation 30.13), the radius r_n (equation 30.10), and the angular momentum L_n (equation 30.8).

3. When electrons are bound to atoms, the energies of the various levels are negative. Therefore, an energy level diagram ranges from zero at the top (where the electron is completely separated from the atom) to the largest negative energy at the bottom (ground state).

4. The quantum numbers n and l are only positive integers. The quantum numbers m_l and m_s can be positive or negative. The quantum numbers n and m_s can never be zero while l and m_l may have zero values.

5. No two electrons in an atom can have the same set of four quantum numbers. This allows us to determine the maximum number of electrons that can fit into an energy level or subshell.

6. In determining the ground state electronic configuration for an atom, it is not necessary to fill all energy states in a given shell of quantum number n before going to a higher n. For example, the 4s subshell is usually filled before the 3d subshell begins to be filled.

EVERYDAY PHYSICS

The introduction of photons, or energy quanta, gave rise to the notion of **quantization**. Something is said to be quantized if it is characterized by discrete, whole numbers rather than a continuous distribution. Examples of quantization are common in everyday life. Your television dial is quantized because you view channels 2, 4, or 7, not channel 3.26. Digital watches display discrete, whole number values indicating the time. The rungs on a ladder or steps are quantized as well. Some things, on the other hand, are not quantized. Watches that have an hour and minute hand can display a continuous range of times. Escalators take you up in a continuous fashion from one floor of a department store to another. Try to observe events and activities which occur discretely or continuously in your everyday life.

CHAPTER QUIZ

1. Which of the following is **not** a feature of the Bohr model of the atom?
   a. a small, positively charged nucleus.
   b. accelerating electrons that do not radiate energy.
   c. electrons in planetary-like orbits.
   d. quantized energy levels.
   e. an electron probability cloud.

2. A hydrogen atom is in the ground state when its electron
   a. is in the middle of the atom
   b. is in the innermost orbit
   c. has been completely ionized
   d. has absorbed a photon

3. Light emitted from excited atoms has frequencies that are related to
   a. the mass of the atom
   b. the color of the atom
   c. the difference in energy levels of the atom
   d. the size of the orbits in the atom

4. According to the modern view (quantum mechanical picture) of the atom,
   a. electrons revolve around the nucleus in well-defined circular orbits.
   b. electrons must be treated as particles, not waves.
   c. emitted photons can have any range of continuous energies.
   d. electrons do not have a localized orbit, but are spatially distributed, like a cloud.

5. According to the Pauli exclusion principle,
   a. no two electrons can the same set of values for the four quantum numbers.
   b. electrons are excluded from entering the nucleus.
   c. electrons must exist in discrete energy states.
   d. electrons cannot be removed from an atom.

6. In the n = 1 state, the energy of the hydrogen atom is - 13.6 eV. What is its energy in the n = 2 state?
   a. - 6.80 eV
   b. - 10.2 eV
   c. - 3.40 eV
   d. - 1.51 eV

7. If r_1 is the radius of the first Bohr orbit, the radius of the fourth Bohr orbit is
   a. $4r_1$
   b. $r_1/4$
   c. $16r_1$
   d. $r_1/16$

8. A neutral atom has a ground state electronic configuration $1s^2 2s^2 2p^6 3s^2 3p^5$. What is its atomic number?
   a. 5
   b. 11
   c. 17
   d. 39

9. A neutral atom has an electronic configuration of $1s^2 2s^2 2p^6 3s^2 3p^6$. What is the electronic configuration for a neutral atom holding one additional electron in its orbit?
   a. $1s^2 2s^2 2p^6 3s^2 3p^6 3d^1$
   c. $1s^2 2s^2 2p^6 3s^2 3p^6 4d^1$
   c. $1s^2 2s^2 2p^6 3s^2 3p^5$
   d. $1s^2 2s^2 2p^6 3s^2 3p^6 4s^1$

10. In the ground state, the possible quantum numbers (n, l, m_l, m_s) for hydrogen are.
    a. 1, 1, 1, 1 respectively
    b. 1, 0, 0, ±1/2 respectively
    c. 1, 0, 0, 0 respectively
    d. 1, 1, 0, ±1/2 respectively

11. The number of electrons that can be accommodated in an l = 6 subshell is
    a. 6
    b. 12
    c. 13
    d. 26

SOLUTIONS AND ANSWERS

Practice Problems

---

1. Using equation (30.10) we have,

$$r_3 = (5.29 \times 10^{-11}\ \text{m})(3)^2/1 = \mathbf{4.76 \times 10^{-10}\ m}.$$

Equation (30.13) gives,

$$E_3 = -(13.6\ \text{eV})(1)^2/(3)^2 = \mathbf{-\ 1.51\ eV}.$$

---

2. The longest wavelength occurs from $n_i = 5$ to $n_f = 4$. Equation (30.14) gives,

$$(1/\lambda) = (1.097 \times 10^7\ \text{m}^{-1})(1/4^2 - 1/5^2) = 2.47 \times 10^5\ \text{m}^{-1},$$

so that λ = **4050 nm.**
The shortest wavelength occurs when $n_i = \infty$. Therefore,

$$(1/\lambda) = (1.097 \times 10^7\ \text{m}^{-1})(1/4^2 - 1/\infty^2) = 6.86 \times 10^5\ \text{m}^{-1},$$

so that λ = **1460 nm.**

---

3. The ionization energy for doubly ionized lithium (Z = 3) can be found using equation (30.13).

$$E_n = -(13.6\ \text{eV})\ Z^2/n^2 = -(13.6\ \text{eV})(3)^2/(1)^2 = \mathbf{122\ eV}.$$

---

4. Using equation (30.14),

$$(1/\lambda) = (1.097 \times 10^7\ \text{m}^{-1})(1/n_f^2 - 1/n_i^2).$$

With λ = 1090 nm = 1.09×10^{-6} m and $n_f = 3$ we obtain n_i = **6**.

---

5. (a) Transitions beginning on n = 4 and ending on n = 1 can take the following paths:

$$n = 4 \rightarrow n = 3 \rightarrow n = 2 \rightarrow n = 1 \text{ (3 spectral lines)},$$

or

$$n = 4 \rightarrow n = 2 \rightarrow n = 1 \text{ (one new line)},$$

or

$$n = 4 \rightarrow n = 1 \text{ (one new line)},$$

and also

$$n = 4 \rightarrow n = 3 \rightarrow n = 1 \text{ (one new line)}.$$

The total number of distinct lines = **6**.

(b) If all transitions are equally probable, begin with 1200 in the n = 4 state. One-third goes to n = 3, one third goes to n = 2 and one third goes to n = 1 (for a total of 400 + 400 + 400 = 1200 photons).
In the n = 3 state, half goes to n = 2 and half to n = 1 (200 + 200 = 400 photons). In the n = 2 state, where there are now 600 electrons, all 600 produce photons in going to the n = 1 state. The total number of photons produced is therefore,

$$1200 + 400 + 600 = \mathbf{2200\ photons}.$$

6.

n	l	m_l	m_s
4	3	3	1/2
4	3	3	-1/2
4	3	2	1/2
4	3	2	-1/2
4	3	1	1/2
4	3	1	-1/2
4	3	0	1/2
4	3	0	-1/2
4	3	-1	1/2
4	3	-1	-1/2
4	3	-2	1/2
4	3	-2	-1/2
4	3	-3	1/2
4	3	-3	-1/2

---

7. The ground state configuration for iron (Z = 26) is

$$1s^2 2s^2 2p^6 3s^2 3p^6 4s^2 3d^6$$

---

8. (a) To find the ionization energy use equation (30.13),

$$E_I = (13.6 \text{ eV})(74)^2 = \mathbf{74\ keV}.$$

(b) Use equation (30.14),

$$(1/\lambda) = (1.097 \times 10^7 \text{ m}^{-1})(74)^2(1/1^2 - 1/2^2) = 4.5 \times 10^{10} \text{ m}^{-1}$$

which gives $\lambda = \mathbf{2.2 \times 10^{-11}\ m}$.

(c) This wavelength corresponds to a photon in the **X-ray** region of the spectrum.

---

9. Use equation (30.17) to obtain

$$\lambda_0 = hc/eV = (6.63 \times 10^{-34} \text{ J s})(3.00 \times 10^8 \text{ m/s})/[(1.60 \times 10^{-19} \text{ C})(25\,000 \text{ V})]$$

$$\lambda_0 = \mathbf{5.0 \times 10^{-11}\ m}.$$

---

10. To account for the shielding effect Z = 78 rather than 79 should be used in equation (30.13). We have,

$$E_n = -(13.6 \text{ eV}) Z^2/n^2 = -(13.6 \text{ eV})(78)^2[1/(1)^2 - 1/(2)^2] = \mathbf{-6.2 \times 10^4\ eV}.$$

Quiz answers

1. e
2. b
3. c
4. d
5. a
6. c
7. c
8. c
9. d
10. b
11. d

MCAT REVIEW PROBLEMS

In a hydrogen atom, the electron has energy

$$E_n = -\frac{E_0}{n^2},$$

where E_0 is a constant, and n is the principle quantum number (i.e., the "shell"). For quantum mechanical reasons, n must be a positive integer. For instance, the ground state corresponds to $n = 1$. The first excited state corresponds to $n = 2$. And so on.

In order to determine the value of E_0, a scientist shines photons ("light particles") of various energies at a cloud of atomic hydrogen. Most of the hydrogen atoms occupy the ground state. A detector records the intensity of light transmitted through that cloud; see figure 1. Figure 2 is a graph of part of the scientist's data, showing the intensity of the transmitted light as a function of the photon energy.

A hydrogen atom's electron is likely to absorb a photon only if the photon gives the electron enough energy to knock it into a higher shell.

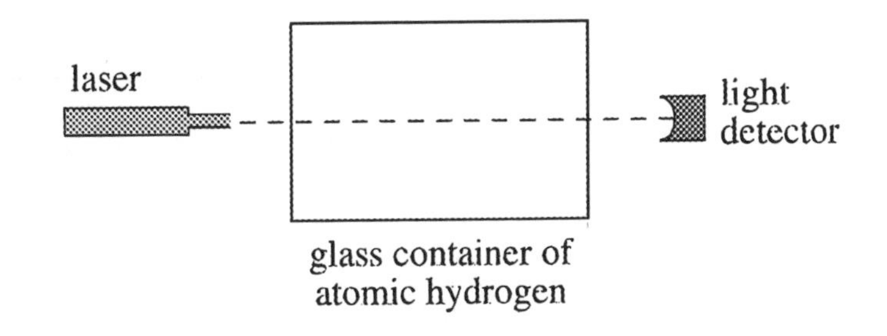

FIGURE 1

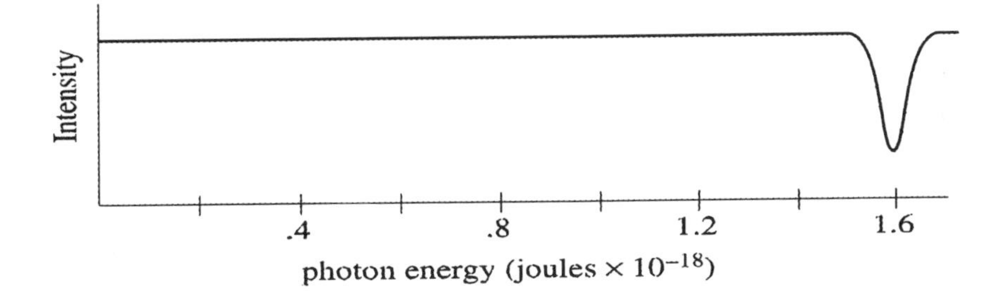

FIGURE 2

---

1. According to this experiment, what is the approximate value of E_0?

   A. 1.6×10^{-18} joules
   B. 2.1×10^{-18} joules
   C. 3.2×10^{-18} joules
   D. 6.4×10^{-18} joules

---

2. If the scientist continues taking data at higher photon energies, he will find the next major "dip" in the intensity graph at what photon energy?

   A. $\frac{1}{9} E_0$ B. $\frac{8}{9} E_0$ C. $3 E_0$ D. $9 E_0$

---

3. When an electron jumps from its ground state to the first excited state, the average value of its *potential* energy:

   A. rises B. stays the same
   C. falls D. We cannot determine the answer without more information.

---

4. A ground-state electron (in a hydrogen atom) absorbs a photon of energy $3E_0$. How much kinetic energy does the electron now have?

   A. E_0 B. $2 E_0$ C. $3 E_0$ D. $4 E_0$

ANSWERS TO MCAT REVIEW PROBLEMS

---

1. **B.** According to figure 2, photons of energy 1.6×10^{-18} joules get absorbed in large numbers; no lower photon energy "works." And according to the passage, substantial absorption occurs only if the photon bumps the ground-state electron into a higher shell. So, a 1.6×10^{-18} joule photon knocks a ground-state electron ($n = 1$) into the first excited state ($n = 2$). In other words, when a ground-state electron absorbs 1.6×10^{-18} joules of energy, it "reaches" the first excited state. Therefore, the *difference* in energy between the ground state and the first excited state must be 1.6×10^{-18} joules. Using $E_n = -\frac{E_0}{n^2}$, we can translate this insight into an equation:

$$\begin{aligned} 1.6\times 10^{-18}\text{ joules} &= E_2 - E_1 \\ &= -\frac{E_0}{2^2} - \left(-\frac{E_0}{1^2}\right) \\ &= E_0\left(1-\frac{1}{4}\right) \\ &= \frac{3}{4}E_0. \end{aligned}$$

Therefore,

$$E_0 = \frac{4}{3}(1.6\times 10^{-18}\text{ joules}) \approx 2.1\times 10^{-18}\text{ joules}.$$

---

2. **B.** As just discussed, a 1.6×10^{-18} joule photon knocks a ground-state electron into the first excited state: $n = 1 \rightarrow n = 2$. The next-lowest energy at which photons get absorbed (in large numbers) corresponds to the $n = 1 \rightarrow n = 3$ transition.

To boost an electron from the ground state to the second excited state, the photon's energy must equal the *difference* in energy between those two shells:

$$\begin{aligned} E_{\text{photon}} &= E_3 - E_1 \\ &= -\frac{E_0}{3^2} - \left(-\frac{E_0}{1^2}\right) \\ &= E_0\left(1-\frac{1}{9}\right) \\ &= \frac{8}{9}E_0. \end{aligned}$$

---

3. **A.** Despite the counterintuitive quantum behavior of electrons, you can employ some intuitive classical reasoning. The ground state ($n = 1$) and the first excited state ($n = 2$) correspond to electron "clouds." On average, the $n = 2$ cloud is "higher" (farther from the nucleus) than the $n = 1$ cloud. Therefore, knocking the electron from $n = 1$ to $n = 2$ corresponds to pushing it farther away from the nucleus.

Since the positive nucleus attracts the negative electron, pushing the electron farther away raises its potential energy, just like lifting a block farther from the Earth raises the block's potential energy.

Mathematical reasoning yields the same answer. As discussed in chapter 19, a point charge generates an electric potential $V = \frac{kq}{r}$. Since the hydrogen nucleus has charge $+e$, it generates a potential $V = \frac{ke}{r}$. Therefore, an electron at distance r from the nucleus has electric potential energy

$$\text{EPE} = q_{\text{electron}}V - (-e)\frac{ke}{r} = -\frac{ke^2}{r}.$$

Consequently, when the electron gets farther from the nucleus, the denominator increases, making the EPE "less negative." In other words, the EPE rises.

---

4. **B.** The ground-state electron is bound to the hydrogen nucleus with energy $-E_0/1^2 = -E_0$. So, E_0 is the "binding energy." In other words, the electron must absorb energy E_0 just to "escape" from the nucleus. More precisely, if it absorbs a photon of energy E_0, the electron breaks free from the nucleus's electrostatic pull; but the electron has no extra energy left over, because it "spent" all of that energy escaping.

Suppose the ground-state electron absorbs a photon of energy $3E_0$. The electron spends the first E_0 of that energy escaping from the nucleus, leaving an extra $2E_0$ for the electron to "keep," in the form of kinetic energy.

Similar reasoning applies to the photoelectric effect, discussed in textbook section 29.3.

CHAPTER 31
Nuclear Physics and Radioactivity

PREVIEW

One of the great successes of physics during the twentieth century was the development of the model of the atom and the subsequent description of its nucleus. In this chapter you will learn about the nucleon inhabitants of the nucleus: the neutron and proton. You will learn how the size of the nucleus depends on the number of nucleons.

Also, you will learn about the force which binds the nucleons together and the energy associated with this binding. If the nucleus is "split", some of this energy can be released. This fact provides the basis for nuclear power.

One-half of radioactive nuclei spontaneously disintegrate after a certain amount of time. You will be able to calculate this half-life in some cases and to determine how much radioactive material has disintegrated after a given time. This idea is used for dating archeological artifacts.

QUICK REFERENCE

Important Terms

Nucleons
Protons and neutrons.

Atomic number (Z)
The total number of protons in a nucleus.

Atomic mass number (A)
The total number of nucleons in a nucleus.

Isotopes
Nuclei which have the same number of protons but different number of neutrons.

Strong nuclear force
One of the four known fundamental forces in nature. The strong force is the force exerted between nucleons which binds them together in a nucleus.

Radioactivity
The spontaneous disintegration or rearrangement of the internal structure of nuclei.

Binding energy
The energy required to break apart a nucleus into separated nucleons.

Mass defect
The difference in the mass of an intact nucleus and total mass of the separated nucleons. This is the mass equivalent of the binding energy of the nucleus.

α particle
A helium nucleus consisting of two protons and two neutrons.

α decay
The radioactive emission of an α particle from a nucleus.

Transmutation
The changing of one element to another by a loss or gain of one or more protons. Often transmutation is caused by α decay.

β decay
The radioactive decay of a nucleus by the emission of an electron or positron.

γ decay
The radioactive decay of a nucleus by emission of a high frequency "gamma" photon.
Neutrino
A particle created during the β decay of a nucleus.
Half-life
The time necessary for one-half of a particular radioactive substance to undergo spontaneous decay.
Radioactive decay series
The sequential decay of one nuclear species after another.
Geiger counter
A common gas filled device used for detecting α, β, and γ emissions of radioactive nuclei.
Scintillation counter
A device consisting of a material which emits light when struck by ionizing radiation. The material is mounted on a photomultiplier tube.
Semiconductor detector
A device which detects ionizing radiation by the electrons and holes produced in a semiconductor material.
Cloud chamber
A device which detects ionizing radiation by its condensation track left in a supercooled gas.
Bubble chamber
A device which detects ionizing radiation by its vapor track left in a superheated liquid.

Equations

The **relationship** between the **atomic number**, Z, number of neutrons, N, and the **atomic mass number**, A, of a nucleus.

$$A = Z + N \tag{31.1}$$

The approximate **radius of a nucleus**

$$r \cong (1.2 \times 10^{-15}\ \text{m})A^{1/3} \tag{31.2}$$

The **binding energy** of a nucleus

$$\text{Binding energy} = (\Delta m)c^2 \tag{31.3}$$

The **activity** of a radioactive sample is the magnitude of

$$\frac{\Delta N}{\Delta t} = -\lambda N \tag{31.4}$$

The **number of radioactive nuclei** present at time, t, in a sample

$$N = N_0 e^{-\lambda t} \tag{31.5}$$

The **half-life** of a radioactive material

$$T_{1/2} = \frac{\ln 2}{\lambda} = \frac{0.693}{\lambda} \tag{31.6}$$

DISCUSSION OF SELECTED SECTIONS

31.1 Nuclear Structure

Example 1
Two isotopes of chlorine occur in nature. The $^{35}_{17}$Cl isotope has an atomic mass of 34.968 85 u and a natural abundance of 75.77%. The $^{37}_{17}$Cl has an atomic mass of 36.965 90 u and a natural abundance of 24.23%. By a calculation of your own, verify that the value of 35.45 u listed in the periodic table is a weighted average of these individual atomic masses.

The weighted average of the masses is

$$(0.7577)(34.968\ 85\ \text{u}) + (0.2423)(36.965\ 90\ \text{u}) = 35.45\ \text{u}$$

Example 2
What is the approximate radius of lead $^{207}_{82}$Pb ?

The atomic mass number of the lead is A = 207. Equation (31.2) gives the approximate radius to be

$$r \approx (1.2 \text{ X } 10^{-15}\ \text{m})\ A^{1/3} = (1.2 \text{ X } 10^{-15}\ \text{m})(207)^{1/3} = 7.1 \text{ X } 10^{-15}\ \text{m}$$

31.3 The Mass Defect of the Nucleus and Nuclear Binding Energy

Relativity tells us that mass and energy are equivalent and related by $E = mc^2$. A nucleus has less energy than the total energy of its separated constituents since some energy is released in the formation of the nucleus (binding energy). Hence, the mass of the nucleus is less than the total mass of its separated nucleons. The difference in mass is called the mass defect and is the mass equivalent of the binding energy of the nucleus. The specific relationship between the binding energy and mass defect is given by equation (31.3).

Note: The total mass of an atom, given in the table in Appendix F, includes the mass of its complement of electrons. The mass of the electrons must be taken into account when calculating mass defects.

Example 3
What is the mass defect and the binding energy for $^{239}_{93}$Np ?

The total mass of the atom including electrons is found in Appendix F to be 239.052 933 u. The atom has 93 protons, 93 electrons and 146 neutrons. The total mass of these separated particles is

$$93(1.007\ 276\ \text{u}) + 93(5.485\ 799 \text{ X } 10^{-4}\ \text{u}) + 146(1.008\ 665\ \text{u}) = 240.992\ 776\ \text{u}.$$

The mass defect of the nucleus is then

$$\Delta m = 240.992\ 776\ \text{u} - 239.052\ 933\ \text{u} = 1.939\ 843\ \text{u}$$

The corresponding binding energy is

$$(\Delta m)c^2 = (1.939\ 843\ \text{u})((931.5\ \text{MeV}/1\ \text{u}) = 1807\ \text{MeV}$$

31.4 Radioactivity

During radioactive decay a nucleus emits an α particle, a β particle or a γ ray. During any of these decay processes, the mass/energy, charge, linear momentum, angular momentum and nucleon number are conserved.

Example 4
What daughter nucleus would be formed if ${}^{238}_{92}U$ could undergo (a) α decay. (b) β decay. (c) γ decay?

(a) Write the reaction as

$$ {}^{238}_{92}U \rightarrow {}^{A}_{Z}X + {}^{4}_{2}He $$

The conservation of nucleon number requires

$$ 238 = A + 4 \rightarrow A = 234 $$

The conservation of charge requires

$$ 92 = Z + 2 \rightarrow Z = 90 $$

Using Appendix F in the text we find that the daughter nucleus is ${}^{234}_{90}Th$.

(b) The reaction is

$$ {}^{239}_{92}U \rightarrow {}^{A}_{Z}X + {}^{0}_{-1}e $$

The conservation of nucleon number gives

$$ 238 = A + 0 \rightarrow A = 238 $$

The conservation of charge requires

$$ 92 = Z + (-1) \rightarrow Z = 93 $$

The periodic chart in Appendix F shows that the daughter nucleus is ${}^{238}_{93}Np$.

(c) The reaction in this case is

$$ {}^{238}_{92}U \rightarrow {}^{A}_{Z}X + \gamma $$

The γ ray does not have a nucleon number or a charge, hence $A = 238$ and $Z = 92$. The daughter nucleus is ${}^{238}_{92}U$.

31.6 Radioactive Decay and Activity

Example 5

The radioactive isotopes of strontium are particularly dangerous to animals since strontium behaves chemically like calcium and is taken up by the body and deposited in bone. Much of the radioactive strontium in the atmosphere is $^{90}_{38}Sr$ which was released by atmospheric nuclear bomb testing during the 1950's and early 1960's. Assume that no strontium has been introduced into the atmosphere for 28.0 years. What percentage of the original radioactive nuclei is still present in the atmosphere? Strontium has a half-life of 28.5 years.

In any given volume of air there are still

$$N = N_0\, e^{-\lambda t}$$

radioactive nuclei left. The percentage left is then

$$\frac{N}{N_0} \text{ X } 100\ \% = 100\ \% \text{ X } e^{-\lambda t}$$

The decay constant, λ, of the strontium isotope is

$$\lambda = (0.693)/T_{1/2} = (0.693)/(28.5 \text{ yr}) = 0.02432 \text{ yr}^{-1}$$

Note that we have carried an extra figure for additional accuracy. The percent of strontium left in the atmosphere is

$$e^{-(0.02432 \text{ yr}^{-1})(28.0 \text{ yr})} \times 100\% = 50.6\%$$

31.7 Radioactive Dating

Example 6

The shroud of Turin was originally believed to be about two thousand years old, but was determined by dating techniques to be about 700 years old. Assuming that the shroud had a carbon-14 activity of 0.23 Bq per gram when it was made, what is its activity now?

The number of nuclei left in a given amount of the shroud is

$$N = N_0\, e^{-\lambda t}$$

The activity, $A = \lambda N$ so

$$A = A_0\, e^{-\lambda t}$$

The decay constant for carbon-14 is

$$\lambda = (0.693)/T_{1/2} = (0.693)/(5730 \text{ yr}) = 1.21 \text{ X } 10^{-4} \text{ yr}^{-1}$$

Now the per gram activity is

$$A = (0.23 \text{ Bq})e^{-(1.21\times10^{-4} \text{ yr}^{-1})(700 \text{ yr})} = 0.21 \text{ Bq}$$

PRACTICE PROBLEMS

1. What is the radius of a nucleus of $^{28}_{14}Si$?

2. What is the binding energy per nucleon of $^{28}_{14}Si$?

3. Polonium $^{214}_{84}Po$ undergoes both α and γ decay. What daughter nucleus is produced in each case?

4. An intermediate nucleus in the decay series of $^{238}_{92}U$ is $^{230}_{90}Th$. After several successive α decays, this intermediate nucleus leads to the isotope $^{214}_{82}Pb$. How many α decays have occurred?

5. The isotope $^{89}_{36}Kr$ has a lifetime of 3.16 min. What percentage of Krypton nuclei are left after 15.0 min?

6. The activity of a sample of a radioactive material is 1.50 Ci and the half-life is 1.6 X 10^3 yr. How many of the radioactive nuclei are left after 250 yr?

7. Plutonium $^{239}_{94}Pu$ (atomic mass = 239.052 157 u) undergoes α decay. Assuming all the released energy is in the form of kinetic energy of the α particle and ignoring the recoil of the daughter nucleus, find the speed of the α particle.

8. One gram of charcoal from an ancient campfire is analyzed for $^{14}_{6}C$ and found to have an activity of 0.11 Bq. How old is the charcoal?

HELPFUL SUGGESTIONS

1. When you are calculating mass defects, you should carry as many significant figures in the atomic masses as possible. This is because the mass defects are often quite small.

2. When calculating the decay constant for use in an exponential, it is desirable to carry an extra figure.

3. Only one-half of a radioactive substance remains after one half-life. After n half-lives only $1/2^n$ of the original amount remains.

EVERYDAY PHYSICS

1. Radioactivity is a concern to many people, particularly those living close to nuclear power plants. Accidental releases of radioactive materials pose a potential hazard to the health of people exposed to the radiation. Additionally, there is the possibility of long term contamination of the area surrounding the site of a nuclear accident. Try to assess and compare the possible long term effects of an accidental release of radioactive gases, such as radon, to the release of heavier radioactive nuclei. Can any clear conclusions be drawn?

2. The potential danger of radioactive radon gas buildup in basements of houses is a cause of concern. (See examples 6 and 7 in the text.) It has been estimated that as many as 50 % of the houses in the United States may have dangerous levels of radon. Inexpensive home test kits are now available for monitoring the radon levels in homes. It might be a good idea to obtain one and determine the radon level in your house.

3. Radon gas itself, apparently, is not as dangerous as its radioactive daughters which attach themselves to particles in the air and can become lodged in the lungs when inhaled. According to Table 31.2 in your text, radon decays by α and γ emissions. What daughters would you expect radon to produce? Do some library research to find out what radioactive daughters radon does produce and their lifetimes.

CHAPTER QUIZ

1. How many neutrons are in the nucleus of $^{260}_{103}Lr$?

   a. 260 b. 103 c. 363 d. 157

2. A certain nucleus has an atomic mass eight times larger than another. The first nucleus has a radius _______ the other.

   a. the same size as b. twice as large as c. four times as large as d. eight times as large as

3. Which of the following is a helium nucleus?

   a. α ray b. β ray c. γ ray d. positron

4. The binding energy of a nucleus is the energy needed to _______.

   a. form it b. tear it apart c. keep it from decaying d. accelerate it to c

5. A nucleus has a mass defect of 1.235 u. What is its binding energy in MeV?

   a. 754.2 b. $1.111 \text{ X } 10^{17}$ c. $1.150 \text{ X } 10^{3}$ d. 931.5

6. A 10.0 g sample of radioactive substance is left to decay for five half-lives? How much of the radioactive substance remains?

   a. Not enough information is given to find an answer.
   b. 0.313 g
   c. 0.625 g
   d. 1.25 g

7. A radioactive nucleus produces a 0.0483 MeV γ ray. How much has the mass of the nucleus changed due to the γ ray emission?

   a. $8.59 \text{ X } 10^{-32}$ kg b. $7.73 \text{ X } 10^{-21}$ kg c. None. The nucleus lost energy not mass.

8. Many of the radiation detectors discussed in the text rely on the fact that the detected particles or rays_______.

   a. have charge
   b. have mass
   c. produce ions in the detector
   d. are actually electrons

9. A radioactive series ultimately results in_______.

   a. hydrogen b. lead c. a stable nucleus d. a β decay

10. In most β^- decay processes a(n) _______ is emitted along with the e^-.

    a. α ray b. γ ray c. neutrino d. antineutrino

SOLUTIONS AND ANSWERS

Practice Problems

---

1. Equation (31.2) gives the approximate radius to be

$$r \approx (1.2 \text{ X } 10^{-15} \text{ m})(28)^{1/3} = \mathbf{3.6 \text{ X } 10^{-15} \text{ m.}}$$

---

2. The mass defect for the nucleus is

$$\Delta m = 14(1.007\,825 \text{ u}) + 14(1.008\,665 \text{ u}) - 27.976\,927 \text{ u} = 0.253\,933 \text{ u}$$

The energy equivalent of this mass defect is

$$\text{Binding energy} = (931.5 \text{ MeV/1 u})(0.253\,933 \text{ u}) = 236.5 \text{ MeV}.$$

There are 28 nucleons, so the energy corresponds to an energy/nucleon of **8.448 MeV/nucleon**.

---

3. For α decay the atomic mass number is conserved so the daughter nucleus has A = 214 - 4 = 210. The charge is also conserved so the daughter has an atomic number of Z = 84 - 2 = 82. The table in Appendix F shows the daughter to be ${}^{210}_{82}\text{Pb}$.
In a gamma decay the nucleus loses only energy, hence the daughter is a de-excited ${}^{214}_{84}\text{Po}$.

---

4. The reaction is

$${}^{230}_{90}\text{Th} \rightarrow {}^{214}_{82}\text{Pb} + N\,{}^{4}_{2}\text{He}$$

where N is the number of a decays which have occurred.

The conservation of charge gives

$$90 = 82 + 2N \Rightarrow N = 4.$$

The conservation of atomic mass number gives

$$230 = 214 + 4N \Rightarrow N = 4.$$

The number of α decays needed is **4.**

---

5. Equation (31.5) gives the percentage of nuclei left.

$$(N/N_0) \text{ X } 100\ \% = e^{-\lambda t} \text{ X } 100\ \%$$

where $\lambda = (0.693)/T_{1/2} = 0.2193 \text{ min}^{-1}$.

The percentage of nuclei left is **3.73 %**.

---

6. Equation (31.4) gives for the initial case Activity = - λ N_0, where $\lambda = (0.693)/T_{1/2}$. The number of nuclei present initially is

$$N_0 = (1.5\ Ci)(3.70 \times 10^{10}\ Bq/\ 1\ Ci)/[(0.693)/(1.6 \times 10^3\ yr)] = 1.3 \times 10^{14}.$$

The number of nuclei remaining after 250 yr is

$$N = (1.3 \times 10^{14})\ e^{-(0.000433)(250)} = \mathbf{1.2 \times 10^{14}}.$$

---

7. The conservation of atomic mass number requires that the daughter have a mass number of 235 and the conservation of charge requires that it have an atomic number of 92. The daughter is uranium. The kinetic energy of the α particle is the equivalent of the total mass defect of the process.

$$M_{lost} = 239.052\ 157\ u - 235.043\ 924\ u - 4.002\ 603\ u = 0.005\ 630\ u.$$

The kinetic energy of the α is KE = $M_{lost}\ c^2$

The speed of the a particle is then

$$v = \sqrt{\frac{2M_{lost}}{M_\alpha}}\ c = \sqrt{\frac{2(0.005\ 630\ u)}{4.002\ 603\ u}}\ (3.00 \times 10^8\ m/s) = \mathbf{1.59 \times 10^7\ m/s}$$

---

8. Assume that the original activity of the radioactive carbon in the wood was 0.23 Bq. The activity at any time is

$$A = A_0\ e^{-\lambda t} \text{ where } \lambda = (0.693)/(5730\ yr) = 1.21 \times 10^{-4}\ yr^{-1}$$

Taking natural logarithm of the above equation and solving for the time yields

$$t = -\ (1/\lambda)\ \ln\ (A/A_0) = -\ (1/1.21 \times 10^{-4}\ yr^{-1})\ \ln\ (0.11\ Bq/0.23\ Bq) = \mathbf{6100\ yr.}$$

Quiz answers

1. d	3. a	5. c	7. a	9. c
2. b	4. b	6. b	8. c	10. d

MCAT REVIEW PROBLEMS

"Beta" radioactive decay refers to two distinct processes. In β^- decay, the nucleus emits an electron. By contrast, in β^+ decay, the nucleus gives off a positron. A positron is the "antiparticle" of an electron. In other words, electrons and positrons have the same mass and spin. But a positron carries charge $+e$ instead of $-e$.

When an electron and positron traveling in opposite directions at the same speed crash into each other, they sometimes annihilate, in accordance with the following reaction:

$$\beta^- + \beta^+ \rightarrow 2\gamma,$$

where γ stands for a gamma ray. A gamma ray is a very energetic photon. All gamma rays are massless and chargeless, but nonetheless carry energy and momentum.

---

1. When an atom undergoes β^+ decay:

   A. A neutron "changes into" a proton.
   B. A proton "changes into" a neutron.
   C. A neutron "changes into" an antiproton.
   D. A proton "changes into" an antineutron.

---

2. An atom of thorium (atomic number 90) undergoes two successive β^- decays. The result is an atom of:

   A. Radium (atomic number 88)
   B. Thorium
   C. Uranium (atomic number 92)
   D. Plutonium (atomic number 94)

---

3. When an electron and positron with equal speeds in opposite directions annihilate each other, they cannot produce just *one* gamma ray, because that would constitute a violation of:

   A. Conservation of charge
   B. Conservation of energy
   C. Conservation of momentum
   D. Conservation of nucleon number

---

4. A container is filled with a radioactive substance for which the half-life is 2 days. A week later, when the container is opened, it contains 5 grams of the substance. Approximately how many grams of the substance were initially placed in the container?

   A. 40 **B.** 60 **C.** 80 **D.** 100

ANSWERS TO MCAT REVIEW PROBLEMS

---

1. **B.** Since a nucleus contains protons and neutrons, with no antiprotons or antineutrons, the answer must be A or B. Intuitively, the proton spits out its positive charge (as a positron), leaving behind a neutron:

$$p \rightarrow n + \beta^+,$$

where p and n denote a proton and neutron. Both sides of the equation have the same charge, $+e$.

By contrast, $n \rightarrow p + \beta^+$ doesn't work, because the left-hand side has charge 0, while the right-hand side has charge $+2e$. Charge conservation makes this reaction "illegal."

---

2. **C.** Since a β^- particle has charge $-e$, the nuclear charge must increase by $+e$ every time it spits out a β^-. Otherwise, charge wouldn't be conserved. So, when the nucleus emits two electrons, the nuclear charge increases by $2e$. In other words, the atomic number increases by 2.

---

3. **C.** You can solve by elimination, or by direct reasoning. Let us review both methods.

Elimination. If one gamma ray were produced instead of two, charge would still be conserved; it would still be zero on both sides of the reaction. And energy conservation could hold, if that one gamma ray carried a "double dose" of energy. Finally, since this reaction involves no nucleons (protons or neutrons), the nucleon number is zero on both sides of the reaction. Therefore, nucleon number is conserved. We've eliminated all the options except C.

Direct reasoning. One gamma ray could *not* conserve linear momentum, for this reason. Initially, the electron and positron have momenta equal in magnitude but opposite in direction. Those momenta cancel:

$$(+)\xrightarrow{\mathbf{p}_1} \quad \xleftarrow{\mathbf{p}_2}(-)$$

$$\mathbf{p}_1 + \mathbf{p}_2 = \mathbf{0}$$

So, the total initial momentum is zero. Therefore, the total final momentum of the gamma rays must be zero. But according to the passage, *one* gamma ray can't have zero momentum. The reaction needs two gamma rays traveling in opposite directions, so that their momenta cancel.

---

4. **B.** Don't try to calculate an exact answer, which turns out to be 57 grams. You can reason more quickly as follows. . .

For this substance, a week (7 days) corresponds to 3.5 half-lives. Over 3 half-lives, the sample reduces to $1/2^3 = 1/8$ of its initial mass. After 4 half-lives, the sample has only $1/2^4 = 1/16$ of its initial mass. Therefore, after 3.5 half-lives, the sample must contain somewhere *between* 1/8 and 1/16 of its initial mass. In other words, 5 grams is somewhere between 1/8 and 1/16 of the initial mass. So, the initial mass is somewhere between 8×5 grams = 40 grams, and 16×5 grams = 80 grams. Only option B fits this description.

By the way, the exact answer is not midway between 40 g and 80 g, because the sample's mass decreases exponentially, not linearly.

CHAPTER 32
Ionizing Radiation, Nuclear Energy, and Elementary Particles

PREVIEW

In this final chapter you will learn about the effects of ionizing radiation on living cells, about nuclear fission and fusion reactions, about nuclear reactors, and about elementary particles including the latest theory of "quarks".

QUICK REFERENCE

Important Terms

Ionizing radiation
Radiation consisting of photons and/or moving particles that have enough energy to remove an electron from an atom or molecule.

Exposure
A measure of the ionization produced in air by X-rays or γ rays. Exposure is measured in roentgens.

Absorbed dose
The amount of energy absorbed from the radiation per unit mass of absorbing material. The SI unit of absorbed dose is the gray (Gy); 1 Gy = 1 J/kg.

Relative biological effectiveness (RBE)
The ratio of the dose of 200-keV X rays needed to produce a certain biological effect to the dose of a given type of radiation required to produce the same effect.

Biologically equivalent dose
The product of the absorbed dose (in rad) and the RBE. The unit is the rem.

Induced nuclear transmutation
The process whereby an incident particle strikes a nucleus and causes the production of a new element.

Nuclear fission
A massive nucleus splits into two less-massive fragments.

Chain reaction
Neutrons released from a fission process induce other nuclei to fission.

Fission reactor
A device that generates energy by a controlled chain reaction.

Nuclear fusion
Two smaller masses combine to form a single nucleus with a larger mass. Energy is released when the binding energy per nucleon is greater for the larger nucleus than for the smaller nuclei.

Elementary particles
The basic building blocks of all matter.

Quarks
Elementary particles which are believed to make up the hadrons.

Equations

$$\text{Exposure (in roentgens)} = \left(\frac{1}{2.58 \times 10^{-4}}\right)\frac{q}{m} \qquad (32.1)$$

$$\text{Absorbed dose} = \frac{\text{Energy absorbed}}{\text{Mass of absorbing material}} \qquad (32.2)$$

$$\begin{matrix}\text{Relative biological}\\ \text{effectiveness (RBE)}\end{matrix} = \frac{\begin{matrix}\text{The dose of 200-keV X rays that}\\ \text{produces a certain biological effect}\end{matrix}}{\begin{matrix}\text{The dose of radiation that}\\ \text{produces the same effect}\end{matrix}} \qquad (32.3)$$

$$\begin{matrix}\text{Biologically equivalent dose}\\ \text{(in rem)}\end{matrix} = \begin{matrix}\text{Absorbed dose}\\ \text{(in rad)}\end{matrix} \times \text{RBE} \qquad (32.4)$$

DISCUSSION OF SELECTED SECTIONS

32.1 Biological Effects of Ionizing Radiation

Ionizing radiation consists of photons and/or moving particles that have sufficient energy to knock an electron out of an atom or molecule, thus forming an ion. The photons usually lie in the ultraviolet, X-ray, and γ-ray regions of the electromagnetic spectrum, while the moving particles can be the α and β particles emitted during radioactive decay. A single high-energy photon or particle can ionize thousands of molecules.

Nuclear radiation is potentially harmful to humans, because the ionization it produces can significantly alter the structure of molecules within a living cell. The alterations cause the cell to malfunction and, if severe enough, can lead to the death of the cell and even the organism itself. Because of these hazards, it is important to learn the fundamentals of radiation exposure, including dose units, and the biological effects of radiation.

Exposure is a measure of the ionization produced in air by X-rays or γ rays, and it is defined as follows: A beam of X rays or γ rays sent through a mass m of dry air at STP will produce ions whose total charge is q. The SI unit of exposure is C/kg. However, for historical reasons, a radiation unit known as the **roentgen (R)**, is still used. The exposure in roentgens is given by

$$\text{Exposure (in roentgens)} = \left(\frac{1}{2.58 \times 10^{-4}}\right)\frac{q}{m} \qquad (32.1)$$

Thus, an exposure of one roentgen produces $q = 2.58 \times 10^{-4}$ coulombs of positive charge in m = 1 kg of dry air, so $1\ R = 2.58 \times 10^{-4}$ C/kg (dry air, STP).

For biological purposes, the **absorbed dose** is a more suitable quantity. It is defined as

$$\text{Absorbed dose} = \frac{\text{Energy absorbed}}{\text{Mass of absorbing material}} \qquad (32.2)$$

The SI unit of absorbed dose is the **gray (Gy)**, which is 1 Gy = 1 J/kg. The absorbed dose is a concept applicable to all types of radiation and absorbing media. A unit also used for absorbed dose is the **rad** (an acronym for radiation absorbed dose). The gray and rad are related by 1 rad = 0.01 Gy.

The amount of biological damage produced by ionizing radiation is different for different types of radiation. For instance, a 1-rad dose of protons is about 10 times as effective in producing biological damage as a 1-rad dose of X rays. To compare the damage done by different types of radiation, the **relative biological effectiveness (RBE)** is used. It is defined by the following equation.

$$\text{Relative biological effectiveness (RBE)} = \frac{\text{The dose of 200-keV X rays that produces a certain biological effect}}{\text{The dose of radiation that produces the same effect}} \qquad (32.3)$$

The RBE depends on the nature of the ionizing radiation and its energy, as well as the type of tissue being irradiated. As seen from Table 32.1, 200-keV X-rays, γ rays, and β^- particles produce less damage than protons, neutrons, and α particles. The RBE is often used in conjunction with the absorbed dose to reflect the damage-producing character of the radiation on tissue. The **biologically equivalent dose** is defined as:

$$\underset{\text{(in rem)}}{\text{Biologically equivalent dose}} = \underset{\text{(in rad)}}{\text{Absorbed dose}} \times \text{RBE} \qquad (32.4)$$

Example 1
What absorbed dose of α particles (RBE = 20) causes as much biological damage as a 60-rad dose of protons (RBE = 10)?

Using equation (32.4) we can obtain the biologically equivalent dose for the protons,

$$\underset{\text{(in rem)}}{\text{Biologically equivalent dose}} = \underset{\text{(in rad)}}{\text{Absorbed dose}} \times \text{RBE} = (60\text{ rad})(10) = 600\text{ rem}$$

For the alpha particles, the biologically equivalent dose is the same (600 rem). Therefore,

$$\text{Absorbed dose} = \frac{\text{Biologically equivalent dose}}{\text{RBE}} = \frac{600\text{ rem}}{20} = 30\text{ rad}$$

32.2 Induced Nuclear Reactions

A **nuclear reaction** occurs when a nucleus, particle, or photon strikes a target nucleus and causes a change in the target nucleus. For example, an α particle (a helium nucleus) strikes a target nitrogen nucleus producing a nucleus of oxygen and a proton. The reaction may be written as

$${}^{4}_{2}\text{He} + {}^{14}_{7}\text{N} \rightarrow {}^{17}_{8}\text{O} + {}^{1}_{1}\text{H}$$

A shorthand notation is often used in writing reactions like the one shown above. The above reaction can be written

$${}^{14}_{7}\text{N}\,(\alpha, p)\,{}^{17}_{8}\text{O}$$

The first and last symbols represent the initial and final nuclei, respectively. The symbols inside the parentheses denote the incident particle (on the left) and the small emitted particle (on the right).

In any nuclear reaction, both the total electric charge of the nucleons and the total number of nucleons are conserved during the process. The fact that these quantities are conserved makes it possible to identify the nucleus produced in a nuclear reaction.

Example 2

Determine the values of the atomic number Z (number of protons) and atomic mass number A (number of nucleons) for the following reactions. Identify the final nucleus and write each reaction in shorthand notation.

$$^{32}_{16}S + ^{1}_{0}n \rightarrow ^{A}_{Z}X + ^{1}_{1}H$$

$$^{40}_{18}Ar + ^{4}_{2}He \rightarrow ^{A}_{Z}X + ^{1}_{1}H$$

$$^{9}_{4}Be + ^{1}_{1}H \rightarrow ^{A}_{Z}X + ^{4}_{2}He$$

The atomic number Z and mass number A must be conserved. Therefore, the left and right hand sides of each reaction must balance. In the first reaction $A = 32 + 1 - 1 = 32$, and $Z = 16 + 0 - 1 = 15$. The element with atomic number $Z = 15$ is P. The reaction may therefore be written in shorthand as

$$^{32}_{16}S\ (n, p)^{32}_{15}P$$

Similarly, for the other two reactions we have

$$^{40}_{18}Ar\ (\alpha, p)^{43}_{19}K \qquad \text{and} \qquad ^{9}_{4}Be\ (p, \alpha)^{6}_{3}Li$$

32.3 Nuclear Fission

In 1939 four German physicists found that a uranium nucleus, after absorbing a neutron, splits into two fragments, each with a smaller mass than the original nucleus. The splitting of a massive nucleus into two less-massive fragments is known as **nuclear fission**. The following reaction shows one possible fission reaction.

$$^{1}_{0}n + ^{235}_{92}U \rightarrow ^{236}_{92}U \rightarrow ^{141}_{56}Ba + ^{92}_{36}Kr + 3^{1}_{0}n$$

In this reaction, a uranium-235 nucleus absorbs a neutron, creating a "compound nucleus", uranium-236. The compound nucleus disintegrates quickly into barium-141, krypton-92, and three neutrons.

The fission of uranium is accompanied by the release of an enormous amount of energy, which appears primarily as kinetic energy of the fission products. An average of 200 MeV of energy is released per fission.

The fact that the uranium fission reaction releases 2 or 3 neutrons makes it possible for a self-sustaining series of fissions to occur. Each neutron released by the fission process can initiate another fission reaction, resulting in the release of still more neutrons, followed by more fissions, and so on. This series of fission reactions is referred to as a **chain reaction**. If the reaction is uncontrolled, it would not be unusual for the number of fissions to increase a thousand fold within a few millionths of a second. With an average energy of 200 MeV per fission being released, an incredible amount of energy can be released in a very short time. This is the basis of the atomic bomb.

By limiting the number of neutrons allowed to participate in further fission reactions, it is possible to control the rate of fissions, and hence the amount of energy released. This controlled nuclear fission chain reaction is the principle behind the nuclear reactors used in the commercial generation of electric power.

32.5 Nuclear Fusion

Another way of generating nuclear energy is to combine two very low-mass nuclei with relatively small binding energies per nucleon, and "fuse" them together into a single, more massive nucleus that has a greater binding energy per nucleon. A substantial amount of energy can be released in such a reaction called a **nuclear fusion** reaction. A typical fusion reaction is shown below

$$^{2}_{1}H + ^{3}_{1}H \rightarrow ^{4}_{2}He + ^{1}_{0}n$$

This reaction involves two isotopes of hydrogen, $^{2}_{1}H$ (deuterium) and $^{3}_{1}H$ (tritium), which fuse to form $^{4}_{2}He$ with the release of a neutron. The amount of energy released in the process can be determined by looking at the initial and final masses of the reacting nuclei. We have

Initial Masses		*Final Masses*	
$^{2}_{1}H$	2.014 102 u	$^{4}_{2}He$	4.002 603 u
$^{3}_{1}H$	3.016 050 u	$^{1}_{0}n$	1.008 665 u
Total	5.030 152 u		5.011 268 u

The **mass defect** is

$$\Delta m = 5.030\ 152\ u - 5.011\ 268\ u = 0.018\ 884\ u.$$

Since 1 u is equivalent to 931.5 MeV, the energy released by the fusion reaction is

$$(931.5\ MeV/1\ u)(0.018\ 884\ u) = 17.59\ MeV.$$

Example 3
In one type of fusion reaction a proton fuses with a neutron to form a deuterium nucleus:

$$^{1}_{1}H + ^{1}_{0}n \rightarrow ^{2}_{1}H.$$

The masses are $^{1}_{1}H$ (1.007 825 u), $^{1}_{0}n$ (1.008 665 u), and $^{2}_{1}H$ (2.014 102 u). How much energy (in MeV) is released by this reaction?

We just need to find the mass deficit between the proton plus neutron, and the deuterium. We have

$$\Delta m = 1.007\ 825\ u + 1.008\ 665\ u - 2.014\ 102\ u = 0.002\ 388\ u.$$

The energy released is therefore,

$$\Delta E = (0.002\ 388\ u)(931.5\ MeV/\ 1\ u) = 2.224\ MeV.$$

PRACTICE PROBLEMS

1. A beam of X rays passes through 7.5×10^{-4} kg of dry air and generates 1.2×10^{13} singly charged ions. What is the exposure in roentgens?

2. A person stands near a radioactive source and receives doses of the following radiation: β^- particles (35 mrad, RBE = 1), protons (7.5 mrad, RBE = 10), and α particles (3.0 mrad, RBE = 15). What is the total biologically equivalent dose?

3. During an X ray examination, a person is exposed to radiation at a rate of 110 milligrays per hour. The exposure time is 0.10 s, and the mass of the exposed tissue is 1.2 kg. Determine the energy absorbed.

4. Write the following reactions in shorthand form.

$$^{14}_{7}N + ^{1}_{0}n \rightarrow ^{14}_{6}C + ^{1}_{1}H$$

$$^{43}_{20}Ca + ^{4}_{2}He \rightarrow ^{46}_{21}Sc + ^{1}_{1}H$$

$$^{55}_{25}Mn + ^{1}_{1}H \rightarrow ^{52}_{24}Cr + ^{4}_{2}He$$

5. Complete the following nuclear reactions, assuming the unknown quantity signified by a question mark is a single entity:

$$^{16}_{8}\mathrm{O}\,(\alpha,\, n)\ ?$$

$$?(p,\, \gamma)\,^{29}_{15}\mathrm{P}$$

$$^{48}_{22}\mathrm{Ti}\,(?,\, \alpha)^{45}_{21}\mathrm{Sc}$$

$$^{35}_{17}\mathrm{Cl}\,(?,\, n)^{38}_{19}\mathrm{K}$$

6. $^{235}_{92}\mathrm{U}$ absorbs a thermal neutron and fissions into rubidium $^{93}_{37}\mathrm{Rb}$ and cesium $^{141}_{55}\mathrm{Cs}$. What other nucleons are produced by the fission, and how many are there?

7. A nuclear reactor which uses $^{235}_{92}\mathrm{U}$ as fuel has an output of 35 MW. If the energy released in the fission of one uranium-235 nucleus is 210 MeV, how many grams of $^{235}_{92}\mathrm{U}$ are consumed in one day?

8. What energy (in MeV) is liberated by the following fission reaction?

$$^{235}_{92}\mathrm{U} + {}^{1}_{0}n \rightarrow {}^{93}_{37}\mathrm{Rb} + {}^{141}_{55}\mathrm{Cs} + 2\,^{1}_{0}n$$

Use the following values for the masses: n = 1.0087 u; U = 235.0439 u; Rb = 92.9217 u; Cs = 140.9195 u.

HELPFUL SUGGESTIONS

1. Nuclear reactions conserve energy (where mass-energy equivalence is assumed), conserve charge (the atomic number Z does not change), and conserve nucleon number (A does not change).

2. Since the mass defect is quite small, the atomic masses of the nuclei in a reaction should be given to 5 or 6 significant figures in order to obtain the mass defect to 2 or 3 significant figures.

3. Nuclear fission occurs when a nucleus splits into two or more smaller nuclei. Energy is released as a result of the difference in the average binding energy per nucleon between the first nucleus and the resulting nuclei.

4. Nuclear fusion, on the other hand, occurs when two very low-mass nuclei with small binding energies per nucleon "fuse" into a single, more massive nucleus that has a greater binding energy per nucleon.

EVERYDAY PHYSICS

1. Some watches and clocks have luminous hands that continuously glow in the dark. The hands contain traces of radium bromide mixed with zinc sulfide. If you have such a watch, examine it with a magnifying glass in the dark. You should see tiny flashes of light which are produced when an alpha particle is ejected from the radium nucleus and strikes a molecule of zinc sulfide.

2. Today, many people may risk exposure to ionizing radiation where they work. Some hazardous environments include nuclear facilities, vehicles which transport nuclear waste, or large hospitals. Alpha rays pose little danger since they are easily stopped by a piece of paper or the outer layer of your skin.
   Beta radiation is about 100 times more penetrating than alpha radiation, and is not stopped by skin tissue or normal clothing. However, materials such as glass, wood, aluminum and plastic will effectively shield out beta rays. Both alpha and beta rays can be damaging if ingested into the body (that is, breathed or eaten).
   Gamma radiation is the most penetrating of the three, about 10 000 times more so than alpha radiation. Gamma rays can only be halted by thick layers of concrete or lead. These rays can easily penetrate the skin and cause severe internal damage.

   Since a number of activities result in radioactive materials becoming airborne, or finding their way into water supplies, everyone should be educated about this subject regardless of where they work or live.

CHAPTER QUIZ

1. Which of the following is a unit that measures the biological effects caused by radiation?
   a. rad b. rem c. roentgen d. curie

2. Which of the following sources of radiation contributes the **least** to the average yearly dose received by humans?
   a. cosmic rays
   b. radioactivity from the earth and air
   c. medical and dental diagnostics
   d. consumer products

3. Which statement correctly relates the atomic number Z, the atomic mass number A, and the neutron number Z for any given atom?
   a. N = A + Z b. N = Z - A c. N = A - Z d. N = 2(A + Z)

4. What is the missing nucleon in the following reaction?

$$^{1}_{0}n + ^{10}_{5}B \rightarrow ^{7}_{3}Li + ?$$

   a. a proton b. a neutron c. an electron d. an alpha particle

5. What is the missing nucleon in the following reaction?

$$^{14}_{7}N\ (\alpha,\ ?)^{17}_{8}O$$

   a. a proton b. a neutron c. an electron d. an alpha particle

6. A single uranium-235 fission reaction can produce a chain reaction because it
   a. emits photons
   b. emits two or three neutrons
   c. emits a lot of energy
   d. converts matter to energy

7. Nuclear fission
   a. is the splitting of heavy nuclei into lighter ones.
   b. is the combining of light nuclei to form heavier ones.
   c. results from a series of alpha decays.
   d. releases much less energy per atom than a chemical process.

8. Which of the following is **not** needed in a fission reactor?
   a. a moderator
   b. neutron absorbing control rods
   c. fissionable fuel
   d. a special source of neutrons

9. A nitrogen $^{14}_{7}N$ nucleus absorbs a deuterium $^{2}_{1}H$ nucleus during a nuclear reaction. What is the name, atomic number, and nucleon number of the compound nucleus?
   a. $^{16}_{6}C$ b. $^{16}_{8}O$ c. $^{12}_{6}C$ d. $^{12}_{8}O$

SOLUTIONS AND ANSWERS

Practice Problems

---

1. The exposure is obtained using equation (32.1),

Exposure = $(1/2.58 \times 10^{-4})(q/m) = (1/2.58 \times 10^{-4})[(1.2 \times 10^{13})(1.6 \times 10^{-19}\ C)/(7.5 \times 10^{-4}\ kg)]$
Exposure = **9.9 R**.

---

2. The biologically equivalent dose (BED) will be the sum of the individual doses. Using equation (32.4) we have

Biol.Equiv.Dose = BED(β) + BED(prot) + BED(α)
Biol.Equiv.Dose = $(35 \times 10^{-3}\ rd)(1) + (7.5 \times 10^{-3}\ rd)(10) + (3.0 \times 10^{-3}\ rd)(15)$

The biologically equivalent dose is therefore, BED = **160 mrem.**

---

3. The absorbed dose is,

$$AD = (110 \times 10^{-3}\ Gy/hr)(1\ hr/3600\ s)(0.10\ s) = 3.1 \times 10^{-6}\ Gy.$$

The energy absorbed is, by equation (32.2),

Energy absorbed = AD x mass of absorbed matter = $(3.1 \times 10^{-6}\ Gy)(1.2\ kg)$ = **3.7×10^{-6} J**

---

4. The shorthand notation is:

$$^{14}_{7}N\ (n, p)\ ^{14}_{6}C$$

$$^{43}_{20}Ca\ (\alpha, p)\ ^{46}_{21}Sc$$

$$^{55}_{25}Mn\ (p, \alpha)\ ^{52}_{24}Cr$$

---

5. Adding up the atomic numbers and atomic masses on each side of the equations yields:

a. A = 16 + 4 - 1 = 19, Z = 8 + 2 - 0 = 10. The unknown quantity has 10 protons and is $^{19}_{10}$**Ne**.

b. A = 29 + 0 - 1 = 28, Z = 15 + 0 - 1 = 14. The unknown quantity has 14 protons and is $^{28}_{14}$**Si**.

c. A = 45 + 4 - 48 = 1, Z = 21 + 2 - 22 = 1. The unknown quantity is therefore $^{1}_{1}$**H**.

d. A = 38 + 1 - 35 = 4, Z = 19 + 0 -17 = 2. The unknown quantity is therefore $^{4}_{2}$**He**.

---

6. If we balance the equation we have

$$A = 235 + 1 - 93 - 141 = 2, Z = 92 + 0 - 37 - 55 = 0.$$

Therefore, since Z = 0 and A = 2, the only possible nucleons are **two neutrons**.

---

7. The power is 35 MW = 35×10^6 J/s. In one day the energy requirement is,

$$E = (3.5 \times 10^7 \text{ J/s})(86\,400 \text{ s}) = 3.0 \times 10^{12} \text{ J}.$$

We know that one gram of uranium-235 has

$$(6.02 \times 10^{23} \text{ mol}^{-1})(1 \text{ g})/(235 \text{ g mol}^{-1}) = 2.56 \times 10^{21} \text{ atoms}.$$

If each atom yields 210 MeV, then 1 g of uranium-235 yields

$$\Delta E = (2.56 \times 10^{21} \text{ atoms})(210 \times 10^6 \text{ eV})(1.60 \times 10^{-19} \text{ J/1 eV}) = 8.6 \times 10^{10} \text{ J}.$$

Finally, the amount of uranium-235 needed in one day is

$$m = (1 \text{ g})(E/\Delta E) = (1 \text{ g})(3.0 \times 10^{12} \text{ J})/(8.6 \times 10^{10} \text{ J}) = \mathbf{35 \text{ g}}.$$

---

8. Find the mass deficit:

$$\Delta m = 235.0439 \text{ u} + 1.0087 \text{ u} - 92.9217 \text{ u} - 140.9195 \text{ u} - 2(1.0087 \text{ u}) = 0.194 \text{ u}.$$

Since 1 u = 931.5 MeV the energy is

$$(931.5 \text{ MeV/1 u})(0.194 \text{ u}) = \mathbf{181 \text{ MeV}}.$$

Quiz answers

1. b
2. d
3. c
4. d
5. a
6. b
7. a
8. d
9. b

MCAT REVIEW PROBLEMS

The compound unstable nucleus $^{236}_{92}U$ often decays in accordance with the following reaction:

$$^{236}_{92}U \rightarrow {}^{140}_{54}Xe + {}^{94}_{38}Sr + \text{other particles}$$

During the reaction, the uranium nucleus "fissions" (splits) into the two smaller nuclei. The reaction is energetically favorable because the smaller nuclei have higher nuclear binding energy per nucleon (although the lighter nuclei have lower total nuclear binding energies, because they contain fewer nucleons).

Inside a nucleus, the nucleons (protons and neutrons) attract each other with a "strong nuclear" force. All nucleons exert approximately the same strong nuclear force on each other. This force holds the nucleus together. Importantly, the strong nuclear force becomes important only when the protons and neutrons are very close together, at intranuclear distances.

---

1. In the nuclear reaction presented above, the "other particles" might be

   A. An alpha particle, which consists of two protons and two neutrons
   B. Two protons
   C. One proton and one neutron
   D. Two neutrons

---

2. Why is a $^{4}_{2}He$ nucleus more stable than a $^{4}_{3}Li$ nucleus?

   A. The strong nuclear force is larger when the neutron-to-proton ratio is higher.
   B. The laws of nuclear physics forbid a nucleus from containing more protons than neutrons.
   C. Forces other than the strong nuclear force make the lithium nucleus less stable.
   D. None of the above.

---

3. A proton and a neutron are both shot at 100 m/s towards a $^{12}_{6}C$ nucleus. Which particle, if either, is more likely to be absorbed by the nucleus?

   A. The proton
   B. The neutron
   C. Both particles are about equally likely to be absorbed.
   D. Neither particle will be absorbed.

---

4. Which of the following graphs might represent the relationship between atomic mass number (i.e., "atomic weight") and the total binding energy of the nucleus, for nuclei heavier than $^{94}_{38}Sr$?

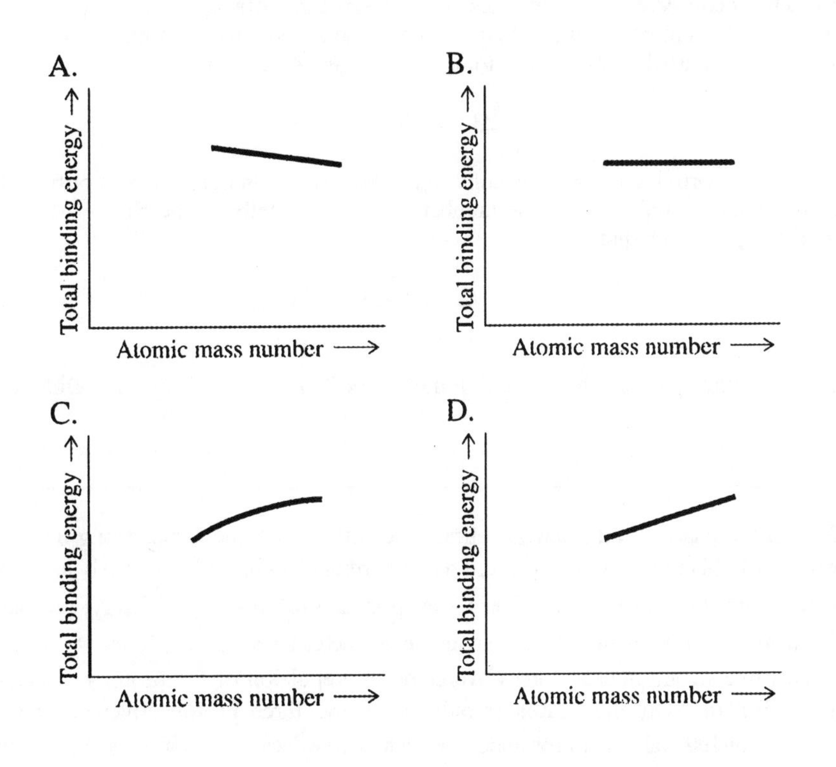

ANSWERS TO MCAT REVIEW PROBLEMS

---

1. **D.** Nuclear reactions conserve total charge, and also conserve the total *approximate* mass (as measured by the atomic mass number). Therefore, since the uranium, xenon, and strontium nuclei have atomic mass 236, 140, and 94, the "other particles" must have total atomic mass A such that

$$236 = 140 + 94 + A.$$

So, $A = 2$. The other particles are two nucleons. This narrows down the answer to options B, C, and D.

For nuclei, the atomic number—i.e., the number of protons—tells us the charge. So, the other particles must have total charge Z such that

$$92 = 54 + 38 + Z,$$

and hence, $Z = 0$.

In summary, the other particles have total atomic mass 2 and total charge 0. Only option D fits this description.

---

2. **C.** According to the passage, all nucleons attract each other with the same strong nuclear force. So, the strong nuclear force holds together three protons and neutron (^4_3Li) just as vigorously as it holds together two protons and two neutrons (^4_2He). But other forces play a role, too. Specifically, protons electrostatically repel other protons. This repulsion "tries" to make a nucleus fly apart. Since ^4_2He contains only two protons, the attractive strong nuclear forces overcome the repulsion of the protons. Therefore, the nucleus holds together. But in ^4_3Li, the mutual repulsion of the three protons overcomes the strong nuclear attractions, and the nucleus falls apart (or undergoes radioactive decay into a more stable nucleus).

---

3. **B.** Once the neutron gets sufficiently close to the nucleus, the strong nuclear force sucks it in. The same reasoning would apply to the proton, except for one thing; the proton feels electrostatically repelled by the six protons already inside the carbon nucleus. This repulsion prevents a 100 m/s proton from getting close enough to the nucleus for the strong force to "kick in." Remember, the strong force becomes important only at tiny distances.

---

4. **C.** Graph A would correctly show the binding energy *per nucleon*. But this differs from the *total* binding energy of the nucleus. The total binding energy takes into account not only the binding energy per nucleon, but also the number of nucleons.

To understand this subtle point, compare $^{94}_{38}\text{Sr}$ to $^{236}_{92}\text{U}$. As the passage states, $^{94}_{38}\text{Sr}$ has a higher binding energy *per nucleon*. Here are the numbers:

$$^{94}_{38}\text{Sr binding energy per nucleon} \approx 9\text{ MeV}.$$
$$^{236}_{92}\text{U binding energy per nucleon} \approx 7\text{ MeV}.$$

But $^{236}_{92}\text{U}$ still has a higher *total* binding energy, simply because it contains more nucleons. Indeed, since

Total binding energy of a nucleus
= binding energy per nucleon × number of nucleons,

the total binding energies are

$$^{94}_{38}\text{Sr}: \quad \frac{9\text{ MeV}}{\text{nucleon}} \times (94\text{ nucleons}) \approx 846\text{ MeV}.$$

$$^{236}_{92}\text{U}: \quad \frac{7\text{ MeV}}{\text{nucleon}} \times (236\text{ nucleons}) \approx 1650\text{ MeV}.$$

So, as atomic mass increases, so does the total binding energy. Uranium has more binding energy, even though it has less binding energy per nucleon.

Crucially, however, the total binding energy does not increase linearly with atomic mass. It increases at a smaller and smaller rate, precisely because the binding energy *per nucleon* decreases. Indeed, since "atomic mass number" specifies the number of nucleons, the binding energy per nucleon is simply the *slope* of a binding energy vs. atomic mass number graph. Between strontium and uranium, this slope decreases, as just explained. Thus, the answer must be C rather than D.

SAMPLE EXAMINATIONS

The following sample examinations may be used to practice for mid-term or final examinations. The questions appear in sequential chapter order so that you may tailor the test to fit your particular course. Keep in mind that these are only sample exams and may not represent the actual exam given by your instructor. The answers are found following Exam 4.

EXAM 1 — CHAPTERS 1 - 9

1. A 44 m high tree casts a 66 m long shadow over level ground. What is the sun's elevation angle above the horizon?
   a. 34° b. 42° c. 48° d. 56°

2. Find the resultant of the following forces:
   20.0 N at 10.0° east of north;
   40.0 N at 45.0° east of north;
   30.0 N at 45.0° north of east.

3. A car travels 60 km due east and then 80 km due north, taking 2 hours for the trip. What is the magnitude of the car's average velocity for the trip?
   a. 30 km/h b. 40 km/h c. 50 km/h d. 70 km/h

4. An object is dropped from the top of a 32-m high building. Neglecting air resistance, how fast is the object moving when it reaches the ground below?
   a. 15 m/s b. 25 m/s c. 64 m/s d. 630 m/s

5. A car traveling at 12.0 m/s sees a traffic light turn red. After 0.500 s have elapsed, the driver applies the brakes and the car decelerates at 6.25 m/s^2. What is the stopping distance of the car, as measured from the point where the driver first notices the red light?

6. A 25-kg mass and a 5-kg mass are simultaneously released from the same vertical height. Neglecting the effects of friction and air resistance
   a. the 25-kg mass has a greater acceleration because it is heavier.
   b. the 25-kg mass has a greater acceleration because it has the greater inertia.
   c. both objects have the same acceleration because they both have the same inertia.
   d. both objects have the same acceleration because they both have the same weight-to-mass ratio.

7. A bullet is fired horizontally from a height of 2.0 m with an initial speed of 310 m/s. How far does the bullet travel?

This situation pertains to questions 8 and 9..
A rocket is launched at an angle of 53° to the horizontal with an initial speed of 1.0 X 10^2 m/s. It moves along a straight line with an acceleration of 30.0 m/s X 10^2 for 3.0 s, the engines then cutoff and it follows a freebody trajectory.

8. What is the maximum altitude reached by the rocket?

9. What is the total time of flight of the rocket?

10. The earth exerts a force, F, on the moon to keep it in orbit. The moon exerts _______ on the earth?
    a. No force. b. an unknown force c. a force, F d. a force, 2F

11. When you stand on the floor you do not expect to fall through it. Which law of physics assures you that the floor will support you?
    a. Newton's first   b. Newton's second   c. Newton's third   d. law of gravity

Questions 12 and 13 pertain to the following situation:

The coefficient of static friction between a 60.0-kg box and the floor is 0.500. A horizontal push is applied to the box and slowly increased.

12. What is the magnitude of the applied force needed to start the box moving?
    a. 30.0 N   b. 294 N   c. 965 N   d. 368 N

13. What is the initial acceleration of the box if the coefficient of kinetic friction between the box and floor is 0.300?
    a. 2.50 m/s^2 b. 1.46 m/s^2 c. 1.00 m/s^2 d. 1.96 m/s^2

14. A 20.0-N box is sliding down a 10.0° inclined plane at a constant speed. What is the coefficient of kinetic friction between the block and plane?
    a. 0.537   b. 0.215   c. 0.791   d. 0.176

Questions 15 and 16 pertain to the following situation:
You are standing in a smoothly operating elevator. Your mass is 60.0 kg and the elevator has an acceleration of 2.0 m/s^2.

15. If the acceleration is upward, what is your apparent weight?
    a. 710 N   b. 470 N   c. 180 N   d. 590 N

16. If the acceleration is downward, what is your apparent weight?
    a. 710 N   b. 470 N   c. 180 N   d. 590 N

Use M = 6.0 kg and m = 4.0 kg with the following diagram and answer questions 17 - 20.

17. What is the acceleration of each mass?

18. What is the tension in the connecting cord?

19. If the system starts from rest, how far will the 6.0-kg mass fall in 2.0 s?

20. What is the velocity of the 4.0 kg mass after 3.0 s have elapsed?

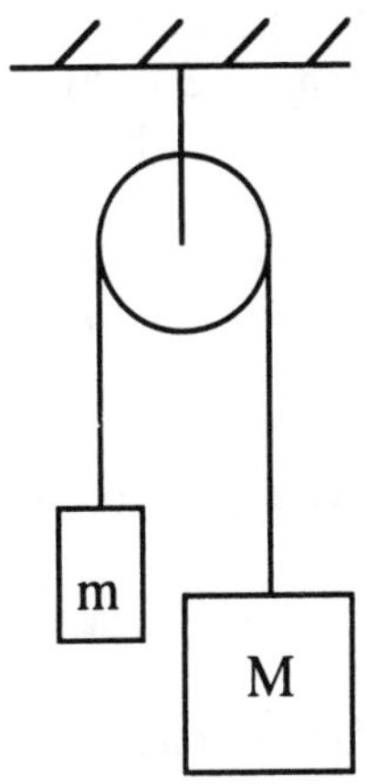

m

M

21. A horizontal force of 5 N accelerates a 4 kg mass from rest at a rate of 0.5 m/s^2 in the positive direction. What friction force acts on the mass?
    a. +3 N   b. - 3 N   c. + 2 N   d. - 2 N

The following pertains to questions 22 - 24.
A 1.0 X 10^3 kg satellite orbits a planet in a circular orbit at an altitude where the acceleration due to gravity is 0.20 m/s^2.

22. What is the speed of the satellite if it is in an orbit of radius 8.0 X 10^3 km?
    a. 41 m/s   b. 54 m/s   c. 1300 m/s   d. 1700 m/s

23. How fast must a 2.0 X 10^3 kg satellite travel in order to maintain the same orbit?
    a. Twice as fast.   b. The same speed.   c. $\sqrt{2}$ times as fast.   d. For times as fast.

24. What is the net force on the 1.0 X 10^3 kg satellite?
    a. 2.0 X 10^2 N   b. 9800 N   c. 0 N   d. 5.0 X 10^2 N

25. What angle of bank should a road curve of radius 1.0 X 10^2 m have for a car to safely negotiate it at a speed of 8.0 X 10^1 km/h without relying on friction.
    a. 27°   b. 45°   c. 62°   d. 31°

The following situation applies to questions 26 - 28.
A balloon lifts its 2.0 X 10^2 kg load at a constant speed. It takes 10.0 minutes for the balloon to attain a height of 1500 m.

26. Find the work done by the balloon in its ascent.
    a. 3.0 X 10^5 J   b. 3.0 X 10^4 J   c. 2.9 X 10^6 J   d. 2.0 X 10^3 J

27. What power is supplied by the balloon to the mass?
    a. 4.9 X 10^3 W   b. 2.0 X 10^4 W   c. 3.0 X 10^5 W   d. 1.8 X 10^8 W

28. What is the change in potential energy of the mass during its ascent?
    a. 3.0 X 10^5 J   b. 2.0 X 10^3 J   c. 2.9 X 10^6 J   d. 4.9 X 10^4 J

29. A 60.0 kg weight is pulled 20.0 m at a constant velocity over a table by a rope which is inclined 20.0° above the surface of the table. The coefficient of kinetic friction between the weight and table is 0.500. How much work is done by the force exerted by the rope on the weight?
    a. 2530 J   b. 1240 J   c. 5290 J   d. 6260 J

30. Two 0.50 kg objects collide head-on with initial speeds of 2.0 m/s each. The objects rebound with a speed of 1.0 m/s and 2.0 m/s, respectively. What quantities (momentum and mechanical energy) are conserved during the collision?

    The following situation applies to questions 31 and 32.

    During a burn a rocket exhausts gases at a rate of 150 kg/s and at a relative speed of 2.0 X 10^3 m/s.

31. What is the thrust of the rocket?

32. What is the acceleration of the rocket when its mass and the mass of the remaining fuel is 6.0 X 10^4 kg?

33. A 2.0-kg mass is moving to the east at a speed of 4.0 m/s and collides head-on with a stationary 2.0-kg mass. The collision is completely inelastic. How much energy is lost during the collision?
    a. 16 J   b. 4.0 J   c. 8.0 J   d. 2.0 J

Questions 34 and 35 refer to the following:
A constant horizontal force of 6.0 N acts for 4.0 s on a 12-kg object which moves on a friction free level surface.

34. What is the object's change in momentum?
    a. 2.0 kg m/s b. 24 kg m/s c. 48 kg m/s d. 72 kg m/s

35. What is the object's change in velocity?
    a. 2.0 m/s b. 24 m/s c. 48 m/s d. 72 m/s

36. An propeller accelerates from rest to a final angular velocity of 1800 rev/min. What is the angular acceleration of the propeller if it took 3.0 s to reach the final speed?

37. A circular saw which has a 0.30-m radius is turning at 1600 rev/min. What is the speed of a tooth on the saw?
    a. 5.0 m/s b. 25 m/s c. 5.0×10^{1} m/s d. 1.0×10^{2} m/s

38. An automobile is traveling on a level road at 60.0 km/h. The diameter of the wheels is 0.750 m. What is the angular velocity of the wheel about its spindle?

Answer questions 39 and 40 based on the following diagram. The uniform rod has a mass of 5.0 kg, and the mass at the end of the rod is M = 10.0 kg.

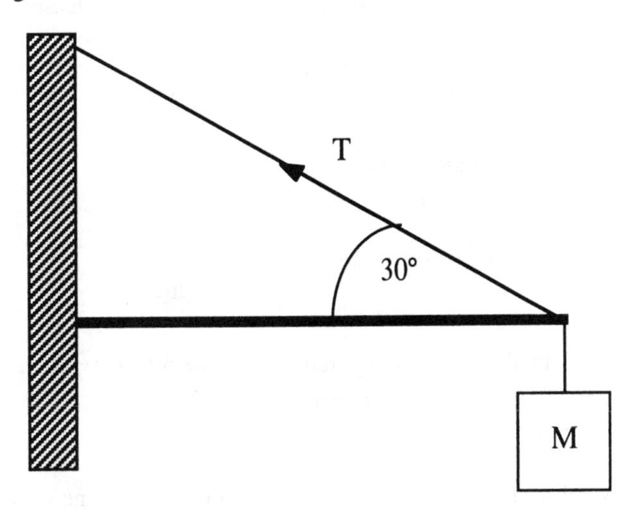

39. What is the tension in the supporting cable?

40. What are the magnitudes of the horizontal and vertical components of the force at the wall?

EXAM 2 — CHAPTERS 10 - 15

1. A 50.0 kg mass is suspended from a steel wire of length 1.5 m and diameter 0.75 mm. By how much does the wire stretch? Young's modulus for steel is $Y = 2.0 \times 10^{11}$ N/m^2.

Questions 2 and 3 refer to the following:
The end of a prong of a tuning fork executes simple harmonic motion with a frequency of 256 Hz and an amplitude of 0.400 mm.

2. What is the maximum speed of the end of the prong?
   a. 0.644 m/s b. 0.512 m/s c. 102 m/s d. 0.721 m/s

3. What is the maximum acceleration of the end of the prong?
   a. 2.60×10^4 m/s^2 b. 1.20×10^3 m/s^2 c. 1.53×10^2 m/s^2 d. 1.03×10^3 m/s^2

Questions 4 - 6 refer to the following:
A 5.0-kg mass is attached to a spring with a force constant of 98 N/m. The mass is resting on a frictionless horizontal plane and a horizontal force of 12.0 N is applied to the mass, and it is then released.

4. What is the amplitude of the simple harmonic motion?
   a. 0.50 m b. 0.12 m c. 0.090 m d. 0.18 m

5. What is the period of simple harmonic motion?
   a. 2.1 s b. 0.51 s c. 1.4 s d. 1.8 s

6. What is the total energy of the simple harmonic oscillator?
   a. 1.5 J b. 0.70 J c. 0.40 J d. 1.2 J

7. The areas of the large and small pistons in a hydraulic press are 50.0 cm^2 and 2.0 cm^2, respectively. What force must be applied to the small piston in order to lift a 4.0×10^3 N load?
   a. 290 N b. 240 N c. 160 N d. 180 N

8. What is the speed at which water flows from a small hole 3.0 m below the water level in a full tank which is 5.0 m high?
   a. 6.0 m/s b. 5.0 m/s c. 7.7 m/s d. 8.3 m/s

9. Water flows through a pipe of cross-sectional area 1.0 cm^2 with a velocity of 50.0 cm/s. What is the velocity if the pipe area expands to 5.0 cm^2?
   a. 250 cm/s b. 0.50 m/s c. 5.0 cm/s d. 0.10 m/s

10. A steel cable supporting a suspension bridge has a length of 5.0×10^2 m at 30.0 °C. What is the change in its length when the temperature is decreased to - 10.0 °C? Take the coefficient of thermal expansion for steel to be $\alpha = 1.2 \times 10^{-5}$ (C°)$^{-1}$.
   a. 0.15 m b. 5.2×10^{-2} m c. 6.4×10^{-3} m d. 0.24 m

11. How much pressure would be required to reduce the volume of a block of brass by a factor of one part in one thousand? The bulk modulus for brass is $B = 6.7 \times 10^{10}$ N/m^2.

12. Two equal masses brought into thermal contact are isolated from their surroundings. One object has an initial temperature of 100.0 °C and the other has an initial temperature of 0.00 °C. The hotter object has a specific heat capacity that is three times greater than that of the cooler object. What is the equilibrium temperature of the system?
    a. 25.0 °C b. 50.0 °C c. 75.0 °C d. None of the above

13. Sixteen grams of water initially at 100 °C are poured into a cavity in a very large block of ice at 0 °C. How many grams of ice will melt?
    a. 2 g b. 20 g c. 200 g d. None of the above

14. A sample of CO_2 (atomic mass = 44 g/mol) contains 12 x 10^{23} molecules. What is the mass of this sample?
    a. 22 g b. 44 g c. 88 g d. 22 kg

15. An aluminum cup having a mass of 250.0 g is filled with 50.0 g of water. Both are initially at 25.0 °C. A 75.0-g piece of iron initially at 350.0 °C is dropped into the water. What is the final equilibrium temperature of the system assuming that no heat was lost to the outside environment?

16. A 1.0 m X 1.5 m window made from a single pane of glass conducts heat energy at the rate of 2100 W when the inside temperature is 23 °C and the outside temperature is 1.0 °C. How thick is the pane of glass?
    a. 0.63 cm b. 0.95 cm c. 2.5 cm d. 1.3 cm

17. What is the mass of the carbon dioxide contained in a 50.0-liter tank at a pressure of 3.0 atmospheres and a temperature of 30.0 °C?
    a. 0.26 kg b. 0.50 kg c. 0.34 kg d. 1.51 kg

Questions 18 - 22 are based on the following situation:
An ideal diatomic gas is confined to a volume of 20.0 liters at 50.0 °C and 1.00 atm. Some heat is slowly added to the gas, which is allowed to expand at constant pressure until its volume is 30.0 L.

18. How many moles of gas are present in the container?

19. What is the final temperature of the gas?

20. How much work is done by the gas during its expansion?

21. How much heat has been supplied to the gas?

22. What is the change in internal energy of the gas?

Questions 23 and 24 deal with the following:
A cylinder contains air at a pressure of 50.0 N/m^2. The original volume at 27.0 °C is 5.00 liters. The air is carried through the following processes:

1. heating at a constant pressure to 227 °C,
2. cooling at constant volume to - 23.0 °C,
3. cooling at constant pressure to - 123 °C,
4. heating at constant volume to 27.0 °C.

23. What are the final pressure and volume of the gas?
    a. 50.0 N/cm^2, 8.3 liters c. 50.0 N/cm^2, 5.0 liters
    b. 25.0 N/cm^2, 8.3 liters d. 25.0 N/cm^2, 5.0 liters

24. Calculate the net work done by the gas.
    a. 4150 J b. 825 J c. 3320 J d. 4980 J

Answer questions 25 and 26 based on the following:
A gas at 27 °C and 1.0 atm is compressed until its volume is one-half of its original value. This compression is done so fast as to be adiabatic. The ratio of specific heats is $\gamma = 1.5$.

25. What is the final pressure of the gas?
    a. 25 atm b. 1.4 atm c. 6.2 atm d. 2.5 atm

26. What is the final temperature of the gas?
    a. 210 °C b. 300 °C c. 250 °C d. 180 °C

27. Which law is a statement of the conservation of energy for a thermodynamic system?
    a. zeroth law of thermodynamics
    b. first law of thermodynamics
    c. second law of thermodynamics
    d. third law of thermodynamics

Questions 28 and 29 are based on the following:
A Carnot heat engine operates between 275 °C and 115 °C. It absorbs 3.75×10^5 J at the higher temperature.

28. What is the efficiency of the engine?

29. How much work per cycle is the engine capable of performing?

30. If 5.0×10^3 J of heat flow spontaneously from a hot reservoir at 227 °C to a cold reservoir at 37 °C, what is the total entropy change for this irreversible process.

EXAM 3 CHAPTERS 16 - 23

Questions 1 - 3 deal with the following:
A traveling wave propagates in the +x direction along a taut string that has a linear density of 5.0 g/m and is under a tension of 2.0×10^2 N. The amplitude of the wave is 8.0 cm and the wavelength is 50.0 cm. At time t = 0, x = 0, the displacement y = 0.

1. What is the propagation velocity of the wave?

2. What are the frequency and period of the wave?

3. Write an expression for the displacement of the wave based on the above description.

4. How much more intense is a 90-dB shout than a 10-dB whisper?
   a. 1×10^4 b. 1×10^6 c. 1×10^8 d. 1×10^{10}

5. How much energy is received 5.0 m away from a sound source which emits an 80.0 dB sound continuously for 10.0 minutes?
   a. 19 J b. 38 J c. 4.0×10^2 J d. 2500 J

6. A train is traveling toward you blowing a 444 Hz note on its whistle. If the train is moving 20.0 m/s, what frequency will you hear? Take the speed of sound to be 343 m/s.
   a. 462 Hz b. 471 Hz c. 418 Hz d. 501 Hz

7. The intensity of a sound wave (at your ear) increases by a factor of 1.0×10^6. How much louder is the sound?
   a. 8.0 times b. 16 times c. 32 times d. 64 times

8. A wave travels down a rope and encounters a fixed end. What happens?
   a. Nothing
   b. It only reverses its direction of travel.
   c. It reverses directions and is phase shifted by 90°.
   d. It reverses directions and is phase shifted by 180°.

9. Standing waves result from
   a. the superposition of identical waves traveling in opposite directions.
   b. the superposition of identical waves traveling in the same direction.
   c. waves that are tired of sitting.

Questions 10 and 11 pertain to the following:
A stretched string, 0.80 m long, has a fundamental frequency of 1.0×10^2 Hz.

10. What is the velocity of the traveling waves whose superposition produces the standing wave?
    a. 45 m/s b. 120 m/s c. 160 m/s d. 220 m

11. What is the wavelength of the standing waves?
    a. 0.40 m b. 0.80 m c. 1.6 m d. 2.4 m

12. Two positive charges exert a mutual repulsive force on each other of 5.00×10^{-3} N when they are separated by 5.00 m. What force do they exert on each other when they are separated by 1.0 m?
    a. 2.50×10^{-2} N b. 0.125 N c. 2.00×10^{-4} N d. 1.00×10^{-3} N

Questions 13 - 15 are based on the following figure:

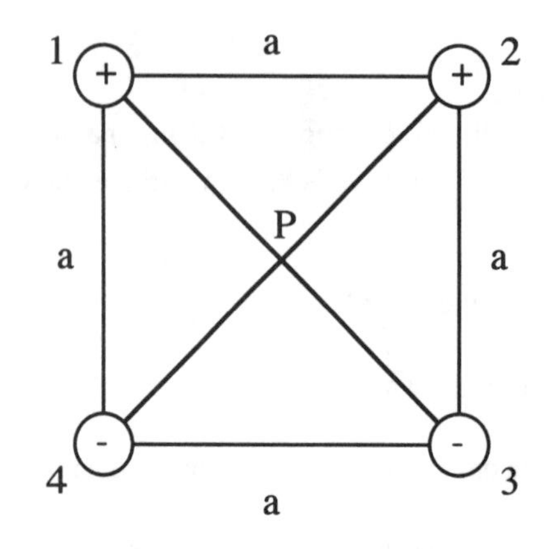

The magnitude of each charge in the above diagram is 4.0 μC. The sides have a = 0.20 m.

13. What are the magnitude and direction of the electric force on charge #4?

14. What are the magnitude and direction of the electric field at the center of the square, point P?

15. What is the electric potential at point P?

16. Two free point charges +q and +4q are a distance d apart. A third point charge is so placed that the entire system is in equilibrium. Find the location, magnitude, and sign of the third charge.
    a. d/2 from + 4q; + q/2
    b. 4d/3 from + q; - 5q/8
    c. d from + 4q; - q
    d. d/3 from + q; - 4q/9

Questions 17 - 20 are based on the following:
A negative charge of 0.50 C is accelerated from point B to point C by an electric field and thereby acquires 1.0 J of energy.

17. What did the work on the charge that it might gain this energy?
    a. A positive charge at infinity.
    b. A negative charge at infinity.
    c. The electric field
    d. The moon.

18. What is the potential difference between points B and C?
    a. 0.50 V
    b. 1.0 V
    c. 1.5 V
    d. 2.0 V

19. Which point is said to be at the higher potential?
    a. B
    b. C
    c. Both are at the same potential.

20. If point C is assigned zero potential, what is the potential at point B?
    a. - 2.0 V
    b. + 2.0 V
    c. + 0.5 V
    d. - 1.0 V

21. If 3.0 A are flowing through a conductor, how much charge will pass through the conductor in 4.0 s?
    a. 3.0 C
    b. 4.0 C
    c. 1.3 C
    d. 12 C

22. The resistance of a copper wire (resistivity $\rho = 1.7 \times 10^{-8}\ \Omega$) is 2.5 Ω. If the wire has a diameter of 2.0 mm, what is the length of the wire?

Questions 23 - 25 refer to the following:
A generator delivers an ac voltage given by $V = 240 \sin 188t$, and produces an rms current of 3.0 A in a purely resistive circuit.

23. What is the rms voltage of the generator?

24. What is the frequency of the generator?

25. What is the average power delivered to the circuit?

26. You are given a number of 10-Ω resistors, each capable of dissipating only 1.0 W. What is the minimum number of such resistors that you can combine in series and parallel combinations to make a 10-Ω resistor capable of dissipating 5.0 W?
   a. 6 b. 7 c. 8 d. 9

27. Which of the following is a true statement concerning a network of parallel resistors connected in parallel with a battery?
   a. The current through each resistor is the same.
   b. The IR drop across each resistor is the same and is equal to the EMF of the battery.
   c. The IR drop across each resistor is different.
   d. The IR drop across each resistor is the same, but is not equal to the EMF of the battery.

28. You are in dire need of a resistor with a value smaller than any that you have on hand. How could you possible use the resistors you have, (a large variety) to satisfy your requirements?
   a. Construct a combination of series resistors.
   b. Construct a parallel combination of resistors.
   c. Use magic.
   d. There is no way.

29. Two parallel wires carry currents in opposite direction. What is the nature of the force between the wires?
   a. repulsive b. attractive c. zero d. Impossible to know.

30. A 10.0 Ω and a 15.0 Ω resistor are connected in parallel with each other and in series with another 10.0 Ω resistor and a 12 V battery. Find the current delivered by the battery.
   a. 0.34 A b. 0.28 A c. 0.75 A d. 1.5 A

31. Voltmeters have a high resistance in order to
   a. protect them from damage.
   b. not disturb the circuit.
   c. allow large currents through the meter.
   d. produce large magnetic fields.

This situation applies to the questions 32 and 33.
An electron travels perpendicular to a constant magnetic field of magnitude 0.010 Wb/m^2 with a speed of 2.0×10^4 m/s.

32. What force does the field exert on the electron?

33. What is the radius of the electron's orbit in the field?

34. The current in a current carrying wire is moving toward you. What is the direction of the magnetic field produced by the wire?
   a. clockwise b. counterclockwise c. alone the wire d. radially outward

This situation applies to questions 35 and 36.
A uniform magnetic field of magnitude 0.20 T makes an angle of 30.0° with the plane of a circular loop of wire of radius 10.0 cm.

35. Calculate the flux through the loop.

36. If the magnetic field decreases to 0 T in 0.10 s, what is the emf induced in the loop?

37. A wire loop is pushed into a region of increasing magnetic field. According to Lenz's law, the magnetic force on the loop attempts to
    a. pull the loop into the region of magnetic field.
    b. push the loop out of the region of magnetic field.
    c. to do nothing.
    d. to move the loop parallel to the magnetic field.

38. The rms value of the alternating current in a 200 Ω resistor is 10 A. What is the rms voltage across this resistor?

39. The 60.0 Hz current (rms) in a circuit containing a 4.00 μF capacitor is 20.0 A. Find the rms voltage across the capacitor.

40. What is the rms current in a 15.0 mH inductor when the rms voltage across the inductor is 120 V?

EXAM 4 CHAPTERS 24 - 32

1. What is the frequency of a radio wave whose wavelength is $\lambda = 20.0$ m?
   a. 6.70×10^4 Hz b. 6.00×10^9 Hz c. 1.50×10^7 Hz d. 2.0×10^{15} Hz

2. An electromagnetic wave has an average intensity of 2.0×10^4 W/m^2. What is the energy density of the wave?
   a. 6.7×10^{-5} J/m^3 b. 6.0×10^{-8} J/m^3 c. 2.0×10^4 J/m^3 d. 7.6×10^{-7} J/m^3

3. The axes of a sheet of polarizing material and an analyzer are oriented 45° to one another. If unpolarized light of intensity I is incident on the system, what is the intensity of the transmitted light?
   a. 0.45 I b. 0.25 I c. 0.80 I d. 0.35 I

Questions 4 - 6 are based on the following:
A concave mirror forms a real image 20.0 cm from the mirror when the object is 50.0 cm away.

4. What is the focal length of the mirror?
   a. + 33.3 cm b. + 70.0 cm c. + 14.3 cm d. - 14.3 cm

5. What is the magnification of the mirror?
   a. - 2.5 b. - 0.40 c. + 2.5 d. + 0.40

6. If the object is moved to a point 20.0 cm from the mirror, what is the image distance?
   a. 15.0 cm b. 33.3 cm c. 50.0 cm d. 70.0 cm

Questions 7 and 8 are based on the following:
A convex mirror has a radius of curvature of 50.0 cm. You stand 1.0 m away and look at yourself in the mirror.

7. Where will you see your image?
   a. 0.33 m behind the mirror
   b. 1.0 m behind the mirror
   c. 0.20 m behind the mirror
   d. 0.25 m in front of the mirror

8. How will your image look?
   a. Larger, with left and right reversed.
   b. Larger, no inversion.
   c. Smaller, but inverted.
   d. Smaller, no inversion.

9. An object is placed 12.0 cm from a converging lens of focal length 8.0 cm. Determine the position of the image.
   a. + 12.0 cm b. +18.0 cm c. - 24.0 cm d. + 24.0 cm

Questions 10 and 11 are based on the following:
A diverging lens has f = - 15.0 cm. An object is placed on the principal axis 12.0 cm from the lens.

10. Describe the position and orientation of the final image.
    a. - 6.67 cm, virtual b. + 6.67 cm, real c. - 3.33 cm, virtual d. +3.33 cm, real

11. What is the magnification in this situation?
    a. + 1.0 b. - 0.56 c. + 0.56 d. - 2.2

12. A person has a near point of 100 cm. What is the power of the correct lenses to prescribe for this person?
    a. 5 diopters b. 4 diopters c. 3 diopters d. 2 diopters

13. A magnifying glass (f = 20 cm) is used to read small print. Where should the magnifying glass be held?
    a. greater than 20 cm from the print
    b. less than 20 cm from the print
    c. exactly 20 cm from the print
    d. it should be in contact with the print.

14. The aperture of a camera lens is set at f/8. What's the focal length of the lens if the aperture diameter is 0.50 cm?
    a. 4.0 cm b. 3.0 cm c. 2.5 cm d. 4.5 cm

15. The objective of a telescope has a focal length of 2.5 m. If this telescope is used with an eyepiece of focal length 10.0 mm, what is the magnification?
    a. 0.25 b. 25 c. 250 d. 2500

16. In the Young's double slit experiment the slit separation is 1.0×10^{-5} m. How far away must one place the screen in order to observe green ($\lambda = 5.0 \times 10^{-7}$ m) spots separated by 1.0 cm?
    a. 5.0 cm b. 1.2 m c. 2.0 m d. 0.20 m

17. A light beam (λ = 500 nm) is passed through a 0.5 m wide slit. Would you expect to observe diffraction?
    a. Yes, because waves always diffract.
    b. Yes, because waves diffract best through large openings.
    c. No, because the slit is too big compared to λ.
    d. No, light is a particle, not a wave.

18. An anti-reflection coating is to applied to a lens. How thick must the coating be in order that green light ($\lambda = 5.00 \times 10^{-7}$ m) will not be reflected?
    a. 5.00×10^{-7} m b. 2.50×10^{-7} m c. 1.25×10^{-7} m d. 7.50×10^{-8} m

19. A laboratory measurement of the coordinates of the ends of a moving meter stick indicates that the "meter stick" is 0.20 m long. What is the velocity of the meter stick?
    a. 0.98c b. 0.50c c. 0.90c d. 0.20c

20. If you were able to convert 1 g of matter into energy, with 100% efficiency, how much energy would result?
    a. 9×10^{19} J b. 9×10^{16} J c. 3×10^{20} J d. 9×10^{13} J

21. An electron is known to have a position that is uncertain to $\Delta x = 3.0 \times 10^{-8}$ cm. What is the uncertainty in its momentum?
    a. greater than 3.5×10^{-24} kg m/s
    b. less than 2.2×10^{-34} kg m/s
    c. zero
    d. greater than 3.0×10^{-20} kg m/s

22. What is the de Broglie wavelength of an electron which is moving at a speed of 2.0×10^{6} m/s?
    a. 1.2×10^{-10} m b. 3.6×10^{-10} m c. 8.5×10^{-9} m d. 1.7×10^{-15} m

23. What is the energy of the hydrogen atom in its n = 3 state?
    a. - 13.6 eV b. - 3.40 eV c. - 1.51 eV d. - 0.94 eV

24. What possible states in a hydrogen atom correspond to n = 2?
    a. $l = 2$; $m_l = 2, 1, 0, -1, -2$; $m_s = \pm 1/2$.
    $l = 0$; $m_l = 0$; $m_s = \pm 1/2$.
    b. $l = 1$; $m_l = 0$; $m_s = \pm 1/2$.
    $l = 0$; $m_l = 1, -1$; $m_s = 0$.
    c. $l = 0$; $m_l = 0$; $m_s = \pm 1/2$.
    $l = 1$; $m_l = 1, 0, -1$; $m_s = \pm 1/2$.

25. A 144-g sample of a radioactive substance is left to decay for five half-lives. How much of the substance remains after this time?
    a. 18 g   b. 9.0 g   c. 36 g   d. 4.5 g

26. A nucleus has a mass defect of 1.025 u. What is the binding energy in MeV?
    a. 954.8   b. 908.8   c. 931.5   d. 985.2

27. In most β^- decay processes, a(n) ____________ is emitted along with an electron.
    a. α particle   b. γ ray   c. neutrino   d. antineutrino

28. Which of the following units is a unit that measures the biological effects caused by radiation?
    a. rd   b. rem   c. roentgen   d. curie

29. The fissioning of uranium-235 can produce a chain reaction because it
    a. emits photons   c. emits a lot of energy
    b. emits two or three neutrons   d. converts mass to energy

30. Nuclear fusion is
    a. when a high-mass nucleus splits into two or more smaller nuclei.
    b. when a low-mass nucleus splits into two high mass nuclei.
    c. when two very low-mass nuclei combine to form a single, more massive nucleus.
    d. when two high-mass nuclei "fuse" into several lower mass nuclei.

ANSWERS TO SAMPLE EXAMINATIONS

EXAM 1

1. a
2. 70.0 N at 8.6° E of N
3. c
4. b
5. 17.5 m
6. d
7. 2.0 X 10^2 m
8. 1500 m
9. 36
10. c
11. c
12. b
13. d
14. d
15. a
16. b
17. 1.96 m/s^2
18. 47.0 N
19. 3.92 m
20. 5.88 m/s
21. a
22. c
23. c
24. a
25. a
26. c
27. a
28. c
29. c
30. neither
31. 3.0 X 10^5 N
32. 5.0 m/s^2
33. c
34. b
35. a
36. 63 rad/s^2
37. c
38. 44.4 rad/s
39. 245 N
40. 212 N, 24.5 N

EXAM 2

1. 8.3 X 10^{-3} m
2. a
3. d
4. b
5. c
6. b
7. c
8. c
9. d
10. d
11. 6.7 X 10^7 Pa
12. c
13. b
14. c
15. 48.5 °C
16. d
17. a
18. 0.745 mol
19. 212 °C
20. 1.00 X 10^3 J
21. 3.52 X 10^3 J
22. 2.52 X 10^3 J
23. d
24. c
25. b
26. a
27. b
28. 0.292
29. 1.10 X 10^5 J
30. 6.1 J/K

EXAM 3

1. 2.0×10^{2} m/s
2. 4.0×10^{3} Hz, 2.5×10^{-3} s
3. $y = 0.08\sin(800\pi t - 4\pi x)$
4. c
5. a
6. b
7. d
8. d
9. a
10. c
11. c
12. b
13. 5.4 N, 116 ° CCW from the line joining the charges.
14. 5.1×10^{6} N/C
15. 0 V
16. d
17. c
18. d
19. b
20. a
21. d
22. 5.0×10^{2} m
23. 170 V
24. 3.0×10^{1} Hz
25. 510 W
26. d
27. b
28. b
29. a
30. c
31. b
32. 3.2×10^{-17} N
33. 1.1×10^{-5} m
34. b
35. 3.14×10^{-3} Wb
36. 3.14×10^{-2} V
37. b
38. 2000 V
39. 1.33×10^{4} V
40. 21.2 A

EXAM 4

1. c
2. a
3. b
4. c
5. b
6. c
7. a
8. d
9. d
10. a
11. c
12. c
13. b
14. a
15. c
16. d
17. c
18. c
19. a
20. d
21. a
22. b
23. c
24. c
25. d
26. a
27. d
28. b
29. b
30. c